Essentials
of Geology

Eighth Edition

Essentials of Geology

Frederick K. Lutgens
Edward J. Tarbuck

Both Illinois Central College, Emeritus

Illustrated by
Dennis Tasa

Prentice Hall
Upper Saddle River, New Jersey 07458

Library of Congress Cataloging-in-Publication Data

Lutgens, Frederick K.
    Essentials of geology / Frederick K. Lutgens, Edward J. Tarbuck ; illustrated by Dennis Tasa. — 8th ed.
        p. cm.
    Includes index.
    ISBN 0-13-008157-4
    1. Geology.   I. Tarbuck, Edward J.
QE26.3 .L87 2003
550—dc21

2001058780

Senior Editor: Patrick Lynch
Editor-in-Chief: Sheri Snavely
Vice President of Production and Manufacturing: David W. Riccardi
Executive Managing Editor: Kathleen Schiaparelli
Assistant Managing Editor: Beth Sweeten
Production Editor: Edward Thomas
Assistant Managing Editor, Science Media: Nicole Bush
Assistant Editor: Amanda Griffith
Marketing Manager: Christine Henry
Manufacturing Manager: Trudy Pisciotti
Assistant Manufacturing Manager: Michael Bell
Art Editor: Adam Velthaus
Director of Creative Services: Paul Belfanti
Director of Design: Carole Anson
Art Director: Maureen Eide
Assistant to Art Director: John Christiana
Interior and Cover Design: Jonathan Boylan
Managing Editor, Audio/Video Assets: Grace Hazeldine
Editorial Assistant: Sean Hale
Photo Research Administrator: Melinda Reo
Photo Researcher: Yvonne Gerin
Copy Editor/Proofreader: Barbara Booth
Image Permission Coordinator: Debbie Hewittson
Color Scanning Supervisor: Joseph Conti
Production Assistant: Nancy Bauer
Composition: Pine Tree Composition
Cover Photo: Wizard Island in Crater Lake National Park, Oregon (photo by Greg Vaughn).
Title Page Photo: Delicate Arch in Arches National Park, Utah, is a product of differential weathering (photo by David Muench Photography, Inc.).

 © 2003, 2000, 1998, 1995 by Prentice-Hall, Inc.
Upper Saddle River, New Jersey 07458

Earlier editions © 1992 by Macmillan Publishing Company,
and © 1989, 1986, 1982 by Merrill Publishing Company.

Printed in the United States of America
10 9 8 7 6 5 4 3 2 1

ISBN 0-13-008157-4

Pearson Education Ltd.
Pearson Education Australia Pty., Limited
Pearson Education Singapore, Pte. Ltd
Pearson Education North Asia Ltd.
Pearson Education Canada, Ltd.
Pearson Educacion de Mexico, S.A. de C.V.
Pearson Education—Japan
Pearson Education Malaysia, Pte. Ltd.

Brief Contents

GEODe III CD-ROM

A copy of the GEODe III CD-ROM is packaged with each copy of ESSENTIALS OF GEOLOGY, Eighth Edition. This dynamic learning aid reinforces key geologic concepts by using tutorials, animations and interactive exercises.

Unit Introduction

A. A View of Earth
B. Earth's Layered Structure
C. Features of the Continents
D. Floor of the Ocean

Unit Earth Materials

A. Minerals
   1. Introduction
   2. Major Mineral Groups
   3. Properties Used to Identify Minerals
   4. Mineral Identification
B. Mineral Review
C. Rock Cycle
D. Igneous Rocks
   1. Introduction
   2. Igneous Textures
   3. Mineral Composition of Igneous Rocks
   4. Naming Igneous Rocks
E. Sedimentary Rocks
   1. Introduction
   2. Types of Sedimentary Rocks
   3. Interpreting Environments
F. Metamorphic Rocks
   1. Introduction
   2. Agents of Metamorphism
   3. Textural and Mineralogical Changes
   4. Common Metamorphic Rocks
G. Rock Review

Unit External Processes

A. External vs. Internal Processes
B. Hydrologic Cycle
C. Running Water
   1. Stream Characteristics
   2. Review—Valleys and Stream-Related Features
D. Groundwater
   1. Groundwater and Its Importance
   2. Springs and Wells
E. Glaciers
   1. Introduction
   2. Budget of a Glacier
   3. Reviewing Glacial Features

F. Deserts
   1. Distribution and Causes of Dry Lands
   2. Common Misconceptions About Deserts
   3. Review of Landforms and Landscapes
G. Coastal Processes
   1. Waves and Beaches
   2. Wave Erosion

Unit Internal Processes

A. Plate Tectonics
   1. Introduction
   2. Plate Boundaries
B. Crustal Deformation
   1. Introduction
   2. Mapping Geologic Structures
   3. Folds
   4. Faults and Fractures (Joints)
   5. Review
C. Earthquakes
   1. What Is an Earthquake?
   2. Seismology
   3. Locating an Earthquake
D. Igneous Activity
   1. The Nature of Volcanic Eruptions
   2. Materials Extruded During an Eruption
   3. Volcanoes
   4. Intrusive Igneous Activity

Unit Geologic Time

A. Geologic Time Scale
B. Relative Dating
C. Radiometric Dating

Unit Landforms of the United States

A. Landform Regions
B. A Geologic Tour

 This GEODe III icon appears throughout the book wherever a text discussion has a corresponding activity on the CD-ROM.

Contents

5   Weathering and Soils   *100*

6   Sedimentary Rocks   *120*

7   Metamorphic Rocks   *145*

8   Mass Wasting: The Work of Gravity   *165*

17 Mountain Building   *362*

18 Geologic Time   *387*

19 Earth History:
A Brief Summary   *412*

Appendix A

Appendix B

Appendix C

Glossary   *444*

Index   *457*

Preface

The Eighth Edition of *Essentials of Geology,* like its predecessors, is a college-level text for students taking their first and perhaps only course in geology. The book is intended to be a meaningful, nontechnical survey for people with little background in science. Usually students are taking this class to meet a portion of their college's or university's general requirements.

In addition to being informative and up-to-date, a major goal of *Essentials of Geology* is to meet the need of beginning students for a readable and user-friendly text, a book that is a highly usable tool for learning the basic principles and concepts of geology.

Distinguishing Features

Readability

The language of this book is straightforward and *written to be understood.* Clear, readable discussions with a minimum of technical language are the rule. The frequent headings and subheadings help students follow discussions and identify the important ideas presented in each chapter. In the Eighth Edition, improved readability was achieved by examining chapter organization and flow, and writing in a more personal style. Large portions of several chapters were substantially rewritten in an effort to make the material more understandable.

Illustrations and Photographs

Geology is highly visual. Therefore, photographs and artwork are a very important part of an introductory book. *Essentials of Geology,* Eighth Edition, contains dozens of new high-quality photographs that were carefully selected to aid understanding, add realism, and heighten the interest of the reader.

The many new and revised pieces of art help to illustrate ideas and concepts even more clearly and realistically than before. The art program was carried out by Dennis Tasa, a gifted artist and respected Earth science illustrator.

Focus on Learning

To assist student learning, every chapter opens with a series of questions. Each question alerts the reader to an important idea or concept in the chapter. When a chapter has been completed, three useful devices help students review. First, a helpful summary—*The Chap-*

ter in Review—recaps all of the major points. Next is a checklist of *Key Terms* with page references. Learning the language of geology helps students learn the material. This is followed by *Questions for Review,* which helps students examine their knowledge of significant facts and ideas. Each chapter closes with a reminder to visit the Website for *Essentials of Geology,* Eighth Edition (http://www.prenhall.com/lutgens). It contains many excellent opportunities for review and exploration.

Maintaining a Focus on Basic Principles and Instructor Flexibility

The main focus of the Eighth Edition remains the same as in the first seven—to foster student understanding of basic geological principles. As much as possible, we have attempted to provide the reader with a sense of the observational techniques and reasoning processes that constitute the discipline of geology.

The organization of the text remains intentionally traditional. Following the overview of geology in Chapter 1, we turn to a discussion of Earth materials and the related processes of volcanism and weathering. Next, we explore the geological work of gravity, water, wind, and ice in modifying and sculpting landscapes. After this look at external processes, we examine Earth's internal structure and the processes that deform rocks and give rise to mountains. Finally, the text concludes with chapters on geologic time and Earth history. This organization accommodates the study of minerals and rocks in the laboratory, which usually comes early in the course.

Realizing that some instructors may prefer to structure their courses somewhat differently, we made each of the chapters self-contained so that they may be taught in a different sequence. Thus, the instructor who wishes to discuss earthquakes, plate tectonics, and mountain building prior to dealing with erosional processes may do so without difficulty. We also chose to introduce plate tectonics in Chapter 1 so that this important and basic theory could be incorporated in appropriate places throughout the text.

More About the Eighth Edition

The Eighth Edition of *Essentials of Geology* represents a thorough revision. *Every* part of the book was examined carefully with the dual goals of keeping topics current and improving the clarity of text discussions. Based on feedback from reviewers and students, we believe we have succeeded.

Here are some examples of what is new in the Eighth Edition of *Essentials of Geology.*

- Fourteen of the special-interest boxes are new. Some highlight recent geologic events, such as the devastating earthquake near Ismit, Turkey, in 1999 (Box 15.2). Others are examples that help develop an awareness of Earth as a system, such as Box 6.1, "The Carbon Cycle and Sedimentary Rocks." Still other new boxes focus on environmental issues and on how geologists study our planet.

- Chapter 1, An Introduction to Geology, offers a new discussion of "The Science of Geology," and revised and expanded treatment of "Earth's Internal Structure." The section on the "Dynamic Earth" is a stronger, more complete first introduction to plate tectonics.

- Chapter 3, Igneous Rocks, and Chapter 7, Metamorphic Rocks, have been significantly revised, rewritten, and reorganized. These chapters now provide stronger technical explanations and easier-to-follow discussions.

- Much of the material in Chapter 4, Volcanoes and Other Igneous Activity, is new. The revision includes expanded treatment of volcanic structures, volcanic hazards, calderas, and the relationship between plate tectonics and igneous activity. These discussions are now clearer and organized in a more logical manner.

- Some of the chapters dealing with external processes have new discussions. Examples include "Causes and Types of Floods" (Chapter 9) and "Interactions Between Groundwater and Streams" (Chapter 10).

- Rewritten and updated discussions on "Intensity Scales" and "Magnitude Scales" appear in Chapter 15, Earthquakes and Earth's Interior.

- Chapter 16, Plate Tectonics, has a reorganized and revised section entitled "Plate Tectonics: The New Paradigm" and completely revised and rewritten sections on "Divergent Plate Boundaries" and "Continental-Continental Convergence." The chapter also includes a new discussion of "Measuring Plate Motion" and a revised look at "The Driving Mechanism."

- Chapter 17, Mountain Building, offers clearer, more logical organization, as well as new material on "Mountain Building at Convergent Boundaries" and "Accretion of Foreign Terranes."

The Teaching and Learning Package

The authors and publisher have been pleased to work with a number of talented people who produced an excellent supplements package. This package includes the traditional supplements that students and professors have come to expect from authors and publishers, as well as some new kinds of supplements that involve electronic media.

For the Student

 Geode III CD-ROM. Each copy of *Essentials of Geology,* Eighth Edition, comes with GEODe III, by Ed Tarbuck, Fred Lutgens, and Dennis Tasa of Tasa Graphic Arts, Inc. GEODe III is a dynamic program that reinforces key concepts by using animations, tutorials, and interactive exercises. This *new version* has been updated and reorganized to provide content that more closely correlates to the text's content. A special GEODe III icon appears throughout the book wherever a text discussion has a corresponding GEODe III activity. This special offering gives students two valuable products (GEODe III and the textbook) for the price of one.

 Companion Website. The Website, specific to the text, contains numerous review exercises (from which students get immediate feedback), exercises to expand one's understanding of geology, and resources for further exploration. This Website provides an excellent platform from which to start using the Internet for the study of geology. Please visit the site at **http://www.prenhall. com/lutgens.**

For the Professor

A wealth of valuable teaching tools are available.

Instructor's Manual. Written by Kenneth Pinzke of Southwestern Illinois College, the Instructor's Manual is intended as a resource for both new and experienced instructors. It includes a variety of lecture outlines, additional source materials, teaching tips, advice about how to integrate visual supplements (including the Web-based resources and *GEODe III*), and various other ideas for the classroom. Also included in every chapter are a wide variety of test questions.

Test Item File. Averages more than 100 multiple-choice, true/false, and short-answer questions per chapter.

Transparencies. Includes more than 300 transparencies, including every Dennis Tasa illustration from the text.

Slides. Includes 150 photos, many from geologist and professional photographer Martin Miller.

Digital Image Gallery for Interactive Teaching DIGIT CD-ROM. All illustrations from the text, the slide-set images, animations from GEODe III, and many text photos in high-resolution JPEG files that can be easily imported into PowerPoint®.

PowerPoint®lecture presentations. Ken Pinzke and Stan Hatfield of Southwestern Illinois College have created compelling, heavily illustrated PowerPoint lecture presentations for each chapter of this text. The presentations are fully customizable or suitable for use as provided. They can be found on the Digital Image Gallery (DIGIT) CD-ROM, in the appropriate chapters.

Prentice Hall Custom Test CD-ROM. Available for Mac and Windows.

New York Times **Themes of the Times—Geology.** This unique newspaper-format supplement features recent articles about geology from the pages of *The New York Times*. This supplement, available at no extra charge from your local Prentice Hall representative, encourages students to make connections between the classroom and the world around them.

Acknowledgments

Writing a college textbook requires the talents and cooperation of many individuals. Working with Dennis Tasa, who is responsible for all of the outstanding illustrations, is always special for us. We not only value his outstanding artistic talents and imagination but his friendship as well. We are grateful to Professor Ken Pinzke at Southwestern Illinois College. In addition to his many helpful suggestions regarding the manuscript, Ken is responsible for the text's Website and Instructor's Manual. Ken is an important part of our team and a valued friend as well. Thanks also to Nancy Lutgens, who helped us greatly in the preparation of this manuscript.

Special thanks goes to those colleagues who prepared in-depth reviews. Their critical comments and thoughtful input helped guide our work and clearly strengthened our text. We wish to thank:

Walter Manger, *University of Arkansas*
John Dembosky, *University of Pittsburgh*
Brendan Schwab, *Humboldt State University*
Kevin McCartney, *University of Maine at Presque Isle*
Cassi Paslick, *Rock Valley College*
Glenn Blaylock, *Utah Valley State College*
Alejandro Amigo, *Sierra College*
Harry Williams, *University of North Texas*

We also want to acknowledge the team of professionals at Prentice Hall. As always, thanks to Paul Corey, President of Prentice Hall's Engineering, Science, and Math division. We sincerely appreciate his continuing strong support for excellence and innovation. Thanks also to our senior editor, Patrick Lynch. His excellent communication skills and helpful and easygoing style contributed greatly to the project. The production team, led by Ed Thomas, has once again done an outstanding job. All are true professionals with whom we are very fortunate to be associated.

Frederick K. Lutgens
Edward J. Tarbuck

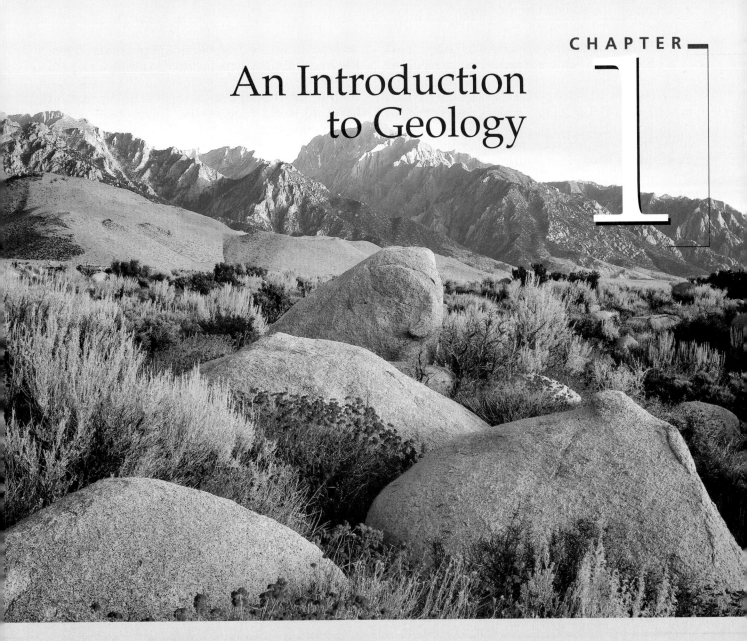

An Introduction to Geology

Focus on Learning

To assist you in learning the important concepts in this chapter, you will find it helpful to focus on the following questions:

• How does physical geology differ from historical geology?

• What is the fundamental difference between uniformitarianism and catastrophism?

• What is relative dating? What are some principles of relative dating?

• How does a scientific hypothesis differ from a scientific theory?

• What are Earth's four major "spheres"?

• Why can Earth be regarded as a system?

• What is the rock cycle? Which geologic interrelationships are illustrated by the cycle?

• What criteria were used to establish Earth's layered structure?

• What is the theory of plate tectonics? How do the three types of plate boundaries differ?

Mount Williamson in California's eastern Sierra Nevada. (Photo by Carr Clifton)

The spectacular eruption of a volcano, the terror brought by an earthquake, the magnificent scenery of a mountain valley, and the destruction created by a landslide all are subjects for the geologist (Figure 1.1). The study of geology deals with many fascinating and practical questions about our physical environment. What forces produce mountains? Will there soon be another great earthquake in California? What was the Ice Age like? Will there be another? How were these ore deposits formed? Should we look for water here? Is strip mining appropriate in this area? Will oil be found if a well is drilled at this location?

The Science of Geology

The subject of this text is **geology,** from the Greek *geo*, "Earth," and *logos*, "discourse." It is the science that pursues an understanding of planet Earth. Geology is traditionally divided into two broad areas—physical and historical. **Physical geology** examines the materials composing Earth and seeks to understand the many processes that operate beneath and upon its surface. The aim of **historical geology,** on the other hand, is to understand the origin of Earth and its development through time. Thus, it strives to establish a chronologi-

cal arrangement of the multitude of physical and biological changes that have occurred in the geologic past. The study of physical geology logically precedes the study of Earth history because we must first understand how Earth works before we attempt to unravel its past.

To understand Earth is challenging because our planet is a dynamic body with many interacting parts and a complex history. Throughout its long existence, Earth has been changing. In fact, it is changing as you read this page and will continue to do so into the foreseeable future. Sometimes the changes are rapid and violent, as when landslides or volcanic eruptions occur. Just as often, change takes place so slowly that it goes unnoticed during a lifetime. Scales of size and space also vary greatly among the phenomena that geologists study. Sometimes they must focus on phenomena that are submicroscopic, and at other times they must deal with features that are continental or global in scale.

Geology is perceived as a science that is done in the out of doors, and rightly so. A great deal of geology is based on observations and experiments conducted in the field. But geology is also done in the laboratory where, for example, the study of various Earth materials provides insights into many basic processes. Frequently geology requires an understanding and

Figure 1.1 On January 13, 2001, a magnitude 7.6 earthquake caused considerable damage in El Salvador. The damage pictured here was caused by a landslide that was triggered by the earthquake. As many as 1000 people were buried under 8 meters (26 feet) of landslide debris. Geologists seek to understand the processes that create such events. (Photo by Reuters/STR/Archive Photos)

application of knowledge and principles from physics, chemistry, and biology. Geology is a science that seeks to expand our knowledge of the natural world and our place in it.

Historical Notes About Geology

The nature of our Earth—its materials and its processes—has been a focus of study for centuries. Writings about fossils, gems, earthquakes, and volcanoes date back to the Greeks, more than 2300 years ago. Certainly, the most influential Greek philosopher was Aristotle. Unfortunately, Aristotle's explanations about the natural world were not derived from keen observations and experiments, as is modern science. Instead, they were arbitrary pronouncements based on the limited knowledge of his day. He believed that rocks were created under the "influence" of the stars and that earthquakes occurred when air in the ground was heated by central fires and escaped explosively! When confronted with a fossil fish, he explained that "a great many fishes live in the earth motionless and are found when excavations are made." Although Aristotle's explanations may have been adequate for his day, they unfortunately continued to be expounded for many centuries, thus thwarting the acceptance of more up-to-date ideas.

Catastrophism

In the mid-1600s James Ussher, Anglican Archbishop of Armagh, Primate of all Ireland, published a major work that had immediate and profound influences. A respected scholar of the Bible, Ussher constructed a chronology of human and Earth history in which he determined that Earth was only a few thousand years old, having been created in 4004 B.C. Ussher's treatise earned widespread acceptance among Europe's scientific and religious leaders, and his chronology was soon printed in the margins of the Bible itself.

During the seventeenth and eighteenth centuries the doctrine of **catastrophism** strongly influenced people's thinking about Earth. Briefly stated, catastrophists believed that Earth's landscapes had been shaped primarily by great catastrophes. Features such as mountains and canyons, which today we know take great periods of time to form, were explained as having been produced by sudden and often worldwide disasters produced by unknown causes that no longer operate. This philosophy was an attempt to fit the rates of Earth processes to the then current ideas on the age of Earth.

The relationship between catastrophism and the age of Earth has been summarized nicely:

That the earth had been through tremendous adventures and had seen mighty changes during its obscure past was plainly evident to every inquiring eye; but to concentrate these changes into a few brief millenniums required a tailor-made philosophy, a philosophy whose basis was sudden and violent change.*

The Birth of Modern Geology

Against this backdrop of Aristotle's views and an Earth created in 4004 B.C., a Scottish physician and gentleman farmer named James Hutton published *Theory of the Earth* in 1795. In this work Hutton put forth a fundamental principle that is a pillar of geology today: **uniformitarianism**. It states that the *physical, chemical, and biological laws that operate today also operated in the geologic past*. In other words, the forces and processes that we observe shaping our planet today have been at work for a very long time. Thus, to understand ancient rocks, we must first understand present-day processes and their results. This idea is commonly stated as *The present is the key to the past*.

Prior to Hutton's *Theory of the Earth*, no one had effectively demonstrated that geological processes can continue over extremely long periods of time. Hutton persuasively argued that forces that appear small could, over long spans of time, produce effects just as great as those resulting from sudden catastrophic events. Hutton carefully cited verifiable observations to support his ideas.

For example, when he argued that mountains are sculpted and ultimately destroyed by weathering and the work of running water, and that their wastes are carried to the oceans by processes that can be observed, Hutton said, "We have a chain of facts which clearly demonstrates . . . that the materials of the wasted mountains have traveled through the rivers"; and further, "There is not one step in all this progress . . . that is not to be actually perceived." He then went on to summarize this thought by asking a question and immediately providing the answer: "What more can we require? Nothing but time."

Today the basic tenets of uniformitarianism are just as viable as in Hutton's day. We realize more strongly than ever that the present gives us insight into the past and that the physical, chemical, and biological laws that govern geological processes remain unchanged through time. However, we also understand that the doctrine should not be taken too literally. To say that geological processes in the past were the same as those occurring today is not to suggest that they always had the same relative importance or operated at precisely the same rate. Moreover, some important geologic processes are not currently observable, but evidence that they occur is well established. For example, we know that Earth has experienced impacts from large meteorites even though we have no human witnesses. Such events altered Earth's crust, modified its climate, and strongly influenced life on the planet.

The acceptance of uniformitarianism meant the acceptance of a very lengthy history for Earth. Although

*H. E. Brown, V. E. Monnett, and J. W. Stovall, *Introduction to Geology* (New York: Blaisdell, 1958).

processes vary in their intensity, they still take a very long time to create or destroy major landscape features (Figure 1.2).

For example, geologists have established that mountains once existed in portions of present-day Minnesota, Wisconsin, and Michigan. Today the region consists of low hills and plains. Erosion (processes that wear land away) gradually destroyed these peaks. Estimates indicate that the North American continent is being lowered at a rate of about 3 centimeters per 1000 years. At this rate, it would take 100 million years for water, wind, and ice to lower mountains that were 3000 meters (10,000 feet) high.

But even this time span is relatively short on the time scale of Earth history, for the rock record contains evidence that shows Earth has experienced many cycles of mountain building and erosion.

In the chapters that follow, we will be examining the materials that compose our planet and the processes that modify it (Box 1.1). It is important to remember that although many features of our physical landscape may seem to be unchanging in terms of the decades over which we observe them, they are nevertheless changing, but on time scales of hundreds, thousands, or even many millions of years. Concerning the ever changing nature of Earth through great expanses of geologic time, James Hutton made a statement that was to become his most famous. In concluding his classic 1788 paper, published in the *Transactions of the Royal Society of Edinburgh*, he stated, "The result, therefore, of our present enquiry is, that we find no vestige of a beginning, no prospect of an end."

Geologic Time

 Geologic Time
↳ Relative Dating

Although Hutton and others recognized that geologic time is exceedingly long, they had no methods to accurately determine the age of Earth. However, in 1896 radioactivity was discovered. Using radioactivity for

Figure 1.2 Over millions of years, weathering, gravity, and the erosional work of the Colorado River and the streams that flow into it have created Arizona's Grand Canyon. Geologic processes often act so slowly that changes may not be visible during an entire human lifetime. The relative ages of the rock layers in the canyon can be determined by applying the law of superposition. The youngest rocks are on top, and the oldest are at the bottom. (Photo by Marc Muench/David Muench Photography, Inc.)

Geology, People, and the Environment

The study of geology sometimes involves interactions between people and the natural environment. Stated another way, many of the problems and issues addressed by geology are of practical value to people.

Natural hazards are a part of living on Earth. Every day they adversely affect millions of people worldwide and are responsible for staggering damages. Among the hazardous Earth processes studied by geologists are volcanoes, floods, earthquakes, and landslides. Of course, geologic hazards are simply *natural* processes. They become hazards only when people try to live where these processes occur (Figure 1.A).

Resources represent another important focus of geology that is of great practical value to people. They include water and soil, a great variety of metallic and nonmetallic minerals, and energy. Together they form the very foundation of modern civilization. Geology deals not only with the formation and occurrence of these vital resources but also with maintaining supplies and with the environmental impact of their extraction and use.

Complicating all environmental issues is rapid world population growth and everyone's aspiration to a better standard of living. Earth is now gaining about 100 million people each year. This means a ballooning demand for resources and a growing pressure for peo-

Figure 1.A   This is an image of Italy's Mt. Vesuvius in September 2000. This major volcano is surrounded by the city of Naples and the Bay of Naples. In 79 A.D. Vesuvius explosively erupted, burying the towns of Pompeii and Herculanaeum in volcanic ash. Will it happen again? Geologic hazards are *natural* processes. They only become hazards when people try to live where these processes occur. (Image courtesy of NASA)

ple to dwell in environments having significant geologic hazards.

Not only do geologic processes have an impact on people, but we humans can dramatically influence geologic processes as well. For example, river flooding is natural, but the magnitude and frequency of flooding can be changed significantly by human activi-

ties such as clearing forests, building cities, and constructing dams. Unfortunately, natural systems do not always adjust to artificial changes in ways that we can anticipate. Thus, an alteration to the environment that was intended to benefit society often has the opposite effect.

---

dating was first attempted in 1905 and has been refined ever since. Geologists are now able to assign fairly accurate dates to events in Earth history. For example, we know that the dinosaurs died out about 65 million years ago. Today the age of Earth is put at about 4.5 billion years.

The concept of geologic time is new to many nongeologists. People are accustomed to dealing with increments of time that are measured in hours, days, weeks, and years. Our history books often examine events over spans of centuries, but even a century is difficult to appreciate fully. For most of us, someone or something that is 90 years old is *very old*, and a 1000-year-old artifact is *ancient*.

By contrast, those who study geology must routinely deal with vast time periods—millions or billions (thousands of millions) of years. When viewed in the

context of Earth's 4.5-billion-year history, a geologic event that occurred 100 million years ago may be characterized as "recent" by a geologist, and a rock sample that has been dated at 10 million years may be called "young." An appreciation for the magnitude of geologic time is important in the study of geology because many processes are so gradual that vast spans of time are needed before significant changes occur.

During the nineteenth century, long before the discovery of radioactivity and the establishment of reliable *numerical* dates, a geologic time scale was developed using principles of relative dating. **Relative dating** means that events are placed in their proper sequence or order without knowing their age in years. This is done by applying principles such as the **law of superposition** which states that in layers of sedimentary rocks or lava flows, the youngest layer is on top, and the oldest is on the bottom (assuming that nothing has turned the layers upside down, which sometimes happens). Arizona's Grand Canyon provides a fine example where the oldest rocks are located in the inner gorge while the youngest rocks are found on the rim (see Figure 1.2). So the law of superposition establishes the *sequence* of rock layers but not, of course, their numerical ages. Today such a proposal appears to be elementary, but 300 years ago it amounted to a major breakthrough in scientific reasoning by establishing a rational basis for relative time measurements.

Fossils, the remains or traces of prehistoric life, were also essential to the development of a geologic time scale (Figure 1.3). Fossils are the basis for the **principle of fossil succession**, which states that fossil organisms succeed one another in a definite and determinable order, and therefore any time period can be recognized by its fossil content. This principle was laboriously worked out over decades by collecting fossils from countless rock layers around the world. Once established, it allowed geologists to identify rocks of the same age in widely separated places and to build the *geologic time scale* shown in Figure 1.4.

Notice that units having the same designations do not necessarily extend for the same number of years. For example, the Cambrian period lasted about 50 million years, whereas the Silurian period spanned only about 26 million years. As we will emphasize again in Chapter 18, this situation exists because the basis for establishing the time scale was not the regular rhythm of a clock, but the changing character of life forms through time. Specific dates were added long after the time scale was established. A glance at Figure 1.4 also reveals that the Phanerozoic eon is divided into many more units than earlier eons even though it encompasses only about 12 percent of Earth history. The meager fossil record for these earlier eons is the primary reason for

the lack of detail on this portion of the time scale. Without abundant fossils, geologists lose their primary tool for subdividing geologic time.

The Nature of Scientific Inquiry

All science is based on the assumption that the natural world behaves in a consistent and predictable manner that is comprehensible through careful, systematic study. The overall goal of science is to discover the underlying patterns in nature and then to use this knowledge to make predictions about what should or should not be expected, given certain facts or circumstances. For example, by knowing how oil deposits form, geologists are able to predict the most favorable sites for exploration and, perhaps as important, how to avoid regions having little or no potential.

The development of new scientific knowledge involves some basic logical processes that are universally accepted. To determine what is occurring in the natural world, scientists collect scientific *"facts"* through observation and measurement. Because some error is inevitable, the accuracy of a particular measurement or observation is always open to question. Nevertheless, these data are essential to science and serve as the springboard for the development of scientific theories.

Hypothesis

Once facts have been gathered and principles have been formulated to describe a natural phenomenon, investigators try to explain how or why things happen in the manner observed. They often do this by constructing a tentative (or untested) explanation, which is called a scientific **hypothesis** or **model**. (The term model, although often used synonymously with hypothesis, is a less precise term because it is sometimes used to describe a scientific theory as well.) It is best if an investigator can formulate more than one hypothesis to explain a given set of observations. If an individual scientist is unable to devise multiple models, others in the scientific community will almost always develop alternative explanations. A spirited debate frequently ensues. As a result, extensive research is conducted by proponents of opposing models, and the results are made available to the wider scientific community in scientific journals.

Before a hypothesis can become an accepted part of scientific knowledge, it must pass objective testing and analysis. (If a hypothesis cannot be tested, it is not scientifically useful, no matter how interesting it might seem.) The verification process requires that *predictions* be made based on the model being considered and that the predictions be tested by comparing them against objective observations of nature. Put another way, hypotheses must fit observations other than those used to formulate them in the first place. Those hypotheses that

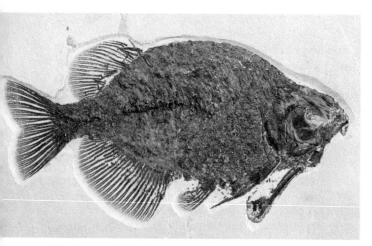

Figure 1.3   Fossils are important tools for the geologist. In addition to being very important in relative dating, fossils can be useful environmental indicators. This fossil fish of Eocene age is from the Green River Formation in Wyoming. (Photo by John Cancalosi/DRK Photo)

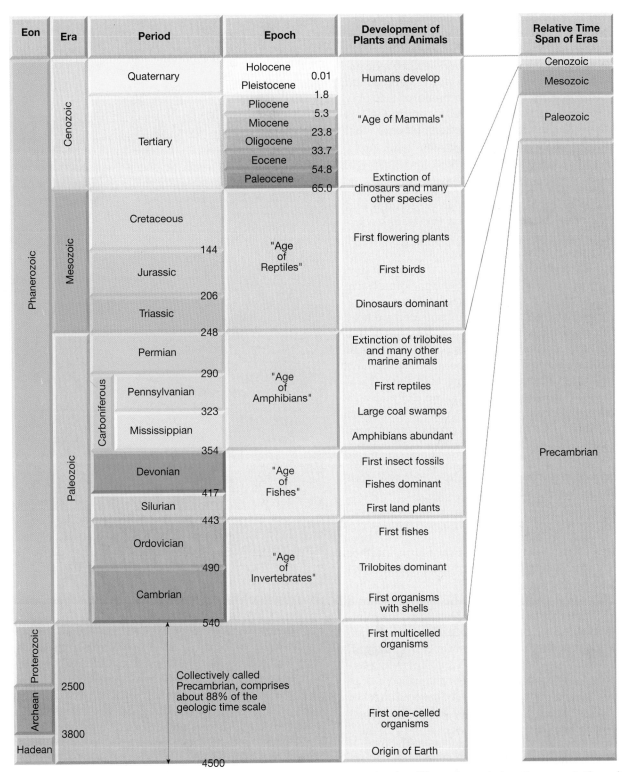

Figure 1.4 The geologic time scale. Numbers on the time scale represent time in millions of years before the present. These dates were added long after the time scale had been established using relative dating techniques. The Precambrian accounts for more than 88 percent of geologic time. (Data from Geological Society of America)

fail rigorous testing are ultimately discarded. The history of science is littered with discarded hypotheses. One of the best known is the Earth-centered model of the universe—a proposal that was supported by the apparent daily motion of the Sun, Moon, and stars around Earth. As the mathematician Jacob Bronowski so ably stated, "Science is a great many things, but in the end they all return to this: Science is the acceptance of what works and the rejection of what does not."

Theory

When a hypothesis has survived extensive scrutiny and when competing models have been eliminated, a hypothesis may be elevated to the status of a scientific **theory**. In everyday language we may say "that's only a theory." But a scientific theory is a well-tested and widely accepted view that the scientific community agrees best explains certain observable facts.

Theories that are extensively documented are held with a very high degree of confidence. Theories of this stature that are comprehensive in scope have a special status. They are called **paradigms** because they explain a large number of interrelated aspects of the natural world. For example, the theory of plate tectonics is a paradigm of the geological sciences that provides the framework for understanding the origin of mountains, earthquakes, and volcanic activity. In addition, plate tectonics explains the evolution of the continents and the ocean basins through time—a topic we will consider later in this chapter.

Scientific Methods

The processes just described, in which scientists gather facts through observations and formulate scientific hypotheses and theories is called the *scientific method*. Contrary to popular belief, the scientific method is not a standard recipe that scientists apply in a routine manner to unravel the secrets of our natural world. Rather, it is an endeavor that involves creativity and insight. Rutherford and Ahlgren put it this way: "Inventing hypotheses or theories to imagine how the world works and then figuring out how they can be put to the test of reality is as creative as writing poetry, composing music, or designing skyscrapers."*

There is not a fixed path that scientists always follow that leads unerringly to scientific knowledge. Nevertheless, many scientific investigations involve the following steps: (1) collection of scientific facts (data) through observation and measurement (Figure 1.5); (2) development of one or more working hypotheses or models to explain these facts; (3) development of ob-

*F. James Rutherford and Andrew Ahlgren, *Science for All Americans* (New York: Oxford University Press, 1990), p. 7.

Figure 1.5 This geologist is taking a lava sample from a "skylight" in a lava tube. Near Kilauea Volcano, Hawaii. (Photo by G. Brad Lewis/Tony Stone Images)

servations and experiments to test the hypotheses; and (4) the acceptance, modification, or rejection of the hypotheses based on extensive testing.

Other scientific discoveries may result from purely theoretical ideas that stand up to extensive examination. Some researchers use high-speed computers to simulate what is happening in the "real" world. These models are useful when dealing with natural processes that occur on very long time scales or take place in extreme or inaccessible locations. Still other scientific advancements have been made when a totally unexpected happening occurred during an experiment. These serendipitous discoveries are more than pure luck; for as Louis Pasteur stated, "In the field of observation, chance favors only the prepared mind."

Scientific knowledge is acquired through several avenues, so it might be best to describe the nature of scientific inquiry as the *methods* of science rather than *the* scientific method. In addition, it should always be remembered that even the most compelling scientific theories are still simplified explanations of the natural world.

A View of Earth

Introduction
↳ A View of Earth

A view such as the one in Figure 1.6A provided the *Apollo 8* astronauts as well as the rest of humanity with a unique perspective of our home. Seen from space, Earth is breathtaking in its beauty and startling in its solitude. Such an image reminds us that our home is, after all, a planet—small, self-contained, and in some ways even fragile.

As we look more closely at our planet from space, it becomes apparent that Earth is much more than rock and soil. In fact, the most conspicuous features in Figure 1.6A are not continents but swirling clouds suspended above the surface and the vast global ocean. These features emphasize the importance of water to our planet.

The closer view of Earth from space shown in Figure 1.6B helps us appreciate why the physical environment is traditionally divided into three major parts: the water portion of our planet, the hydrosphere; Earth's gaseous envelope, the atmosphere; and, of course, the solid Earth. It needs to be emphasized that our environment is highly integrated and is not dominated by rock, water, or air alone. Rather, it is characterized by continuous interactions as air comes in contact with rock, rock with water, and water with air. Moreover, the biosphere, which is the totality of all plant and animal life on our planet, interacts with each of the three physical realms and is an equally integral part of the planet. Thus, Earth can be thought of as consisting of four major spheres: the hydrosphere, atmosphere, solid Earth, and biosphere.

The interactions among Earth's four spheres are uncountable. Figure 1.7 provides us with one easy-to-visualize example. The shoreline is an obvious meeting place for rock, water, and air. In this scene, ocean waves that were created by the drag of air moving across the water are breaking against the rocky shore. The force of the water can be powerful, and the erosional work that is accomplished can be great.

Hydrosphere

Earth is sometimes called the *blue* planet. Water more than anything else makes Earth unique. The **hydrosphere** is a dynamic mass of liquid that is continually on the move, evaporating from the oceans to the atmosphere, precipitating back to the land, and running back to the ocean again. The global ocean is certainly the most prominent feature of the hydrosphere, blanketing

A.

B.

Figure 1.6 **A**. View that greeted the *Apollo 8* astronauts as their spacecraft emerged from behind the Moon. (NASA Headquarters) **B**. Africa and Arabia are prominent in this image of Earth taken from *Apollo 17*. The tan cloud-free zones over the land coincide with major desert regions. The band of clouds across central Africa is associated with a much wetter climate that in places sustains tropical rain forests. The dark blue of the oceans and the swirling cloud patterns remind us of the importance of the oceans and the atmosphere. Antarctica, a continent covered by glacial ice, is visible at the South Pole. (NASA/Science Source/Photo Researchers, Inc.)

Atmosphere

Earth is surrounded by a life-giving gaseous envelope called the **atmosphere.** Compared with the solid Earth, the atmosphere is thin and tenuous. One half lies below an altitude of 5.6 kilometers (3.5 miles), and 90 percent occurs within just 16 kilometers (10 miles) of Earth's surface. By comparison, the radius of the solid Earth (distance from the surface to the center) is about 6400 kilometers (nearly 4000 miles)! Despite its modest dimensions, this thin blanket of air is an integral part of the planet. It not only provides the air that we breathe but also acts to protect us from the Sun's intense heat and dangerous ultraviolet radiation. The energy exchanges that continually occur between the atmosphere and the surface and between the atmosphere and space produce the effects we call weather and climate.

If, like the Moon, Earth had no atmosphere, our planet would not only be lifeless but many of the processes and interactions that make the surface such a dynamic place could not operate. Without weathering and erosion, the face of our planet might more closely resemble the lunar surface, which has not changed appreciably in nearly 3 billion years.

Biosphere

The **biosphere** includes all life on Earth. It is concentrated near the surface in a zone that extends from the ocean floor upward for several kilometers into the atmosphere. Plants and animals depend on the physical environment for the basics of life. However, organisms do not just respond to their physical environment. Indeed, the biosphere powerfully influences the other three spheres. Without life, the makeup and nature of the solid Earth, hydrosphere, and atmosphere would be very different.

The Solid Earth

Lying beneath the atmosphere and the oceans is the solid Earth. Much of our study of the solid Earth focuses on the more accessible surface features. Fortunately, many of these features represent the outward expressions of the dynamic behavior of Earth's interior. By examining the most prominent surface features and their global extent, we can obtain clues to the dynamic processes that have shaped our planet. A first look at the structure of Earth's interior and at the major surface features of the solid Earth will come later in the chapter.

Earth as a System

As we study Earth, it becomes clear that our planet can be viewed as a dynamic body with many separate but interacting parts or *spheres.* The hydrosphere, atmosphere, biosphere, and solid Earth, and all of their com-

Figure 1.7   The shoreline is one obvious meetingplace for the hydrosphere, atmosphere, and solid Earth. In this scene, ocean waves that were created by the force of moving air break against the rocky shore. The force of the water can be powerful, and the erosional work that is accomplished can be great. (Photo by Galen Rowell)

nearly 71 percent of Earth's surface to an average depth of about 3800 meters (12,500 feet). It accounts for about 97 percent of Earth's water. However, the hydrosphere also includes the freshwater found underground and in streams, lakes, and glaciers. Moreover, water is an important component of all living things.

Although these latter sources constitute just a tiny fraction of the total, they are much more important than their meager percentage indicates. In addition to providing the freshwater that is so vital to life on land, streams, glaciers, and groundwater are responsible for creating many of our planet's varied landforms.

ponents, can be studied separately. However, the parts are not isolated. Each is related in some way to the others to produce a complex and continuously interacting whole that we call the *Earth system*.

A **system** is a group of interacting, or interdependent parts that form a complex whole. Most of us hear and use the term frequently. We may service our car's cooling *system*, make use of the city's transportation *system*, and be a participant in the political *system*. A news report might inform us of an approaching weather *system*. Further, we know that Earth is just a small part of a large system known as the *solar system*.

The parts of the Earth system are linked, so a change in one part can produce changes in any or all of the other parts. For example, when a volcano erupts, lava from Earth's interior may flow out at the surface and block a nearby valley. This new obstruction influences the region's drainage system by creating a lake or causing streams to change course. The large quantities of volcanic ash and gases that can be emitted during an eruption might be blown high into the atmosphere and influence the amount of solar energy that can reach Earth's surface. The result could be a drop in air temperatures over the entire hemisphere.

Where the surface is covered by lava flows or a thick layer of volcanic ash, existing soils are buried. This causes the soil-forming processes to begin anew to transform the new surface material into soil (Figure 1.8). The soil that eventually forms will reflect the interactions among many parts of the Earth system—the volcanic parent material, the type and rate of weathering, and the impact of biological activity. Of course, there would also be significant changes in the biosphere. Some organisms and their habitats would be eliminated by the lava and ash, whereas new settings for life, such as the lake, would be created. The potential climate change could also impact sensitive life forms.

The Earth system is characterized by processes that vary on spatial scales from fractions of millimeters to thousands of kilometers. Time scales for Earth's processes range from milliseconds to billions of years. As we learn about Earth, it becomes increasingly clear that despite significant separations in distance or time, many processes are connected, and a change in one component can influence the entire system.

The Earth system is powered by energy from two sources. The Sun drives external processes that occur in the atmosphere, hydrosphere, and at Earth's surface. Weather and climate, ocean circulation, and erosional processes are driven by energy from the Sun. Earth's interior is the second source of energy. Heat remaining from when our planet formed, and heat that is continuously generated by radioactive decay powers the internal processes that produce volcanoes, earthquakes, and mountains.

Humans are *part of* the Earth system, a system in which the living and nonliving components are entwined and interconnected. Therefore, our actions produce changes in all of the other parts. When we burn gasoline and coal, build seawalls along the shoreline, dispose of our wastes, and clear the land, we cause other parts of the system to respond, often in unforeseen ways. Throughout this book you will learn about many of Earth's subsystems: the hydrologic system, the tectonic (mountain-building) system, and the rock cycle, to name a few. Remember that these components *and we humans* are all part of the complex interacting whole we call the Earth system.

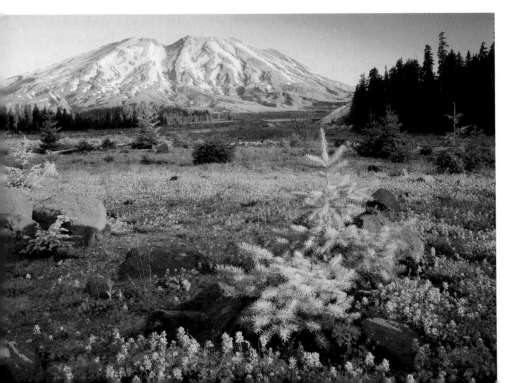

Figure 1.8 When Mount St. Helens erupted in May 1980, the area shown here was buried by a volcanic mudflow. Now plants are reestablished and new soil is forming. (Jack W. Dykinga Associates)

The Rock Cycle: Part of the Earth System

GEODe Earth Materials
↳ The Rock Cycle

The Earth system has a nearly endless array of cycles or loops in which matter is recycled over and over again. Viewed over long time spans, the rocks of the solid Earth are constantly forming, changing, and re-forming. The loop that involves the processes by which one rock changes to another is called the **rock cycle.** This cycle allows us to view many of the interrelationships among different parts of the Earth system. It helps us to understand the origin of igneous, sedimentary, and metamorphic rocks and to see that each type is linked to the others by the processes that act upon and within the planet. Learn the rock cycle well; you will be examining its interrelationships in greater detail throughout this book.

The Basic Cycle

We will begin our look at the rock cycle at the bottom of Figure 1.9. **Magma** is molten material that forms in certain environments in Earth's interior where temperatures and pressures are such that rock melts. Once formed, magma migrates upward into Earth's outer layer or crust. Eventually magma cools and solidifies. This process, called *crystallization*, may occur either beneath the surface or, following a volcanic eruption, at the surface. In either situation the resulting rocks are called **igneous rocks.**

If igneous rocks are exposed at the surface, they will undergo *weathering*, in which the day-in and day-out influences of the atmosphere slowly disintegrate and decompose rocks. The materials that result are often moved downslope by gravity before being picked up and transported by a number of erosional agents, such as running water, glaciers, wind, or waves. Eventually these particles and dissolved substances, called **sediment,** are deposited. Although most sediment ultimately comes to rest in the ocean, other sites of deposition include river floodplains, desert basins, swamps, and dunes.

Next the sediments undergo *lithification,* a term meaning "conversion into rock." Sediment is usually lithified into **sedimentary rock** when compacted by the weight of overlying layers or when cemented as percolating water fills the pores with mineral matter.

If the resulting sedimentary rock is buried deep within Earth and involved in the dynamics of mountain building, or intruded by a mass of magma, it will be subjected to great pressures and/or intense heat. The sedimentary rock will react to the changing environment and turn into the third rock type, **metamorphic rock.** When metamorphic rock is subjected to additional pressure changes or to still higher temperatures,

it will melt, creating magma, which will eventually crystallize into igneous rock.

Processes driven by heat from Earth's interior are responsible for creating igneous and metamorphic rocks. Weathering and erosion—external processes powered by a combination of energy from the Sun and gravity—produce the sediment from which sedimentary rocks form.

Alternative Paths

The paths shown in the basic cycle are not the only ones that are possible. To the contrary, other paths are just as likely to be followed as those described in the preceding section. These alternatives are indicated by the blue arrows in Figure 1.9.

Igneous rocks, rather than being exposed to weathering and erosion at Earth's surface, may remain deeply buried. Eventually these masses may be subjected to the strong compressional forces and high temperatures associated with mountain building. When this occurs, they are transformed directly into metamorphic rocks.

Metamorphic and sedimentary rocks, as well as sediment, do not always remain buried. Rather, overlying layers may be stripped away, exposing the once buried rock. When this happens, the material is attacked by weathering processes and turned into new raw materials for sedimentary rocks.

Although rocks might seem to be unchanging masses, the rock cycle clearly shows that they are not. The changes, however, take time—great amounts of time.

Earth's Internal Structure

GEODe Introduction
↳ Earth's Layered Structure

Earth's interior consists of three layers defined by their chemical composition—the crust, mantle, and core. These layers formed early in Earth's history.* In addition to these compositionally distinct layers, Earth can be divided into layers based on physical properties. The physical properties used to define such zones include whether the layer is solid or liquid and how weak or strong it is. Knowledge of both types of layered structures is essential to our understanding of basic geologic processes, such as volcanism, earthquakes, and mountain building (Figure 1.10).

Layers Defined by Composition

Crust. The **crust,** Earth's comparatively thin, rocky outer skin, is generally divided into oceanic and continental crust. The oceanic crust is roughly 7 kilometers (5 miles) thick and composed of dark igneous rocks

*For more on this, see "Early Evolution of Earth" in Chapter 19.

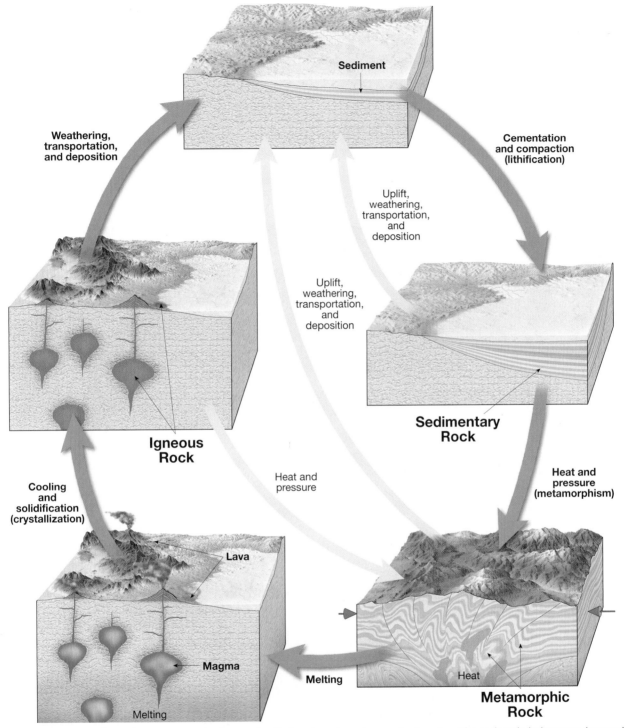

Figure 1.9 Viewed over long spans, rocks are constantly forming, changing, and reforming. The rock cycle helps us understand the origin of the three basic rock groups. Arrows represent processes that link each group to the others.

called *basalt*. By contrast, the continental crust averages 35–40 kilometers (25 miles) thick but may exceed 70 kilometers (40 miles) in some mountainous regions. Unlike the oceanic crust, which has a relatively homogeneous chemical composition, the continental crust consists of many rock types. The upper crust has an average composition of a *granitic rock* called *granodiorite*, whereas the composition of the lowermost continental crust is more akin to basalt (see Chapter 3). Continental rocks have an average density of about 2.7 g/cm^3, and some have been discovered that are 4 billion years old. The rocks of the oceanic crust are

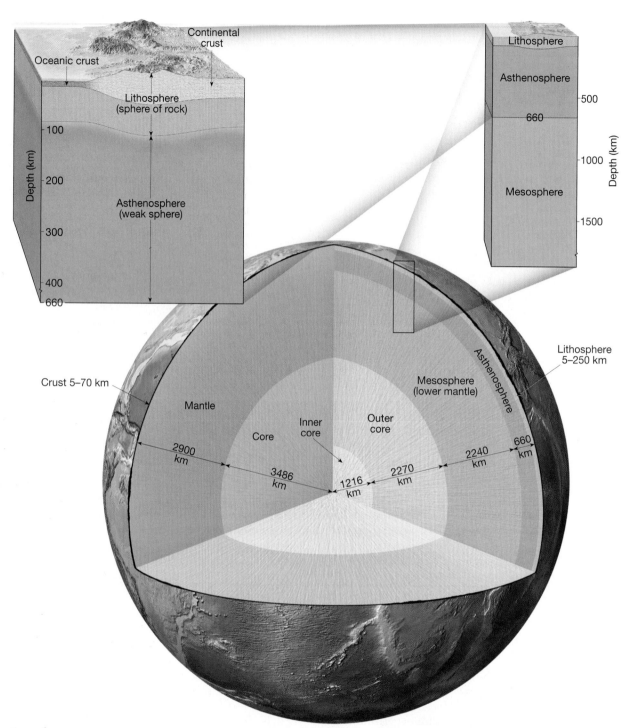

Figure 1.10 Views of Earth's layered structure. The left side of the large cross section shows that Earth's interior is divided into three different layers based on compositional differences—the crust, mantle, and core. The right side of the large cross section depicts the five main layers of Earth's interior based on physical properties and hence mechanical strength—the lithospere, asthenosphere, mesosphere, outer core, and inner core. The block diagrams above the large cross section show an enlarged view of the upper portion of Earth's interior.

younger (180 million years or less) and more dense (about 3.0 g/cm^3) than continental rocks.*

Mantle. Over 82 percent of Earth's volume is contained in the **mantle,** a solid, rocky shell that extends to a depth of 2900 kilometers (1800 miles). The boundary between the crust and mantle represents a marked change in chemical composition. The dominant rock type in the uppermost mantle is *peridotite*, which has a density of 3.3 g/cm^3. At greater depth peridotite changes by assuming a more compact crystalline structure and hence a greater density.

Core. The composition of the **core** is thought to be an iron-nickel alloy with minor amounts of oxygen, silicon, and sulfur—elements that readily form compounds with iron. At the extreme pressure found in the core, this iron-rich material has an average density of nearly 11 g/cm^3 and approaches 14 times the density of water at Earth's center.

Layers Defined by Physical Properties

Earth's interior is characterized by a gradual increase in temperature, pressure, and density with depth. Estimates put the temperature at a depth of 100 kilometers at between 1200°C and 1400°C, whereas the temperature at Earth's center may exceed 6700°C. Clearly, Earth's interior has retained much of the energy acquired during its formative years, despite the fact that heat is continuously flowing toward the surface, where it is lost to space. The increase in pressure with depth causes a corresponding increase in rock density.

The gradual increase in temperature and pressure with depth affects the physical properties and hence the mechanical behavior of Earth materials. When a substance is heated, its chemical bonds weaken and its mechanical strength (resistance to deformation) is reduced. If the temperature exceeds the melting point of an Earth material, the material's chemical bonds break and melting ensues. If temperature were the only factor that determined whether a substance melted, our planet would be a molten ball covered with a thin, solid outer shell. However, pressure also increases with depth and tends to increase rock strength. Furthermore, because melting is accompanied by an increase in volume, it occurs at higher temperatures at depth because of greater confining pressure. Thus, depending on the physical environment (temperature and pressure), a particular Earth material may behave like a brittle solid, deform in a puttylike manner, or even melt and become liquid.

Earth can be divided into five main layers based on physical properties and hence mechanical strength—the *lithosphere, asthenosphere, mesosphere (lower mantle), outer core,* and *inner core.*

Lithosphere and Asthenosphere. Based on physical properties, Earth's outermost layer consists of the crust and uppermost mantle and forms a relatively cool, rigid shell. Although this layer is composed of materials with markedly different chemical compositions, it tends to act as a unit that exhibits rigid behavior—mainly because it is cool and thus strong. This layer, called the **lithosphere** (*sphere of rock*), averages about 100 kilometers in thickness but may be more than 250 kilometers thick below the older portions of the continents (Figure 1.10). Within the ocean basins the lithosphere is only a few kilometers thick along the oceanic ridges but increases to perhaps 100 kilometers in regions of older and cooler oceanic crust.

Beneath the lithosphere, in the upper mantle (to a depth of about 660 kilometers) lies a soft, comparatively weak layer known as the **asthenosphere** ("weak sphere"). The top portion of the asthenosphere has a temperature/pressure regime that results in a small amount of melting. Within this very weak zone the lithosphere is mechanically detached from the layer below. The result is that the lithosphere is able to move independently of the asthenosphere, a fact we will consider in the next section.

It is important to emphasize that the strength of various Earth materials is a function of both their composition and of the temperature and pressure of their environment. You should not get the idea that the entire lithosphere behaves like a brittle solid similar to rocks found on the surface. Rather, the rocks of the lithosphere get progressively hotter and weaker (more easily deformed) with increasing depth. At the depth of the uppermost asthenosphere, the rocks are close enough to their melting temperatures (some melting may actually occur) that they are very easily deformed. Thus, the uppermost asthenosphere is weak because it is near its melting point, just as hot wax is weaker than cold wax.

Mesosphere or Lower Mantle. Below the zone of weakness in the uppermost asthenosphere, increased pressure counteracts the effects of higher temperature, and the rocks gradually strengthen with depth. Between the depths of 660 kilometers and 2900 kilometers a more rigid layer, called the **mesosphere** (*middle sphere*) or **lower mantle,** is found. Despite their strength, the rocks of the mesosphere are still very hot and capable of very gradual flow.

Inner and Outer Core. The core, which is composed mostly of an iron-nickel alloy, is divided into two regions that exhibit very different mechanical strengths.

---

*Liquid water has a density of 1 g/cm^3; therefore, the density of basalt is three times that of water.

A. Divergent boundary

B. Convergent boundary

Figure 1.11   Mosaic of rigid plates that constitute Earth's outer shell. (After W. B. Hamilton, U.S. Geological Survey)

The **outer core** is a *liquid layer* 2270 kilometers (1410 miles) thick. It is the movement of metallic iron within this zone that generates Earth's magnetic field. The **inner core** is a sphere having a radius of 1216 kilometers (754 miles). Despite its higher temperature, the material in the inner core is stronger (because of immense pressure) than the outer core and behaves like a *solid*.

Dynamic Earth

GEODe II   Internal Processes
    ↳ Plate Tectonics

Earth is a dynamic planet! If we could go back in time a few hundred million years, we would find the face of face has remained essentially unchanged—only a few craters have been added.

The Theory of Plate Tectonics

Within the past several decades a great deal has been learned about the workings of our dynamic planet. This period has seen an unequaled revolution in our understanding of Earth. The revolution began in the early part of the twentieth century with the radical proposal of *continental drift*—the idea that the continents moved about the face of the planet. This proposal contradicted the established view that the continents and ocean basins are permanent and stationary features on the face of Earth. For that reason, the notion of drifting continents was received with great skepticism and even ridicule. More than 50 years passed before enough data were gathered to transform this controversial hypothesis into a sound theory that wove together the basic processes known to operate on Earth. The theory that finally emerged, called the **theory of plate tectonics**, provided geologists with the first comprehensive model of Earth's internal workings.*

According to the plate tectonics model, Earth's rigid outer shell (*lithosphere*) is broken into numerous slabs called **plates**, which are in motion and are continually changing shape and size. As shown in Figure 1.11, seven major lithospheric plates are recognized. They are the North American, South American, Pacific, African, Eurasian, Australian, and Antarctic plates. Intermediate-size plates include the Caribbean, Nazca, Philippine, Arabian, Cocos, and Scotia plates. In addition, over a dozen smaller plates have been identified, but are not shown in Figure 1.11. Note that several large plates include an entire continent plus a large area of seafloor (for example, the South American plate). However, none of the plates are defined entirely by the margins of a single continent.

The lithospheric plates move relative to each other at a very slow but continuous rate that averages about 5 centimeters (2 inches) a year. This movement is ultimately driven by the unequal distribution of heat within Earth. Hot material found deep in the mantle moves slowly upward and serves as one part of our planet's internal convective system. Concurrently, cooler, denser slabs of lithosphere descend back into the mantle, setting Earth's rigid outer shell in motion. Ultimately, the titanic, grinding movements of Earth's lithospheric plates generate earthquakes, create volcanoes, and deform large masses of rock into mountains.

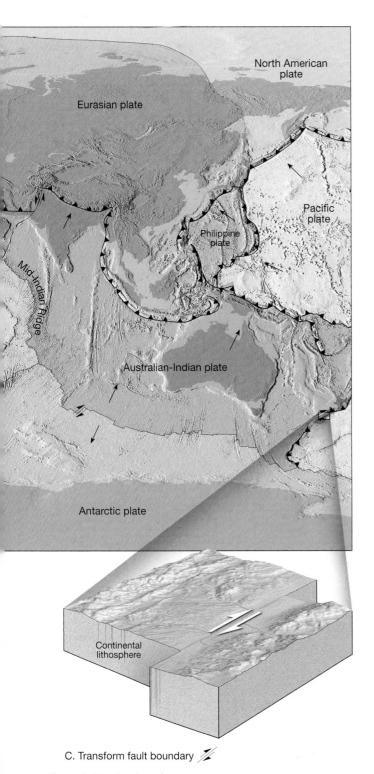

C. Transform fault boundary

Figure 1.11   Continued.

our planet dramatically different from what we see today. There would be no Mount St. Helens, Rocky Mountains, or Gulf of Mexico. Moreover, we would find continents having different sizes and shapes and located in different positions than today's landmasses. In contrast, over the past few billion years the Moon's sur-

*Plate tectonics can be defined as the composite of various ideas explaining the observed motion of Earth's lithosphere through the mechanisms of subduction and seafloor spreading, which in turn generate Earth's major features, including continents and ocean basins.

Plate Boundaries

Lithospheric plates move as coherent units relative to all other plates. Although the interiors of plates may experience some deformation, all major interactions among individual plates (and therefore most deformation) occurs along their *boundaries*. In fact, the first attempts to outline plate boundaries were made using locations of earthquakes. Later work showed that plates are bounded by three distinct types of boundaries, which are differentiated by the type of relative movement they exhibit. These boundaries are depicted at the bottom of Figure 1.11 and are briefly described here:

1. **Divergent boundaries**—where plates move apart, resulting in upwelling of material from the mantle to create new seafloor (Figure 1.11A).
2. **Convergent boundaries**—where plates move together, resulting in the subduction (consumption) of oceanic lithosphere into the mantle (Figure 1.11B). Convergence can also result in the collision of two continental margins to create a major mountain system.
3. **Transform fault boundaries**—where plates grind past each other without the production or destruction of lithosphere (Figure 1.11C).

If you examine Figure 1.11, you can see that each large plate is bounded by a combination of these boundaries. Movement along one boundary requires that adjustments be made at the others.

Divergent Boundaries. Plate spreading (divergence) occurs mainly at the oceanic ridge. As plates pull apart, the fractures created are immediately filled with molten rock that wells up from the asthenosphere below (Figure 1.12). This hot material slowly cools to hard rock, producing new slivers of seafloor. This happens again and again over millions of years, adding thousands of square kilometers of new seafloor.

This mechanism has created the floor of the Atlantic Ocean during the past 160 million years and is appropriately called **seafloor spreading** (Figure 1.12). The rate of seafloor spreading varies considerably from one spreading center to another. Spreading rates of only 2.5 centimeters (1 inch) per year are typical in the North Atlantic, whereas much faster rates (20 centimeters, 8 inches per year) have been measured along the East Pacific Rise. Even the most rapid rates of spreading are slow on the scale of human history. Nevertheless, the slowest rate of lithosphere production is rapid enough to have created all of Earth's ocean basins over the last 200 million years. In fact, none of the ocean floor that has been dated exceeds 180 million years in age.

Along divergent boundaries where molten rock emerges, the oceanic lithosphere is elevated, because it is hot and occupies more volume than do cooler rocks. Worldwide, this ridge extends for over 70,000 kilometers through all major ocean basins (Figure 1.13). As new lithosphere is formed along the oceanic ridge, it is slowly yet continually displaced away from the zone of upwelling along the ridge axis. Thus, it begins to cool and contract, thereby increasing in density. This thermal contraction accounts for the greater ocean depths that exist away from the ridge. In addition, cooling causes the mantle rocks below the oceanic crust to strengthen,

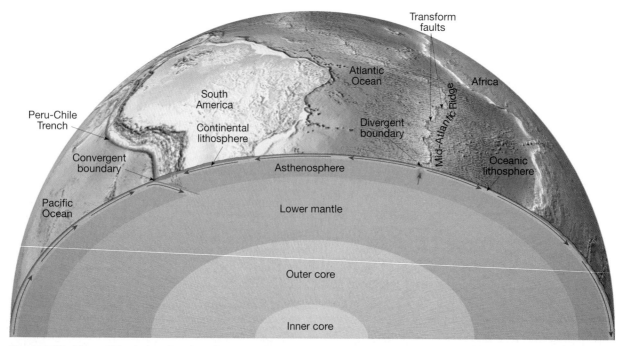

Figure 1.12   View of Earth showing the relationship between divergent, convergent, and transform fault boundaries.

Figure 1.13 Major physical features of the continents and ocean basins. Note the underwater mountain chains that mark diverging plate boundaries in the Atlantic (Mid-Atlantic Ridge) and Pacific (East Pacific Rise). Also note the trenches created where two plates converge, such as the Peru–Chile trench.

thereby adding to the plate's thickness. Stated another way, the thickness of oceanic lithosphere is age dependent. The older (cooler) it is, the greater its thickness.

Convergent Boundaries.   Although new lithosphere is constantly being added at the oceanic ridges, the planet is not growing in size—its total surface area remains constant. To accommodate the newly created lithosphere, older oceanic plates return to the mantle along *convergent boundaries*. As two plates slowly converge, the leading edge of one slab is bent downward, allowing it to slide beneath the other. The surface expression produced by the descending plate is an ocean *trench*, like the Peru–Chile trench illustrated in Figures 1.12 and 1.13.

Plate margins where oceanic crust is being consumed are called **subduction zones.** Here, as the subducted plate moves downward, it enters a high-temperature, high-pressure environment. Some subducted materials, as well as more voluminous amounts of the asthenosphere located above the subducting slab, melt and migrate upward into the overriding plate. Occasionally this molten rock may reach the surface, where it gives rise to explosive volcanic eruptions like Mount St. Helens in 1980 (see Figure 4.1). However, much of this molten rock never reaches the surface; rather, it solidifies at depth and acts to thicken the crust.

Whenever slabs of continental lithosphere and oceanic lithosphere converge, the continental plate being less dense remains "floating," while the denser oceanic lithosphere sinks into the asthenosphere (Figure 1.12). The classic convergent boundary of this type occurs along the western margin of South America where the Nazca plate descends beneath the adjacent continental block. Here subduction along the Peru–Chile trench gave rise to the Andes Mountains, a linear chain of deformed rocks capped with numerous volcanoes—a number of which are still active.

The simplest type of convergence occurs where one oceanic plate is thrust beneath another. Here subduction results in the production of magma in a manner similar to that in the Andes, except volcanoes grow from the floor of the ocean rather than on a continent. If this activity is sustained, it will eventually build a chain of volcanic structures that emerge from the sea as a *volcanic island arc* (see Figure 4.28, upper left). Most volcanic island arcs are found in the Pacific Ocean, as exemplified by the Aleutian, Mariana, and Tonga islands.

As we saw earlier, when an oceanic plate is subducted beneath continental lithosphere, an Andean-type mountain range develops along the margin of the continent. However, if the subducting plate also contains continental lithosphere, continued subduction eventually brings the two continents together (Figure 1.14). Whereas oceanic lithosphere is relatively dense and sinks into the asthenosphere, continental lithosphere is buoyant, which prevents it from being subducted to any great depth. The result is a collision between the two continental blocks (Figure 1.14). Such a collision occurred when the subcontinent of India "rammed" into Asia and produced the Himalayas—the most spectacular mountain range on Earth. During this collision, the continental crust buckled, fractured, and was generally shortened and thickened. In addition to the Himalayas, several other major mountain systems, including the Alps, Appalachians, and Urals, formed during continental collisions along convergent plate boundaries.

Transform Fault Boundaries. Transform fault boundaries are located where plates grind past each other without either generating new lithosphere or consuming old lithosphere. These faults form in the direction of plate movement and were first discovered in association with offsets in oceanic ridges (see Figure 1.13).

Although most transform faults are located along oceanic ridges, some slice through the continents. Two examples are the earthquake-prone San Andreas Fault of California and the Alpine Fault of New Zealand. Along the San Andreas Fault the Pacific plate is moving toward the northwest, relative to the adjacent North American plate (see Figure 1.11). The movement along this boundary does not go unnoticed. As these plates pass, strain builds in the rocks on opposite sides of the fault. Occasionally the rocks adjust, releasing energy in the form of a great earthquake of the type that devastated San Francisco in 1906.

Changing Boundaries. Although the total surface area of Earth does not change, individual plates may diminish or grow in area depending on the distribution of convergent and divergent boundaries. For example, the Antarctic and African plates are almost entirely bounded by spreading centers and hence are growing larger. By contrast, the Pacific plate is being subducted along much of its perimeter and is therefore diminishing in area. At the current rate, the Pacific would close completely in 300 million years—but this is unlikely because changes in plate boundaries will probably occur before that time.

New plate boundaries are created in response to changes in the forces acting on the lithosphere. For example, a relatively new divergent boundary is located

Figure 1.14 When two plates containing continental lithosphere collide, complex mountains are formed. The formation of the Himalayas represents a relatively recent example.

in Africa, in a region known as the East African Rift Valleys (see Figure 4.28, lower right). If spreading continues in this region, the African plate will split into two plates, separated by a new ocean basin. At other locations plates carrying continental crust are moving toward each other. Eventually these continents may collide and be sutured together. Thus, the boundary that once separated these plates disappears, and two plates become one.

As long as temperatures within the interior of our planet remain significantly higher than those at the surface, material within Earth will continue to circulate. This internal flow, in turn, will keep the rigid outer shell of Earth in motion. Thus, while Earth's internal heat engine is operating, the positions and shapes of the continents and ocean basins will change and Earth will remain a dynamic planet.

In the remaining chapters we will examine in more detail the workings of our dynamic planet in light of the plate tectonics theory.

The Chapter in Review

- *Geology* means "the study of Earth." The two broad areas of the science of geology are (1) *physical geology*, which examines the materials composing Earth and the processes that operate beneath and upon its surface; and (2) *historical geology*, which seeks to understand the origin of Earth and its development through time.

- During the seventeenth and eighteenth centuries, *catastrophism* influenced the formulation of explanations about Earth. Catastrophism states that Earth's landscapes developed over short time spans primarily as a result of great catastrophes. By contrast, *uniformitarianism*, one of the fundamental principles of modern geology advanced by *James Hutton* in the late 1700s, states that the physical, chemical, and biological laws that operate today have also operated in the geologic past. The idea is often summarized as *"The present is the key to the past."* Hutton argued that processes that appear to be slow-acting could, over long spans of time, produce effects that are just as great as those resulting from sudden catastrophic events.

- Using the principles of *relative dating*, the placing of events in their proper sequence or order without knowing their age in years, scientists developed a geologic time scale during the nineteenth century. Relative dates can be established by applying such principles as the *law of superposition* and the *principle of fossil succession.*

- All science is based on the assumption that the natural world behaves in a consistent and predictable manner. The process by which scientists gather facts and formulate scientific *hypotheses* and *theories* is called the *scientific method*. To determine what is occurring in the natural world, scientists often (1) collect facts, (2) develop a scientific hypothesis, (3) construct experiments to test the hypothesis, and (4) accept, modify, or reject the hypothesis on the basis of extensive testing. Other discoveries represent purely theoretical ideas that have stood up to extensive examination. Still other scientific advancements have been made when a totally unexpected happening occurred during an experiment.

- Earth's physical environment is traditionally divided into three major parts: the solid Earth; the water portion of our planet, the *hydrosphere*; and Earth's gaseous envelope, the *atmosphere*. In addition, the *biosphere*, the totality of life on Earth, interacts with each of the three physical realms and is an equally integral part of Earth.

- Although each of Earth's four spheres can be studied separately, they are all related in a complex and continuously interacting whole that we call the *Earth system*. Changing one part of the Earth system can produce changes in any or all of the other parts.

- The two sources of energy that power the Earth system are (1) the Sun, which drives the external processes that occur in the atmosphere, hydrosphere, and at Earth's surface, and (2) heat from Earth's interior that powers the internal processes that produce volcanoes, earthquakes, and mountains.

- The *rock cycle* is one of the many cycles or loops of the Earth system in which matter is recycled. The rock cycle is a means of viewing many of the interrelationships of geology. It illustrates the origin of the three basic rock groups and the role of various geologic processes in transforming one rock type into another.

- Earth's internal structure is divided into layers based on differences in chemical composition and on the basis of changes in physical properties. Compositionally, Earth is divided into a thin outer *crust*, a solid rocky *mantle*, and a dense *core*. Based on physical properties, the layers of Earth are (1) the *lithosphere*—the cool, rigid outermost layer that averages about 100 kilometers thick, (2) the *asthenosphere*, a relatively weak layer located in the mantle beneath the lithosphere, (3) the more rigid *mesosphere*, where rocks are very hot and capable of very gradual flow, (4) the liquid *outer core*, where Earth's magnetic field is generated, and (5) the solid *inner core*.

- The *theory of plate tectonics* provides a comprehensive model of Earth's internal workings. It holds that Earth's rigid outer lithosphere consists of several segments called *plates* that are slowly and continually in motion relative to one another. Most earthquakes, volcanic activity, and mountain building are associated with the movements of these plates.

- The three distinct types of plate boundaries are (1) *divergent boundaries*, where plates move apart; (2) *convergent boundaries*, where plates move together, causing one to go beneath another, or where plates collide, which occurs when the leading edges are made of continental crust; and (3) *transform fault boundaries*, where plates slide past one another.

Key Terms

asthenosphere (p. 15)
atmosphere (p. 10)
biosphere (p. 10)
catastrophism (p. 3)
convergent boundary
   (p. 18)
core (p. 15)
crust (p. 2)
divergent boundary (p. 18)
fossil (p. 6)
fossil succession, principle
   of (p. 6)

geology (p. 2)
historical geology (p. 2)
hydrosphere (p. 9)
hypothesis (p. 6)
igneous rock (p. 12)
inner core (p. 16)
lithosphere (p. 15)
lower mantle (p. 15)
magma (p. 12)
mantle (p. 15)
mesosphere (p. 15)
metamorphic rock (p. 12)

model (p. 6)
outer core (p. 16)
paradigm (p. 8)
physical geology (p. 2)
plate (p. 17)
plate tectonics, theory of
   (p. 17)
relative dating (p. 5)
rock cycle (p. 12)
seafloor spreading (p. 18)
sediment (p. 12)
sedimentary rock (p. 12)

subduction zone (p. 19)
superposition, law of
   (p. 5)
system (p. 11)
theory (p. 8)
transform fault boundary
   (p. 18)
uniformitarianism (p. 3)

Questions for Review

1. Geology is traditionally divided into two broad areas. Name and describe these two subdivisions.

2. Briefly describe Aristotle's influence on the science of geology.

3. How did the proponents of catastrophism perceive the age of Earth?

4. Describe the doctrine of uniformitarianism. How did the advocates of this idea view the age of Earth?

5. About how old is Earth?

6. The geologic time scale was established without the aid of radiometric dating. What principles were used to develop the time scale?

7. How is a scientific hypothesis different from a scientific theory?

8. List and briefly describe the four spheres that constitute our natural environment.

9. What are the two sources of energy for the Earth system?

10. Using the rock cycle, explain the statement "One rock is the raw material for another."

11. List and briefly describe Earth's compositional layers.

12. Contrast the lithosphere and the asthenosphere.

13. With which type of plate boundary is each of the following associated: subduction zone, San Andreas Fault, seafloor spreading, and Mount St. Helens?

Web Resources

The *Essentials of Geology* Website uses the resources and flexibility of the Internet to aid in your study of the topics in this chapter. Written and developed by geology instructors, this site will help improve your understanding of geology. Visit **www.prenhall.com/lutgens** and click on the cover of *Essentials of Geology, 8/e* to find:

- On-line review quizzes.
- Web-based critical thinking and writing exercises.
- Links to chapter-specific Web resources.
- Internet-wide key term searches.

www.prenhall.com/lutgens

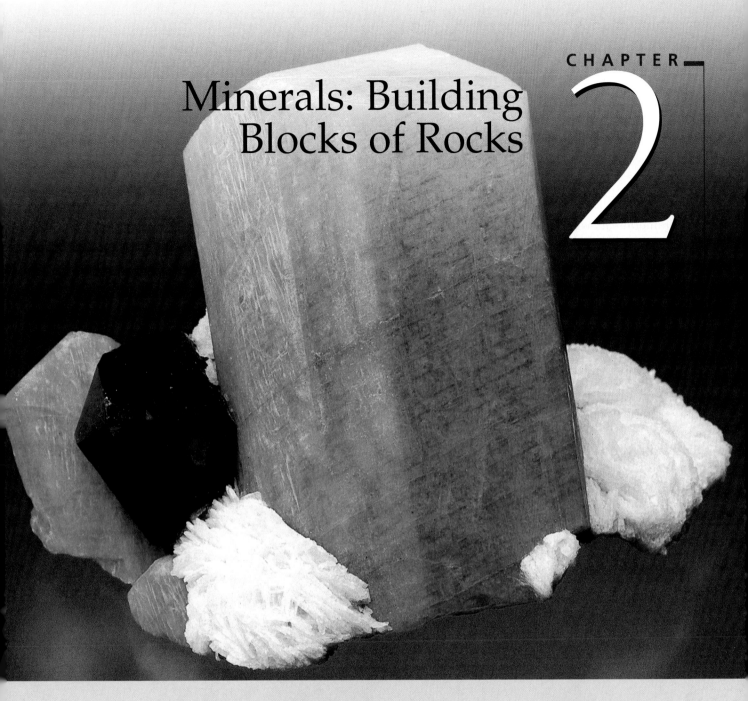

Minerals: Building Blocks of Rocks

Focus on Learning

To assist you in learning the important concepts in this chapter, you will find it helpful to focus on the following questions:

- What are minerals, and how are they different from rocks?

- What is the basic structure of an atom? How do atoms bond together?

- How do isotopes of the same element vary from one another, and why are some isotopes radioactive?

- What are some of the physical and chemical properties of minerals? How can these properties be used to distinguish one mineral from another?

- What is the most abundant mineral group? What do all minerals within this group have in common?

- What are some important nonsilicate minerals?

- When is the term ore used in reference to a mineral?

Crystals of microcline (green), a variety of feldspar. (Photo by Jeff Scovil) arth's crust is the source of a wide variety of useful and essential minerals (Figure 2.1). In fact, practically every manufactured product contains materials obtained from minerals. Most people are familiar with the common uses of many basic metals, including aluminum in beverage cans, copper in electrical wiring, gold in jewelry, and silicon in computer chips. But few are aware that pencil lead does not contain the metal lead but is really made of the greasy-feeling mineral *graphite*. Baby powder is ground-up rock made of the mineral *talc*, with perfume added. Drill bits impregnated with pieces of *diamond* (a mineral) are used by dentists to drill through tooth enamel. The common mineral *quartz* is the main ingredient in glass and is the source of silicon for computer chips. As the material demands of modern society grow, the need to locate additional supplies of useful minerals also grows and becomes more challenging as the easily mined sources become depleted.

In addition to the economic uses of rocks and minerals, every process studied by geologists is in some way dependent upon the properties of these basic Earth

materials. Events such as volcanic eruptions, mountain building, weathering and erosion, and even earthquakes involve rocks and minerals. Consequently, a basic knowledge of Earth materials is essential to an understanding of all geologic phenomena.

Minerals: The Building Blocks of Rocks

Earth Materials
↳ Minerals

We begin our discussion of Earth materials with an overview of **mineralogy** (the study of minerals), because minerals are the building blocks of rocks. Geologists define **mineral** as any naturally occurring inorganic solid that possesses an orderly internal structure and a definite chemical composition. Thus, for any Earth material to be considered a mineral, it must exhibit the following characteristics:

A.   B.   C.

D.   E.   F.

G.   H.   I.

Figure 2.1 Mineral samples. **A.** Quartz; **B.** Olivine (variety Fosterite); **C.** Fluorite; **D.** Realgar; **E.** Beryl (variety aquamarine); **F.** Bornite and Chalcopyrite; **G.** Native copper; **H.** Gold nugget; **I.** Cut diamond. (Photos A–F by Dennis Tasa; G by E. J. Tarbuck; H and I by Dane Pendland, courtesy of Smithsonian Institution)

1. It must occur naturally.
2. It must be inorganic.
3. It must be a solid.
4. It must possess an orderly internal structure—that is, its atoms must be arranged in a definite pattern.
5. It must have a definite chemical composition that may vary within specified limits.

When the term *mineral* is used by geologists, only those substances that meet these criteria are considered minerals. Consequently, synthetic diamonds and a wide variety of other useful materials produced by chemists are not considered minerals. Further, the gemstone *opal* is classified as a *mineraloid* rather than a mineral, because it lacks an orderly internal structure.

Rocks, on the other hand, are more loosely defined. Simply, a **rock** is any solid mass of mineral, or mineral-like, matter that occurs naturally as part of our planet. A few rocks are composed almost entirely of one mineral. A common example is the sedimentary rock *limestone*, which consists of impure masses of the mineral calcite. However, most rocks, like the common rock granite (shown in Figure 2.2), occur as aggregates of several kinds of minerals. Here, the term *aggregate* implies that the minerals are joined in such a way that the properties of each mineral are retained. Note that you can easily identify the mineral constituents of the sample of granite shown in Figure 2.2.

A few rocks are composed of nonmineral matter. These include the volcanic rocks *obsidian* and *pumice*, which are noncrystalline glassy substances, and *coal*, which consists of solid organic debris.

Although this chapter deals primarily with the nature of minerals, keep in mind that most rocks are simply aggregates of minerals. Because the properties of rocks are determined largely by the chemical composition and internal structure of those minerals contained within them, we will first consider these Earth materials. Subsequent chapters will consider the major rock types.

The Composition and Structure of Minerals

Each of Earth's nearly 4000 minerals is uniquely defined by its chemical composition and internal structure. In other words, every sample of the same mineral contains the same elements joined together in a consistent, repeating pattern. We will first review the basic building blocks of minerals, the *elements*, and then examine how elements bond together to form mineral structures.

Granite
(Rock)

Quartz
(Mineral)

Hornblende
(Mineral)

Feldspar
(Mineral)

Figure 2.2 Rocks are aggregates of one or more minerals. (All photos by E. J. Tarbuck)

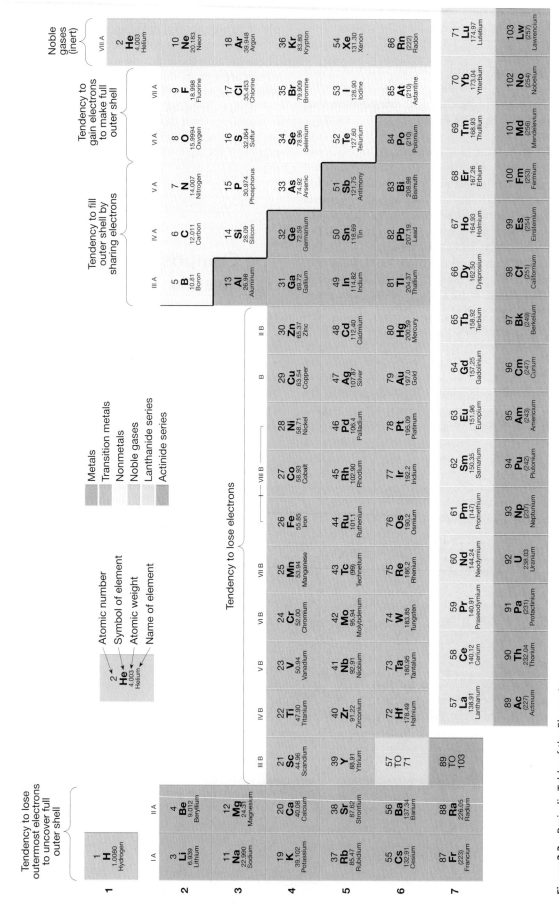

Figure 2.3 Periodic Table of the Elements.

At present, 112 elements are known. Of these, only 92 are naturally occurring (Figure 2.3). Some minerals, such as gold and sulfur, are made entirely from one element. But most minerals are a combination of two or more elements joined to form a chemically stable compound. To understand better how elements combine to form molecules and compounds, we must first consider the **atom,** the smallest part of matter that still retains the characteristics of an element. It is this extremely small particle that does the combining.

Atomic Structure

Two simplified models illustrating basic atomic structure are shown in Figure 2.4. Note that atoms have a central region called the **nucleus,** which contains very dense **protons** (particles with positive electrical charges) and equally dense **neutrons** (particles with neutral electrical charges). Surrounding the nucleus are very light particles called **electrons,** which travel at high speeds and are negatively charged. For convenience, we often diagram atoms showing the electrons in orbits around the nucleus, like the orbits of the planets around the Sun (Figure 2.4A). However, electrons *do not* travel in the same plane like planets. Further, because of their rapid motion, electrons create spherically shaped negatively charged zones around the nucleus called **energy levels,** or **shells.** Hence, a more realistic picture of the atom can be obtained by envisioning cloudlike shells of fast-moving electrons surrounding a central nucleus (Figure 2.4B). As we shall see, an important fact about these shells is that each can accommodate a specific number of electrons.

The number of protons found in an atom's nucleus determines the **atomic number** and name of the element. For example, all atoms with six protons are carbon atoms, those with eight protons are oxygen atoms, and so forth. Because atoms have the same number of electrons as protons, the atomic number also equals the number of electrons surrounding the nucleus (Table 2.1). Moreover, because neutrons have no charge, the positive charge of the protons is exactly balanced by the negative charge of the electrons. Consequently, atoms are electrically neutral. Thus, an **element** is a large collection of electrically neutral atoms, all having the same atomic number.

The simplest element, hydrogen, is composed of atoms that have only one proton in the nucleus and one electron surrounding the nucleus. Each successively heavier atom has one more proton and one more electron, in addition to a certain number of neutrons (Table 2.1). Studies of electron configurations have shown that each electron is added in a systematic fashion to a particular energy level or shell. In general, electrons enter higher energy levels after lower energy levels have been filled to capacity.* The first principal shell holds a maximum of two electrons, while each of the higher shells holds eight or more electrons. As we shall see, it is generally the outermost electrons, also referred to as **valence electrons,** that are involved in chemical bonding.

How Atoms Bond Together

Elements combine with one another to form a wide variety of more complex substances. The strong attractive force linking atoms together is called a *chemical bond.* When chemical bonding joins two or more elements together in definite proportions, the substance is called a **compound.** Most minerals are chemical compounds.

Why do elements join together to form compounds? From experimentation it has been learned that the forces holding the atoms together are electrical. Further, it is known that chemical bonding results in a change in the electron configuration of the bonded atoms. As we noted earlier, it is the valence electrons (outer-shell electrons) that are generally involved in chemical bonding. Other than the first shell, which contains two electrons, *a stable configuration occurs when the valence shell contains eight electrons.* Only the so-called noble gases, such as neon and argon, have a complete outermost electron shell. Therefore, the noble gases are the least chemically reactive; and thus their designation as inert. However, all other atoms seek a valence shell containing eight electrons, like the noble gases. Simply stated, atoms combine to form compounds and molecules in order to obtain the stable electron configuration of the noble gases. To accomplish this, an atom can either gain, lose, or share electrons with one or more atoms. The result of this process is the formation of an electrical "glue" that bonds the atoms. In summary, *most atoms are chemically reactive and bond together in order to achieve the stable noble-gas configuration while retaining overall electrical neutrality.*

Ionic Bonds.   Perhaps the easiest type of bond to visualize is an **ionic bond.** In ionic bonding, one or more valence electrons are transferred from one atom to another. Simply, one atom gives up its valence electrons, and the other uses them to complete its outer shell. A common example of ionic bonding is sodium (Na) and chlorine (Cl), joining to produce sodium chloride (common table salt). This is shown in Figure 2.5. Notice that sodium gives up its single outer electron to chlorine. As a result, sodium achieves a stable configuration having eight electrons in its outermost shell. By acquiring the electron that sodium loses, chlorine, which has seven valence electrons, gains the eighth electron needed to complete its outermost shell. Thus, through the transfer of a single electron, both the sodium and chlorine atoms have acquired the stable noble-gas configuration.

---

*This principle holds for the first 18 elements.

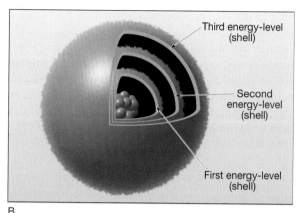

Figure 2.4 Two models of the atom. **A.** A very simplified view of the atom, which consists of a central nucleus, composed of protons and neutrons, encircled by high-speed electrons. **B.** Another model of the atoms showing spherically shaped electron clouds (energy level shells). Note that these models are not drawn to scale. Electrons are minuscule in size compared to protons and neutrons, and the relative space between the nucleus and electron shells is much greater than illustrated.

Once electron transfer takes place, atoms are no longer electrically neutral. By giving up one electron, a neutral sodium atom (11 protons/11 electrons) becomes *positively charged* (11 protons/10 electrons). Similarly, by acquiring one electron, the neutral chlorine atom (17 protons/17 electrons) becomes *negatively charged* (17 protons/18 electrons). Atoms such as these, which have an electrical charge because of the unequal numbers of electrons and protons, are called **ions.**

We know that particles (ions) with like charges repel and those with unlike charges attract. Thus, an *ionic bond* is the attraction of oppositely charged ions to one another producing an electrically neutral compound. Figure 2.6 illustrates the arrangement of sodium and chloride ions in ordinary table salt. Notice that salt consists of alternating sodium and chloride ions, positioned in such a manner that each positive ion is attracted to and surrounded on all sides by negative ions, and vice versa. This arrangement maximizes the attraction between ions with unlike charges while minimizing the repulsion between ions with like charges. Thus, *ionic compounds consist of an orderly arrangement of oppositely charged ions assembled in a definite ratio that provides overall electrical neutrality.*

The properties of a chemical compound are *dramatically different* from the properties of the elements comprising it. For example, chlorine is a green, poisonous gas that is so toxic it was used as a chemical weapon during World War I. Sodium is a soft, silvery metal that reacts vigorously with water and, if held in your hand, could burn it severely. Together, however, these atoms

Table 2.1 Atomic number and distribution of electrons

Element	Symbol	Atomic Number	Number of Electrons in Each Shell			
			1st	2nd	3rd	4th
Hydrogen	H	1	1			
Helium	He	2	2			
Lithium	Li	3	2	1		
Beryllium	Be	4	2	2		
Boron	B	5	2	3		
Carbon	C	6	2	4		
Nitrogen	N	7	2	5		
Oxygen	O	8	2	6		
Fluorine	F	9	2	7		
Neon	Ne	10	2	8		
Sodium	Na	11	2	8	1	
Magnesium	Mg	12	2	8	2	
Aluminum	Al	13	2	8	3	
Silicon	Si	14	2	8	4	
Phosphorus	P	15	2	8	5	
Sulfur	S	16	2	8	6	
Chlorine	Cl	17	2	8	7	
Argon	Ar	18	2	8	8	
Potassium	K	19	2	8	8	1
Calcium	Ca	20	2	8	8	2

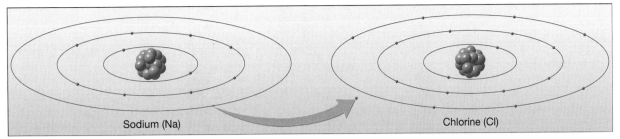

Figure 2.5 Chemical bonding of sodium and chlorine atoms to produce sodium chloride. Through the transfer of one electron in the outer shell of a sodium atom to the outer shell of a chlorine atom, the sodium becomes a positive ion and chlorine a negative ion.

produce the compound sodium chloride (table salt), a clear crystalline solid that is essential for human life. This example also illustrates an important difference between a rock and a mineral. Most *minerals* are *chemical compounds* with unique properties that are very different from the elements that comprise them. A *rock*, on the other hand, is a *mixture* of minerals, with each mineral retaining its own identity.

Covalent Bonds. Not all atoms combine by transferring electrons to form ions. Other atoms *share* electrons. For example, the gaseous elements oxygen (O_2), hydrogen (H_2), and chlorine (Cl_2) exist as stable molecules consisting of two atoms bonded together by sharing their valence electrons.

Figure 2.7 illustrates the sharing of a pair of electrons between two chlorine atoms to form a molecule of chlorine gas (Cl_2). By overlapping their outer shells, these chlorine atoms share a pair of electrons. Thus, each chlorine atom has acquired, through cooperative action, the needed eight electrons to complete its outer shell. The bond produced by the sharing of electrons is called a **covalent bond.**

A common analogy may help you visualize a covalent bond. Imagine two people at opposite ends of a dimly lit room, each reading under a separate lamp. By moving the lamps to the center of the room, they are able to combine their light sources so each can see better. Just as the overlapping light beams meld, the shared electrons that provide the electrical glue in covalent bonds are indistinguishable from each other. The most common mineral group, the silicates, contains the element silicon, which readily forms covalent bonds with oxygen.

Other Bonds. As you might suspect, many chemical bonds are actually hybrids. They consist to some degree of electron sharing, as in covalent bonding, and to some degree of electron transfer, as in ionic bonding. Furthermore, both ionic and covalent bonds may occur within the same compound. This occurs in many silicate minerals, where silicon and oxygen atoms are covalently bonded to form the basic building block common to all silicates. These structures in turn are ionically bonded to metallic ions, producing various electrically neutral chemical compounds.

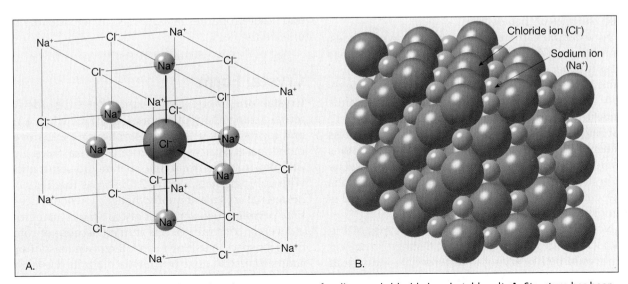

Figure 2.6 Schematic diagrams illustrating the arrangement of sodium and chloride ions in table salt. **A.** Structure has been opened up to show arrangement of ions. **B.** Actual ions are closely packed.

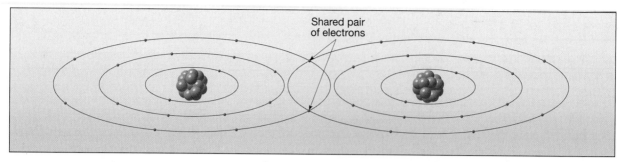

Figure 2.7 Illustration of the sharing of a pair of electrons between two chlorine atoms to form a chlorine molecule.

Another chemical bond exists in which valence electrons are free to migrate from one ion to another. The mobile valence electrons serve as the electrical glue. This type of electron sharing is found in metals such as copper, gold, aluminum, and silver and is called **metallic bonding.** Metallic bonding accounts for the high electrical conductivity of metals, the ease with which metals are shaped, and numerous other special properties of metals.

Isotopes and Radioactive Decay

Subatomic particles are so incredibly small that a special unit was devised to express their mass. A proton or a neutron has a mass just slightly more than one *atomic mass unit*, whereas an electron is only about one two-thousandth of an atomic mass unit. Thus, although electrons play an active role in chemical reactions, they do not contribute significantly to the mass of an atom.

The **mass number** of an atom is simply the total of its neutrons and protons in the nucleus. Atoms of the same element always have the same number of protons but they commonly have varying numbers of neutrons. This means that an element can have more than one mass number. These variants of the same element are called **isotopes** of that element.

For example, carbon has three well-known isotopes. One has a mass number of 12 (carbon-12), another has a mass number of 13 (carbon-13), and the third, carbon-14, has a mass number of 14. All atoms of the same element must have the same number of protons (atomic number), and carbon always has six. Hence, carbon-12 must have *six* protons plus six neutrons to give it a mass number of 12, whereas carbon-14 must have six protons plus *eight* neutrons to give it a mass number of 14. The *average* atomic mass of any random sample of carbon is much closer to 12 than 13 or 14, because carbon-12 is the more abundant isotope.

Note that in a chemical sense all isotopes of the same element are nearly identical. To distinguish among them would be like trying to differentiate individual members from a group of similar objects, all having the same shape, size, and color, with some being only

slightly heavier. Further, different isotopes of an element are generally found together in the same mineral.

Although the nuclei of most atoms are stable, some elements do have isotopes in which the nuclei are unstable. Unstable isotopes, such as carbon-14, disintegrate through a process called **radioactive decay.** During radioactive decay unstable nuclei spontaneously break apart, giving off subatomic particles and/or electromagnetic energy similar to X-rays. The rate at which the unstable nuclei decay is steady and measurable, thus making such isotopes useful "clocks" for dating the events of Earth history. A discussion of radioactive decay and its applications in dating past geological events can be found in Chapter 18.

Properties of Minerals

Minerals occur in a fascinating palette of colors, shapes, and lusters. They vary in hardness and may even have a distinctive taste or scent. We will now look at these and other properties that are used to help identify minerals in the field.

Crystal Form

Crystal form is the external expression of the orderly internal arrangement of atoms. Generally, when a mineral forms without space restrictions, it will develop individual crystals with well-formed crystal faces. Figure 2.8A illustrates this for pyrite, an iron-and-sulfur mineral with cubic crystals. Figure 2.8B shows the distinctive hexagonal crystals of quartz that grow when space and time permit. However, most of the time crystal growth is severely constrained. It is stunted because of competition for space, resulting in an intergrown mass of small, jammed crystals, none of which exhibits its crystal form. Thus, most inorganic solid objects are composed of crystals, but they are not clearly visible to the unaided eye.

A.

B.

Figure 2.8 Crystal form is the external expression of a mineral's orderly internal structure. **A.** Pyrite, commonly known as "fool's gold," often forms cubic crystals that may exhibit parallel lines called striations. (Photo by E. J. Tarbuck) **B.** Quartz sample that exhibits well-developed hexagonal (six-sided) crystals with pyramidal-shaped ends. (Photo by Breck P. Kent)

Luster

Luster is the appearance or quality of light reflected from the surface of a mineral. Minerals that have the appearance of metals, regardless of color, are said to have a *metallic luster* like the pyrite crystals in Figure 2.8A. Minerals with a *nonmetallic luster* are described by various adjectives, including vitreous (glassy), pearly, silky, resinous, and earthy (dull). Some minerals appear somewhat metallic in luster and are said to be submetallic.

Color

Although **color** is an obvious feature of a mineral, it is often an unreliable diagnostic property. Slight impurities in the common mineral quartz, for example, give it a variety of colors, including pink, purple (amethyst), white, and even black (see Figure 2.20).

Streak

Streak is the color of a mineral in its powdered form, which is a much more consistent indication of color. Streak is obtained by rubbing the mineral across a piece of unglazed porcelain termed a *streak plate*. Whereas the color of a mineral may vary from sample to sample, the streak usually does not, and is therefore the more reliable property. Streak can also be an aid in distinguishing minerals with metallic lusters from those having nonmetallic lusters. Metallic minerals generally have a dense, dark streak, whereas minerals with nonmetallic lusters do not.

Hardness

One of the most useful diagnostic properties is **hardness,** a measure of the resistance of a mineral to abrasion or scratching. This property is determined by rubbing the mineral to be identified against another mineral of known hardness, or vice versa. One will scratch the other (unless they have the same hardness).

Geologists use a standard hardness scale, called the **Mohs scale.** It consists of 10 minerals arranged in order from 1 (softest) to 10 (hardest), as shown in Table 2.2. Any mineral of unknown hardness can be rubbed against these to determine its hardness. In the field, other handy objects work, too. For example, your fingernail has a hardness of 2.5, a copper penny 3, and a piece of glass 5.5. The mineral gypsum, which has a

Table 2.2 Mohs scale of mineral hardness

Relative	Scale	Mineral	Hardness of Some Common Objects
Hardest	10	Diamond	
	9	Corundum	
	8	Topaz	
	7	Quartz	
	6	Potassium feldspar	5.5 Glass, Pocketknife
	5	Apatite	
	4	Fluorite	
	3	Calcite	3 Copper penny
	2	Gypsum	2.5 Fingernail
Softest	1	Talc	

hardness of 2, can be easily scratched with your fingernail. On the other hand, the similar-looking mineral calcite, which has a hardness of 3, will scratch your fingernail but will not scratch glass. Quartz, the hardest of the common minerals, will scratch a glass plate. You can see that by rubbing an unknown mineral against objects of known hardness, you can quickly narrow its hardness range.

Cleavage

In the crystal structure of a mineral, some bonds may be weaker than others. These bonds are where a mineral will break when it is stressed. **Cleavage** is the tendency of a mineral to break along planes of weak bonding. Not all minerals have definite planes of weak bonding, but those that possess cleavage can be identified by the distinctive smooth surfaces produced when the mineral is broken.

The simplest type of cleavage is exhibited by the micas (Figure 2.9). Because the micas have weak bonds in one direction, they cleave to form thin, flat sheets. Some minerals have several cleavage planes that produce smooth surfaces when broken, whereas others exhibit poor cleavage, and still others have no cleavage at all. When minerals break evenly in more than one direction, cleavage is described by the number of planes exhibited and the angles at which they meet (Figure 2.10).

Do not confuse cleavage with crystal form. When a mineral exhibits cleavage, it will break into pieces that have the same *geometry as each other*. By contrast, the quartz crystals shown in Figure 2.8B do not have cleavage. If broken, they fracture into shapes that do not resemble each other or the original crystals.

Figure 2.9 The thin sheets shown here were produced by splitting a mica (muscovite) crystal parallel to its perfect cleavage. (Photo by Breck P. Kent)

Fracture

Minerals that do not exhibit cleavage when broken, such as quartz, are said to **fracture.** Even fracturing has variety: Minerals that break into smooth curved surfaces like those seen in broken glass have a *conchoidal fracture* (Figure 2.11). Others break into splinters or fibers, but most minerals fracture irregularly.

Specific Gravity

Specific gravity compares the weight of a mineral to the weight of an equal volume of water. For example, if a mineral weighs three times as much as an equal volume of water, its specific gravity is 3. With a little practice, you can estimate the specific gravity of minerals by hefting them in your hand. For example, if a mineral feels as heavy as the common rocks you have handled, its specific gravity will probably be somewhere between 2.5 and 3. Some metallic minerals have a specific gravity two or three times that of common rock-forming minerals. Galena, which is an ore of lead, has a specific gravity of roughly 7.5, whereas the specific gravity of 24-karat gold is approximately 20 (Figure 2.12).

Other Properties of Minerals

In addition to the properties already discussed, some minerals can be recognized by other distinctive properties. For example, halite is ordinary salt, so it is quickly identified with your tongue. Thin sheets of mica will bend and elastically snap back. Gold is malleable and can be easily shaped. Talc and graphite both have distinctive feels; talc feels soapy and graphite feels greasy. A few minerals, such as magnetite, have a high iron content and can be picked up with a magnet, while some varieties (lodestone) are natural magnets and will pick up small iron-based objects such as pins and paper clips. Some minerals exhibit special optical properties. For example, when a transparent piece of calcite is placed over printed material, the letters appear twice. This optical property is known as *double refraction*. In addition, the streak of many sulfur-bearing minerals smells like rotten eggs.

One very simple chemical test involves placing a drop of dilute hydrochloric acid from a dropper bottle on a freshly broken mineral surface. Certain minerals, called *carbonates*, will effervesce (fizz) with hydrochloric acid. This test is useful in identifying the mineral calcite, which is a common carbonate mineral.

In summary, a number of special physical and chemical properties are useful in identifying certain minerals. These include taste, smell, elasticity, malleability, feel, magnetism, double refraction, and chemical reaction with hydrochloric acid. Remember that every one of these properties depends on the composi-

Figure 2.10 Smooth surfaces produced when a mineral with cleavage is broken. The sample on the left (fluorite) exhibits four planes of cleavage (eight sides), whereas the other two samples exhibit three planes of cleavage (six sides). Also notice that the mineral in the center (halite) has cleavage planes that meet at 90-degree angles, whereas the mineral on the right (calcite) has cleavage planes that meet at 75-degree angles. (Photo by E. J. Tarbuck)

tion (elements) of a mineral and its structure (how the atoms are arranged).

Mineral Groups

GEODe Earth Materials
↳ Minerals

Nearly 4000 minerals have been named, and several new ones are identified each year. Fortunately, for students who are beginning to study minerals, no more than a few dozen are abundant! Collectively, these few make up most of the rocks of Earth's crust and, as such, are classified as the rock-forming minerals. It is also interesting to note that only eight elements make up the bulk of these minerals and represent over 98 percent (by weight) of the continental crust (Table 2.3). These elements, in order of abundance, are oxygen (O), silicon (Si), aluminum (Al), iron (Fe), calcium (Ca), sodium (Na), potassium (K), and magnesium (Mg).

The two most abundant elements are silicon and oxygen, which combine to form the framework of the most common mineral group, the **silicates.** Perhaps the next most common mineral group is the carbonates, of which calcite is the most prominent member. Other common rock-forming minerals include gypsum and halite.

We will first discuss the most common mineral group, the silicates, and then consider some of the other prominent mineral groups.

The Silicates

Every silicate mineral contains the elements oxygen and silicon. Moreover, except for a few minerals such as quartz, the crystalline structure of most silicate minerals contains one or more of the other common elements of Earth's crust. These elements give rise to the great variety of silicate minerals and their varied properties.

The Silicon–Oxygen Tetrahedron. All silicates have the same fundamental building block, the **silicon–**

Figure 2.11 Conchoidal fracture. The smooth curved surfaces result when minerals break in a glasslike manner. (Photo by E. J. Tarbuck)

Figure 2.12 Galena is lead sulfide. Like other metallic ores, it has a relatively high specific gravity and therefore feels heavy when hefted. (Photo by E. J. Tarbuck)

Table 2.3 Relative abundance of the most common elements in the continental crust

Element	Approximate Percentage by Weight
Oxygen (O)	46.6
Silicon (Si)	27.7
Aluminum (Al)	8.1
Iron (Fe)	5.0
Calcium (Ca)	3.6
Sodium (Na)	2.8
Potassium (K)	2.6
Magnesium (Mg)	2.1
All others	1.5
Total	100

Source: Data from Brian Mason.

oxygen tetrahedron. This structure consists of four oxygen ions surrounding a much smaller silicon ion (Figure 2.13). The silicon–oxygen tetrahedron is a complex ion (SiO_4^{-4}) with a charge of −4.

In nature, one of the simplest ways in which these tetrahedra join together to become neutral compounds is through the addition of positively charged ions (Figure 2.14). In this way, a chemically stable structure is produced, consisting of individual tetrahedra linked together by various positively charged ions.

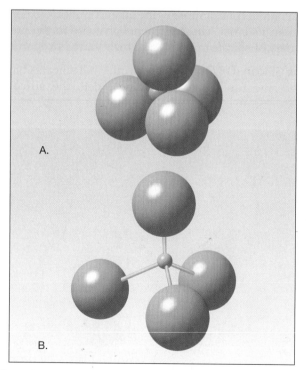

Figure 2.13   Two representations of the silicon–oxygen tetrahedron. **A.** The four large spheres represent oxygen atoms and the blue sphere represents a silicon atom. The spheres are drawn in proportion to the radii of the atoms. **B.** A model of the tetrahedron using rods to depict the bonds that connect the atoms.

Other Silicate Structures. In addition to positive ions providing the opposite electrical charge needed to bind the tetrahedra, the tetrahedra may link with themselves in a variety of configurations. For example, the tetrahedra may be joined to form *single chains,* *double chains,* or *sheet structures,* as shown in Figure 2.15. The joining of tetrahedra in each of these configurations results from the sharing of oxygen atoms between silicon atoms in adjacent tetrahedra.

To understand better how this sharing takes place, select one of the silicon ions (small blue spheres) near the middle of the single-chain structure shown in Figure 2.15. Notice that this silicon ion is completely surrounded by four larger oxygen ions (you are looking *through* one of the four to see the blue silicon ion). Also notice that, of the four oxygen ions, two are joined to other silicon ions, whereas the other two are not shared in this manner. *It is the linkage across the shared oxygen ions that joins the tetrahedra into a chain structure.* Now, examine a silicon ion near the middle of the sheet structure and count the number of shared and unshared oxygen ions surrounding it. The increase in the degree of sharing accounts for the sheet structure. Other silicate structures exist, and the most common has all of the oxygen ions shared to produce a complex three-dimensional framework.

By now you can see that the ratio of oxygen ions to silicon ions differs in each of the silicate structures. In the isolated tetrahedron, there are four oxygen ions for every silicon ion. In the single chain, the oxygen-to-silicon ratio is 3:1, and in the three-dimensional framework this ratio is 2:1. Consequently, as more of the oxygen ions are shared, the percentage of silicon in the structure increases. Silicate minerals are therefore described as having a high or low silicon content based on their ratio of oxygen to silicon. This difference in silicon content is important, as we shall see in a later chapter when we consider the formation of igneous rocks.

Joining Silicate Structures. Most silicate structures, including single chains, double chains, and sheets, are not neutral chemical compounds. Thus, like the individual tetrahedra, they are all neutralized by the inclusion of positively charged ions that bond them together into a variety of complex crystalline configurations. The ions that most often link silicate structures are those of the elements iron (Fe), magnesium (Mg), potassium (K), sodium (Na), aluminum (Al), and calcium (Ca).

Notice in Figure 2.14 that each of these ions has a particular atomic size and a particular charge. Generally, ions of approximately the same size are able to substitute freely for one another. For instance, ions of iron (Fe^{2+}) and magnesium (Mg^{2+}) are nearly the same size and substitute for each other without altering the mineral structure. This also holds true for calcium and sodium ions, which can occupy the same site in a crystalline

Negative Ion (Anion)	Positive Ions (Cations)		
	Si⁴⁺ 0.39	Al³⁺ 0.51	
	Fe³⁺ 0.64	Mg²⁺ 0.66	Fe²⁺ 0.74
O²⁻ 1.40	Na¹⁺ 0.97	Ca²⁺ 0.99	K¹⁺ 1.33

Figure 2.14 Relative sizes and electrical charges of the eight most common elements in Earth's crust. These are the most common ions in rock-forming minerals. Ionic radii are expressed in angstroms (1 angstrom equals 10^{-8} cm).

structure. In addition, aluminum (Al) often substitutes for silicon in the silicon–oxygen tetrahedron.

Because of the ability of silicate structures to readily accommodate different ions at a given bonding site, individual specimens of a particular mineral may contain varying amounts of certain elements. A mineral of this type is often expressed by a chemical formula that uses parentheses to show the variable component. A good example is the mineral olivine, $(Mg, Fe)_2SiO_4$, which is magnesium/iron silicate. As you can see from the formula, it is the iron (Fe^{2+}) and magnesium (Mg^{2+}) ions in olivine that freely substitute for each other. At one extreme, olivine may contain iron without any magnesium $(Fe_2SiO_4$, or iron silicate) and at the other, iron is totally lacking $(Mg_2SiO_4$, or magnesium silicate). Between these end members, any ratio of iron to magnesium is possible. Thus, olivine, as well as many other silicate minerals, is actually a *family* of minerals with a range of composition between the two end members.

In certain substitutions, the ions that interchange do not have the same electrical charge. For instance, when calcium (Ca^{2+}) substitutes for sodium (Na^{1+}), the structure gains a positive charge. In nature, one way in which this substitution is accomplished, while still maintaining overall electrical neutrality, is the simultaneous substitution of aluminum (Al^{3+}) for silicon (Si^{4+}). This particular double substitution occurs in a mineral called *plagioclase feldspar*. It is a member of the most

A. Single chains   B. Double chains   C. Sheet structures

Figure 2.15 Three types of silicate structures. **A.** Single chains. **B.** Double chains. **C.** Sheet structures.

abundant family of minerals found in Earth's crust. The end members of this particular feldspar series are a calcium–aluminum silicate (anorthite, $CaAl_2Si_2O_8$) and a sodium–aluminum silicate (albite, $NaAlSi_3O_8$).

We are now prepared to review silicate structures in light of what we know about chemical bonding. An examination of Figure 2.14 shows that among the major constituents of the silicate minerals, only oxygen is a negatively charged ion. Because oppositely charged ions attract (and similarly charged ions repel), the chemical bonds that hold silicate structures together form between oxygen and positively charged ions. Thus, the positively charged ions arrange themselves so that they can be as close as possible to oxygen while remaining as far apart from *each other* as possible. Because of its small size and high charge (-4), the silicon (Si) ion forms the strongest bonds with oxygen. Aluminum (Al), although not as strongly bonded to oxygen as silicon, is more strongly bonded than calcium (Ca), magnesium (Mg), iron (Fe), sodium (Na), and potassium (K). In many ways, aluminum plays a role similar to silicon by being the central ion in the basic tetrahedral structure.

Most silicate minerals consist of a basic framework composed of either a single silicon or aluminum ion surrounded by four negatively charged oxygen ions. These tetrahedra often link together to form a variety of other silicate structures (chains, sheets, etc.) through shared oxygen atoms. Finally, positively charged ions bond with the oxygen atoms of these silicate structures to create the more complex crystalline structures that characterize the silicate minerals.

Common Silicate Minerals

To reiterate, the silicates are the most abundant mineral group and have the silicate ion (SiO_4^{4-}) as their basic building block. The major silicate groups and common examples are given in Figure 2.16. The feldspars are by far the most plentiful silicate group, comprising more than 50 percent of Earth's crust. Quartz, the second most abundant mineral in the continental crust, is the only common mineral made completely of silicon and oxygen.

Notice in Figure 2.16 that each mineral group has a particular silicate structure and that a relationship exists between the internal structure of a mineral and the cleavage it exhibits. Because the silicon–oxygen bonds are strong, silicate minerals tend to cleave between the silicon–oxygen structures rather than across them. For example, the micas have a sheet structure and thus tend to cleave into flat plates (see Figure 2.9). Quartz, which has equally strong silicon–oxygen bonds in all directions, has no cleavage but fractures instead.

Most silicate minerals form (crystallize) as molten rock is cooling. This cooling can occur at or near Earth's surface (low temperature and pressure) or at great depths (high temperature and pressure). The environment during crystallization and the chemical composition of the molten rock determine to a large degree which minerals are produced. For example, the silicate mineral olivine crystallizes at high temperatures, whereas quartz crystallizes at much lower temperatures.

In addition, some silicate minerals form at Earth's surface from the weathered products of other silicate minerals. Still other silicate minerals are formed under the extreme pressures associated with mountain building. Each silicate mineral, therefore, has a structure and a chemical composition that *indicate the conditions under which it formed*. Thus, by carefully examining the mineral constituents of rocks, geologists can often determine the circumstances under which the rocks formed.

We will now examine some of the most common silicate minerals, which we divide into two major groups on the basis of their chemical makeup.

The Dark Silicate Minerals.   The **dark** (or **ferromagnesian**) **silicates** are those minerals containing ions of iron (iron = *ferro*) and/or magnesium in their structure. Because of their iron content, ferromagnesian silicates are dark in color and have greater specific gravity, between 3.2 and 3.6, than do nonferromagnesian silicates. The most common dark silicate minerals are olivine, the pyroxenes, the amphiboles, and biotite (dark-colored mica).

Olivine is a family of high-temperature silicate minerals that are black to olive green in color and have a glassy luster and a conchoidal fracture. Although not abundant in continental rocks, olivine is thought to be a major component of Earth's upper mantle.

The *pyroxenes* are a group of chemically complex minerals. The most common member, augite, is a black, opaque mineral that is found in basalt, a common igneous rock of the oceanic crust and volcanic areas on the continents.

Hornblende is the most common member of a group of silicate minerals called *amphiboles* (Figure 2.17). Hornblende is usually dark green to black in color and very similar in appearance to augite. Predominately found in continental rocks of igneous origin, hornblende makes up the dark portion of an otherwise light-colored rock.

Biotite is the dark iron-rich member of the mica family. Like other micas, biotite possesses a sheet structure that gives it excellent cleavage in one direction. Biotite also has a shiny black appearance that helps to distinguish it from the other dark ferromagnesian minerals. Like hornblende, biotite is a common constituent of continental rocks, including the igneous rock granite.

The Light Silicate Minerals.   As the name implies, the **light** (or **nonferromagnesian**) **silicates** are generally light in color and have a specific gravity of about 2.7, which is considerably less than the ferromagnesian silicates. As indicated earlier, these differences

Mineral		Idealized Formula	Cleavage	Silicate Structure	
Olivine		$(Mg, Fe)_2SiO_4$	None	Single tetrahedron	
Pyroxene group (Augite)		$(Mg,Fe)SiO_3$	Two planes at right angles	Single chains	
Amphibole group (Hornblende)		$Ca_2(Fe,Mg)_5Si_8O_{22}(OH)_2$	Two planes at 60° and 120°	Double chains	
Micas	Biotite	$K(Mg,Fe)_3AlSi_3O_{10}(OH)_2$	One plane	Sheets	
	Muscovite	$KAl_2(AlSi_3O_{10})(OH)_2$			
Feld-spars	Orthoclase (Potassium feldspar)	$KAlSi_3O_8$	Two planes at 90°	Three-dimensional networks	
	Plagioclase	$(Ca,Na)AlSi_3O_8$			
Quartz		SiO_2	None		

Figure 2.16 Common silicate minerals. Note that the complexity of the silicate structure increases down the chart.

are mainly attributable to the presence or absence of iron and magnesium. The light silicates contain varying amounts of aluminum, potassium, calcium, and sodium rather than iron and magnesium.

The most common mineral group found in Earth's crust are the nonferromagnesians known as *feldspars*. Because the feldspars form (crystallize) under a wide range of pressures and temperatures, they are a major constituent of most igneous rocks. All of the feldspars have similar physical properties. They have two planes of cleavage meeting at or near 90-degree angles, are relatively hard (6 on the Mohs scale), and have a luster that ranges from glassy to pearly.

Two different feldspar structures exist. *Orthoclase feldspar* is a common member of a group of feldspar minerals containing potassium ions in its structure and is, therefore, often referred to as *potassium feldspar*. The other group, called *plagioclase feldspar*, contains both sodium and calcium ions that freely substitute for one another depending on the environment during crystallization.

Potassium feldspar is usually light cream to salmon pink in color (Figure 2.18). The plagioclase feldspars, in contrast, range in color from white to medium gray. However, color should not be used to distinguish these groups. The only sure way to physically distinguish the

Figure 2.17   Hornblende amphibole. Hornblende is a common dark silicate mineral having two cleavage directions that intersect at roughly 60° and 120°. (Photo by E. J. Tarbuck)

Figure 2.19   These parallel lines, called *striations*, are a distinguishing characteristic of the plagioclase feldspars. (Photo by E. J. Tarbuck)

feldspars is to look for a multitude of fine parallel lines, called *striations*. Striations are found on some cleavage planes of plagioclase feldspar but are not present on orthoclase feldspar (Figure 2.19).

Quartz is the only common silicate mineral consisting entirely of silicon and oxygen. As such, the term *silica* is applied to quartz, which has the chemical formula SiO_2. Quartz is hard, resistant to weathering, and does not have cleavage. When broken, quartz generally exhibits conchoidal fracture. In a pure form, quartz is clear and, if allowed to solidify without interference, will form hexagonal crystals that develop pyramid-shaped ends (see Figure 2.8B). However, like most other clear minerals, quartz is often colored (Figure 2.20) by the inclusion of various ions (impurities) and forms without developing good crystal faces. The most common varieties of quartz are milky (white), smoky (gray), rose (pink), amethyst (purple), and rock crystal (clear).

Figure 2.18   Sample of orthoclase feldspar. (Photo by E. J. Tarbuck)

Muscovite is a common member of the mica family. It is light in color and has a pearly luster. Like other micas, muscovite has excellent cleavage in one direction (see Figure 2.9). In thin sheets muscovite is clear, a property that accounts for its use as window glass during the Middle Ages. Because muscovite is very shiny, it can often be identified by the sparkle it gives to rocks.

Important Nonsilicate Minerals

Other mineral groups can be considered scarce when compared to the silicates, although many are important economically. Table 2.4 lists examples of oxides, sulfides, sulfates, halides, native elements, and carbonates of economic value. A discussion of a few of the more common nonsilicate rock-forming minerals follows.

The carbonate minerals are much simpler structurally than the silicates. This mineral group is composed of the carbonate ion (CO_3^{2-}) and one or more kinds of positive ions. The two most common carbonate minerals are *calcite*, $CaCO_3$ (calcium carbonate) and *dolomite*, $CaMg(CO_3)_2$ (calcium/magnesium carbonate).

Because these minerals are similar both physically and chemically, they are difficult to distinguish from each other. Both have a vitreous luster, a hardness between 3 and 4, and nearly perfect rhombic cleavage (see Figure 2.10, right side). They can, however, be distinguished by using dilute hydrochloric acid. Calcite reacts vigorously with this acid, whereas dolomite reacts much more slowly. Calcite and dolomite are usually found together as the primary constituents in the sedimentary rocks limestone and dolostone. When calcite is the dominant mineral, the rock is called *limestone*, whereas *dolostone* results from a predominance of

Figure 2.20 Quartz. Some minerals, such as quartz, occur in a variety of colors. These samples include crystal quartz (colorless), amethyst (purple), citrine (yellow), and smoky quartz (gray to black). (Photo by E. J. Tarbuck)

dolomite. Limestone has numerous economic uses, including as a road aggregate, as building stone, and as the main ingredient in portland cement.

Two other nonsilicate minerals frequently found in sedimentary rocks are *halite* and *gypsum*. Both minerals are commonly found in thick layers, which are the last vestiges of ancient seas that have long since evaporated (Figure 2.21). Like limestone, both are important

nonmetallic resources. Halite is the mineral name for common table salt (NaCl). Gypsum ($CaSO_4 \cdot 2H_2O$), which is calcium sulfate with water bound into the structure, is the mineral of which plaster and other similar building materials are composed.

In addition, a number of other minerals are prized for their economic value (see Table 2.4). Included in this group are the ores of metals, such as hematite (iron),

Table 2.4 Common nonsilicate mineral groups

Mineral Group	Name	Chemical Formula	Economic Use
Oxides	Hematite	Fe_2O_3	Ore of iron, pigment
	Magnetite	Fe_3O_4	Ore of iron
	Corundum	Al_2O_3	Gemstone, abrasive
	Ice	H_2O	Solid form of water
	Uraninite	UO_2	Ore of uranium
Sulfides	Galena	PbS	Ore of lead
	Sphalerite	ZnS	Ore of zinc
	Pyrite	FeS_2	Sulfuric acid production
	Chalcopyrite	$CuFeS_2$	Ore of copper
	Cinnabar	HgS	Ore of mercury
Sulfates	Gypsum	$CaSO_4 \cdot 2H_2O$	Plaster
	Anhydrite	$CaSO_4$	Plaster
	Barite	$BaSO_4$	Drilling mud
Native elements	Gold	Au	Trade, jewelry
	Copper	Cu	Electrical conductor
	Diamond	C	Gemstone, abrasive
	Sulfur	S	Sulfa drugs, chemicals
	Graphite	C	Pencil lead, dry lubricant
	Silver	Ag	Jewelry, photography
	Platinum	Pt	Catalyst
Halides	Halite	NaCl	Common salt
	Fluorite	CaF_2	Used in steel making
	Sylvite	KCl	Fertilizer
Carbonates	Calcite	$CaCO_3$	Portland cement, lime
	Dolomite	$CaMg(CO_3)_2$	Portland cement, lime

Figure 2.21 Thick bed of halite (salt) at an underground mine near Lansing, New York. (Photo by Michael J. Okoniewski/Liaison Agency, Inc.)

sphalerite (zinc), and galena (lead); the native (free-occurring, not in compounds) elements, including gold, silver, carbon (diamonds); and a host of others, such as fluorite, corundum, and uraninite (a uranium source).

Mineral Resources

Earth's crust and oceans are the source of a wide variety of useful and essential substances (see Box 2.1). In

Once considered safe enough to use in toothpaste, asbestos is now one of the most feared contaminants on Earth. Although health concerns arose two decades earlier, the asbestos panic began in 1986 when the Environmental Protection Agency (EPA) instituted the Asbestos Hazard Emergency Response Act. It required inspection of all public and private schools for asbestos. This brought asbestos to public attention and raised parental fears that their children could contract asbestos-related cancers because of high levels of airborne fibers in schools. Since that time, billions of dollars have been spent on asbestos testing and removal.

What Is Asbestos?

Asbestos is not a single material. Rather, asbestos is a general term for a group of silicate minerals that *readily separate into thin, strong fibers* (Figure 2.A). Because these fibers are flexible, heat resistant, and relatively chemically inert, they have many uses. Asbestos has been widely used to strengthen concrete, make fireproof fabrics, and insulate boilers and pipes that carry hot fluids. It was a component in floor tiles and the major ingredient in automobile brake linings. Wall coatings rich in asbestos fibers were used extensively during the U.S. building boom of the 1950s and early 1960s.

← 2 cm →

Figure 2.A Asbestos. This sample is a fibrous form of the mineral serpentine. (Photo by E. J. Tarbuck)

Most asbestos comes from three minerals. *Chrysotile* (white asbestos) is a fibrous form of the mineral serpentine. It is the only type of asbestos mined in North America and once made up 95 percent of world production. *Crocidolite* (blue asbestos) and *amosite* (brown asbestos) are currently mined in South Africa and make up about 5 percent of world production.

Exposure and Risk

There is no question that prolonged exposure to air laden with certain types of asbestos dust in an unregulated workplace can be dangerous. When the thin, rodlike fibers are inhaled into the lungs, they are neither easily broken down nor easily expelled, but can remain *for life*. Three lung diseases can result: (1) as-fact, practically every manufactured product contains materials obtained from minerals. Table 2.4 lists important mineral groups in this category.

Mineral resources are the endowment of useful minerals ultimately available commercially. Resources include already identified deposits from which minerals can be extracted profitably, called **reserves,** as well as known deposits that are not yet economically or technologically recoverable. Deposits inferred to exist, but not yet discovered, are also considered mineral resources.

The term **ore** is used to denote those useful metallic minerals that can be mined at a profit. In common usage, the term ore is also applied to some nonmetallic minerals such as fluorite and sulfur. However, materials used for such purposes as building stone, road aggregate, abrasives, ceramics, and fertilizers are not usually called ores; rather, they are classified as industrial rocks and minerals.

Recall that more than 98 percent of Earth's crust is composed of only eight elements, and except for oxygen and silicon, all other elements make up a relatively small fraction of common crustal rocks (see Table 2.3). Indeed, the natural concentrations of many elements are exceedingly small. A deposit containing the average percentage of a valuable element is worthless because the cost of extracting it greatly exceeds the value of the material recovered.

To be considered of value, an element must be concentrated above the level of its average crustal abundance. Generally, the lower the crustal abundance, the greater the concentration. For example, copper makes up about 0.0135 percent of the crust. However, for a material to be considered as copper ore, it must contain a concentration that is about 100 times this amount. Aluminum, in contrast, represents 8.13 percent of the crust and must be concentrated to only about four times its

---

bestosis, a scarring of tissue that decreases the lung's ability to absorb oxygen; (2) *mesothelioma,* a rare tumor that develops in the chest or gut; and (3) *lung cancer.*

Evidence that incriminates blue asbestos and brown asbestos comes from health studies at mines in South Africa and western Australia. Miners and millworkers showed an extremely high incidence of mesothelioma, sometimes after less than a year of exposure.

However, the U.S. Geological Survey concludes that the risks from the most widely used form of asbestos (chrysotile or white asbestos) are minimal to nonexistent. They cite studies of miners of white asbestos in Canada and in northern Italy, where mortality rates from mesothelioma and lung cancer differ very little from the general public.

Another study was conducted on miners' wives in the area of Thetford Mines, Quebec, once the largest chrysotile mine in the world. For many years there were no dust controls on mining and milling operations, so these women were exposed to extremely high levels of airborne asbestos. Nevertheless, they exhibited *below-normal* levels of the diseases thought to be associated with asbestos exposure.

The various types of asbestos fibers differ in their chemical composition, shape, and durability. The thin, rodlike blue asbestos and brown asbestos

fibers, which can easily penetrate the lining of the lungs, are certainly pathogenic. But chrysotile fibers are curly and can be expelled more easily than can the rodlike fibers.

Further, if inhaled, chrysotile fibers break down within a year. This is not the case for the other forms of asbestos or for fiberglass, which is frequently used as a substitute for asbestos. These differences are thought to explain the fact that the mortality rates for chrysotile workers differ very little from the rates for the general population. Although the EPA has lifted its ban on asbestos, very little of this once exalted mineral is presently used in the United States.

A Risk in U.S. Schools?

Does asbestos present a risk to the nation's students? Available data indicate that the levels of airborne asbestos in schools are approximately 0.01 of the permissible exposure levels for the U.S. workplace. In addition, the indoor concentrations of the most biologically active fibers are comparable to outdoor levels. With few exceptions, the type of asbestos found in schools is chrysotile, which in low concentrations has been shown to be relatively harmless.

A comparison of the risk from asbestos exposure in schools to other risks in society is shown in Table 2.A. Playing high school football is more than 100 times riskier than exposure to airborne asbestos. These data clearly demonstrate that the asbestos panic was and is unwarranted. Misguided public policies, no matter how well intentioned, are expensive.

Table 2.A Estimates of risk from asbestos exposure in schools in comparison with other risks

Cause	Annual Death Rate (per million)
Asbestos exposure in schools	0.005–0.093
Whooping cough vaccination (1970–80)	1–6
Aircraft accidents (1979)	6
High school football (1970–80)	10
Drowning (ages 5–14)	27
Motor vehicle accident, Pedestrian (ages 5–14)	32
Home accidents (ages 1–14)	60
Long-term smoking	1200

Figure 2.22 Aerial view of Bingham Canyon copper mine near Salt Lake City, Utah. Although the amount of copper in the rock is less than 1 percent, the huge volumes of material removed and processed each day (about 200,000 tons) yield enough metal to be profitable. (Photo by Michael Collier)

average crustal percentage before it can be extracted profitably.

It is important to realize that a deposit may become profitable to extract or lose its profitability because of economic changes. If demand for a metal increases and prices rise sufficiently, the status of a previously unprofitable deposit changes, and it becomes an ore. The status of unprofitable deposits may also change if a technological advance allows the ore to be extracted at a lower cost than before.

Conversely, changing economic factors can turn a once profitable ore deposit into an unprofitable deposit that can no longer be called an ore. This situation was illustrated at the copper mining operation located at Bingham Canyon, Utah, one of the largest open-pit mines on Earth (Figure 2.22). Mining was halted there in 1985 because outmoded equipment had driven the cost of extracting the copper beyond the current selling price. The owners responded by replacing an antiquated 1000-car railroad with conveyor belts and pipelines for transporting the ore and waste. These devices achieved a cost reduction of nearly 30 percent and returned this mining operation to profitability.

Over the years, geologists have been keenly interested in learning how natural processes produce localized concentrations of essential minerals. One well-established fact is that occurrences of valuable mineral resources are closely related to the rock cycle. That is, the mechanisms that generate igneous, sedimentary, and metamorphic rocks, including the processes of weathering and erosion, play a major role in producing concentrated accumulations of useful elements.

Moreover, with the development of the theory of plate tectonics, geologists have added another tool for understanding the processes by which one rock is transformed into another. As these rock-forming processes are examined in the following chapters, we will consider their role in producing some of our important mineral resources.

The Chapter in Review

- A *mineral* is a naturally occurring inorganic solid possessing a definite chemical structure that gives it a unique set of physical properties. Most *rocks* are aggregates composed of two or more minerals.

- The building blocks of minerals are *elements*. An *atom* is the smallest particle of matter that still retains the characteristics of an element. Each atom has a *nucleus*, which contains *protons* (particles with positive electrical charges) and *neutrons* (particles with neutral electrical charges). Orbiting the nucleus of an atom in regions called *energy levels*, or *shells*, are *electrons*, which have negative electrical charges. The number of protons in an atom's nucleus determines its *atomic number* and the name of the element. An element is a large collection of electrically neutral atoms, all having the same atomic number.

- Atoms combine with each other to form more complex substances called *compounds*. Atoms bond together by either gaining, losing, or sharing electrons with other atoms. In an *ionic bond*, one or more electrons are transferred from one atom to another, giving the atom a net positive or negative charge. The resulting electrically charged atom is called an *ion*. Ionic compounds consist of an orderly arrangement of oppositely charged ions assembled in a definite ratio that provides overall electrical neutrality. Another type of bond, the *covalent bond*, is produced when atoms share electrons.

- *Isotopes* are variants of the same element, but with a different *mass number* (the total number of neutrons plus protons found in an atom's nucleus). Some isotopes are unstable and disintegrate naturally through a process called *radioactivity*.

- The properties of minerals include *crystal form, luster, color, streak, hardness, cleavage, fracture,* and *specific gravity*. In addition, a number of special physical and chemical properties (*taste, smell, elasticity, malleability, feel, magnetism, double refraction,* and *chemical reaction to hydrochloric acid*) are useful in identifying certain minerals. Each mineral has a unique set of properties that can be used for identification.

- Of the nearly 4000 minerals, no more than a few dozen make up most of the rocks of Earth's crust and, as such, are classified as *rock-forming minerals*. Eight elements (oxygen, silicon, aluminum, iron, calcium, sodium, potassium, and magnesium) make up the bulk of these minerals and represent over 98 percent (by weight) of Earth's continental crust.

- The most common mineral group is the *silicates*. All silicate minerals have the negatively charged *silicon–oxygen tetrahedron* as their fundamental building block. In some silicate minerals the tetrahedra are joined in chains (the pyroxene and amphibole groups); in others, the tetrahedra are arranged into sheets (the micas, biotite, and muscovite), or three-dimensional networks (the feldspars and quartz). Binding the tetrahedra and various silicate structures are often the positive ions of iron, magnesium, potassium, sodium, aluminum, and calcium. Each silicate mineral has a structure and a chemical composition that indicates the conditions under which it was formed.

- The *nonsilicate* mineral groups, which contain several economically important minerals, include the *oxides* (e.g., the mineral hematite, mined for iron), *sulfides* (e.g., the mineral sphalerite, mined for zinc, and the mineral galena, mined for lead), *sulfates, halides,* and *native elements* (e.g., gold and silver). The more common nonsilicate rock-forming minerals include the *carbonate minerals*, calcite and dolomite. Two other nonsilicate minerals frequently found in sedimentary rocks are halite and gypsum.

- *Mineral resources* are the endowment of useful minerals ultimately available commercially. Resources include already identified deposits from which minerals can be extracted profitably, called *reserves*, as well as known deposits that are not yet economically or technologically recoverable. Deposits inferred to exist, but not yet discovered, are also considered mineral resources. The term *ore* is used to denote those useful metallic minerals that can be mined for a profit, as well as some nonmetallic minerals, such as fluorite and sulfur, that contain useful substances.

Key Terms

atom (p. 26)	ferromagnesian silicates	mineral (p. 24)	reserve (p. 41)
atomic number (p. 26)	(p. 36)	mineral resource (p. 41)	rock (p. 25)
cleavage (p. 32)	fracture (p. 32)	mineralogy (p. 24)	shell (p. 26)
color (p. 31)	hardness (p. 31)	Mohs scale (p. 31)	silicates (p. 33)
compound (p. 26)	ion (p. 28)	neutron (p. 26)	silicon–oxygen
covalent bond (p. 29)	ionic bond (p. 26)	nonferromagnesian	tetrahedron (p. 33)
crystal form (p. 30)	isotope (p. 30)	silicates (p. 36)	specific gravity (p. 32)
dark silicates (p. 36)	light silicates (p. 36)	nucleus (p. 26)	streak (p. 31)
electron (p. 26)	luster (p. 31)	ore (p. 41)	valence electron (p. 26)
element (p. 26)	mass number (p. 30)	proton (p. 26)	
energy-levels (p. 26)	metallic bond (p. 30)	radioactive decay (p. 30)	

Questions for Review

1. Define the term *rock*.

2. List the three main particles of an atom, and explain how they differ from one another.

3. If the number of electrons in an atom is 35 and its mass number is 80, calculate the following:
   a. the number of protons
   b. the atomic number
   c. the number of neutrons

4. What is the significance of valence electrons?

5. Briefly distinguish between ionic and covalent bonding.

6. What occurs in an atom to produce an ion?

7. What is an isotope?

8. Although all minerals have an orderly internal arrangement of atoms (crystalline structure), most mineral samples do not demonstrate their crystal form. Why?

9. Why might it be difficult to identify a mineral by its color?

10. If you found a glassy-appearing mineral while rock hunting and had hopes that it was a diamond, what simple test might help you make a determination?

11. Explain the use of corundum as given in Table 2.4 in terms of the Mohs hardness scale.

12. Gold has a specific gravity of almost 20. If a 25-liter pail of water weighs about 25 kilograms, how much would a 25-liter pail of gold weigh?

13. Explain the difference between the terms *silicon* and *silicate*.

14. What do ferromagnesian minerals have in common? List examples of ferromagnesian minerals.

15. What do muscovite and biotite have in common? How do they differ?

16. Should color be used to distinguish between orthoclase and plagioclase feldspar? What is the best means of distinguishing between the two types of feldspar?

17. Each of the following statements describes a silicate mineral or mineral group. In each case, provide the appropriate name.
    a. The most common member of the amphibole group.
    b. The most common nonferromagnesian member of the mica family.
    c. The only silicate mineral made entirely of silicon and oxygen.
    d. A high-temperature silicate with a name that is based on its color.
    e. Characterized by striations.
    f. Originates as a product of chemical weathering.

18. What simple test can be used to distinguish calcite from dolomite?

19. Contrast *resource* and *reserve*.

20. What might cause a mineral deposit that had not been considered an ore to be reclassified as an ore?

Web Resources

The *Essentials of Geology* Website uses the resources and flexibility of the Internet to aid in your study of the topics in this chapter. Written and developed by geology instructors, this site will help improve your understanding of geology. Visit **www.prenhall.com/lutgens** and click on the cover of *Essentials of Geology, 8/e* to find:

- On-line review quizzes.
- Web-based critical thinking and writing exercises.
- Links to chapter-specific Web resources.
- Internet-wide key term searches.

www.prenhall.com/lutgens

Igneous Rocks

Focus on Learning

To assist you in learning the important concepts in this chapter, you will find it helpful to focus on the following questions:

- How are igneous rocks formed?
- How does magma differ from lava?
- What two criteria are used to classify igneous rocks?
- How does the rate of cooling of magma influence the crystal size of minerals in igneous rocks?
- How is the mineral makeup of an igneous rock related to Bowen's reaction series?
- In what ways are granitic rocks different from basaltic rocks?
- How are economic deposits of gold, silver, and many other metals formed?

Igneous rocks exposed at Devil's Tower National Monument, Wyoming. (Photo by Carr Clifton)

Igneous rocks make up the bulk of Earth's crust. In fact, with the exception of Earth's core, the remaining solid portion of our planet is basically a huge mass of igneous rock that is partially mantled with a thin veneer of sedimentary and metamorphic rocks. Consequently, a basic knowledge of igneous rocks is essential to our understanding of the structure, composition, and internal workings of our planet.

Magma: The Parent Material of Igneous Rock

In our discussion of the rock cycle, it was pointed out that **igneous rocks** (from the Latin *ignis*, or fire) form as molten rock cools and solidifies. Abundant evidence supports the fact that the parent material for igneous rocks, called *magma*, is formed by a process called *partial melting*. Partial melting occurs at various levels within Earth's crust and upper mantle to depths of perhaps 250 kilometers (about 150 miles). We will explore the origin of magma later in this chapter.

Once formed, a magma body buoyantly rises toward the surface because it is less dense than the surrounding rocks. Occasionally molten rock breaks through, producing a spectacular volcanic eruption. Magma that reaches Earth's surface is called **lava.** Sometimes magma is explosively ejected from a vent, producing a catastrophic eruption. However, not all eruptions are violent; many volcanoes emit quiet outpourings of very fluid lava (Figure 3.1).

Igneous rocks that form when molten rock solidifies *at the surface* are classified as **extrusive,** or **volcanic** (after the fire god Vulcan). Extrusive igneous rocks are abundant in western portions of the Americas, including the volcanic cones of the Cascade Range and the extensive lava flows of the Columbia Plateau. In addition, many oceanic islands, typified by the Hawaiian chain, are composed almost entirely of volcanic igneous rocks.

Magma that loses its mobility before reaching the surface eventually crystallizes at depth. Igneous rocks that *form at depth* are termed **intrusive,** or **plutonic** (after Pluto, the god of the lower world in classical mythology). Intrusive igneous rocks would never outcrop at the surface if portions of the crust were not uplifted and the overlying rocks stripped away by erosion. (When a mass of crustal rock is exposed—not covered with soil—it is called an *outcrop.*) Exposures of intrusive igneous rocks occur in many places, including Mount Washington, New Hampshire, Stone Mountain, Georgia, the Black Hills of South Dakota, and Yosemite National Park, California.

The Nature of Magma

Magma is completely or partly molten material, which on cooling solidifies to form an igneous rock. Most magmas consist of three distinct parts—a liquid component, a solid component, and a gaseous phase.

Figure 3.1 Fluid basaltic lava erupting and flowing down the slopes of Hawaii's Kilauea Volcano toward the sea. (Photo by G. Brad Lewis)

The liquid portion, called **melt,** is composed of mobile ions of those elements commonly found in Earth's crust. Melt is made up mostly of ions of silicon and oxygen that readily combine to form silica (SiO_2), as well as lesser amounts of aluminum, potassium, calcium, sodium, iron, and magnesium.

The solid components (if any) in magma are silicate minerals that have already crystallized from the melt. As a magma body cools, the size and number of crystals increases. During the last stage of cooling, a magma body is mostly a crystalline solid with only minor amounts of melt.

Water vapor (H_2O), carbon dioxide (CO_2), and sulfur dioxide (SO_2) are the most common gases found in magma and are confined by the immense pressure exerted by the overlying rocks. These gaseous components, called **volatiles,** are dissolved within the melt. (Volatiles are those materials that will readily vaporize—form a gas—at surface pressures.) Volatiles remain part of the magma until it either moves near the surface (low-pressure environment) or until the magma body is essentially crystallized, at which time any remaining volatiles freely migrate away.

From Magma to Crystalline Rock

As a magma cools, the ions in the melt begin to lose their mobility and arrange themselves into orderly crystalline structures. This process, called **crystallization,** generates various silicate minerals that reside within the remaining melt.

Before we examine how magma crystallizes, let us first examine how a simple crystalline solid melts. In any crystalline solid, the ions are arranged in a closely packed regular pattern. However, they are not without some motion. They exhibit a sort of restricted vibration about fixed points. As the temperature rises, the ions vibrate more rapidly and consequently collide with ever increasing vigor with their neighbors. Thus, heating causes the ions to occupy more space, which in turn causes the solid to expand. When the ions are vibrating rapidly enough to overcome the force of the chemical bonds, the solid begins to melt. At this stage the ions are able to slide past one another, and their orderly crystalline structure disintegrates. Thus, melting converts a solid consisting of tight, uniformly packed ions into a liquid composed of unordered ions moving randomly about.

In the process of crystallization, cooling reverses the events of melting. As the temperature of the liquid drops, the ions pack closer and closer together as they slow their rate of movement. When cooled sufficiently, the forces of the chemical bonds will again confine the ions to an orderly crystalline arrangement.

When magma cools, it is generally the silicon and oxygen atoms that link together first to form silicon-oxygen tetrahedra, the basic building blocks of the silicate minerals. As a magma continues to lose heat to its surroundings, the tetrahedra join with each other and with other ions to form embryonic crystal nuclei. Slowly each nucleus grows as ions lose their mobility and join the crystalline network.

The earliest formed minerals have space to grow and tend to have better developed crystal faces than do the later ones that fill the remaining space. Eventually all of the melt is transformed into a solid mass of interlocking silicate minerals that we call an *igneous rock* (Figure 3.2).

As you will see later, the crystallization of magma is much more complex than just described. Whereas a single compound, such as water, crystallizes at a specific temperature, solidification of magma with its diverse chemistry spans a temperature range of 200°C. During crystallization, the composition of the melt con-

A.   **B.**

Figure 3.2 **A.** Close-up of interlocking crystals in a coarse-grained igneous rock. The largest crystals are about 1 centimeter in length. **B.** Photomicrograph of interlocking crystals in a coarse-grained igneous rock. (Photos by E. J. Tarbuck) tinually changes as ions are selectively removed and incorporated into the earliest formed minerals. If the melt should separate from the earliest formed minerals, its composition will be different from that of the original magma. Thus, a single magma may generate rocks with widely differing compositions. As a consequence, a great variety of igneous rocks exist. We will return to this important idea later in the chapter.

Although the crystallization of magma is complex, it is nevertheless possible to classify igneous rocks based on their mineral composition and the conditions under which they formed. Their environment during crystallization can be roughly inferred from the size and arrangement of the mineral grains, a property called *texture*. Consequently, *igneous rocks are most often classified by their texture and mineral composition*. We will consider these two rock characteristics in the following sections.

Igneous Textures

 Earth Materials
↳ Igneous Rocks

The term **texture**, when applied to an igneous rock, is used to describe the overall appearance of the rock based on the size, shape, and arrangement of its interlocking crystals (Figure 3.3). Texture is an important characteristic because it reveals a great deal about the environment in which the rock formed. This fact allows geologists to make inferences about a rock's origin while working in the field where sophisticated equipment is not available.

Factors Affecting Crystal Size

Three factors contribute to the textures of igneous rocks: (1) *the rate at which magma cools*; (2) *the amount of silica present*; and (3) *the amount of dissolved gases in the magma*. Of these, the rate of cooling is the dominant factor, but like all generalizations, this one has numerous exceptions.

As a magma body loses heat to its surroundings, the mobility of its ions decreases. A very large magma body located at great depth will cool over a period of perhaps tens or hundreds of thousands of years. Initially, relatively few crystal nuclei form. Slow cooling permits ions to migrate freely until they eventually join one of the existing crystalline structures. Consequently, slow cooling promotes the growth of fewer but larger crystals.

On the other hand, when cooling occurs more rapidly—for example, in a thin lava flow—the ions quickly lose their mobility and readily combine to form crystals. This results in the development of numerous embryonic nuclei, all of which compete for the available ions. The result is a solid mass of small intergrown crystals.

When molten material is quenched quickly, there may not be sufficient time for the ions to arrange into a crystalline network. Rocks that consist of unordered ions are referred to as **glass.**

Types of Igneous Textures

As you saw, the effect of cooling on rock textures is fairly straightforward. Slow cooling promotes the growth of large crystals, whereas rapid cooling tends to generate smaller crystals. We will consider the other factors affecting crystal growth as we examine the major textural types.

Aphanitic (fine-grained) Texture. Igneous rocks that form at the surface or as small masses within the upper crust where cooling is relatively rapid possess a very fine-grained texture termed **aphanitic.** By definition, the crystals that make up aphanitic rocks are so small that individual minerals can only be distinguished with the aid of a microscope (Figure 3.3A). Because mineral identification is not possible, we commonly characterize fine-grained rocks as being light, intermediate, or dark in color. Using this system of grouping, light-colored aphanitic rocks are those containing primarily light-colored nonferromagnesian silicate minerals and so forth (see the section titled "Common Silicate Minerals" in Chapter 2).

Commonly seen in many aphanitic rocks are the voids left by gas bubbles that escape as lava solidifies. These spherical or elongated openings are called *vesicles*, and the rocks that contain them are said to have a **vesicular texture.** Rocks that exhibit a vesicular texture usually form in the upper zone of a lava flow, where cooling occurs rapidly enough to "freeze" the lava, thereby preserving the openings produced by the expanding gas bubbles (Figure 3.4).

Phaneritic (coarse-grained) Texture. When large masses of magma slowly solidify far below the surface, they form igneous rocks that exhibit a coarse-grained texture described as **phaneritic.** These coarse-grained rocks consist of a mass of intergrown crystals that are roughly equal in size and large enough so that the individual minerals can be identified without the aid of a microscope (Figure 3.3B). (Geologists often use a small magnifying lens to aid in identifying coarse-grained minerals.) Because phaneritic rocks form deep within Earth's crust, their exposure at Earth's surface results only after erosion removes the overlying rocks that once surrounded the magma chamber.

Porphyritic Texture. A large mass of magma located at depth may require tens to hundreds of thousands of years to solidify. Because different minerals crystallize at different temperatures (as well as at differing rates),

A. Aphanitic

B. Phaneritic

C. Porphyritic

D. Glassy

Figure 3.3   Igneous rock textures. **A.** Aphanitic (fine-grained). **B.** Phaneritic (coarse-grained). **C.** Porphyritic (large crystals embedded in a matrix). **D.** Glassy (cooled too rapidly to form crystals). (Photos by E. J. Tarbuck)

it is possible for some crystals to become quite large before others even begin to form. If magma containing some large crystals should change environments—for example, by erupting at the surface—the remaining liquid portion of the lava would cool relatively quickly. The resulting rock, which has large crystals embedded in a matrix of smaller crystals, is said to have a **porphyritic texture** (Figure 3.3C). The large crystals in such a rock are referred to as **phenocrysts,** whereas the matrix of smaller crystals is called **groundmass.** A rock with such a texture is termed a **porphyry.**

Glassy Texture.   During some volcanic eruptions, molten rock is ejected into the atmosphere, where it is quenched quickly. Rapid cooling of this type may generate rocks having a **glassy texture.** As we indicated

earlier, glass results when unordered ions are "frozen" before they are able to unite into an orderly crystalline structure. *Obsidian*, a common type of natural glass, is similar in appearance to a dark chunk of manufactured glass (Figure 3.3D).

Lava flows composed of obsidian several tens of feet thick occur in some places (Figure 3.5). Thus, rapid cooling is not the only mechanism by which a glassy texture can form. As a general rule, magmas with a high silica content tend to form long, chainlike structures before crystallization is complete. These structures in turn impede ionic transport and increase the magma's viscosity. (*Viscosity* is a measure of a fluid's resistance to flow.)

Granitic magma, which is rich in silica, may be extruded as an extremely viscous mass that eventually solidifies to form obsidian. By contrast, basaltic

Figure 3.4   Scoria is a volcanic rock that exhibits a vesicular texture. Vesicles are small holes left by escaping gas bubbles. (Photo by E. J. Tarbuck)

magma, which is low in silica, forms very fluid lavas that upon cooling usually generate fine-grained crystalline rocks. However, the surface of basaltic lava may be quenched rapidly enough to form a thin, glassy skin. Moreover, Hawaiian volcanoes sometimes generate lava fountains, which spray basaltic lava tens of meters into the air. Such activity can produce strands of volcanic glass called *Pele's hair*, after the Hawaiian goddess of volcanoes.

Pyroclastic (fragmental) Texture.   Some igneous rocks are formed from the consolidation of individual rock fragments that are ejected during a violent volcanic eruption. The ejected particles might be very fine ash, molten blobs, or large angular blocks torn from the walls of the vent during the eruption. Igneous

rocks composed of these rock fragments are said to have a **pyroclastic** or **fragmental texture.**

Because pyroclastic rocks are made of individual particles or fragments rather than interlocking crystals, their textures often appear to be more similar to sedimentary rocks than to other igneous rocks.

Pegmatitic Texture.   Under special conditions, exceptionally coarse-grained igneous rocks, called **pegmatites,** may form. These rocks, which are composed of interlocking crystals all larger than a centimeter in diameter, are said to have a **pegmatitic texture.** Most pegmatites are found around the margins of large plutons as small masses or thin veins that commonly extend into the adjacent host rock.

Pegmatites form in the late stages of crystallization, when water and other volatiles, such as chlorine, fluorine, and sulfur, make up an unusually high percentage of the melt. Because ion migration is enhanced in these fluid-rich environments, the crystals that form are abnormally large. Thus, the large crystals in pegmatites are not the result of inordinately long cooling histories; rather, they are the consequence of the fluid-rich environment in which crystallization takes place.

The composition of most pegmatites is similar to that of granite. Thus, pegmatites contain large crystals of quartz, feldspar, and muscovite. However, some contain significant quantities of comparatively rare and hence valuable minerals (see section titled "Mineral Resources and Igneous Processes" at the end of this chapter).

Igneous Compositions

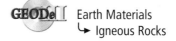

GEODe Earth Materials
   ↳ Igneous Rocks

Igneous rocks are mainly composed of silicate minerals. Furthermore, the mineral makeup of a particular igneous rock is ultimately determined by the chemical

Figure 3.5   This obsidian flow was extruded from a vent along the south wall of Newberry Caldera, Oregon. Note the road for scale. (Photo by E. J. Tarbuck)

composition of the magma from which it crystallizes. Recall that magma is composed largely of the eight elements that are the major constituents of the silicate minerals. Chemical analysis shows that silicon and oxygen (usually expressed as the silica [SiO_2] content of a magma) are by far the most abundant constituents of igneous rocks. These two elements, plus ions of aluminum (Al), calcium (Ca), sodium (Na), potassium (K), magnesium (Mg), and iron (Fe), make up roughly 98 percent by weight of most magmas. In addition, magma contains small amounts of many other elements, including titanium and manganese, and trace amounts of much rarer elements such as gold, silver, and uranium.

As magma cools and solidifies, these elements combine to form two major groups of silicate minerals. The *dark* (or *ferromagnesian*) *silicates* are rich in iron and/or magnesium and comparatively low in silica. *Olivine, pyroxene, amphibole,* and *biotite mica* are the common dark silicate minerals of Earth's crust. By contrast, the *light* (or *nonferromagnesian*) *silicates* contain greater amounts of potassium, sodium, and calcium rather than iron and magnesium. As a group, these minerals are richer in silica than the dark silicates. The light silicates include *quartz, muscovite mica*, and the most abundant mineral group, the *feldspars*. The feldspars make up over

40 percent of most igneous rocks. Thus, in addition to feldspar, igneous rocks contain some combination of the other light and/or dark silicates listed here.

Granitic versus Basaltic Compositions

Despite their great compositional diversity, igneous rocks (and the magmas from which they form) can be divided into broad groups according to their proportions of light and dark minerals (Figure 3.6). Near one end of the continuum are rocks composed almost entirely of light-colored silicates—quartz and feldspar. Igneous rocks in which these are the dominant minerals have a **granitic composition**. Geologists also refer to granitic rocks as being **felsic,** a term derived from *f*eldspar and *si*lica (quartz). In addition to quartz and feldspar, most granitic rocks contain about 10 percent dark silicate minerals, usually biotite mica and amphibole. Granitic rocks are rich in silica (about 70 percent) and are major constituents of the continental crust.

Rocks that contain substantial dark silicate minerals and calcium-rich plagioclase feldspar (but no quartz) are said to have a **basaltic composition** (Figure 3.6). Because basaltic rocks contain a high percentage of ferro-

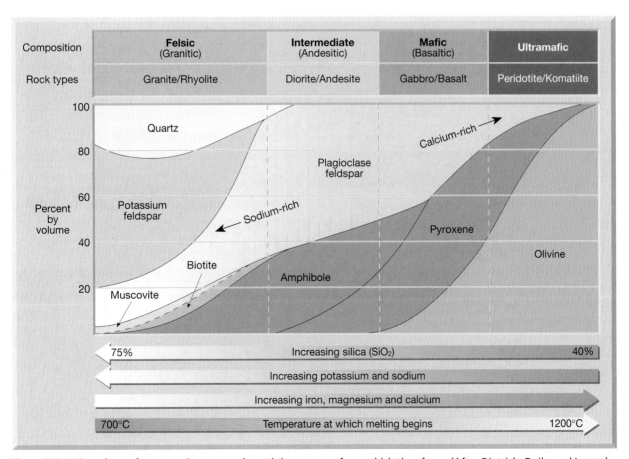

Figure 3.6 Mineralogy of common igneous rocks and the magmas from which they form. (After Dietrich, Daily, and Larsen)

magnesian minerals, geologists also refer to them as **mafic** (from *ma*gnesium and *f*errum, the Latin name for iron). Because of their iron content, mafic rocks are typically darker and have a greater density than granitic rocks. Basaltic rocks make up the ocean floor as well as many of the volcanic islands located within the ocean basins. Basalt is also found on the continents.

Other Compositional Groups

As you can see in Figure 3.6, rocks with a composition between granitic and basaltic rocks are said to have an **intermediate,** or **andesitic, composition** after the common volcanic rock *andesite.* Intermediate rocks contain at least 25 percent dark silicate minerals, mainly amphibole, pyroxene, and biotite mica with the other dominant mineral being plagioclase feldspar. This important category of igneous rocks is associated with volcanic activity that is typically confined to the margins of the continents.

Another important igneous rock, *peridotite,* contains mostly olivine and pyroxene and thus falls on the opposite side of the compositional spectrum from granitic rocks (see Figure 3.6). Because peridotite is composed almost entirely of ferromagnesian minerals, its chemical composition is referred to as **ultramafic.** Although ultramafic rocks are rare at Earth's surface, peridotite is believed to be the main constituent of the upper mantle.

Silica Content as an Indicator of Composition

An important aspect of the chemical composition of igneous rocks is their silica (SiO_2) content. Recall that silicon and oxygen are the two most abundant elements in igneous rocks. Typically, the silica content of crustal rocks ranges from a low of about 45 percent in ultramafic rocks to a high of over 70 percent in felsic rocks (Figure 3.6). The percentage of silica in igneous rocks actually varies in a systematic manner that parallels the abundance of other elements. For example, rocks comparatively low in silica contain large amounts of iron, magnesium, and calcium. By contrast, rocks high in silica contain very small amounts of those elements but are enriched instead in sodium and potassium. Consequently, the chemical makeup of an igneous rock can be inferred directly from its silica content.

Further, the amount of silica present in magma strongly influences its behavior. Granitic magma, which has a high silica content, is quite viscous (thick) and exists as a liquid at temperatures as low as 700°C. On the other hand, basaltic magmas are low in silica and are

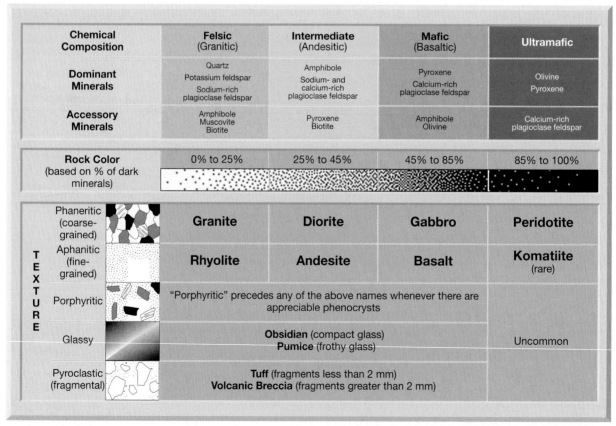

Figure 3.7   Classification of the major groups of igneous rocks based on their mineral composition and texture. Phaneritic (coarse-grained) rocks are plutonic, solidifying deep underground. Aphanitic (fine-grained) rocks are volcanic, or solidify as shallow, thin plutons.

generally more fluid. Further, basaltic magmas crystallize at higher temperatures than granitic magmas and are completely solid when cooled to 1000°C.

In summary, igneous rocks can be divided into broad groups according to the proportions of light and dark minerals they contain. Granitic (felsic) rocks, which are almost entirely composed of the light-colored minerals quartz and feldspar, are at one end of the compositional spectrum (Figure 3.6). Basaltic (mafic) rocks, which contain abundant dark silicate minerals in addition to plagioclase feldspar, make up the other major igneous rock group of Earth's crust. Between these groups are rocks with an intermediate (andesitic) composition, while ultramafic rocks, which totally lack light-colored minerals, lie at the other end of the compositional spectrum from granitic rocks.

Naming Igneous Rocks

 Earth Materials
  ↳ Igneous Rocks

As was stated previously, igneous rocks are most often classified, or grouped, on the basis of their texture and mineral composition (Figure 3.7). The various igneous textures result mainly from different cooling histories, whereas the mineral composition of an igneous rock is the consequence of the chemical makeup of its parent magma. Because igneous rocks are classified on the basis of their mineral composition and texture, two rocks may have similar mineral constituents but have different textures and hence different names. For example, *granite*, a coarse-grained plutonic rock, has a fine-grained volcanic equivalent called *rhyolite*. Although these rocks are mineralogically the same, they have different cooling histories and do not look at all alike (Figure 3.8).

Felsic (Granitic) Igneous Rocks

Granite.   Granite is perhaps the best known of all igneous rocks (Figure 3.8A). This is partly because of its natural beauty, which is enhanced when it is polished, and partly because of its abundance in the continental crust. Slabs of polished granite are commonly used for tombstones and monuments and as building stones. Well-known areas in the United States where granite is quarried include Barre, Vermont; Mount Airy, North Carolina; and Saint Cloud, Minnesota.

Granite is a phaneritic rock composed of about 25 percent quartz and roughly 65 percent feldspar, mostly potassium- and sodium-rich varieties. Quartz crystals, which are roughly spherical in shape, are often glassy and clear to light gray in color. By contrast, feldspar crystals are not as glassy, are generally white to gray or salmon pink in color, and exhibit a rectangular rather than spherical shape. When potassium feldspar is dominant and dark pink in color, granite appears almost

A. Granite

Close up

B. Rhyolite

Close up

Figure 3.8   **A.** Granite, one of the most common phaneritic igneous rocks. **B.** Rhyolite, the aphanitic equivalent of granite, is less abundant. (Photos by E. J. Tarbuck)

reddish. This variety is popular as a building stone. However, the feldspar grains are often white to gray, so that when mixed with lesser amounts of dark silicates, granite appears light gray in color.

Other minor constituents of granite are muscovite and some dark silicates, particularly biotite and amphibole. Although the dark components generally make up less than 10 percent of most granites, dark minerals appear to be more prominent than their percentage would indicate.

Granite may also have a porphyritic texture. These specimens contain feldspar crystals of a centimeter or more in length that are scattered among the coarse-grained groundmass of quartz and amphibole.

Granite and other related crystalline rocks are often the by-products of mountain building. Because granite is very resistant to weathering, it frequently forms the core of eroded mountains. For example, Pikes Peak in the Rockies, Mount Rushmore in the Black Hills, the White Mountains of New Hampshire, Stone Mountain

in Georgia, and Yosemite National Park in the Sierra Nevada are all areas where large quantities of granite are exposed at the surface (Figure 3.9).

Granite is a very abundant rock. However, it has become common practice among geologists to apply the term granite to any coarse-grained intrusive rock composed predominantly of light silicate materials. We will follow this practice for the sake of simplicity. You should keep in mind that this use of the term granite covers rocks having a wide range of mineral compositions.

Rhyolite. Rhyolite is the extrusive equivalent of granite and, like granite, is composed essentially of the light-colored silicates (Figure 3.8B). This fact accounts for its color, which is usually buff to pink or occasionally very light gray. Rhyolite is aphanitic and frequently contains glass fragments and voids, indicating rapid cooling in a surface environment. When rhyolite contains phenocrysts, they are small and composed of

Figure 3.9 El Capitan, a large igneous monolith located in Yosemite National Park, California. Granite bedrock is exposed over much of the park. (Photo by Carr Clifton)

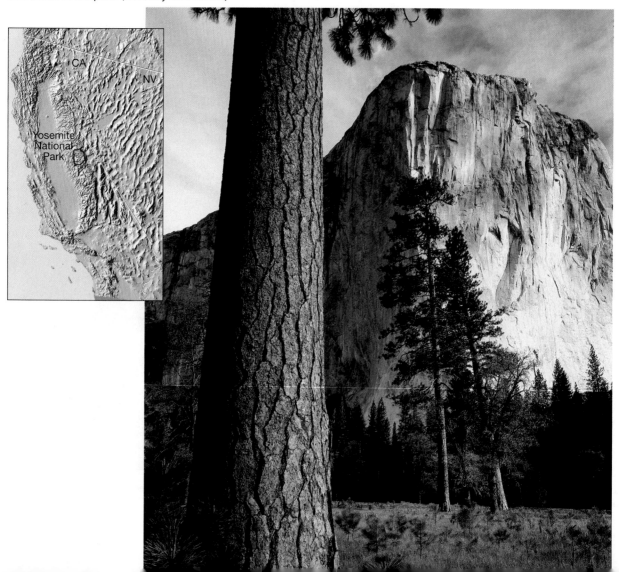

either quartz or potassium feldspar. In contrast to granite, which is widely distributed as large plutonic masses, occurences of rhyolite are less common and generally less voluminous. Yellowstone Park is one well-known exception. Here rhyolite lava flows and thick ash deposits of similar composition are extensive.

Obsidian.   *Obsidian* is a dark-colored glassy rock that usually forms when silica-rich lava is quenched quickly (Figure 3.10A). In contrast to the orderly arrangement of ions characteristic of minerals, *the ions in glass are unordered*. Consequently, glassy rocks such as obsidian are not composed of minerals in the same sense as most other rocks.

Although usually black or reddish-brown in color, obsidian has a high silica content (Figure 3.10A). Thus, its composition is more akin to the light-colored igneous rocks such as granite than to the dark rocks of basaltic composition. By itself, silica is clear like window glass; the dark color results from the presence of metallic ions. If you examine a thin edge of a piece of obsidian, it will be nearly transparent. Because of its excellent conchoidal fracture and ability to hold a sharp, hard edge, obsidian was a prized material from which Native Americans chipped arrowheads and cutting tools.

Pumice.   *Pumice* is a volcanic rock that, like obsidian, has a glassy texture. Usually found with obsidian, pumice forms when large amounts of gas escape through lava to generate a gray, frothy mass (Figure 3.10B). In some samples, the voids are quite noticeable, whereas in others the pumice resembles fine shards of intertwined glass. Because of the large percentage of voids, many samples of pumice will float when placed in water. Oftentimes flow lines are visible in pumice, indicating that some movement occurred before solidification was complete. Moreover, pumice and obsidian can often be found in the same rock mass, where they exist in alternating layers.

Intermediate (Andesitic) Igneous Rocks

Andesite.   Andesite is a medium-gray fine-grained rock of volcanic origin. Its name comes from South America's Andes Mountains, where numerous volcanoes are composed of this rock type. In addition to the volcanoes of the Andes, many of the volcanic structures encircling the Pacific Ocean are of andesitic composition. Andesite commonly exhibits a porphyritic texture (Figure 3.11). When this is the case, the phenocrysts are often light, rectangular crystals of plagioclase feldspar or black, elongated amphibole crystals. Andesite often resembles rhyolite, so their identification usually requires microscopic examination to verify the abundance, or lack, of quartz crystals. Andesite contains minor amounts of quartz, whereas rhyolite is composed of about 25 percent quartz.

Diorite.   *Diorite* is the plutonic equivalent of andesite. It is a coarse-grained intrusive rock that looks some-

A.

B.

Figure 3.10   Igneous rocks that exhibit a glassy texture. **A.** Obsidian, a glassy volcanic rock. **B.** Pumice, a glassy rock containing numerous vesicles. (Photos by E. J. Tarbuck)

A. Andesite porphyry

B. Close up

Figure 3.11 Andesite porphyry. **A.** Hand sample of andesite porphyry, a common volcanic rock. **B.** Photomicrograph of a thin section of andesite porphyry to illustrate texture. Notice that the few large crystals (phenocrysts) are surrounded by much smaller crystals (groundmass). (Photo courtesy of E. J. Tarbuck)

Close up

Figure 3.12 Diorite is a phaneritic igneous rock of intermediate composition. The white crystals are plagioclase feldspar and the black crystals are amphibole and biotite. (Photo courtesy of E. J. Tarbuck)

what similar to gray granite. However, it can be distinguished from granite by the absence of visible quartz crystals and because it contains a higher percentage of dark silicate minerals. The mineral makeup of diorite is primarily sodium-rich plagioclase feldspar and amphibole, with lesser amounts of biotite. Because the light-colored feldspar grains and dark amphibole crystals appear to be roughly equal in abundance, diorite has a salt-and-pepper appearance (Figure 3.12).

Mafic (Basaltic) Igneous Rocks

Basalt. *Basalt* is a very dark green to black fine-grained volcanic rock composed primarily of pyroxene and calcium-rich plagioclase feldspar, with lesser amounts of olivine and amphibole present (Figure 3.13A). When porphyritic, basalt commonly contains small light-colored feldspar phenocrysts or glassy-appearing olivine phenocrysts embedded in a dark groundmass.

Basalt is the most common extrusive igneous rock. Many volcanic islands, such as the Hawaiian Islands and Iceland, are composed mainly of basalt (see Figure 3.1). Further, the upper layers of the oceanic crust con-

sist of basalt. In the United States, large portions of central Oregon and Washington were the sites of extensive basaltic outpourings (see Figure 4.20). At some locations these once fluid basaltic flows have accumulated to thicknesses approaching 3 kilometers.

Gabbro. Gabbro is the intrusive equivalent of basalt (Figure 3.13B). Like basalt, it is very dark green to black in color and composed primarily of pyroxene and calcium-rich plagioclase feldspar. Although gabbro is not a common constituent of the continental crust, it undoubtedly makes up a significant percentage of the oceanic crust. Here large portions of the magma found in underground reservoirs that once fed basalt flows eventually solidified at depth to form gabbro.

Pyroclastic Rocks

Pyroclastic rocks are composed of fragments ejected during a volcanic eruption. One of the most common pyroclastic rocks, called *tuff*, is composed mainly of tiny ash-size fragments that were later cemented together. In situations where the ash particles remained hot enough to fuse, the rock is called *welded tuff*. Although welded tuff consists mostly of tiny glass shards, it may

Figure 3.13   These dark-colored mafic rocks are composed primarily of pyroxene and calcium-rich plagioclase. **A.** Basalt is aphanitic and a very common extrusive rock. **B.** Gabbro, the phaneritic equivalent of basalt, is less abundant. (Photos by E. J. Tarbuck)

contain walnut-size pieces of pumice and other rock fragments.

Welded tuffs blanket vast portions of once volcanically active areas of the western United States. Some of these tuff deposits are hundreds of feet thick and extend for tens of miles from their source. Most formed millions of years ago as volcanic ash spewed from large volcanic structures (calderas) in an avalanche style, spreading laterally at speeds approaching 100 kilometers per hour. Early investigators of these deposits incorrectly classified them as rhyolite lava flows. Today we know that this silica-rich lava is too viscous (thick) to flow more than a few miles from a vent.

Unlike most igneous rock names, such as granite and basalt, the term tuff does not imply mineral composition. Thus, it is frequently used with a modifier, as, for example, rhyolite tuff.

Origin of Magma

The origin of magma has been a controversial topic in geology almost from the very beginning of the science. How can magma form in Earth's mantle, which is composed essentially of solid rock? How do magmas of different compositions form? Why do volcanoes in the deep-ocean basins primarily extrude basaltic lava, whereas those on the continental margins adjacent to oceanic trenches extrude mainly andesitic lava? These are some of the questions we will address in the following section.

Generating Magma from Solid Rock

Based on available scientific evidence, *Earth's crust and mantle are composed primarily of solid, not molten, rock.* Although the outer core is a fluid, its iron-rich material is very dense and remains deep within Earth. So, what is the source of magma that produces igneous activity?

Geologists conclude that magma originates when essentially solid rock, located in the crust and upper mantle, melts. The most obvious way to generate magma from solid rock is to raise the temperature above the rock's melting point.

Role of Heat.   What source of heat is sufficient to melt rock? Workers in underground mines know that temperatures get higher as they go deeper. Although the rate of temperature change varies from place to place, it *averages* between 20°C and 30°C per kilometer in the *upper* crust. This change in temperature with depth is known as the **geothermal gradient.** Estimates indicate that the temperature at a depth of 100 kilometers ranges between 1200°C and 1400°C. At these high temperatures, rocks in the lower crust and upper mantle are near but somewhat below their melting points. Thus, they are very hot but still essentially solid.

There are several ways that enough additional heat can be generated within the crust or upper mantle to produce some magma. First, at subduction zones friction generates heat as huge slabs of crust slide past one another. Second, crustal rocks are heated as they descend into the mantle during subduction. Third, hot mantle rocks can rise and intrude crustal rocks. Although all of these processes generate some magma, the quantities produced are comparatively small and the distribution is limited. As you will see, the vast bulk of magma forms without the aid of an additional heat source.

Role of Pressure.   If temperature were the only factor that determined whether or not rock melts, our planet would be a molten ball covered with a thin, solid outer shell. This, of course, is not the case. The reason is that pressure also increases with depth.

Melting, which is accompanied by an increase in volume, *occurs at higher temperatures at depth* because of greater confining pressure. Consequently, an increase in confining pressure causes an increase in the rock's melting temperature. Conversely, reducing confining pressure lowers a rock's melting temperature. When confining pressure drops enough, **decompression melting** is triggered. This may occur when rock *ascends* as a result of convective upwelling, thereby moving into zones of lower pressure. (Recall that even though the mantle is a *solid, it does flow* at very slow rates over time

scales of millions of years.) This process is responsible for generating magma along oceanic ridges where plates are rifting apart (Figure 3.14).

Role of Volatiles.   Another important factor affecting the melting temperature of rock is its water content. Water and other volatiles act as salt does to melt ice. That is, volatiles cause rock to melt at lower temperatures. Further, the effect of volatiles is magnified by increased pressure. Consequently, "wet" rock buried at depth has a much lower melting temperature than does "dry" rock of the same composition and under the same confining pressure. Therefore, in addition to a rock's composition, its temperature, depth (confining pressure), and water content determine whether it exists as a solid or liquid.

Volatiles play an important role in generating magma in regions where cool slabs of oceanic lithosphere descend into the mantle (Figure 3.15). As an oceanic plate sinks, both heat and pressure drive water from the subducting crustal rocks. These volatiles, which are very mobile, migrate into the wedge of hot mantle that lies above. This process is believed to lower the melting temperature of mantle rock sufficiently to generate some melt. Laboratory studies have shown that the melting point of basalt can be lowered by as much as 100°C by the addition of only 0.1 percent water.

When enough mantle-derived basaltic magma forms, it will buoyantly rise toward the surface. In a continental setting, basaltic magma may "pond" be-

Figure 3.14 As hot mantle rock ascends, it continually moves into zones of lower pressure. This drop in confining pressure can trigger melting, even without additional heat.

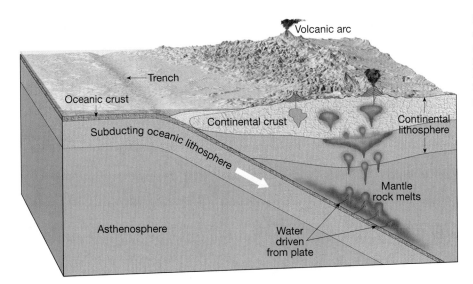

Figure 3.15 As an oceanic plate descends into the mantle, water and other volatiles are driven from the subducting crustal rocks. These volatiles lower the melting temperature of mantle rock sufficiently to generate melt.

neath crustal rocks, which have a lower density and are already near their melting temperature. This may result in some melting of the crust and the formation of a secondary, silica-rich magma.

In summary, magma can be generated under three sets of conditions: (1) *heat* may be added; for example, a magma body from a deeper source intrudes and melts crustal rock; (2) *a decrease in pressure* (without the addition of heat) can result in *decompression melting*; and (3) the *introduction of volatiles* (principally water) can lower the melting temperature of mantle rock sufficiently to generate magma.

How Magmas Evolve

Because a large variety of igneous rocks exists, it is logical to assume that an equally large variety of magmas must also exist. However, geologists have observed that a single volcano may extrude lavas exhibiting quite different compositions. Data of this type led them to examine the possibility that magma might change (evolve) and thus become the parent to a variety of igneous rocks. To explore this idea, a pioneering investigation into the crystallization of magma was carried out by N. L. Bowen in the first quarter of the twentieth century.

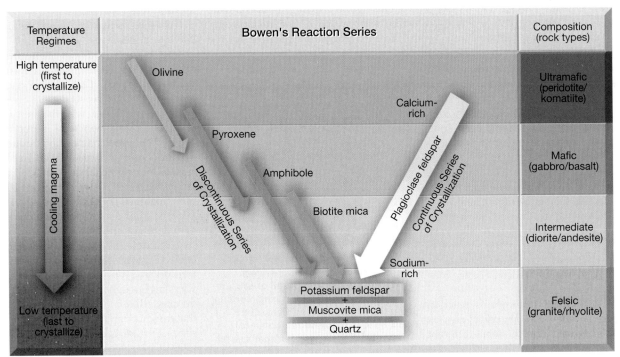

Figure 3.16 Bowen's reaction series shows the sequence in which minerals crystallize from a magma. Compare this figure to the mineral composition of the rock groups in Figure 3.7. Note that each rock group consists of minerals that crystallize in the same temperature range.

Bowen's Reaction Series and the Composition of Igneous Rocks

Recall that ice freezes at a single temperature whereas magma crystallizes through at least 200°C of cooling. In a laboratory setting Bowen and his coworkers demonstrated that as a basaltic magma cools, minerals tend to crystallize in a systematic fashion based on their melting points. As shown in Figure 3.16, the first mineral to crystallize from a basaltic magma is the ferromagnesian mineral olivine. Further cooling generates calcium-rich plagioclase feldspar as well as pyroxene, and so forth down the diagram.

During the crystallization process, the composition of the liquid portion of the magma continually changes. For example, at the stage when about a third of the

BOX 3.1

A Closer Look at Bowen's Reaction Series

Although it is highly idealized, Bowen's reaction series provides us with a visual representation of the order in which minerals crystallize from a magma of average composition (see Figure 3.16). This model assumes that the magma cools slowly at depth in an otherwise unchanging environment. Notice that Bowen's reaction series is divided into two branches—a discontinuous series and a continuous series.

Discontinuous Reaction Series. The upper left branch of Bowen's reaction series indicates that as a magma cools, olivine is the first mineral to crystallize. Once formed, olivine will chemically react with the remaining melt to form the mineral pyroxene (see Figure 3.16). In this reaction, olivine, which is composed of individual silicon–oxygen tetrahedra, incorporates more silica into its structure, thereby linking its tetrahedra into single-chain structures of the mineral pyroxene. (Note: Pyroxene has a lower crystallization temperature than olivine and is more stable at lower temperatures.) As the magma body cools further, the pyroxene crystals will in turn react with the melt to generate the double-chain structure of amphibole. This reaction will continue until the last mineral in this series, biotite mica, crystallizes. In nature, these reactions do not usually run to completion, so that various amounts of each of the minerals in the series may exist at any given time, and some minerals such as biotite may never form.

This branch of Bowen's reaction series is called a *discontinuous reaction series* because at each step a different silicate structure emerges. Olivine, the first mineral in the sequence, is composed of isolated tetrahedra, whereas pyroxene is composed of single chains, amphibole of double chains, and biotite of sheet structures.

Continuous Reaction Series. The right branch of the reaction series, called the *continuous reaction series*, illustrates that calcium-rich plagioclase feldspar crystals react with the sodium ions in the melt to become progressively more sodium-rich (see Figure 3.16). Here the sodium ions diffuse into the feldspar crystals and displace the calcium ions in the crystal lattice. Often, the rate of cooling occurs rapidly enough to prohibit a complete replacement of the calcium ions by sodium ions. In these instances, the feldspar crystals will have calcium-rich interiors surrounded by zones that are progressively richer in sodium (Figure 3.A).

During the last stage of crystallization, after much of the magma has solidified, potassium feldspar forms.

Figure 3.A   Photomicrograph of a zoned plagioclase feldspar crystal. After this crystal (composed of calcium-rich feldspar) solidified, further cooling resulted in sodium ions displacing calcium ions. Because replacement was not complete, this feldspar crystal has a calcium-rich interior surrounded by zones that are progressively richer in sodium. (Photo courtesy of E. J. Tarbuck)

(Muscovite will form in pegmatites and other plutonic igneous rocks that crystallize at considerable depth.) Finally, if the remaining melt has excess silica, the mineral quartz will form.

Testing Bowen's Reaction Series. During an eruption of Hawaii's Kilauea volcano in 1965, basaltic lava poured into a pit crater, forming a lava lake that became a natural laboratory for testing Bowen's reaction series. When the surface of the lava lake cooled enough to form a crust, geologists drilled into the magma and periodically removed samples that were quenched to preserve the melt, and minerals that were growing within it. By sampling the lava at successive stages of cooling, a history of crystallization was recorded.

As Bowen's reaction series predicts, olivine crystallized early but later ceased to form, and was partly reabsorbed into the cooling melt. (In a larger magma body that cooled more slowly, we would expect most, if not all, of the olivine to react with the melt and change to pyroxene.) Most important, the melt changed composition throughout the course of crystallization. In contrast to the original basaltic lava, which contained about 50 percent silica (SiO_2), the final melt contained more than 75 percent silica and had a composition similar to granite.

Although the lava in this setting cooled rapidly compared to rates experienced in deep magma chambers, it was slow enough to verify that minerals do crystallize in a systematic fashion that roughly parallels Bowen's reaction series. Further, had the melt been separated at any stage in the cooling process, it would have formed a rock with a composition much different from the original lava.

magma has solidified, the melt will be nearly depleted of iron, magnesium, and calcium because these elements are constituents of the earliest-formed minerals. The removal of these elements from the melt will cause it to become enriched in sodium and potassium. Further, because the original basaltic magma contained about 50 percent silica (SiO_2), the crystallization of the earliest-formed mineral, olivine, which is only about 40 percent silica, leaves the remaining melt richer in SiO_2. Thus, the silica component of the melt becomes enriched as the magma evolves.

Bowen also demonstrated that if the solid components of a magma remain in contact with the remaining melt, they will chemically react and evolve into the next mineral in the sequence shown in Figure 3.16. For this reason, this arrangement of minerals became known as **Bowen's reaction series** (see Box 3.1). As you will see, in some natural settings the earliest-formed minerals can be separated from the melt, thus halting any further chemical reaction.

The diagram of Bowen's reaction series in Figure 3.16 depicts the sequence that minerals crystallize from a magma of average composition under laboratory conditions. Evidence that this highly idealized crystallization model approximates what can happen in nature comes from the analysis of igneous rocks. In particular, we find that minerals that form in the same general temperature regime on Bowen's reaction series are found together in the same igneous rocks. For example, notice in Figure 3.16 that the minerals quartz, potassium feldspar, and muscovite, which are located in the same region of Bowen's diagram, are typically found together as major constituents of the plutonic igneous rock granite.

Magmatic Differentiation. Bowen demonstrated that minerals crystallize from magma in a systematic fashion. But how do Bowen's findings account for the great diversity of igneous rocks? It has been shown that at one or more stages during crystallization, a separation of the solid and liquid components of a magma can occur. One example is called **crystal settling.** This process occurs when the earlier-formed minerals are denser (heavier) than the liquid portion and sink toward the bottom of the magma chamber, as shown in Figure 3.17. When the remaining melt solidifies—either in place or in another location if it migrates into fractures in the surrounding rocks—it will form a rock with a chemical composition much different from the parent magma (Figure 3.17). The formation of one or more secondary magmas from a single parent magma is called **magmatic differentiation.**

At any stage in the evolution of a magma, the solid and liquid components can separate into two chemically distinct units. Further, magmatic differentiation within the secondary melt can generate additional chemically distinct fractions. Consequently, magmatic differentiation and separation of the solid and liquid

A.

B.

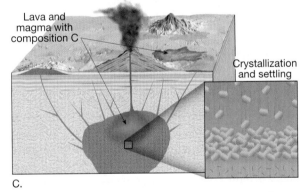
C.

Figure 3.17 Illustration of how a magma evolves as the earliest-formed minerals (those richer in iron, magnesium, and calcium) crystallize and settle to the bottom of the magma chamber, leaving the remaining melt richer in sodium, potassium, and silica (SiO_2). **A.** Emplacement of a magma body and associated igneous activity generates rocks having a composition similar to that of the initial magma. **B.** After a period of time, crystallization and settling changes the composition of the melt, while generating rocks having a composition quite different than the original magma. **C.** Further magmatic differentiation results in another more highly evolved melt with its associated rock types.

components at various stages of crystallization can produce several chemically diverse magmas and ultimately a variety of igneous rocks (see Figure 3.17).

Assimilation. Once a magma body forms, its composition can change through the incorporation of foreign material. For example, as magma migrates upward, it may incorporate some of the surrounding host rock, a process called **assimilation** (Figure 3.18). This process may operate in a near-surface environment where rocks are brittle. As the magma pushes upward, stress causes numerous cracks in the overlying rock. The force of the injected magma is often sufficient to dislodge blocks of "foreign" rock and incorporate them into the magma body. In deeper environments, the magma may be hot enough to simply melt and assimilate some of the surrounding host rock, which is near its melting temperature.

Magma Mixing.   Another means by which the composition of a magma body can be altered is called **magma mixing.** This process occurs whenever one magma body intrudes another (Figure 3.18). Once combined, convective flow may stir the two magmas and generate a fluid with an intermediate composition. Magma mixing may occur during the ascent of two chemically distinct magma bodies as the more buoyant mass overtakes the slower moving mass.

Partial Melting and Magma Formation

Recall that the crystallization of a magma occurs over a temperature range of at least 200°C. As you might expect, melting, the reverse process, spans a similar temperature range. As rock begins to melt, those minerals with the lowest melting temperatures are the first to melt. Should melting continue, minerals with higher melting points begin to melt and the composition of the magma steadily approaches the overall composition of the rock from which it was derived. Most often, however, melting is not complete. The incomplete melting of rocks is known as **partial melting,** a process that produces most, if not all, magma.

Notice in Figure 3.16 that rocks with a felsic composition are composed of minerals with the lowest melting (crystallization) temperatures—namely, quartz and potassium feldspar. Also note that as we move up Bowen's reaction series, the minerals have progressively higher melting temperatures and that olivine, which is found at the top, has the highest melting point. When a rock undergoes partial melting, it will form a melt that is enriched in ions from minerals with the lowest melting temperatures. The unmelted crystals are those of minerals with higher melting temperatures. Separation of these two fractions would yield a melt with a chemical composition that is richer in silica and nearer to the felsic (granitic) end of the spectrum than the rock from which it was derived.

In summary, a large number of processes dominated by magmatic differentiation and partial melting, as well as magma mixing and contamination by crustal rocks, account for the great diversity of igneous rocks and the magmas from which they formed.

Figure 3.18 This illustration shows three ways that the composition of a magma body may be altered: magma mixing; assimilation of host rock; and crystallization and settling (magmatic differentiation).

Mineral Resources and Igneous Processes

Some of the most important accumulations of metals, such as gold, silver, copper, mercury, lead, platinum, and nickel, are produced by igneous processes (Table 3.1). These mineral resources result from processes that concentrate desirable materials to the extent that extraction is economically feasible.

The igneous processes that generate some of these metal deposits are quite straightforward. For example, as a large basaltic magma body cools, the heavy minerals that crystallize early tend to settle to the lower portion of the magma chamber. This type of magmatic segregation serves to concentrate selected metals producing major deposits of chromite (ore of chromium), magnetite, and platinum. Layers of chromite, interbedded with other heavy minerals, are mined at Montana's Stillwater Complex, whereas the Bushveld Complex in South Africa contains over 70 percent of the world's known reserves of platinum.

Magmatic segregation is also important in the late stages of the magmatic process. This is particularly true of granitic magmas in which the residual melt can become enriched in rare elements and some heavy metals. Further, because water and other volatile substances do not crystallize along with the bulk of the magma body, these fluids make up a high percentage of the melt during the final phase of solidification. Crystallization in a fluid-rich environment, where ion migration is enhanced, results in the formation of crystals several centimeters, or even a few meters, in length. The resulting rocks, called *pegmatites,* are composed of these unusually large crystals (Figure 3.19).

Feldspar masses the size of houses have been quarried from a pegmatite located in North Carolina. Gigantic hexagonal crystals of muscovite measuring a few meters across have been found in Ontario, Canada. In the Black Hills, spodumene crystals as thick as telephone poles have been mined (Figure 3.19). The largest of these was more than 12 meters (40 feet) long. Not all pegmatites contain such large crystals, but these exam-

Table 3.1   Occurrences of metallic minerals

Metal	Principal Ores	Geological Occurrences
Aluminum	Bauxite	Residual product of weathering
Chromium	Chromite	Magmatic segregation
Copper	Chalcopyrite	Hydrothermal deposits; contact metamorphism; enrichment by
	Bornite	weathering processes
	Chalcocite	
Gold	Native gold	Hydrothermal deposits; placers
Iron	Hematite	Banded sedimentary formations; magmatic segregation
	Magnetite	
	Limonite	
Lead	Galena	Hydrothermal deposits
Magnesium	Magnesite	Hydrothermal deposits
	Dolomite	
Manganese	Pyrolusite	Residual product of weathering
Mercury	Cinnabar	Hydrothermal deposits
Molybdenum	Molybdenite	Hydrothermal deposits
Nickel	Pentlandite	Magmatic segregation
Platinum	Native platinum	Magmatic segregation; placers
Silver	Native silver	Hydrothermal deposits; enrichment by weathering processes
	Argentite	
Tin	Cassiterite	Hydrothermal deposits; placers
Titanium	Ilmenite	Magmatic segregation; placers
	Rutile	
Tungsten	Wolframite	Pegmatites; contact metamorphic deposits; placers
	Scheelite	
Uranium	Uraninite (pitchblende)	Pegmatites; sedimentary deposits
Zinc	Sphalerite	Hydrothermal deposits

Figure 3.19 This pegmatite in the Black Hills of South Dakota was mined for its large crystals of spodumene, an important source of lithium. Arrows are pointing to impressions left by crystals. Note person in upper center of photo for scale. (Photo by James G. Kirchner)

ples emphasize the special conditions that must exist during their formation.

Most pegmatites are granitic in composition and consist of unusually large crystals of quartz, feldspar, and muscovite. Feldspar is used in the production of ceramics, and muscovite is used for electrical insulation and glitter. Further, pegmatites often contain some of the least abundant elements. Minerals containing the elements lithium, cesium, uranium, and the rare earths are occasionally found. Moreover, some pegmatites contain semiprecious gems such as beryl, topaz, and tourmaline. Most pegmatites are located within large igneous masses or as dikes or veins that cut into the host rock that surrounds the magma chamber (Figure 3.20).

Not all late-stage magmas produce pegmatites, nor do all have a granitic composition. Rather, some magmas become enriched in iron or occasionally copper. For example, at Kirava, Sweden, magma composed of over 60 percent magnetite solidified to produce one of the largest iron deposits in the world.

Among the best known and most important ore deposits are those generated from hot, ion-rich fluids called **hydrothermal** (hot-water) **solutions.*** Included

---

*Because these hot, ion-rich fluids tend to chemically alter the host rock, this process is called *hydrothermal metamorphism* and is discussed in Chapter 7.

Figure 3.20 Illustration of the relationship between a parent igneous body and the associated pegmatite and hydrothermal deposits.

Figure 3.21 Light-colored vein deposits emplaced along a series of fractures in dark-colored metamorphic rock. Diablo Lake Overlook, North Cascades National Park, Washington. (Photo by James E. Patterson)

in this group are the gold deposits of the Homestake mine in South Dakota, the lead, zinc, and silver ores near Coeur d'Alene, Idaho, the silver deposits of the Comstock Lode in Nevada, and the copper ores of Michigan's Keweenaw Peninsula.

The majority of hydrothermal deposits originate from hot, metal-rich fluids that are remnants of late-stage magmatic processes. During solidification, liquids plus various metallic ions accumulate near the top of the magma chamber. Because of their mobility, these ion-rich solutions can migrate great distances through the surrounding rock before they are eventually deposited, usually as sulfides of various metals (Figure 3.20). Some of these fluids move along openings such as fractures or bedding planes, where they cool and precipitate metallic ions to produce *vein deposits* (Figure 3.21). Many of the most productive deposits of gold, silver, and mercury occur as hydrothermal vein deposits.

Another important type of accumulation generated by hydrothermal activity is called a *disseminated deposit*. Rather than being concentrated in narrow veins and dikes, these ores are distributed as minute masses throughout the entire rock mass. Much of the world's copper is extracted from disseminated deposits, in-

cluding those at Chuquicamata, Chile, and the huge Bingham Canyon copper mine in Utah (see Figure 2.22). Because these accumulations contain only 0.4 to 0.8 percent copper, between 125 and 250 kilograms of ore must be mined for every kilogram of metal recovered. The environmental impact of these large excavations, including the problems of waste disposal, is significant.

Another economically important mineral with an igneous origin is diamond. Although best known as gems, diamonds are used extensively as abrasives. Diamonds are thought to originate at depths of nearly 200 kilometers, where the confining pressure is great enough to generate this high-pressure form of carbon. Once crystallized, the diamonds are carried upward through pipe-shaped conduits that increase in diameter toward the surface. In diamond-bearing pipes, nearly the entire pipe contains diamond crystals that are disseminated throughout an ultramafic rock called *kimberlite*. The most productive kimberlite pipes are those in South Africa. The only equivalent source of diamonds in the United States is located near Murfreesboro, Arkansas, but this deposit is exhausted and serves today merely as a tourist attraction.

The Chapter in Review

- *Igneous rocks* form when *magma cools* and solidifies. *Extrusive*, or *volcanic*, igneous rocks result when *lava* cools at the surface. Magma that solidifies at depth produces *intrusive*, or *plutonic*, igneous rocks.

- As magma cools, the ions that compose it arrange themselves into orderly patterns during a process called *crystallization*. Slow cooling results in the formation of rather large crystals. Conversely, when cooling occurs rapidly, the outcome is a solid mass consisting of tiny intergrown crystals. When molten material is quenched instantly, a mass of unordered atoms, referred to as *glass*, forms.

- Igneous rocks are most often classified by their *texture* and *mineral composition*.

- The texture of an igneous rock refers to the overall appearance of the rock based on the size and arrangement of its interlocking crystals. The most important factor affecting texture is the rate at which magma cools. Common igneous rock textures include *aphanitic*, with grains too small to be distinguished without the aid of a microscope; *phaneritic*, with intergrown crystals that are roughly equal in size and large enough to be identified with the unaided eye; *porphyritic*, which has large crystals (*phenocrysts*) interbedded in a matrix of smaller crystals (*groundmass*); and *glassy*.

- The mineral composition of an igneous rock is the consequence of the chemical makeup of the parent magma and the environment of crystallization. Igneous rocks are divided into broad compositional groups based on the percentage of dark and light silicate minerals they contain. *Felsic rocks* (e.g., granite and rhyolite) are composed mostly of the light-colored silicate minerals potassium feldspar and quartz. Rocks of *intermediate* composition (e.g., andesite and diorite) contain plagioclase feldspar and amphibole. *Mafic rocks* (e.g., basalt and gabbro) contain abundant olivine, pyroxene, and calcium feldspar. They are high in iron, magnesium, and calcium, low in silicon, and are dark gray to black in color.

- The mineral makeup of an igneous rock is ultimately determined by the chemical composition of the magma from which it crystallizes. N. L. Bowen discovered that as magma cools in the laboratory, those minerals with higher melting points crystallize before minerals with lower melting points. *Bowen's reaction series* illustrates the sequence of mineral formation within magma.

- During the crystallization of magma, if the earlier-formed minerals are denser than the liquid portion, they will settle to the bottom of the magma chamber during a process called *crystal settling*. Owing to the fact that crystal settling removes the earlier-formed minerals, the remaining melt will form a rock with a chemical composition much different from the parent magma. The process of developing more than one magma type from a common magma is called *magmatic differentiation*.

- Once a magma body forms, its comparison can change through the incorporation of foreign material, a process termed *assimilation*, or by *magma mixing*.

- Magma originates from essentially solid rock of the crust and mantle. In addition to a rock's composition, its temperature, depth (confining pressure), and water content determine whether it exists as a solid or liquid. Thus, magma can be generated by *raising a rock's temperature*, as occurs when a hot mantle plume "ponds" beneath crustal rocks. A *decrease in pressure* can cause *decompression melting*. Further, the *introduction of volatiles* (water) can lower a rock's melting point sufficiently to generate magma. Because melting is generally not complete, a process called *partial melting* produces a melt made of the lowest-melting-temperature minerals, which are higher in silica than the original rock. Thus, magmas generated by partial melting are nearer to the felsic end of the compositional spectrum than are the rocks from which they formed.

- Some of the most important accumulations of metals, such as gold, silver, lead, and copper, are produced by igneous processes. The best-known and most important ore deposits are generated from *hydrothermal* (hot-water) *solutions*. Hydrothermal deposits are thought to originate from hot, metal-rich fluids that are remnants of late-stage magmatic processes. These ion-rich solutions move along fractures or bedding planes, cool, and precipitate the metallic ions to produce *vein deposits*. In a *disseminated deposit* (e.g., much of the world's copper deposits), the ores from hydrothermal solutions are distributed as minute masses throughout the entire rock mass.

Key Terms

andesitic composition
(p. 52)
aphanitic texture (p. 48)
assimilation (p. 62)
basaltic composition (p. 51)
Bowen's reaction series
(p. 61)
crystallization (p. 47)
crystal settling (p. 61)
decompression melting
(p. 58)
extrusive (p. 46)

felsic (p. 51)
fragmental texture (p. 50)
geothermal gradient
(p. 57)
glass (p. 48)
glassy texture (p. 49)
groundmass (p. 49)
granitic (p. 51)
hydrothermal solutions
(p. 64)
igneous rocks (p. 46)
intermediate (p. 52)

intrusive (p. 46)
lava (p. 46)
mafic (p. 52)
magma (p. 46)
magma mixing (p. 62)
magmatic differentiation
(p. 61)
melt (p. 47)
partial melting (p. 62)
pegmatite (p. 50)
pegmatitic texture (p. 50)

phaneritic texture (p. 48)
phenocryst (p. 49)
plutonic (p. 46)
porphyritic texture (p. 49)
porphyry (p. 49)
pyroclastic texture (p. 50)
texture (p. 48)
ultramafic (p. 52)
vesicular texture (p. 48)
volatiles (p. 47)
volcanic (p. 46)

Questions for Review

1. What is magma?

2. How does lava differ from magma?

3. How does the rate of cooling influence the crystallization process?

4. In addition to the rate of cooling, what two other factors influence the crystallization process?

5. The classification of igneous rocks is based largely on two criteria. Name these criteria.

6. The statements that follow relate to terms describing igneous rock textures. For each statement, identify the appropriate term.
  a. Openings produced by escaping gases.
  b. Obsidian exhibits this texture.
  c. A matrix of fine crystals surrounding phenocrysts.
  d. Crystals are too small to be seen without a microscope.
  e. A texture characterized by two distinctly different crystal sizes.
  f. Coarse-grained, with crystals of roughly equal size.
  g. Exceptionally large crystals exceeding 1 centimeter in diameter.

7. Why are the crystals in pegmatites so large?

8. What does a porphyritic texture indicate about an igneous rock?

9. How are granite and rhyolite different? In what way are they similar?

10. Compare and contrast each of the following pairs of rocks:
  a. granite and diorite
  b. basalt and gabbro
  c. andesite and rhyolite

11. How does tuff differ from other igneous rocks, such as granite and basalt?

12. What is the geothermal gradient?

13. Describe the three conditions that cause rock to melt.

14. What is magmatic differentiation? How might this process lead to the formation of several different igneous rocks from a single magma?

15. Relate the classification of igneous rocks to Bowen's reaction series.

16. What is partial melting?

17. How does the composition of a melt produced by partial melting compare with the composition of the parent rock?

18. List two types of hydrothermal deposits.

Web Resources

The *Essentials of Geology* Website uses the resources and flexibility of the Internet to aid in your study of the topics in this chapter. Written and developed by geology instructors, this site will help improve your understanding of geology. Visit **www.prenhall.com/lutgens** and click on the cover of *Essentials of Geology, 8/e* to find:

- On-line review quizzes.
- Web-based critical thinking and writing exercises.
- Links to chapter-specific Web resources.
- Internet-wide key term searches.

www.prenhall.com/lutgens

4 Volcanoes and Other Igneous Activity

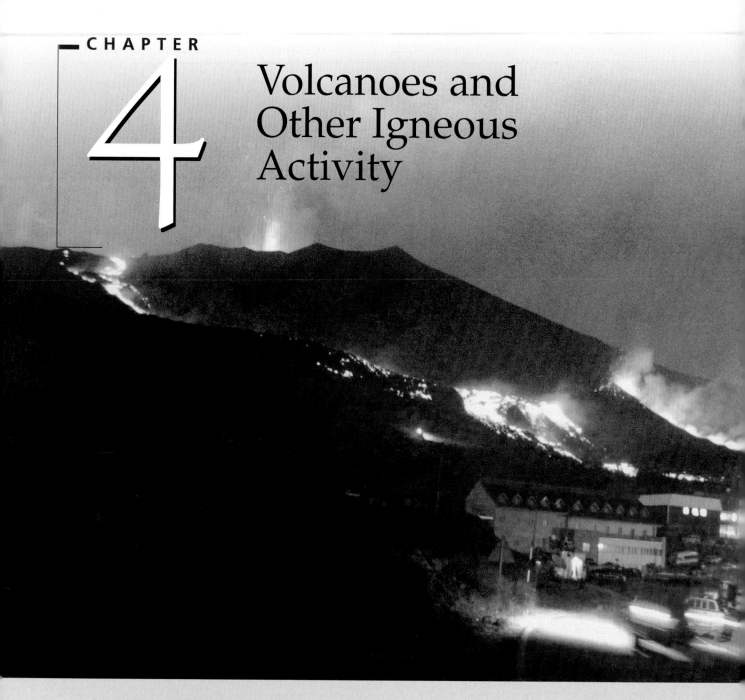

Focus on Learning

To assist you in learning the important concepts in this chapter, you will find it helpful to focus on the following questions:

- What primary factors determine the nature of volcanic eruptions? How do these factors affect a magma's viscosity?

- What materials are associated with a volcanic eruption?

- What are the eruptive patterns and basic characteristics of the three types of volcanoes generally recognized by volcanologists?

- What criteria are used to classify intrusive igneous bodies? What are some of these features?

- What is the relation between volcanic activity and plate tectonics?

Glowing lava spews from Italy's Mount Etna, one of Europe's most active volcanoes, July 27, 2001. (Photo by Sipa Press)

On Sunday, May 18, 1980, the largest volcanic eruption to occur in North America in historic times transformed a picturesque volcano into a decapitated remnant (Figure 4.1). On this date in southwestern Washington State, Mount St. Helens erupted with tremendous force. The blast blew out the entire north flank of the volcano, leaving a gaping hole. In one brief moment, a prominent volcano whose summit had been more than 2900 meters (9500 feet) above sea level was lowered by more than 400 meters (1350 feet).

The event devastated a wide swath of timber-rich land on the north side of the mountain (Figure 4.2). Trees within a 400-square-kilometer area lay intertwined and flattened, stripped of their branches and appearing from the air like toothpicks strewn about. The accompanying mudflows carried ash, trees, and water-saturated rock debris 29 kilometers (18 miles) down the Toutle River. The eruption claimed 59 lives, some dying from the intense heat and the suffocating cloud of ash and gases, others from being hurled by the blast, and still others from entrapment in the mudflows.

The eruption ejected nearly a cubic kilometer of ash and rock debris. Following the devastating explosion, Mount St. Helens continued to emit great quantities of hot gases and ash. The force of the blast was so strong that some ash was propelled more than 18,000 meters (over 11 miles) into the stratosphere. During the next few days, this very fine-grained material was carried around Earth by strong upper-air winds. Measurable deposits were reported in Oklahoma and Minnesota, with crop damage into central Montana. Meanwhile, ash fallout in the immediate vicinity exceeded 2 meters in depth. The air over Yakima, Washington (130 kilometers to the east), was so filled with ash that residents experienced midnight-like darkness at noon.

Not all volcanic eruptions are as violent as the 1980 Mount St. Helens event. Some volcanoes, such as Hawaii's Kilauea volcano, generate relatively quiet outpourings of fluid lavas. These "gentle" eruptions are not without some fiery displays; occasionally fountains of incandescent lava spray hundreds of meters into the air (Figure 4.3). Such events, however, typically pose minimal threat to human life and property, and the lava generally falls back into a lava pool.

Testimony to the quiet nature of Kilauea's eruptions is the fact that the Hawaiian Volcanoes Observatory has operated on its summit since 1912. This, despite the fact that Kilauea has had more than 50 eruptive phases since record-keeping began in 1823. Further, the longest and largest of Kilauea's eruptions began in 1983 and remains active, although it has received only modest media attention.

Why do volcanoes like Mount St. Helens erupt explosively, whereas others like Kilauea are relatively quiet? Why do volcanoes occur in chains like the Aleutian Islands or the Cascade Range? Why do some volca-noes form on the ocean floor, while others occur on the continents? This chapter will deal with these and other questions as we explore the nature and movement of magma and lava.

The Nature of Volcanic Eruptions

GEODe Internal Processes
    ↳ Igneous Activity

Volcanic activity is commonly perceived as a process that produces a picturesque, cone-shaped structure that periodically erupts in a violent manner, like Mount St. Helens (Box 4.1). Although some eruptions may be very explosive, many are not. What determines whether a volcano extrudes magma violently or "gently"? The primary factors include the magma's *composition*, its *temperature*, and the amount of *dissolved gases* it contains. To varying degrees, these factors affect the magma's mobility, or **viscosity.** The more viscous the material, the greater its resistance to flow. (For example, syrup is more viscous than water.) Magma associated with an explosive eruption may be five times more viscous than magma that is extruded in a quiescent manner.

Factors Affecting Viscosity

The effect of temperature on viscosity is easily seen. Just as heating syrup makes it more fluid (less viscous), the mobility of lava is strongly influenced by temperature. As lava cools and begins to congeal, its mobility decreases and eventually the flow halts.

A more significant factor influencing volcanic behavior is the chemical composition of the magma. This was discussed in Chapter 3 with the classification of igneous rocks. Recall that a major difference among various igneous rocks is their silica (SiO_2) content (Table 4.1). Magmas that produce mafic rocks such as basalt contain about 50 percent silica, whereas magmas that produce felsic rocks (granite and its extrusive equivalent, rhyolite) contain over 70 percent silica. The intermediate rock types—andesite and diorite—contain about 60 percent silica.

A magma's viscosity is directly related to its silica content. In general, the more silica in magma, the greater its viscosity. The flow of magma is impeded because silica structures link together into long chains, even before crystallization begins. Consequently, because of their high silica content, rhyolitic (felsic) lavas are very viscous and tend to form comparatively short, thick flows. By contrast, basaltic (mafic) lavas, which contain less silica, tend to be quite fluid and have been known to travel distances of 150 kilometers (90 miles) or more before congealing.

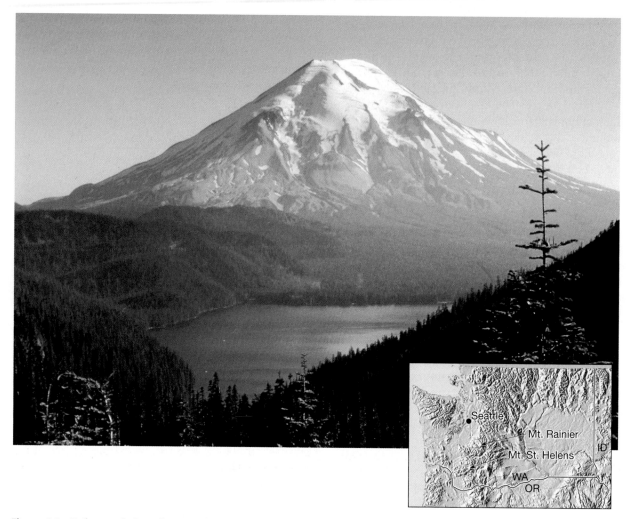

Figure 4.1 Before-and-after photographs show the transformation of Mount St. Helens caused by the May 18, 1980, eruption. The dark area in the "after" photo is debris-filled Spirit Lake, partially visible in the "before" photo. (Photos courtesy of U.S. Geological Survey)

Figure 4.2 Douglas fir trees were snapped off or uprooted by the lateral blast of Mount St. Helens on May 18, 1980. (Photo by Krafft-Explorer/Photo Researchers, Inc.)

Importance of Dissolved Gases

The gas content of a magma also affects its mobility. Dissolved gases tend to increase the fluidity of magma. Of far greater consequence is the fact that escaping gases provide enough force to propel molten rock from a volcanic vent.

The summits of volcanoes often begin to inflate months or even years before an eruption occurs. This indicates that magma is migrating into a shallow reservoir located within the cone. During this phase **volatiles** (the gaseous component of magma consisting mostly

Figure 4.3 Fluid lava emitted from a flank eruption on Kilauea, Hawaii. (Photo by Paul Chesley/Tony Stone Images)

of water) tend to migrate upward and accumulate near the top of the magma chamber. Thus, the upper portion of a magma body is enriched in dissolved gases.

When an eruption starts, gas-charged magma moves from the magma chamber and rises through the volcanic conduit or vent. As the magma nears the surface, the confining pressure is greatly reduced. This reduction in pressure allows the dissolved gases to be released suddenly, just as opening a warm soda bottle allows carbon dioxide gas bubbles to escape. At temperatures of 1000°C and low near-surface pressures, these gases will expand to occupy hundreds of times their original volume.

Very fluid basaltic magmas allow the expanding gases to migrate upward and escape from the vent with relative ease. As they escape, the gases may propel incandescent lava hundreds of meters into the air, producing lava fountains (see Figure 4.3). Although spectacular, such fountains are mostly harmless and not generally associated with major explosive events that cause great loss of life and property. Rather, eruptions of fluid basaltic lavas, such as those that occur in Hawaii, are generally quiescent.

At the other extreme, highly viscous magmas explosively expel jets of hot ash-laden gases that evolve into buoyant plumes called **eruption columns** that extend thousands of meters into the atmosphere (Figure 4.4). Prior to an explosive eruption, an extended period of *magmatic differentiation* occurs in which iron-rich minerals crystallize and settle out, leaving the upper part of the magma enriched in silica and dissolved gases. As this volatile-rich magma moves up the volcanic vent toward the surface, these gases begin to collect as tiny bubbles. For reasons that are still poorly understood, at some height in the conduit, this mixture is transformed into a gas jet containing tiny magma fragments that are explosively ejected from the volcano. This type of explosive eruption is exemplified by Mount Pinatubo in the Philippines (1991) and Mount St. Helens (1980).

71

Anatomy of an Eruption

The events leading to the May 18, 1980, eruption of Mount St. Helens began about two months earlier as a series of minor Earth tremors centered beneath the awakening mountain (Figure 4.A, part A). The tremors were caused by the upward movement of magma within the mountain. The first volcanic activity took place a week later, when a small amount of ash and steam rose from the summit. Over the next several weeks, sporadic eruptions of varied intensity occurred. Prior to the main eruption, the primary concern had been the potential hazard of mudflows. These moving lobes of saturated soil and rock are created as ice and snow melt from the heat emitted from magma within the volcano.

The only warning of a potential eruption was a bulge on the volcano's north flank (Figure 4.A, part B). Careful monitoring of this dome-shaped structure indicated a very slow but steady growth rate of a few meters per day. If the growth rate of the bulge changed appreciably, an eruption might quickly follow. Unfortunately, no such variation was detected prior to the explosion. In fact, the seismic activity decreased during the two days preceding the huge blast.

Dozens of scientists were monitoring the mountain when it exploded. "Vancouver, Vancouver, this is it!" was the only warning—and last words from one scientist—that preceded the unleashing of tremendous quantities of pent-up gases. The trigger was a medium-sized earthquake. Its vibrations sent the north slope of the cone plummeting into the Toutle River, removing the overburden that had trapped the magma below (Figure 4.A, part C). With the pressure reduced, the water in the magma vaporized and expanded, causing the mountainside to rupture like an overheated steam boiler. Because the eruption originated around the bulge, several hundred meters below the summit, the initial blast was directed laterally rather than vertically. Had the full

A. March 20, 1980. Intrusion of magma generates earthquake.

B. April 23, 1980. Continued intrusion of magma produces bulge.

Bulge

C. May 18, 1980

Lateral blast

Landslide debris

Full vertical eruption develops

D. May 18, 1980

Figure 4.A   Idealized diagrams showing the events in the May 18, 1980, eruption of Mount St. Helens. A. First, a sizable earthquake recorded on Mount St. Helens indicates that renewed volcanic activity is possible. B. Alarming growth of a bulge on the north flank suggests increasing magma pressure below. C. Triggered by an earthquake, a giant landslide reduced the confining pressure on the magma body and initiated an explosive lateral blast. D. Within seconds a large vertical eruption sent a column of volcanic ash to an altitude of about 18 kilometers (11 miles). This phase of the eruption continued for over nine hours.

force of the eruption been upward, far less destruction would have occurred.

Mount St. Helens is one of 15 large volcanoes and innumerable smaller ones that comprise the Cascade Range, which extends from British Columbia to northern California. Eight of the largest volcanoes have been active in the past few hundred years. Of the remaining seven active volcanoes, the most likely to erupt again are Mount Baker and Mount Rainier in Washington, Mount Shasta and Lassen Peak in California, and Mount Hood in Oregon.

Table 4.1 Magmas have different compositions, which cause their properties to vary

Composition	Silica Content	Viscosity	Gas Content	Tendency to Form Pyroclastics	Volcanic Landform
Mafic (Basaltic) magma	Least (~50%)	Least	Least (1–2%)	Least	Shield Volcanoes Basalt Plateaus Cinder Cones
Intermediate (Andesitic) magma	Intermediate (~60%)	Intermediate	Intermediate (3–4%)	Intermediate	Composite Cones
Felsic (Rhyolitic) magma	Most (~70%)	Greatest	Most (4–6%)	Greatest	Volcanic Domes Pyroclastic Flows

As magma in the upper portion of the vent is ejected, the pressure on the molten rock directly below drops. Thus, rather than a single "bang," volcanic eruptions are really a series of explosions. This process might logically continue until the magma chamber is emptied, much like a geyser empties itself of water (see Chapter 10). However, this generally does not happen. The soluble gases in a viscous magma migrate upward quite slowly. Only within the uppermost portion of the magma body does the gas content build sufficiently to trigger explosive eruptions. Thus, an explosive event is commonly followed by the quiet emission of "degassed" lavas. However, once this eruptive phase ceases, the process of gas buildup begins anew. This time lag may partially explain the sporadic eruptive patterns of volcanoes that eject viscous lavas.

To summarize, the viscosity of magma, plus the quantity of dissolved gases and the ease with which they can escape, determines the nature of a volcanic eruption. We can now understand the "gentle" volcanic eruptions of hot, fluid lavas in Hawaii and the explosive and sometimes catastrophic eruptions of viscous lavas from volcanoes such as Mount St. Helens.

Materials Extruded During an Eruption

 Internal Processes
    ↳ Igneous Activity

Volcanoes extrude lava, large volumes of gas, and pyroclastic materials (broken rock, lava "bombs," fine ash, and dust). In this section we will examine each of these materials.

Lava Flows

Because of their low silica content, hot basaltic lavas are usually very fluid. They flow in thin, broad sheets or streamlike ribbons. On the island of Hawaii, such lavas

Figure 4.4 Eruption column formed by the 1986 eruption of Mount Augustine, Cook Inlet, Alaska. Volcanoes that border the Pacific Ocean are fed largely by magmas that have intermediate or felsic compositions. These silica-rich magmas often erupt explosively, generating large plumes of volcanic dust and ash. (Photo by Steve Kaufman/DRK)

have been clocked at speeds of 30 kilometers per hour down steep slopes, but flow rates of 10 to 300 meters per hour are more common. Further, basaltic lavas have been known to travel distances of 150 kilometers (90 miles) or more before congealing. In contrast, the movement of silica-rich (rhyolitic) lava may be too slow to perceive.

When fluid basaltic lavas of the Hawaiian type congeal, they commonly form a relatively smooth skin that wrinkles as the still-molten subsurface lava continues to advance (Figure 4.5A). These are known as **pahoehoe flows** (pronounced *pah-hoy-hoy*) and resemble the twisting braids in ropes.

Another common type of basaltic lava, called **aa** (pronounced *ah-ah*), has a surface of rough, jagged blocks with dangerously sharp edges and spiny projections (Figure 4.5B). Active aa flows are relatively cool and thick and advance at rates from 5 to 50 meters per hour. Moreover, gases escaping from the surface produce numerous voids and sharp spines in the congealing lava. As the molten interior advances, the outer crust is broken further, giving the flow the appearance of an advancing mass of lava rubble.

The lava that flowed from the Mexican volcano Parícutin and buried the city of San Juan Parangaricutiro was of the aa type (see Figure 4.12). At times one of the flows from Parícutin moved only 1 meter per day, but it continued to advance day in and day out for more than three months.

Hardened lava flows commonly contain tunnels that once were horizontal conduits carrying lava from the volcanic vent to the flow's leading edge. These openings develop in the interior of a flow where temperatures remain high long after the surface congeals. Under these conditions, the still-molten lava within the conduits continues its forward motion, leaving behind the cavelike voids called **lava tubes.** Lava tubes are important because they allow fluid lavas to advance great distances from their source.

Gases

Magmas contain varying amounts of dissolved gases (*volatiles*) held in the molten rock by confining pressure, just as carbon dioxide is held in soft drinks. As with soft drinks, as soon as the pressure is reduced, the gases begin to escape. Obtaining gas samples from an erupting volcano is difficult and dangerous, so geologists usually estimate the amount of gas originally contained within the magma.

The gaseous portion of most magmas makes up from 1 to 6 percent of the total weight, with most of this in the form of water vapor. Although the percentage may be small, the actual quantity of emitted gas can exceed thousands of tons per day.

The composition of volcanic gases is important because they contribute significantly to the gases that make up our planet's atmosphere. Analyses of samples taken during Hawaiian eruptions indicate that the gases are about 70 percent water vapor, 15 percent carbon dioxide, 5 percent nitrogen, 5 percent sulfur dioxide, and lesser amounts of chlorine, hydrogen, and argon. Sulfur compounds are easily recognized by their pungent odor. Volcanoes are a natural source of air pollution, including sulfur dioxide, which readily combines with water to form sulfuric acid.

In addition to propelling magma from a volcano, gases play an important role in creating the narrow conduit that connects the magma chamber to the surface. First, high temperatures and the buoyant force

A.

B.

Figure 4.5 **A.** Typical pahoehoe (ropy) lava flow, Kilauea, Hawaii. (Photo by Doug Perrine/DRK Photo)
B. Typical slow-moving aa flow. (Photo by J. D. Griggs/U.S. Geological Survey)

from the magma body cracks the rock above. Then, hot blasts of high-pressure gases expand the cracks and develop a passageway to the surface. Once the passageway is completed, the hot gases armed with rock fragments erode its walls, producing a larger conduit. Because these erosive forces are concentrated on any protrusion along the pathway, the volcanic pipes that are produced have a circular shape. As the conduit enlarges, magma moves upward to produce surface activity. Following an eruptive phase, the volcanic pipe often becomes choked with a mixture of congealed magma and debris that was not thrown clear of the vent. Before the next eruption, a new surge of explosive gases may again clear the conduit.

Occasionally, eruptions emit colossal amounts of volcanic gases that rise high into the atmosphere, where they may reside for several years. Some of these eruptions may have an impact on Earth's climate, a topic we will consider later in the chapter.

Pyroclastic Materials

When basaltic lava is extruded, dissolved gases escape quite freely and continually. These gases propel incandescent blobs of lava to great heights (Figure 4.6). Some of this ejected material may land near the vent and build a cone-shaped structure, whereas smaller particles will be carried great distances by the wind. By contrast, viscous (rhyolitic) magmas are highly charged with gases, and upon release they expand a thousandfold as they blow pulverized rock, lava, and glass fragments from the vent. The particles produced in both of these situations are referred to as **pyroclastic materials.** These ejected fragments range in size from very fine dust and sand-sized volcanic ash (less than 2 millimeters) to pieces that weigh several tons.

Ash and *dust* particles are produced from gas-laden viscous magma during an explosive eruption (see Figure 4.4). As magma moves up in the vent, the gases rapidly expand, generating a froth of melt that might resemble froth that flows from a just opened bottle of champagne. As the hot gases expand explosively, the froth is blown into very fine glassy fragments. When the hot ash falls, the glassy shards often fuse to form a rock called *welded tuff*. Sheets of this material, as well as ash deposits that later consolidate, cover vast portions of the western United States.

Also common are pyroclasts that range in size from small beads to walnuts termed *lapilli* ("little stones"). These ejecta are commonly called *cinders* (2–64 millimeters). Particles larger than 64 millimeters (2.5 inches) in diameter are called *blocks* when they are made of hardened lava, and *bombs* when they are ejected as incandescent lava. Because bombs are semimolten upon ejection, they often take on a streamlined shape as they hurtle through the air (Figure 4.6). Because of their size, bombs and blocks usually fall on the slopes of a cone; however, they are occasionally propelled far from the volcano by the force of escaping gases. For instance, bombs 6 meters (20 feet) long and weighing about 200 tons were blown 600 meters (2000 feet) from the vent during an eruption of the Japanese volcano Asama.

Figure 4.6 Volcanic bombs forming during an eruption of Hawaii's Kilauea volcano. Ejected lava fragments take on a streamlined shape as they sail through the air. The bomb in the insert is about 10 centimeters long. (Photo by Arthur Roy/National Audubon Society; inset photo by E. J. Tarbuck)

← 2 cm →

Volcanic Structures and Eruptive Styles

 Internal Processes
↳ Igneous Activity

The popular image of a volcano is that of a solitary, graceful, snowcapped cone, such as Mt. Hood in Oregon or Japan's Fujiyama. These picturesque, conical mountains are produced by volcanic activity that occurred intermittently over thousands, or even hundreds of thousands, of years. However, many volcanoes do not fit this image. Some volcanoes are only 30 meters (100 feet) high and formed during a single eruptive phase that may have lasted only a few days. Further, numerous volcanic landforms are not "volcanoes" at all. For example, Alaska's Valley of Ten Thousand Smokes is a flat-topped deposit consisting of 15 cubic kilometers of ash that erupted in less than 60 hours and blanketed a section of river valley to a depth of 200 meters (600 feet).

Volcanic landforms come in a wide variety of shapes and sizes, and each structure has a unique eruptive history. Nevertheless, volcanologists have been able to classify volcanic landforms and determine their eruptive patterns. In this section we will consider the general anatomy of a volcano and look at three major volcanic types: shield volcanoes, cinder cones, and composite cones. This discussion will be followed by an overview of other significant volcanic landforms.

Anatomy of a Volcano

Volcanic activity frequently begins when a fissure (crack) develops in the crust as magma moves forcefully toward the surface. As the gas-rich magma moves up this linear fissure, its path is usually localized into a circular **conduit**, or **pipe**, that terminates at a surface opening called a **vent** (Figure 4.7). Successive eruptions of lava, pyroclastic material, or frequently a combination of both often separated by long periods of inactivity eventually build the structure we call a **volcano.**

Located at the summit of many volcanoes is a steep-walled depression called a **crater.** Craters are constructional features that were built upward as ejected fragments collected around the vent to form a doughnut-like structure. Some volcanoes have multiple summit craters, whereas others have very large, more or less circular depressions called *calderas.* Calderas are large collapse structures that may or may not form in association with a volcano. (We will consider the formation of various types of calderas later.)

During early stages of growth, most volcanic discharges come from a central summit vent. As a volcano matures, material also tends to be emitted from fissures that develop along the flanks, or base, of the volcano. Continued activity from a flank eruption may produce a small **parasitic cone.** Mount Etna in Italy, for example, has more than 200 secondary vents, some of which have built cones. Many of these vents, however, emit only gases and are appropriately called **fumaroles.**

The form of a particular volcano is largely determined by the composition of the contributing magma. As you will see, fluid Hawaiian-type lavas tend to produce broad structures with gentle slopes, whereas more viscous silica-rich lavas (and some gas-rich basaltic lavas) tend to generate cones with moderate to steep slopes.

Shield Volcanoes

Shield volcanoes are produced by the accumulation of fluid basaltic lavas and exhibit the shape of a broad, slightly domed structure that resembles a warrior's shield (Figure 4.8). Most shield volcanoes have grown up from the deep-ocean floor to form islands or seamounts. For example, the islands of the Hawaiian chain, Iceland, and the Galapagos are either a single shield volcano or the coalescence of several shields. Some shield volcanoes, however, occur on continents.

Extensive study of the Hawaiian Islands confirms that each shield was built from a myriad of basaltic lava flows averaging a few meters thick. Also, these islands consist of only about 1 percent pyroclastic ejecta.

Figure 4.7 Anatomy of a "typical" composite cone (see also Figures 4.8 and 4.11 for a comparison with a shield and cinder cone, respectively).

Flank eruption   Summit caldera

Central vent

Magma chamber

Oceanic crust

Figure 4.8 Mauna Loa is one of five shield volcanoes that together make up the island of Hawaii. Shield volcanoes are built primarily of fluid basaltic lava flows and contain only a small percentage of pyroclastic materials. (Photo by Greg Vaughn)

Mauna Loa is one of five overlapping shield volcanoes that together comprise the Big Island of Hawaii. From its base on the floor of the Pacific Ocean to its summit, Mauna Loa is over 9 kilometers (6 miles) high, exceeding that of Mount Everest. This massive pile of basaltic rock has a volume of 40,000 cubic kilometers that was extruded over a period of nearly a million years. For comparison, the volume of material composing Mauna Loa is roughly 200 times greater than the amount composing a large composite cone such as Mt. Rainier (Figure 4.9). Most shields, however, are more modest in size. For example, the classic Icelandic shield, Skjalbreidur, rises to a height of only about 600 meters (2000 feet) and is 10 kilometers (6 miles) across its base.

Young shields, particularly those located in Iceland, emit very fluid lava from a central summit vent and have sides with gentle slopes that vary from 1 to 5 degrees. Mature shields, as exemplified by Mauna Loa,

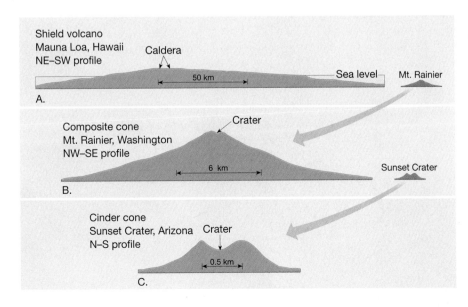

Shield volcano
Mauna Loa, Hawaii
NE–SW profile

Caldera

50 km

Sea level

Mt. Rainier

A.

Composite cone
Mt. Rainier, Washington
NW–SE profile

Crater

6 km

Sunset Crater

B.

Cinder cone
Sunset Crater, Arizona
N–S profile

Crater

0.5 km

C.

Figure 4.9 Profiles of volcanic landforms. **A.** Profile of Mauna Loa, Hawaii, the largest shield volcano in the Hawaiian chain. Note size comparison with Mt. Rainier, Washington, a large composite cone. **B.** Profile of Mt. Rainier, Washington. Note how it dwarfs a typical cinder cone. **C.** Profile of Sunset, Arizona, a typical steep-sided cinder cone.

have steeper slopes in the middle sections (about 10 degrees), while their summits and flanks are comparatively flat. During the mature stage, lavas are discharged from the summit vents, as well as from rift zones that develop along the slopes. Most lava is discharged as the fluid pahoehoe type, but as these flows cool downslope, many change into a clinkery aa flow. Once an eruption is well established, a large fraction of the lava (perhaps 80 percent) flows through a well-developed system of lava tubes. This greatly increases the distance lava can travel before it solidifies. Thus, lava emitted near the summit often reaches the sea, thereby adding to the width of the cone at the expense of its height.

Another feature common to a mature, active shield volcano is a large steep-walled caldera that occupies its summit. Calderas form when the roof of the volcano collapses as magma from the central magma reservoir migrates to the flanks, often feeding fissure eruptions. Mauna Loa's summit caldera measures 2.6 by 4.5 kilometers (1.6 by 2.8 miles) and has a depth that averages about 150 meters (500 feet).

In their final stage of growth, the activity on mature shields is more sporadic and pyroclastic ejections are more prevalent. Further, the lavas increase in viscosity, resulting in thicker, shorter flows. These eruptions tend to steepen the slope of the summit area, which often becomes capped with clusters of cinder cones. This explains why Mauna Kea, a very mature volcano that has not erupted in historic times, has a steeper summit than Mauna Loa, which erupted as recently as 1984. Astronomers are so certain that Mauna Kea is "over the hill" that they have built an elaborate observatory on its summit, housing some of the world's finest (and most expensive) telescopes.

Kilauea, Hawaii: Eruption of a Shield Volcano. Kilauea, the most active and intensely studied shield volcano in the world, is located on the island of Hawaii in the shadow of Mauna Loa. More than 50 eruptions have been witnessed here since record keeping began in 1823. Several months before each eruptive phase, Kilauea inflates as magma gradually migrates upward and accumulates in a central reservoir located a few kilometers below the summit. For up to 24 hours in advance of an eruption, swarms of small earthquakes warn of the impending activity.

Most of the activity on Kilauea during the past 50 years occurred along the flanks of the volcano in a region called the East Rift Zone. A rift eruption here in 1960 engulfed the coastal village of Kapoho, located nearly 30 kilometers (20 miles) from the source. The longest and largest rift eruption ever recorded on Kilauea began in 1983 and continues to this day, with no signs of abating. The first discharge began along a 6-kilometer (4-mile) fissure where a 100-meter- (300-foot-) high "curtain of fire" formed as red-hot lava was ejected skyward (Figure 4.10). When the activity became localized, a cinder and spatter cone given the Hawaiian name *Puu Oo* was built. Over the next three years the general eruptive pattern consisted of short periods (hours to days) when fountains of gas-rich lava sprayed skyward (see Figure 4.3). Each event was followed by nearly a month of inactivity.

Figure 4.10 Lava extruded along the East Rift Zone, Kilauea, Hawaii. (Photo by Greg Vaughn)

By the summer of 1986 a new vent opened up 3 kilometers downrift. Here smooth-surfaced pahoehoe lava formed a lava lake. Occasionally the lake overflowed, but more often lava escaped through tunnels to feed pahoehoe flows that moved down the southeastern flank of the volcano toward the sea. These flows destroyed nearly a hundred rural homes, covered a major roadway, and eventually reached the sea. Lava has been intermittently pouring into the ocean ever since, adding new land to the island of Hawaii.

Cinder Cones

As the name suggests, **cinder cones** (also called **scoria cones**) are built from ejected lava fragments that take on the appearance of cinders or clinkers as they begin to harden while in flight. These fragments range in size from fine ash to bombs but consist mostly of pea- to walnut-size lapilli. Usually a product of relatively gas-rich basaltic magma, cinder cones consist of rounded to irregular fragments that are markedly vesicular (containing voids) and have a black to reddish-brown color. Recall that these vesicular rock fragments are called *scoria*. Occasionally an eruption of silica-rich magma will generate a light-colored cinder cone composed of ash and pumice fragments. Although cinder cones are composed mostly of loose pyroclastic material, they sometimes extrude lava. On such occasions the discharges come from vents located at or near the base rather than from the summit crater.

Cinder cones have a very simple, distinctive shape determined by the slope that loose pyroclastic material maintains as it comes to rest (Figure 4.11). Because cinders have a high angle of repose (the steepest angle at which material remains stable), young cinder cones are steep-sided, having *slopes* between 30 and 40 degrees. In addition, cinder cones have large, deep craters in relation to the overall size of the structure. Although relatively symmetrical, many cinder cones are elongated, and higher on the side that was downwind during the eruptions.

Cinder cones are usually the product of a single eruptive episode that sometimes lasts only a few weeks and rarely exceeds a few years. Once this event ceases, the magma in the pipe connecting the vent to the magma chamber solidifies, and the volcano never erupts again. As a consequence of this short life span, cinder cones are small, usually between 30 meters (100 feet) and 300 meters (1000 feet), and rarely exceed 700 meters (2100 feet) in height (see Figure 4.9).

Cinder cones are found by the thousands all around the globe. Some are located in volcanic fields like the one located near Flagstaff, Arizona, which consists of about 600 cones. Others are parasitic cones of larger volcanoes. Mount Etna, for example, has dozens of cinder cones dotting its flanks.

Parícutin: Life of a Garden-Variety Cinder Cone. One of the very few volcanoes studied by geologists from beginning to end is the cinder cone called Parícutin, located about 320 kilometers (200 miles) west of Mexico City. In 1943 its eruptive phase began in a cornfield owned by Dionisio Pulido, who witnessed the event as he prepared the field for planting.

For two weeks prior to the first eruption, numerous Earth tremors caused apprehension in the nearby village of Parícutin. Then on February 20 sulfurous gases began billowing from a small depression that had been in the cornfield for as long as people could remember. During the night, hot, glowing rock fragments were ejected from the vent, producing a spectacular fireworks display. Explosive discharges continued, throwing hot fragments and ash occasionally as high as 6000 meters (20,000 feet) above the crater rim. Larger fragments fell near the crater, some remaining incandescent as they rolled down the slope. These built an aesthetically pleasing cone, while finer ash fell over a much larger area, burning and eventually covering the village of Parícutin. In the first day the cone grew to 40 meters (130 feet), and by the fifth day it was over 100 meters (330 feet) high. Within the first year more than 90 percent of the total ejecta had been discharged.

The first lava flow came from a fissure that opened just north of the cone, but after a few months flows began to emerge from the base of the cone itself. In June 1944 a clinkery aa flow 10 meters (30 feet) thick moved over much of the village of San Juan Parangaricutiro, leaving only the church steeple exposed (Figure 4.12). After nine years of intermittent pyroclastic explosions and nearly continuous discharge of lava from vents at its base, the activity ceased almost as quickly as it had begun. Today, Parícutin is just another one of the scores of cinder cones dotting the landscape in this region of Mexico. Like the others, it will not erupt again.

Composite Cones

Earth's most picturesque yet potentially dangerous volcanoes are **composite cones,** or **stratovolcanoes** (Figure 4.13). Most are located in a relatively narrow zone that rims the Pacific Ocean, appropriately called the *Ring of Fire* (see Figure 4.27). This active zone includes a chain of continental volcanoes that are distributed along the west coast of South and North America, including the large cones of the Andes and the Cascade Range of the western United States and Canada. The latter group includes Mount St. Helens, Mount Rainier, and Mount Garibaldi. The most active regions in the Ring of Fire are located along curved belts of volcanic islands situated adjacent to the deep ocean trenches of the northern and western Pacific. This nearly continuous chain of volcanoes stretches from the Aleutian Islands to Japan

Figure 4.11 SP Crater, a cinder cone north of Flagstaff, Arizona. Note the road across the lava flow for scale. Cinder cones are built from ejected lava fragments and are usually less than 300 meters (1000 feet) high. (Photo by Michael Collier)

and the Philippines and ends on the North Island of New Zealand.

The classic composite cone is a large, nearly symmetrical structure composed of both lava and pyroclastic deposits. Just as shield volcanoes owe their shape to fluid basaltic lavas, composite cones reflect the nature of the erupted material. For the most part, composite cones are the product of gas-rich magma having an andesitic composition. (Composite cones may also emit various amounts of material having a basaltic and/or rhyolitic composition.) Relative to shields, the silica-rich magmas typical of composite cones generate thick viscous lavas that travel short distances. In addition, composite cones may generate explosive eruptions that eject huge quantities of pyroclastic material.

The growth of a "typical" composite cone begins with both pyroclastic material and lava being emitted from a central vent. As the structure matures, lavas tend to flow from fissures that develop on the lower flanks of the cone. This activity may alternate with explosive eruptions that eject pyroclastic material from the summit crater. Sometimes both activities occur simultaneously.

A conical shape, with a steep summit area and more gradually sloping flanks, is typical of many large composite cones. This classic profile, which adorns calendars and postcards, is partially a consequence of the way viscous lavas and pyroclastic ejecta contribute to the growth of the cone. Coarse fragments ejected from the summit crater tend to accumulate near their source.

80

Because of their high angle of repose, coarse materials contribute to the steep slopes of the summit area. Finer ejecta, on the other hand, are deposited as a thin layer over a large area. This acts to flatten the flank of the cone. In addition, during the early stages of growth, lavas tend to be more abundant and flow greater distances from the vent than do later lavas. This contributes to the cone's broad base. As the volcano matures, the short flows that come from the central vent serve to armor and strengthen the summit area. Consequently,

Figure 4.13 Mount Shasta, California, one of the largest composite cones in the Cascade Range. (Photo by John S. Shelton)

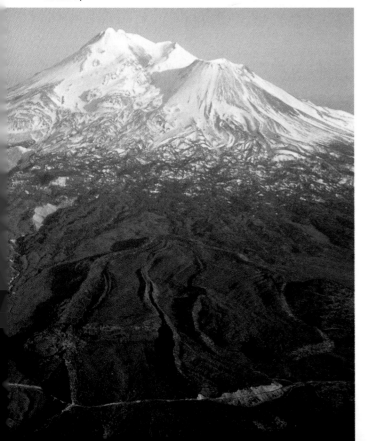

steep slopes exceeding 40 degrees are sometimes possible. Two of the most perfect cones—Mount Mayon in the Philippines and Fujiyama in Japan—exhibit the classic form we expect of a composite cone, with its steep summit and gently sloping flanks.

Despite their symmetrical form, most composite cones have a complex history. Huge mounds of volcanic debris surrounding many cones provide evidence that, in the distant past, a large section of the volcano slid downslope as a massive landslide. Others develop horseshoe-shaped depressions at their summits as a result of explosive eruptions, or as occurred during the 1980 eruption of Mount St. Helens, a combination of a landslide and the eruption of 0.6 cubic kilometer of magma left a gaping void on the north side of the cone. Often, so much rebuilding has occurred since these eruptions that no trace of the amphitheater-shaped scar remains.

Vesuvius in Italy provides us with another example of the complex history of a volcanic region. This young volcano was built on the same site where an older cone was destroyed by an eruption in A.D. 79. In the following section we will look at another aspect of composite cones: their destructive nature.

Living in the Shadow of a Composite Cone

More than 50 volcanoes have erupted in the United States in the past 200 years (Figure 4.14). Fortunately, the most explosive of these eruptions, except for Mount St. Helens in 1980, occurred in sparsely inhabited regions of Alaska. On a global scale, numerous destructive eruptions have occurred during the past few thousand years, a few of which may have influenced the course of human civilization.

Figure 4.14 Of the 13 potentially active volcanoes in the Cascade Range, 11 have erupted in the past 4000 years and 7 in just the past 200 years. More than 100 eruptions, most of which were explosive, have occurred in the past 4000 years. Mount St. Helens is the most active volcano in the Cascades. Its eruptions have ranged from relatively quiet outflows of lava to explosive events much larger than that of May 18, 1980. Each eruption symbol in the diagram represents from one to several dozen eruptions closely spaced in time. (After U.S. Geological Survey)

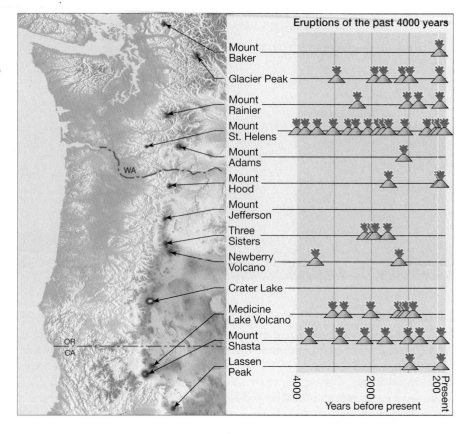

Eruption of Vesuvius A.D. 79

In addition to producing some of the most violent volcanic activity, composite cones can erupt unexpectedly. One of the best documented of these events was the A.D. 79 eruption of the Italian volcano we now call Vesuvius. Prior to this eruption, Vesuvius had been dormant for centuries and had vineyards adorning its sunny slopes. On August 24, however, the tranquility ended, and in less than 24 hours the city of Pompeii (near Naples) and more than 2000 of its 20,000 residents perished. Some were entombed beneath a layer of pumice nearly 3 meters (10 feet) thick, while others were encased within a layer of hardened ash. They remained this way for nearly 17 centuries, until the city was partially excavated, giving archaeologists a superbly detailed picture of ancient Roman life.

By reconciling historical records with detailed scientific studies of the region, volcanologists have pieced together the chronology of the destruction of Pompeii. The eruption most likely began as steam discharges on the morning of August 24. By early afternoon fine ash and pumice fragments formed a tall eruptive cloud emanating from Vesuvius. Shortly thereafter, debris from this cloud began to shower Pompeii, located 9 kilometers (6 miles) downwind of the volcano. Undoubtedly, many people fled during this early phase of the eruption. For the next several hours pumice fragments as large as 5 centimeters (2 inches) fell on Pompeii. One historical record of this eruption states that people located more distant than Pompeii tied pillows to their heads in order to fend off the flying fragments.

The pumice fall continued for several hours, accumulating at the rate of 12 to 15 centimeters (5 to 6 inches) per hour. Most of the roofs in Pompeii eventually gave way. Despite the accumulation of more than 2 meters of pumice, many of the people that had not evacuated Pompeii were probably still alive the morning of August 25. Then suddenly and unexpectedly a surge of searing hot dust and gas swept rapidly down the flanks of Vesuvius. This blast killed an estimated 2000 people who had somehow managed to survive the pumice fall. Some may have been killed by flying debris, but most died of suffocation as a result of inhaling ash-laden gases. Their remains were quickly buried by the falling ash, which rain cemented into a hard mass before their bodies had time to decay. The subsequent decomposition of the bodies produced cavities in the hardened ash that replicated the form of the entombed bodies, preserving in some cases even facial expressions. Nineteenth-century excavators found these cavities and created casts of the corpses by pouring plaster of Paris into the voids (Figure 4.15). Some of the plaster casts show victims trying to cover their mouth in an effort to draw what would be their last breath.

Figure 4.15 Plaster casts of several victims of the A.D. 79 eruption of Mount Vesuvius that destroyed the Italian city of Pompeii. (Photo by Leonard von Matt/Photo Researchers, Inc.)

Nuée Ardente: A Deadly Pyroclastic Flow

Among the most devastating phenomena associated with composite cones are **pyroclastic flows** that consist of hot gases infused with incandescent ash and larger rock fragments. The most destructive of these fiery flows, called **nuée ardentes** (also referred to as *glowing avalanches*), are capable of racing down steep volcanic slopes at speeds that can approach 200 kilometers (125 miles) per hour (Figure 4.16).

The ground-hugging portion of a glowing avalanche is rich in particulate matter, which is suspended by jets of buoyant gases passing upward through the flow. Some of these gases have escaped from newly erupted volcanic fragments. In addition, air that is overtaken and trapped by an advancing flow may be heated sufficiently to provide buoyancy to the particulate matter of the nuée ardente. Thus, these flows, which can include large rock fragments in addition to ash, travel downslope in a nearly frictionless environment. This helps to explain why some nuée ardente deposits are found more than 100 kilometers (60 miles) from their source.

The pull of gravity is the force that causes these heavier-than-air flows to sweep downslope much like a snow avalanche. Some pyroclastic flows result when a powerful eruption blasts pyroclastic material laterally out the side of a volcano. Probably more often, nuée ardentes form from the collapse of tall eruption columns that form over a volcano during an explosive event. Once gravity overcomes the initial upward thrust provided by the escaping gases, the ejecta begin to fall. Massive amounts of incandescent blocks, ash, and pumice fragments that fall onto the summit area begin to cascade downslope under the influence of gravity. The largest fragments have been observed bouncing

Figure 4.16 Nuée ardente races down the slope of Mount St. Helens on August 7, 1980, at speeds in excess of 100 kilometers (60 miles) per hour. (Photo by Peter W. Lipman, U.S. Geological Survey)

down the flanks of a cone, while the finer materials travel rapidly as an expanding tongue-shaped cloud.

The Destruction of St. Pierre. In 1902 an infamous nuée ardente from Mount Pelée, a small volcano on the Caribbean island of Martinique, destroyed the port town of St. Pierre. The destruction happened in moments and was so devastating that almost all of St. Pierre's 28,000 inhabitants were killed. Only one person on the outskirts of town—a prisoner protected in a dungeon—and a few people on ships in the harbor were spared (Figure 4.17).

Shortly after this calamitous eruption, scientists arrived on the scene. Although St. Pierre was mantled by only a thin layer of volcanic debris, they discovered that masonry walls nearly a meter thick were knocked over like dominoes; large trees were uprooted and cannons were torn from their mounts. A further reminder of the destructive force of this nuée ardente is preserved in the ruins of the mental hospital. One of the immense steel chairs that had been used to confine alcoholic patients can be seen today, contorted, as though it were made of plastic.

Lahars: Mudflows on Active and Inactive Cones

In addition to violent eruptions, large composite cones may generate a type of mudflow referred to by its Indonesian name **lahar.** These destructive mudflows occur when volcanic debris becomes saturated with water and rapidly moves down steep volcanic slopes, generally following gullies and stream valleys. Some lahars are triggered when large volumes of ice and snow melt during an eruption. Others are generated when heavy rainfall saturates weathered volcanic deposits. Thus, lahars can occur even when a volcano is *not* erupting.

When Mount St. Helens erupted in 1980, several lahars formed. These mudflows and accompanying flood waters raced down the valleys of the north and south forks of the Toutle River at speeds exceeding 30 kilometers per hour. Water levels in the river rose to 4 meters (13 feet) above flood stage, destroying or severely damaging nearly all the homes and bridges along the impacted area. Fortunately, the area was not densely populated.

In 1985 deadly lahars were produced during a small eruption of Nevado del Ruiz, a 5300 meter (17,400 foot) volcano in the Andes Mountains of Colombia. Hot, pyroclastic material melted ice and snow that capped the mountain (Nevado means *snow* in Spanish) and sent torrents of ash and debris down three major river valleys that flank the volcano. Reaching speeds of 100 kilometers (60 miles) per hour, these mudflows tragically took 25,000 lives.

Mount Rainier, Washington, is considered by many to be America's most dangerous volcano because, like Nevado del Ruiz, it has a thick year-round mantle of snow and ice. Adding to the risk is the fact that 100,000 people live in the valleys around Rainier, and many homes are built on lahars that flowed down the volcano

hundreds or thousands of years ago. A future eruption, or perhaps just a period of heavy rainfall, may produce lahars that will likely take similar paths.

Other Volcanic Landforms

The most obvious volcanic structure is a cone. But other distinctive and important landforms are also associated with volcanic activity.

Calderas

Calderas are large collapse depressions having a more or less circular form. Their diameters exceed one kilometer, and many are tens of kilometers across. (Those less than a kilometer across are called *collapse pits.*) Most calderas are formed by one of the following processes: 1) the collapse of the summit of a large composite volcano following an explosive eruption of silica-rich pumice and ash fragments; 2) the collapse of the top of a shield volcano caused by subterranean drainage from a central magma chamber; and 3) the collapse of a large area, independent of any preexisting volcanic structures, caused by the discharge of colossal volumes of silica-rich pumice and ash along ring fractures.

Crater Lake–Type Calderas. Crater Lake, Oregon, is located in a caldera that has a maximum diameter of 10 kilometers (6 miles) and is 1175 meters (over 3800 feet) deep. This caldera formed about 7000 years ago when a composite cone, later named Mount Mazama, violently extruded 50 to 70 cubic kilometers of pyroclastic material (Figure 4.18). With the loss of support, 1500 meters (nearly a mile) of the summit of this once

Figure 4.17 St. Pierre as it appeared shortly after the eruption of Mount Pelée, 1902. (Reproduced from the collection of the Library of Congress)

Eruption of
Mount Mazama

Partialy emptied
magma chamber

Collapse of
Mount Mazama

Formation of Crater Lake and Wizard Island

Figure 4.18 Sequence of events that formed Crater Lake, Oregon. About 7000 years ago a violent eruption partly emptied the magma chamber, causing the summit of former Mount Mazama to collapse. Rainfall and groundwater contributed to form Crater Lake, the deepest lake in the United States. Subsequent eruptions produced the cinder cone called Wizard Island. (After H. Williams, *The Ancient Volcanoes of Oregon*, p. 47. Courtesy of the University of Oregon)

prominent cone collapsed. After the collapse, rainwater filled the caldera (Figure 4.19). Later volcanic activity built a small cinder cone in the lake. Today this cone, called Wizard Island, provides a mute reminder of past activity.

Hawaiian-Type Calderas. Although most calderas are produced by *collapse following an explosive eruption*, some are not. For example, Hawaii's active shield volcanoes, Mauna Loa and Kilauea, both have large calderas at their summits. Kilauea's measures 3.3 by 4.4 kilometers (about 2 by 3 miles) and is 150 meters (500 feet) deep. Each caldera formed by gradual subsidence of the summit as magma slowly drained laterally from the central magma chamber to a rift zone, often producing flank eruptions.

Yellowstone-Type Calderas. Although the 1980 eruption of Mount St. Helens was spectacular, it pales by comparison to what happened 630,000 years ago in the region now occupied by Yellowstone National Park. Here, approximately 1000 cubic kilometers of pyroclastic material erupted, eventually producing a caldera 70 kilometers (43 miles) across. This event produced showers of ash as far away as the Gulf of Mexico. Vestiges of this activity are the many hot springs and geysers in the region.

Unlike calderas associated with composite cones, these depressions are so large and poorly defined that many remained undetected until high-quality aerial, or satellite, images became available. One of these, the LaGarita caldera, located in the San Juan Mountains of southern Colorado, is about 32 kilometers (20 miles) wide and 80 kilometers (50 miles) long. Despite modern mapping techniques, the entire outline of this structure is still uncertain.

The formation of a large Yellowstone-type caldera begins when a silica-rich (rhyolitic) magma body is emplaced near the surface, upwarping the overlying rocks. Next, ring fractures develop in the roof providing a pathway to the surface for the highly viscous gas-rich magma. This initiates an explosive eruption of colossal proportions ejecting huge volumes (usually exceeding 100 cubic kilometers) of pyroclastic materials, mainly in the form of ash and pumice fragments. Typically, these materials form a pyroclastic flow that spreads across the landscape at speeds that may exceed 100 kilometers (60 miles) per hour, destroying most living things in its path. After coming to rest, the hot fragments of ash and pumice fuse together, forming a welded tuff that closely resembles a solidified lava flow. Finally, with the loss of support, the roof of the magma chamber collapses, generating a large caldera.

Calderas of the type located on the Yellowstone Plateau of northwestern Wyoming are the largest volcanic structures on Earth. Some geologists have compared their destructive force with that of the impact of a small asteroid. Fortunately, no eruption of this type has occurred in historic times. Other examples of large calderas located in the United States are California's Long Valley Caldera and Valles Caldera located west of Los Alamos, New Mexico.

Fissure Eruptions and Lava Plateaus

We think of volcanic eruptions as building a cone or shield from a central vent. But by far the greatest volume of volcanic material is extruded from fractures in

Figure 4.19 Crater Lake in Oregon occupies a caldera about 6 miles in diameter. Wizard Island is a cinder cone located within the caldera. (Photo by Greg Vaughn/Tom Stack and Associates)

the crust called **fissures**. Rather than building a cone, these long, narrow cracks may emit a low-viscosity basaltic lava, blanketing a wide area.

The extensive Columbia Plateau in the northwestern United States was formed this way (Figure 4.20). Here, numerous **fissure eruptions** extruded very fluid

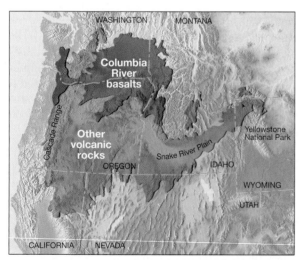

Figure 4.20 Volcanic areas that compose the Columbia Plateau in the Pacific Northwest. The Columbia River basalts cover an area of nearly 200,000 square kilometers (80,000 square miles). Activity here began about 17 million years ago as lava began to pour out of large fissures, eventually producing a basalt plateau with an average thickness of more than 1 kilometer. (After U.S. Geological Survey)

basaltic lava (Figure 4.21). Successive flows, some 50 meters (160 feet) thick, buried the existing landscape as they built a lava plateau nearly a mile thick. The fluid nature of the lava is evident, because some remained molten long enough to flow 150 kilometers (90 miles) from their sources. The term **flood basalts** appropriately describes these flows. Massive accumulations of basaltic lava, similar to those of the Columbia Plateau, occur worldwide. One of the largest is the Deccan Traps, a thick sequence of flat-lying basalt flows covering nearly 500,000 square kilometers (195,000 square miles) of west central India. When the Deccan Traps formed about 66 million years ago, nearly 2 million cubic kilometers of lava were extruded in less than 1 million years. Another huge deposit of flood basalts, called the Ontong Java Plateau, is found on the floor of the Pacific Ocean.

Iceland, which is located astride the Mid-Atlantic Ridge, regularly experiences fissure eruptions. The largest Icelandic eruptions in historic times occurred in 1783 and were named the Laki eruptions. A rift 25 kilometers (15 miles) long generated over 20 separate vents, which initially extruded sulfurous gases and ash deposits that built several small cinder cones. This activity was followed by huge outpourings of very fluid basaltic lava. The total volume of lava extruded by the Laki eruptions was in excess of 12 cubic kilometers. Volcanic gases stunted grasslands and directly killed most of Iceland's livestock. The ensuing famine caused 10,000 deaths.

Figure 4.21 Basaltic fissure eruption. **A.** Lava fountaining from a fissure and formation of fluid lava flows called flood basalts. **B.** Photo of basalt flows near Idaho Falls. (Photo by John S. Shelton)

Volcanic Pipes and Necks

Most volcanoes are fed magma through short conduits, called *pipes*, that connect a magma chamber to the surface. In rare circumstances, pipes may extend tubelike to depths exceeding 200 kilometers (125 miles). When this occurs, the ultramafic magmas that migrate up these structures produce rocks that are thought to be samples of the mantle that have undergone very little alteration during their ascent. Geologists consider these unusually deep conduits to be "windows" into Earth, for they allow us to view rock normally found only at great depth.

The best-known volcanic pipes are the diamond-bearing structures of South Africa. Here, the rocks filling the pipes originated at depths of at least 150 kilometers (90 miles), where pressure is high enough to generate diamonds and other high-pressure minerals.

The task of transporting essentially unaltered magma (along with diamond inclusions) through 150 kilometers of solid rock is exceptional. This fact accounts for the scarcity of natural diamonds.

Volcanoes on land are continually being lowered by weathering and erosion. Cinder cones are easily eroded, because they are composed of unconsolidated materials. However, all volcanoes will eventually succumb to relentless erosion over geologic time. As erosion progresses, the rock occupying the volcanic pipe is often more resistant and may remain standing above the surrounding terrain long after most of the cone has vanished. Shiprock, New Mexico, is such a feature and is called a **volcanic neck** (Figure 4.22). This structure, higher than many skyscrapers, is but one of many such landforms that protrude conspicuously from the red desert landscapes of the American Southwest.

Intrusive Igneous Activity

GEODe Internal Processes
↳ Igneous Activity

Although volcanic eruptions can be among the most violent and spectacular events in nature and therefore worthy of detailed study, most magma is emplaced at depth. Thus, an understanding of intrusive igneous activity is as important to geologists as the study of volcanic events.

The structures that result from the emplacement of igneous material at depth are called **plutons,** named for Pluto, the god of the lower world in classical mythology. Because all plutons form out of view beneath Earth's surface, they can be studied only after uplifting and erosion have exposed them. The challenge lies in reconstructing the events that generated these structures millions or even hundreds of millions of years ago.

For the sake of clarity, we have separated our discussions of volcanism and plutonic activity. Keep in mind, however, that these diverse processes occur simultaneously and involve basically the same materials.

Nature of Plutons

Plutons are known to occur in a great variety of sizes and shapes. Some of the most common types are illustrated in Figure 4.23. Notice that some of these structures have a tabular (tabletop) shape, whereas others are quite massive. Also, observe that some of these bodies cut across existing structures, such as layers of sedimentary rock; others form when magma is injected between sedimentary layers. Because of these differences, intrusive igneous bodies are generally classified according to their shape as either **tabular** or **massive** and by their orientation with respect to the host rock. Plutons are said to be **discordant** if they cut across ex-

Figure 4.22 Shiprock, New Mexico, is a volcanic neck. This structure, which stands over 420 meters (1380 feet) high, consists of igneous rock that crystallized in the vent of a volcano that has long since been eroded away. The tabular structure in the background is a dike that served to feed lava flows along the flanks of the once active volcano. (Photo by Michael Collier)

isting structures, and **concordant** if they form parallel to features such as sedimentary strata. As you can see in Figure 4.23A, plutons are closely associated with volcanic activity. Many of the largest intrusive bodies are the remnants of magma chambers that once fed ancient volcanoes.

Dikes

Dikes are tabular discordant bodies that are produced when magma is injected into fractures. The force exerted by the emplaced magma can be great enough to separate the walls of the fracture further. Once crystallized, these sheetlike structures have thicknesses ranging from less than a centimeter to more than a kilometer. The largest have lengths of hundreds of kilometers. Most dikes, however, are a few meters thick and extend laterally for no more than a few kilometers.

Dikes are often found in groups that once served as vertically oriented pathways followed by molten rock that fed ancient lava flows. The parent pluton is generally not observable. Some dikes are found radiating, like spokes on a wheel, from an eroded volcanic neck (see Figure 4.22). In these situations the active ascent of magma is thought to have generated fissures in the volcanic cone out of which lava flowed.

Sills and Laccoliths

Sills and laccoliths are concordant plutons that form when magma is intruded in a near-surface environment. They differ in shape and usually differ in composition.

Sills. **Sills** are tabular plutons formed when magma is injected along sedimentary bedding surfaces (Figure 4.24). Horizontal sills are the most common, although all orientations, even vertical, are known to exist. Because of their relatively uniform thickness and large

areal extent, sills are likely the product of very fluid magmas. Magmas having a low silica content are more fluid, so most sills are composed of the rock basalt.

The emplacement of a sill requires that the overlying sedimentary rock be lifted to a height equal to the thickness of the sill. Although this is a formidable task, in shallow environments it often requires less energy than forcing the magma up the remaining distance to the surface. Consequently, sills form only at shallow depths, where the pressure exerted by the weight of overlying rock layers is low. Although sills are intruded between layers, they can be locally discordant. Large sills frequently cut across sedimentary layers and resume their concordant nature at a higher level.

One of the largest and most studied of all sills in the United States is the Palisades Sill. Exposed for 80 kilometers along the west bank of the Hudson River in southeastern New York and northeastern New Jersey, this sill is about 300 meters thick. Because of its resistant nature, the Palisades Sill forms an imposing cliff that can be seen easily from the opposite side of the Hudson.

In many respects, sills closely resemble buried lava flows. Both are tabular and often exhibit columnar jointing (Figure 4.25). **Columnar joints** form as igneous rocks cool and develop shrinkage fractures that produce elongated, pillarlike columns. Further, because sills generally form in near-surface environments and may be only a few meters thick, the emplaced magma often cools quickly enough to generate an aphanitic texture.

Laccoliths. **Laccoliths** are similar to sills because they form when magma is intruded between sedimentary layers in a near-surface environment. However, the magma that generates laccoliths is more viscous. This less fluid magma collects as a lens-shaped mass that arches the overlying strata upward (see Figure 4.23). Consequently, a laccolith can occasionally be de-

Cinder
cones

Composite
cones

Fissure
eruption

Magma
chamber

A. Implacement of igneous structures

Figure 4.23 Illustrations showing basic igneous structures. **A.** This block diagram shows the relationship between volcanism and intrusive igneous activity. **B.** This view illustrates the basic intrusive igneous structures, some of which have been exposed by erosion long after their formation. **C.** After millions of years of uplifting and erosion, a batholith is exposed at the surface.

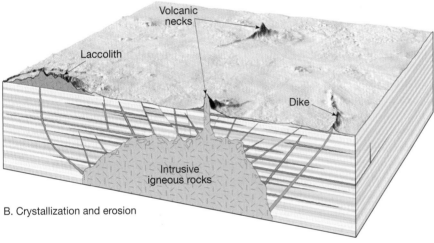

Volcanic
necks

Laccolith

Dike

Intrusive
igneous rocks

B. Crystallization and erosion

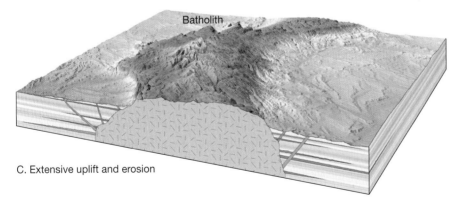

Batholith

C. Extensive uplift and erosion

tected because of the dome-shaped bulge it creates at the surface.

Most large laccoliths are probably not much wider than a few kilometers. The Henry Mountains in southeastern Utah are largely composed of several laccoliths believed to have been fed by a much larger magma body emplaced nearby.

Batholiths

By far the largest intrusive igneous bodies are **batholiths.** Most often, batholiths occur in groups that form linear structures several hundreds of kilometers long and up to 100 kilometers wide, as shown in Figure 4.26. The Idaho batholith, for example, encompasses an area of more than 40,000 square kilometers and consists

Figure 4.24   Salt River Canyon, Arizona. The dark, essentially horizontal band is a sill of basaltic composition that intruded horizontal layers of sedimentary rock. (Photo by E. J. Tarbuck)

of many plutons. Indirect evidence gathered from gravitational studies indicates that batholiths are also very thick, possibly extending dozens of kilometers into the crust. Based on the amount exposed by erosion, some batholiths are at least several kilometers thick.

By definition, a plutonic body must have a surface exposure greater than 100 square kilometers (40 square miles) to be considered a batholith. Smaller plutons of this type are termed **stocks.** Many stocks appear to be portions of batholiths that are not yet fully exposed.

Batholiths usually consist of rock types having chemical compositions toward the granitic end of the spectrum, although diorite is commonly found. Smaller batholiths can be rather simple structures composed almost entirely of one rock type. However, studies of large batholiths have shown that they consist of several distinct plutons that were intruded over a period of millions of years. The plutonic activity that created the Sierra Nevada batholith, for example, occurred nearly continuously over a 130-million-year period that ended about 80 million years ago during the Cretaceous period.

Batholiths may compose the core of mountain systems. Here, uplifting and erosion have removed the surrounding rock, thereby exposing the resistant igneous body. Some of the highest peaks in the Sierra Nevada, such as Mount Whitney, are carved from such a granitic mass.

Large expanses of granitic rock also occur in the stable interiors of the continents, such as the Canadian Shield of North America. These relatively flat exposures are the remains of ancient mountains that have long since been leveled by erosion. Thus, the rocks that make up the batholiths of youthful mountain ranges, such as the Sierra Nevada, were generated near the top of a magma chamber, whereas in shield areas, the roots of former mountains and, thus, the lower portions of batholiths, are exposed.

Plate Tectonics and Igneous Activity

Geologists have known for decades that the global distribution of volcanism is not random. Of the more than 800 active volcanoes* that have been identified, most are located along the margins of the ocean basins—most notably within the circum-Pacific belt known as the *Ring of Fire* (Figure 4.27). This group of volcanoes consists mainly of composite cones that emit volatile-rich magma having an intermediate (andesitic) composition that occasionally produces awe-inspiring eruptions.

The volcanoes comprising a second group emit very fluid basaltic lavas and are confined to the deep ocean basins, including well-known examples on Hawaii and Iceland. In addition, this group contains many active submarine volcanoes that dot the ocean floor; particularly notable are the innumerable small seamounts that lay along the axis of the oceanic ridge. At these depths the pressures are so great that seawater does not boil explosively, even in contact with hot lavas. Thus, first-hand knowledge of these eruptions is limited, coming mainly from deep-diving submersibles.

A third group includes those volcanic structures that are irregularly distributed in the interiors of the continents. None are found in Australia nor in the eastern two-thirds of North and South America. Africa is notable because it has many potentially active volcanoes including Mount Kilimanjaro, the highest point on the continent (5895 meters, 19,454 feet). Volcanism on continents is the most diverse, ranging from eruptions of very fluid basaltic lavas, like those that generated the Columbia River basalts to explosive eruptions of silica-rich rhyolitic magma as occurred in Yellowstone.

Until the late 1960s, geologists had no explanation for the apparently haphazard distribution of continental volcanoes, nor were they able to account for the almost continuous chain of volcanoes that circles the margin of the Pacific basin. With the development of the theory of plate tectonics, the picture was greatly clarified. Recall that all primary (unaltered) magma originates in the upper mantle and that the mantle is essentially solid, *not molten* rock. The basic connection between plate tectonics and volcanism is that *plate motions provide the mechanisms by which mantle rocks melt to generate magma.*

We will examine three zones of igneous activity and their relationship to plate boundaries. These active areas are located 1) along convergent plate boundaries where

*For our purposes, active volcanoes include those with eruptions that have been dated. At least 700 other cones exhibit geologic evidence that they have erupted within the past 10,000 years and are regarded as potentially active. Further innumerable active submarine volcanoes are hidden from view in the depths of the ocean and are not counted in these numbers.

Figure 4.25 Columnar jointing in basalt, Giants Causeway National Park, Northern Ireland. These five- to seven-sided columns are produced by contraction and fracturing that occurs as a lava flow or sill gradually cools. (Photo by Tom Till)

Equally spaced centers of contraction

plates move toward each other and one sinks beneath the other; 2) along divergent plate boundaries, where plates move away from each other and new seafloor is created; and 3) areas within the plates proper that are not associated with any plate boundary. (It should be noted that volcanic activity rarely occurs along transform plate boundaries.) These three volcanic settings are depicted in Figure 4.28. (If you are unclear as to how magma is generated, we suggest that you study the section entitled "Origin of Magma" located at the end of Chapter 3 before proceeding.)

Igneous Activity at Convergent Plate Boundaries

Recall that at convergent plate boundaries, slabs of oceanic crust are bent as they descend into the mantle, generating an oceanic trench. As a slab sinks deeper into the mantle, the increase in temperature and pressure drives volatiles (mostly H_2O) from the oceanic crust. These mobile fluids migrate upward into the wedge-shaped piece of mantle located between the subducting slab and overriding plate (Figure 4.28). Once the sinking slab reaches a depth of about 100 to 150 kilometers, these water-rich fluids reduce the melting point of hot mantle rock sufficiently to trigger some melting. The partial melting of mantle rock (principally peridotite) generates magma with a basaltic composition. After a sufficient quantity of magma has accumulated, it slowly migrates upward.

Volcanism at a convergent plate margin results in the development of a linear or slightly curved chain of volcanoes called a *volcanic arc*. These volcanic chains develop roughly parallel to the associated trench—at distances of 200 to 300 kilometers (100 to 200 miles). Volcanic arcs can be constructed on oceanic, or continental, lithosphere. Those that develop within the ocean and grow large enough for their tops to rise above the surface are labeled *island archipelagos* in most atlases. Geologists prefer the more descriptive term **volcanic island arcs,** or simply **island arcs** (see Figure 4.28). Several young volcanic island arcs of this type border the western Pacific basin, including the Aleutians, the Tongas, and the Marianas.

The early stage of island arc volcanism is typically dominated by the eruption of fluid basalts that build numerous shieldlike structures on the ocean floor. Because this activity begins at great depth, volcanic cones must extrude a great deal of lava before their tops rise above the sea to form islands. This cone-building activity, coupled with massive basaltic intrusions as well as magma that is added to the underside of the crust, tends to thicken the arc crust through time. As a result, mature volcanic arcs are underlain by a comparatively thick crust that impedes the upward flow of the mantle-derived basalts. This in turn provides time for magmatic differentiation to occur, in which heavy iron-rich minerals crystallize and settle out, leaving the melt enriched in silica (see Chapter 3). Consequently, as the arc matures, the magmas that reach the surface tend to erupt silica-rich andesites and even

91

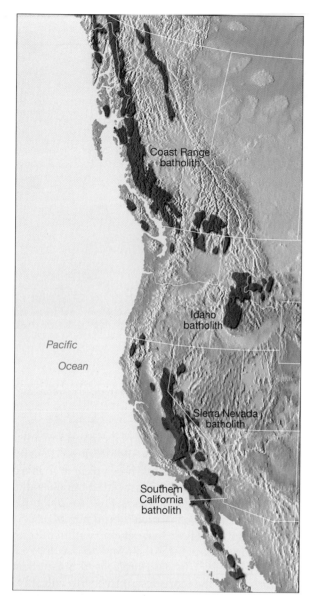

Figure 4.26 Granitic batholiths that occur along the western margin of North America. These gigantic, elongated bodies consist of numerous plutons that were emplaced during the last 150 million years of Earth history.

may become highly evolved as it rises through continental crust. Stated another way, the primary magmas generated in the mantle may change from a comparatively dry, fluid basaltic magma to a viscous andesitic or rhyolitic magma having a high concentration of volatiles as it moves up through the continental crust. The volcanic chain of the Andes Mountains along the western edge of South America is perhaps the best example of a mature continental volcanic arc.

Since the Pacific basin is essentially bordered by convergent plate boundaries (and associated subduction zones), it is easy to see why the irregular belt of explosive volcanoes we call the Ring of Fire formed in this region. The volcanoes of the Cascade Range in the northwestern United States, including Mount Hood, Mount Rainier, and Mount Shasta, are included in this group (Box 4.2).

Igneous Activity at Divergent Plate Boundaries

The greatest volume of magma (perhaps 60 percent of Earth's total yearly output) is produced along the oceanic ridge system in association with seafloor spreading (Figure 4.28). Here, below the ridge axis where the lithospheric plates are being continually pulled apart, the solid yet mobile mantle responds to the decrease in overburden and rises upward to fill in the rift. Recall from Chapter 3 that as rock rises, it experiences a decrease in confining pressure and undergoes melting without the addition of heat. This process, called *decompression melting*, is the most common process by which mantle rocks melt.

Partial melting of mantle rock at spreading centers produces basaltic magma having a composition that is surprisingly similar to that generated along convergent plate boundaries. Because this newly formed basaltic magma is less dense than the mantle rock from which it was derived, it rises faster than the mantle.

Collecting in reservoirs located just beneath the ridge crest, about 10 percent of this magma eventually migrates upward along fissures to erupt as flows on the ocean floor. This activity continuously adds new basaltic rock to the plate margins, temporarily welding them together, only to break again as spreading continues. Along some ridges, outpouring of bulbous pillow lavas build numerous small seamounts. At other locations erupted lavas produce fluid flows that create more subdued topography.

Although most spreading centers are located along the axis of an oceanic ridge, some are not. In particular, the East African Rift is a site where continental lithosphere is being ripped apart. Here, magma is generated by decompression melting in the same manner it is produced along the oceanic ridge system. Vast outpourings of fluid basaltic lavas are common in this region. The East African Rift zone also contains some

some rhyolites. In addition, magmatic differentiation tends to concentrate the available volatiles (water) into the more silica-rich components of these magmas. Because they emit viscous volatile-rich magma, island arc volcanoes typically have explosive eruptions.

Volcanism associated with convergent plate boundaries may also develop where slabs at oceanic lithosphere are subducted under continental lithosphere to produce a **continental volcanic arc** (Figure 4.28). The mechanisms that generate these mantle-derived magmas are essentially the same as those operating at island arcs. The major difference is that continental crust is much thicker and is composed of rocks having a higher silica content than oceanic crust. Hence through the assimilation of silica-rich crustal rocks, plus extensive magmatic differentiation, a mantle-derived magma

Figure 4.27   Locations of some of Earth's major volcanoes.

large composite cones, as exemplified by Mount Kilimanjaro. Like composite cones that form along convergent plate boundaries, these volcanoes form when mantle-derived basalts evolve into volatile-rich andesitic magma as they migrate up through thick silica-rich rocks of the continent.

Intraplate Igneous Activity

We know why igneous activity is initiated along plate boundaries, but why do eruptions occur in the interiors of plates? Hawaii's Kilauea is considered the world's most active volcano, yet it is situated thousands of kilometers from the nearest plate boundary in the middle of the vast Pacific plate. Other sites of **intraplate volcanism** (meaning "within the plate") include the Canary Islands, Yellowstone, and several volcanic centers that you may be surprised to learn are located in the Sahara Desert of northern Africa.

We now recognize that most intraplate volcanism occurs where a mass of hotter than normal mantle material called a **mantle plume** ascends toward the surface (Figure 4.29). Although the depth at which (at least some) mantle plumes originate is still hotly debated, many appear to form deep within Earth at the core–mantle boundary. These plumes of solid yet mobile mantle rock rise toward the surface in a manner similar to the blobs that form within a lava lamp. (These are the gaudy lamps that contain two immiscible liquids in a glass container. As the base of the lamp is heated, the denser liquid at the bottom becomes buoyant and forms blobs that rise to the top.) Like the blobs in a lava lamp, a mantle plume has a bulbous head that draws out a narrow stalk beneath it as it rises. Once the plume head nears the top

of the mantle, decompression melting generates basaltic magma that may eventually trigger volcanism at the surface. The result is a localized volcanic region a few hundred kilometers across called a **hot spot** (Figure 4.29). More than 100 hot spots have been identified, and most have persisted for millions of years. The land surface around hot spots is often elevated, showing that it is buoyed up by a plume of warm low-density material. Furthermore, by measuring the heat flow in these regions, geologists have determined that the mantle beneath hot spots must be 100–150°C hotter than normal.

The volcanic activity on the island of Hawaii, with its outpourings of basaltic lava, is certainly the result of hot-spot volcanism. Where a mantle plume has persisted for long periods of time, a chain of volcanic structures may form as the overlying plate moves over it. In the Hawaiian Islands, hot-spot activity is currently centered on Kilauea. However, over the past 80 million years the same mantle plume generated a chain of volcanic islands (and seamounts) that extend thousands of kilometers from the Big Island in a northwestward direction across the Pacific.

Mantle plumes are also thought to be responsible for the vast outpourings of basaltic lava that create large basalt plateaus such as the Columbia Plateau in the northwestern United States, India's Deccan Plateau, and the Ontong Java Plateau in the western Pacific. The most widely accepted explanation for these eruptions, which emit extremely large volumes of basaltic magma over relatively short time intervals, involves a plume with a substantial-sized head. Sometimes called *superplumes*, these structures may have heads that are hundreds of kilometers in diameter connected to a long, narrow tail rising from the core–mantle boundary (Figure 4.29).

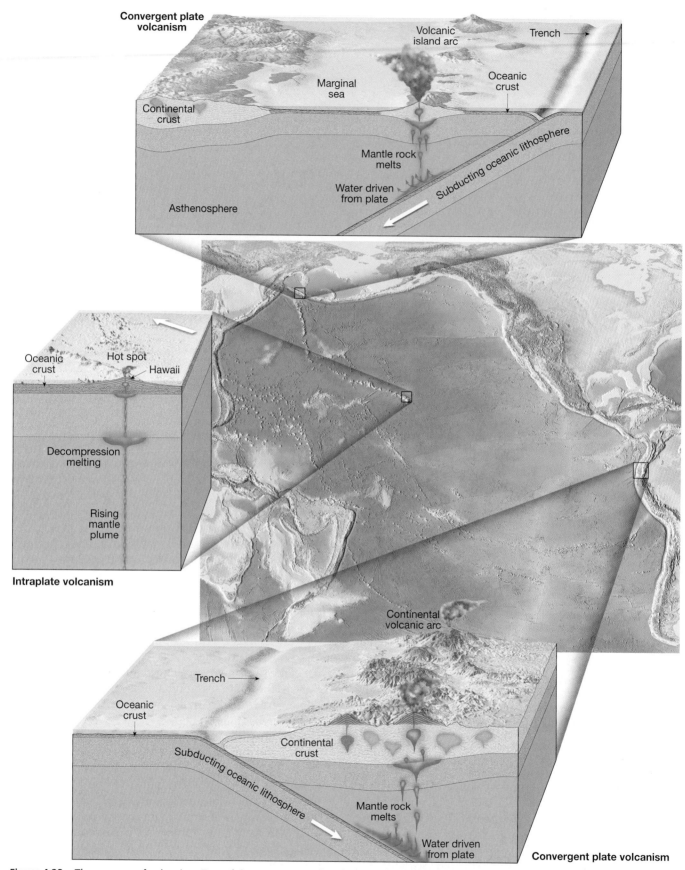

Figure 4.28 Three zones of volcanism. Two of these zones are plate boundaries, and the third is the interior area of the plates.

Divergent plate volcanism

Oceanic crust

Magma chamber

Decompression melting

Asthenosphere

Intraplate volcanism

Flood basalts

Hot spot

Continental crust

Decompression melting

Rising mantle plume

Rift valley

Continental crust

Decompression melting

Divergent plate volcanism

BOX 4.2

Can Volcanoes Change Earth's Climate?

One example of the interplay between different parts of the Earth system is the relationship between volcanic activity and changes in climate. We know that changes in the composition of the atmosphere can have a significant impact on climate. Moreover, we know that volcanic eruptions can emit large quantities of gases and particles into the atmosphere, thus altering its composition. So do volcanic eruptions actually influence Earth's climate?

The idea that explosive eruptions might alter Earth's climate goes back many years. In fact, Benjamin Franklin speculated that the eruption of a large volcano in Iceland might have been responsible for the unusually cold winter of 1783–84. The basic premise is that suspended volcanic material reflects a portion of the incoming solar radiation, which in turn drops temperatures in the lowest layer of the atmosphere (this layer, called the *troposphere*, extends from Earth's surface to a height of about 10 kilometers).

Perhaps the most notable cool period linked to a volcanic event is the "year without a summer" that followed the 1815 eruption of Indonesia's Mount Tambora. The cold temperatures of spring and summer were unprecedented and were followed by an early fall in many parts of the Northern Hemisphere, including eastern North America and Western Europe. The abnormal temperatures and shortened growing season led to crop failures, which in turn led to famine in some regions.

When Mount St. Helens erupted in 1980, there was almost immediate speculation: Can an eruption such as this change our climate? There is no doubt that the large quantity of volcanic ash emitted by the explosive eruption had local and regional effects for a short period (Figure 4.B). Still, studies indicated

Figure 4.B   When Mount St. Helens erupted on May 18, 1980, huge quantities of volcanic ash were blown into the atmosphere. This satellite image was taken less than eight hours after the eruption. The ash cloud had already spread as far as western Montana. Volcanic ash has little long-term impact on global climate because it settles quickly from the air. A more significant factor affecting climate is the quantity of sulfur dioxide gas emitted during an eruption. (Photo courtesy of the National Environmental Satellite Service)

that any longer-term lowering of hemispheric temperatures was negligible. The cooling was so slight, probably less than 0.1°C (0.2°F), that it could not be distinguished from other natural temperature fluctuations.

Two years of monitoring and studies following the 1982 eruption of Mexico's El Chichón volcano indicated that its cooling effect on global mean temperature was greater than that of Mount St. Helens, on the order of 0.3 to 0.5°C (0.5 to 0.9°F). The eruption of El Chichón was

less explosive than the Mount St. Helens blast, so why did it have a greater impact on global temperatures? The reason is that the material emitted by Mount St. Helens was largely fine ash that settled out in a relatively short time. El Chichón, on the other hand, emitted far greater quantities of sulfur dioxide gas (an estimated 40 times more) than Mount St. Helens. This gas combines with water vapor to produce a dense cloud of tiny sulfuric-acid particles in the stratosphere (the atmospheric layer between about 10 and 50 kilometers above the surface). The particles, called *aerosols*, may take up to several years to settle out completely. They lower the troposphere's mean temperature because they reflect solar radiation back to space.

The Philippines volcano, Mount Pinatubo, erupted explosively in June 1991, injecting 25 million to 30 million tons of sulfur dioxide into the stratosphere. The event provided scientists with an opportunity to study the climatic impact of a major explosive volcanic eruption using NASA's spaceborne Earth Radiation Budget Experiment. During the next year the haze of tiny aerosols lowered global temperatures by 0.5°C (0.9°F).

We now understand that volcanic clouds that remain in the stratosphere for a year or more are composed largely of sulfuric-acid droplets and not of volcanic ash, as was once thought. Thus, the volume of ash emitted during an explosive event is *not* an accurate criterion for predicting the global atmospheric effects of an eruption. Furthermore, the preceding examples illustrate that the impact on climate of a single volcanic eruption, no matter how explosive, is relatively small and short-lived.*

*It should be noted that massive outpourings of lava that produced lava plateaus may have released enormous quantities of CO_2, which in turn caused warming in the geologic past.

Upon reaching the base of the lithosphere, the temperature of the material in the plume is estimated to be 200–300°C warmer than the surrounding rock. Thus, as much as 10 to 20 percent of the mantle material making up the plume head rapidly melts. It is this melting that triggers the burst of volcanism that emits volumi-

nous outpourings of lava to form a huge basalt plateau in a matter of a million or so years (Figure 4.29). The comparatively short initial eruptive phase is followed by tens of millions of years of less voluminous activity, as the plume tail slowly rises to the surface. Thus, extending away from most large flood basalt provinces is

Figure 4.29 Model of a mantle plume and associated hot-spot volcanism. **A.** A rising mantle plume with large bulbous head and narrow tail. **B.** Rapid decompression melting of the head of a mantle plume produces vast outpourings of basalt. **C.** Less voluminous activity caused by the plume tail produces a linear volcanic chain on the seafloor.

a chain of volcanic structures, similar to the Hawaiian chain, that terminates over an active hot spot marking the current position of the tail of the plume.

Based on the current state of knowledge, it appears that hot-spot volcanism, with its associated mantle plumes, is responsible for most intraplate volcanism. However, there are some widely scattered volcanic regions located far from any plate boundary that are not linked to hot-spot volcanism. Well-known examples are found in the Basin and Range province of the western United States and northwestern Mexico. We will consider the cause of volcanism in this region in Chapter 17. A few other volcanic regions defy explanation. Thus, the natural world still has some secrets that remain to be explained by future generations of Earth scientists.

The Chapter in Review

- The primary factors that determine the nature of volcanic eruptions include the magma's *composition*, its *temperature*, and the *amount of dissolved gases* it contains. As lava cools, it begins to congeal, and as *viscosity* increases, its mobility decreases. The *viscosity of magma is directly related to its silica content*. Rhyolitic (felsic) lava, with its high silica content (about 70 percent), is very viscous and forms short, thick flows. Basaltic (mafic) lava, with a lower silica content (about 50 percent), is more fluid and may travel a long distance before congealing. Dissolved gases tend to increase the fluidity of magma and, as they expand, provide the force that propels molten rock from the vent of a volcano.

- The materials associated with a volcanic eruption include (1) *lava flows* (*pahoehoe* flows, which resemble twisted braids; and *aa* flows, consisting of rough, jagged blocks; both form from basaltic lavas); (2) *gases* (primarily *water vapor*); and (3) *pyroclastic material* (pulverized rock and lava fragments blown from the volcano's vent, which include *ash, pumice, lapilli, cinders, blocks,* and *bombs.*)

- Successive eruptions of lava from a central vent result in a mountainous accumulation of material known as a *volcano*. Located at the summit of many volcanoes is a steep-walled depression called a *crater*. *Shield cones* are broad, slightly domed volcanoes built primarily of fluid, basaltic lava. *Cinder cones* have steep slopes composed of pyroclastic material. *Composite cones*, or *stratovolcanoes*, are large, nearly symmetrical structures built of interbedded lavas and pyroclastic deposits. Composite cones produce some of the most violent volcanic activity. Often associated with a violent eruption is a *nuée ardente*, a fiery cloud of hot gases infused with incandescent ash that races down steep volcanic slopes. Large composite cones may also generate a type of mudflow known as a *lahar*.

- Most volcanoes are fed by *conduits* or *pipes*. As erosion progresses, the rock occupying the pipe is often more resistant and may remain standing above the surrounding terrain as a *volcanic neck*. The summits of some volcanoes have large, nearly circular depressions called *calderas* that result from collapse following an explosive eruption. Calderas also form on shield volcanos by subterranean drainage from a central magma chamber, and the largest calderas form by the discharge of colossal volumes of silica-rich pumice along ring fractures. Although volcanic eruptions from a central vent are the most familiar, by far the largest amounts of volcanic material are extruded from cracks in the crust called *fissures*. The term *flood basalts* describes the fluid, waterlike, basaltic lava flows that cover an extensive region in the northwestern United States known as the Columbia Plateau. When silica-rich magma is extruded, *pyroclastic flows* consisting largely of ash and pumice fragments usually result.

- Intrusive igneous bodies are classified according to their *shape* and by their *orientation with respect to the host rock*, generally sedimentary rock. The two general shapes are *tabular* (sheetlike) and *massive*. Intrusive igneous bodies that cut across existing sedimentary beds are said to be *discordant*; those that form parallel to existing sedimentary beds are *concordant*.

- *Dikes* are tabular, discordant igneous bodies produced when magma is injected into fractures that cut across rock layers. Tabular, concordant bodies, called *sills*, form when magma is injected along the bedding surfaces of sedimentary rocks. In many respects, sills closely resemble buried lava flows. *Laccoliths* are similar to sills but form from less fluid magmas that collect as lens-shaped masses that arch the overlying strata upward. *Batholiths*, the largest intrusive igneous bodies with surface exposures of more than 100 square kilometers (40 square miles), frequently make up the cores of mountains.

- *Most active volcanoes are associated with plate boundaries.* Active areas of volcanism are found along oceanic ridges where seafloor spreading is occurring (*divergent plate boundaries*), in the vicinity of ocean trenches where one plate is being subducted beneath another (*convergent plate boundaries*), and in the interiors of plates themselves (*intraplate volcanism*). Rising plumes of hot mantle rock are the source of most intraplate volcanism.

Key Terms

aa flow (p. 74)
batholith (p. 89)
caldera (p. 84)
cinder cone (p. 79)
columnar joints (p. 88)
composite cone (p. 79)
concordant (p. 88)
conduit (p. 76)
continental volcanic arc
    (p. 92)
crater (p. 76)
dike (p. 88)

discordant (p. 87)
eruption column (p. 71)
fissure (p. 86)
fissure eruption (p. 86)
flood basalt (p. 86)
fumarole (p. 76)
hot spot (p. 93)
intraplate volcanism (p. 93)
island arc (p. 91)
laccolith (p. 88)
lahar (p. 84)

lava tube (p. 74)
mantle plume (p. 93)
massive (p. 87)
nuée ardente (p. 83)
pahoehoe flow (p. 74)
parasitic cone (p. 76)
pipe (p. 76)
pluton (p. 87)
pyroclastic flow (p. 83)
pyroclastic material (p. 75)
scoria cone (p. 79)

shield volcano (p. 76)
sill (p. 88)
stock (p. 90)
stratovolcano (p. 79)
tabular (p. 87)
vent (p. 76)
viscosity (p. 69)
volcanic island arc (p. 91)
volcanic neck (p. 87)
volcano (p. 76)
volatiles (p. 71)

Questions for Review

1. What event triggered the May 18, 1980, eruption of Mount St. Helens? (See Box 4.1.)

2. List three factors that determine the nature of a volcanic eruption. What role does each play?

3. Why is a volcano that is fed by highly viscous magma likely to be a greater threat than a volcano supplied with very fluid magma?

4. Describe pahoehoe and aa lava.

5. List the main gases released during a volcanic eruption. Why are gases important in eruptions?

6. How do volcanic bombs differ from blocks of pyroclastic debris?

7. Compare a volcanic crater to a caldera.

8. Compare and contrast the three main types of volcanoes (size, composition, shape, and eruptive style).

9. Name a prominent volcano for each of the three types.

10. Briefly compare the eruptions of Kilauea and Parícutin.

11. Contrast the destruction of Pompeii with the destruction of St. Pierre (time frame, volcanic material, and nature of destruction).

12. Describe the formation of Crater Lake. Compare it to the caldera found on shield volcanoes such as Kilauea.

13. What are the largest volcanic structures on Earth?

14. What is Shiprock, New Mexico, and how did it form?

15. How do the eruptions that created the Columbia Plateau differ from eruptions that create volcanic peaks?

16. Extensive pyroclastic flow deposits are most often associated with which volcanic structures?

17. Describe each of the four intrusive features discussed in the text (dike, sill, laccolith, and batholith).

18. Why might a laccolith be detected at Earth's surface before being exposed by erosion?

19. What is the largest of all intrusive igneous bodies? Is it tabular or massive? Concordant or discordant?

20. Volcanism at divergent plate boundaries is associated with which rock type? What causes rocks to melt in these regions?

21. What is the Ring of Fire?

22. What type of plate boundary is associated with the Ring of Fire?

23. Are volcanoes in the Ring of Fire generally described as quiescent or violent? Name a volcano that would support your answer.

24. Describe the process that generates magma along convergent plate boundaries.

25. What is the source of magma for intraplate volcanism?

26. What is meant by hot-spot volcanism?

27. How do geologists identify hot spots other than volcanism?

28. The Hawaiian Islands and Yellowstone are associated with which of the three zones of volcanism? Cascade Range? Flood basalt provinces?

Web Resources

 The *Essentials of Geology* Website uses the resources and flexibility of the Internet to aid in your study of the topics in this chapter. Written and developed by geology instructors, this site will help improve your understanding of geology. Visit **www.prenhall.com/lutgens** and click on the cover of *Essentials of Geology, 8/e* to find:

- On-line review quizzes.
- Web-based critical thinking and writing exercises.
- Links to chapter-specific Web resources.
- Internet-wide key term searches.

www.prenhall.com/lutgens

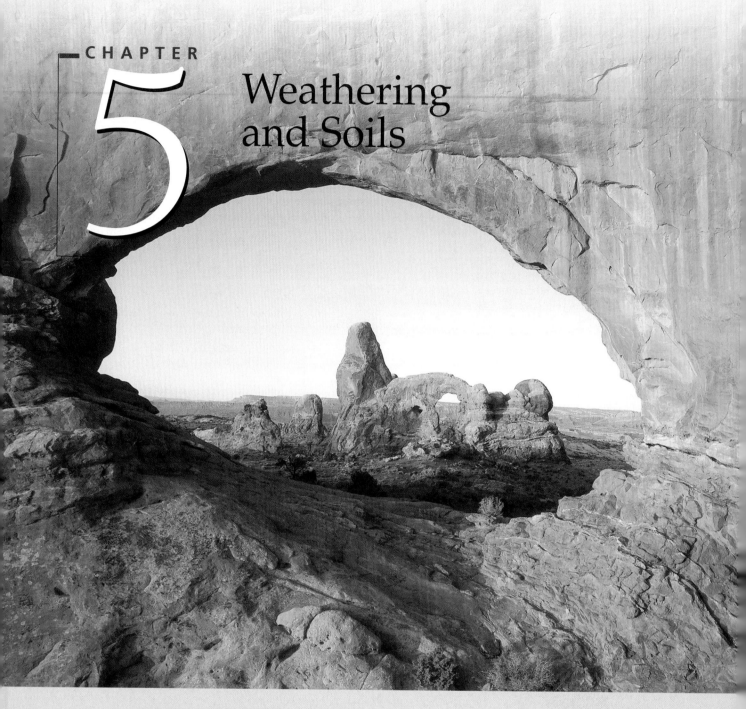

5 Weathering and Soils

Focus on Learning

To assist you in learning the important concepts in this chapter, you will find it helpful to focus on the following questions:

- What are Earth's external processes, and what roles do they play in the rock cycle?

- What are the two types of weathering? In what ways are they different?

- What factors determine the rate at which rock weathers?

- What is soil? What are the factors that control soil formation?

- What factors influence natural rates of soil erosion? What impact have humans had?

- How is weathering related to the formation of ore deposits?

Differential weathering is responsible for much of the spectacular scenery in Arches National Park, Utah. (Photo by David Muench)

Earth's surface is constantly changing. Rock is disintegrated and decomposed, moved to lower elevations by gravity, and carried away by water, wind, or ice. In this manner Earth's physical landscape is sculptured. This chapter focuses on the first step of this never-ending process—weathering. What causes solid rock to crumble, and why does the type and rate of weathering vary from place to place? Soil, an important product of the weathering process and a vital resource, is also examined.

Earth's External Processes

 Earth Materials
↳ External vs Internal Processes

Weathering, mass wasting, and erosion are called **external processes** because they occur at or near Earth's surface and are powered by energy from the Sun. External processes are a basic part of the rock cycle because they are responsible for transforming solid rock into sediment.

To the casual observer, the face of Earth may appear to be without change, unaffected by time. In fact, 200 years ago, most people believed that mountains, lakes, and deserts were permanent features of an Earth that was thought to be no more than a few thousand years old. Today, we know that Earth is 4.5 billion years old and that mountains eventually succumb to weathering and erosion, lakes fill with sediment or are drained by streams, and deserts come and go with changes in climate.

Earth is a dynamic body. Some parts of Earth's surface are gradually elevated by mountain building and volcanic activity. These **internal processes** derive their energy from Earth's interior. Meanwhile, opposing external processes are continually breaking rock apart and moving the debris to lower elevations (Figure 5.1). The latter processes include:

1. **Weathering**—the physical breakdown (disintegration) and chemical alteration (decomposition) of rocks at or near Earth's surface.
2. **Mass wasting**—the transfer of rock and soil downslope under the influence of gravity.
3. **Erosion**—the physical removal of material by mobile agents such as water, wind, or ice.

In this chapter we will focus on rock weathering and the products generated by this activity. However, weathering cannot be easily separated from mass wast-

Figure 5.1 Slopes are places where materials are continually moving from higher to lower elevations. Weathering begins the process by attacking the solid rock exposed at the surface. Next, gravity moves the weathered debris downslope. This step, termed mass wasting, may range from a slow and gradual creep to a thundering landslide. Eventually, the material that was once high up the slope reaches the stream at the bottom. The moving water then transports the debris away. (Photo by Carr Clitton)

ing and erosion, because as weathering breaks rocks apart, mass wasting and erosion remove the rock debris. This transport of material by mass wasting and erosion further disintegrates and decomposes the rock.

Weathering

 Earth Materials
↳ Sedimentary Rocks

Weathering goes on all around us, but it seems like such a slow and subtle process that it is easy to underestimate its importance. However, it is worth remembering that weathering is a basic part of the rock cycle and thus a key process in the Earth system.

All materials are susceptible to weathering. Consider, for example, the fabricated product concrete, which closely resembles the sedimentary rock called conglomerate. A newly poured concrete sidewalk has a smooth, fresh, unweathered look. However, not many years later, the same sidewalk will appear chipped, cracked, and rough, with pebbles exposed at the surface. If a tree is nearby, its roots may heave and buckle the concrete as well. The same natural processes that eventually break apart a concrete sidewalk also act to disintegrate rock.

Weathering occurs when rock is mechanically fragmented (disintegrated) and/or chemically altered (decomposed). **Mechanical weathering** is accomplished by physical forces that break rock into smaller and smaller pieces without changing the rock's mineral composition. **Chemical weathering** involves a chemical transformation of rock into one or more new compounds. These two concepts can be illustrated with a piece of paper. The paper can be disintegrated by tearing it into smaller and smaller pieces, whereas decomposition occurs when the paper is set afire and burned.

Why does rock weather? Simply, weathering is the response of Earth materials to a changing environment. For instance, after millions of years of uplift and erosion, the rocks overlying a large intrusive igneous body may be removed, exposing it at the surface. The mass of

crystalline rock, which formed deep below ground where temperatures and pressures are much greater than at the surface, is now subjected to a very different and comparatively hostile surface environment. In response, this rock mass will gradually change. This transformation of rock is what we call *weathering*.

In the following sections, we will discuss the various modes of mechanical and chemical weathering. Although we will consider these two processes separately, keep in mind that they usually work simultaneously in nature.

Mechanical Weathering

 Earth Materials
↳ Sedimentary Rocks

When a rock undergoes *mechanical weathering*, it is broken into smaller and smaller pieces, each retaining the characteristics of the original material. The end result is many small pieces from a single large one. Figure 5.2 shows that breaking a rock into smaller pieces increases the surface area available for chemical attack. Hence, by breaking rocks into smaller pieces, mechanical weathering increases the amount of surface area available for chemical weathering.

In nature, three physical processes are especially important in breaking rocks into smaller fragments: frost wedging, expansion resulting from unloading, and biological activity.

Frost Wedging

Repeated cycles of freezing and thawing represent an important process of mechanical weathering. Liquid water has the unique property of expanding about 9 percent when it freezes. This increase in volume occurs because, as ice forms, the water molecules arrange themselves into a very open crystalline structure. As a result, when water freezes, it expands and exerts a tremendous outward force. This can be verified by fill-

Figure 5.2 Chemical weathering can occur only to those portions of a rock that are exposed to the elements. Mechanical weathering breaks rock into smaller and smaller pieces, thereby increasing the surface area available for chemical attack.

4 square units ×
6 sides ×
1 cube =
_____
24 square units

1 square unit ×
6 sides ×
8 cubes =
_____
48 square units

.25 square unit ×
6 sides ×
64 cubes =
_____
96 square units

ing a container with water and freezing it. The formation of ice will rupture the container.

In nature, water works its way into cracks in rock and, upon freezing, expands and enlarges these openings. After many freeze-thaw cycles, the rock is broken into angular fragments. This process is appropriately called **frost wedging** (Figure 5.3). Frost wedging is most

Figure 5.3 Frost wedging. As water freezes, it expands, exerting a force great enough to break rock. When frost wedging occurs in a setting such as this, the broken rock fragments fall to the base of the cliff and create a cone-shaped accumulation known as talus. (Photo by Tom & Susan Bean, Inc.)

pronounced in mountainous regions where a daily freeze-thaw cycle often exists. Here, sections of rock are wedged loose and may tumble into large piles called **talus slopes** that often form at the base of steep rock outcrops (Figure 5.3).

Unloading

When large masses of igneous rock, particularly granite, are exposed by erosion, concentric slabs begin to break loose. The process generating these onionlike layers is called **sheeting.** It is thought that this occurs, at least in part, because of the great reduction in pressure when the overlying rock is eroded away, a process called *unloading.* Accompanying this unloading, the outer layers expand more than the rock below and thus separate from the rock body (Figure 5.4). Continued weathering eventually causes the slabs to separate and spall off, creating **exfoliation domes.** Excellent examples of exfoliation domes include Stone Mountain, Georgia, and Half Dome (Figure 5.4C) and Liberty Cap in Yosemite National Park.

Deep underground mining provides us with another example of how rocks behave once the confining pressure is removed. Large rock slabs sometimes explode off the walls of newly cut mine tunnels because of the abruptly reduced pressure. Evidence of this type, plus the fact that fracturing occurs parallel to the floor of a quarry when large blocks of rock are removed, strongly supports the process of unloading as the cause of sheeting.

Biological Activity

Weathering is also accomplished by the activities of organisms, including plants, burrowing animals, and humans. Plant roots in search of minerals and water grow into fractures, and as the roots grow, they wedge the rock apart (Figure 5.5). Burrowing animals further break down the rock by moving fresh material to the surface, where physical and chemical processes can more effectively attack it. Decaying organisms also produce acids, which contribute to chemical weathering. Where rock has been blasted in search of minerals or for road construction, the impact of humans is particularly noticeable.

Chemical Weathering

 Earth Materials
↳ Sedimentary Rocks

Chemical weathering involves the complex processes that break down rock components and internal structures

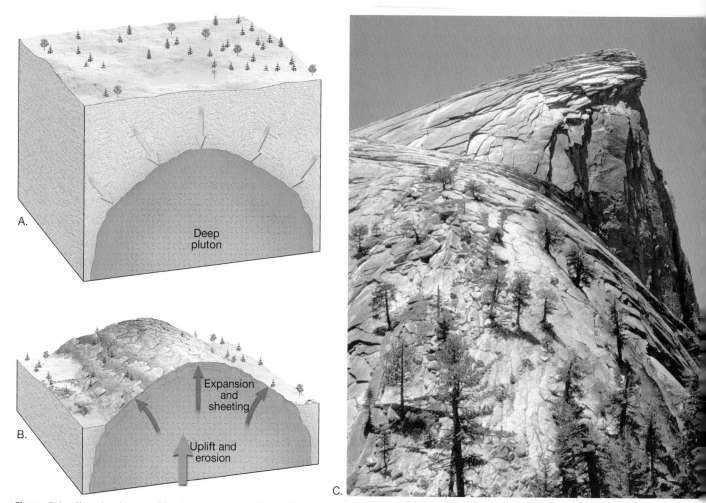

Figure 5.4 Sheeting is caused by the expansion of crystalline rock as erosion removes the overlying material. When the deeply buried pluton in **A** is exposed at the surface following uplift and erosion in **B**, the igneous mass fractures into thin slabs. The photo in **C** is of the summit of Half Dome in Yosemite National Park, California. It is an exfoliation dome and illustrates the onionlike layers created by sheeting. (Photo by Breck P. Kent)

of minerals. Such processes convert the constituents to new minerals or release them to the surrounding environment. During this transformation, the original rock decomposes into substances that are stable in the surface environment. Consequently, the products of chemical weathering will remain essentially unchanged as long as they remain in an environment similar to the one in which they formed.

Water and Carbonic Acid

Water is by far the most important agent of chemical weathering. Although pure water is nonreactive, a small amount of dissolved material is generally all that is needed to activate it. Oxygen dissolved in water will oxidize some materials. For example, when an iron nail is found in moist soil, it will have a coating of rust (iron oxide), and if the time of exposure has been long, the nail will be so weak that it can be broken as easily as a toothpick. When rocks containing iron-rich minerals ox-

idize, a yellow to reddish-brown rust will appear on the surface (Figure 5.6).

Carbon dioxide (CO_2) dissolved in water (H_2O) forms carbonic acid (H_2CO_3), the same weak acid produced when soft drinks are carbonated. Rain dissolves some carbon dioxide as it falls through the atmosphere, and additional amounts released by decaying organic matter are acquired as the water percolates through the soil. Carbonic acid ionizes to form the very reactive hydrogen ion (H^+) and the bicarbonate ion (HCO_3^-).

How Granite Weathers

To illustrate how rock chemically weathers when attacked by carbonic acid, we will consider the weathering of granite, the most abundant continental rock. Recall that granite consists mainly of quartz and potassium feldspar. The weathering of the potassium feldspar component of granite takes place as follows:

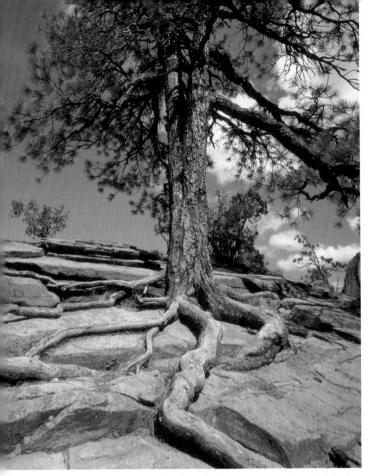

Figure 5.5 Root wedging widens fractures in rock and aids the process of mechanical weathering. (Photo by Tom and Susan Bean, Inc.)

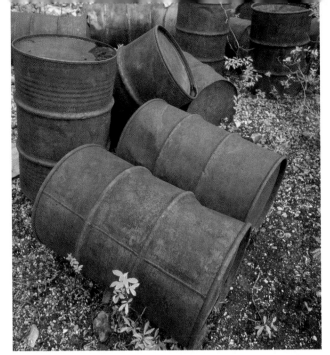

Figure 5.6 Iron reacts with oxygen to form iron oxide as seen on these rusted barrels. (Photo by Stephen J. Krasemann/DRK Photo)

$$2\,KAlSi_3O_8 \;+\; 2(H^+ + HCO_3^-) \;+\; H_2O \rightarrow$$

potassium   carbonic acid   water
feldspar

$$Al_2Si_2O_5(OH)_4 \;+\; 2K^+ \;+\; 2HCO_3^- \;+\; 4SiO_2$$

clay mineral   potassium   bicarbonate   silica
ion   ion

in solution

In this reaction, the hydrogen ions (H^+) attack and replace potassium ions (K^+) in the feldspar structure, thereby disrupting the crystalline network. Once removed, the potassium is available as a nutrient for plants or becomes the soluble salt potassium bicarbonate $(KHCO_3)$, which may be incorporated into other minerals or carried to the ocean in dissolved form by streams.

The most abundant products of the chemical breakdown of feldspar are residual clay minerals. Clay minerals are the end products of weathering and are very stable under surface conditions. Consequently, clay minerals make up a high percentage of the inorganic material in soils. Moreover, the most abundant sedimentary rock, shale, contains a high proportion of clay minerals.

In addition to the formation of clay minerals during the weathering of feldspar, some silica is removed from the feldspar structure and is carried away by groundwater (water beneath Earth's surface). This dissolved silica will eventually precipitate to produce nodules of chert or flint, or it will fill in the pore spaces between sediment grains, or it will be carried to the ocean, where microscopic animals will remove it from the water to build hard silica shells.

To summarize, the weathering of potassium feldspar generates a residual clay mineral, a soluble salt (potassium bicarbonate), and some silica, which enters into solution.

Quartz, the other main component of granite, is *very resistant* to chemical weathering; it remains substantially unaltered when attacked by weak acidic solutions. As a result, when granite weathers, the feldspar crystals dull and slowly turn to clay, releasing the once interlocked quartz grains, which still retain their fresh, glassy appearance. Although some quartz remains in the soil, much is eventually transported to the sea or to other sites of deposition, where it becomes the main constituent of such features as sandy beaches and sand dunes. In time these quartz grains may become lithified to form the sedimentary rock sandstone.

Weathering of Silicate Minerals

Table 5.1 lists the weathered products of some of the most common silicate minerals. Remember that silicate minerals make up most of Earth's crust and that these minerals are composed essentially of only eight elements. When chemically weathered, silicate minerals yield sodium, calcium, potassium, and magnesium ions

Table 5.1 Products of weathering.

Mineral	Residual Products	Material in Solution
Quartz	Quartz grains	Silica
Feldspars	Clay minerals	Silica, K^+, Na^+, Ca^{2+}
Amphibole (hornblende)	Clay minerals Limonite Hematite	Silica, Ca^{2+}, Mg^{2+}
Olivine	Limonite Hematite	Silica, Mg^{2+}

that form soluble products, which may be removed by groundwater. The element iron combines with oxygen, producing relatively insoluble iron oxides, which give soil a reddish-brown or yellowish color. Under most conditions the three remaining elements—aluminum, silicon, and oxygen—join with water to produce residual clay minerals. However, even the highly insoluble clay minerals are very slowly removed by subsurface water.

Spheroidal Weathering

In addition to altering the internal structure of minerals, chemical weathering also causes physical changes. For instance, when angular rock masses are attacked by water that enters along joints, the rocks tend to take on a spherical shape. Gradually the corners and edges of the angular blocks become more rounded. The corners are attacked most readily because of their greater surface area, as compared to the edges and faces. This process, called **spheroidal weathering,** gives the weathered rock a more rounded or spherical shape (Figure 5.7A).

Sometimes during the formation of spheroidal boulders, successive shells separate from the rock's main body (Figure 5.7B). Eventually the outer shells break off, allowing the chemical-weathering activity to penetrate deeper into the boulder. This spherical scaling results because, as the minerals in the rock weather to clay, they increase in size through the addition of water to their structure. This increased bulk exerts an outward force that causes concentric layers of rock to break loose and fall off. Hence, chemical weathering does produce forces great enough to cause mechanical weathering.

This type of spheroidal weathering, in which shells spall off, should not be confused with the phenomenon of sheeting discussed earlier. In sheeting, the fracturing occurs as a result of unloading, and the rock layers that separate from the main body are largely unaltered at the time of separation.

Rates of Weathering

Several factors influence the type and rate of rock weathering. We have already seen how mechanical weathering affects the rate of weathering. By breaking

A.

B.

Figure 5.7 A. Spheroidal weathering is evident in this exposure of granite in California's Joshua Tree National Monument. Because the rocks are attacked more vigorously on the corners and edges, they take on a spherical shape. The lines visible in the rock are called *joints*. Joints are important rock structures that allow water to penetrate and start the weathering process long before the rock is exposed. (Photo by E. J. Tarbuck) **B.** Sometimes successive shells are loosened as the weathering process continues to penetrate ever deeper into the rock. (Photo by Martin Schmidt, Jr.)

rock into smaller pieces, the amount of surface area exposed to chemical weathering is increased. Other important factors examined here include rock characteristics and climate.

Rock Characteristics

Rock characteristics encompass all of the chemical traits of rocks, including mineral composition and solubility. In addition, any physical features such as joints (cracks) can be important because they influence the ability of water to penetrate rock.

The variations in weathering rates due to the mineral constituents can be demonstrated by comparing old headstones made from different rock types. Headstones of granite, which are composed of silicate minerals, are relatively resistant to chemical weathering. We can see this by examining the inscriptions on the headstones shown in Figure 5.8. In contrast, the marble headstone shows signs of extensive chemical alteration over a relatively short period. Marble is composed of calcite (calcium carbonate), which readily dissolves even in a weakly acidic solution.

The silicates, the most abundant mineral group, weather in essentially the same order as their order of crystallization. By examining Bowen's reaction series (see Figure 3.16, p. 59), we see that olivine crystallizes first and is therefore the least resistant to chemical weathering, whereas quartz, which crystallizes last, is the most resistant.

Climate

Climatic factors, particularly temperature and moisture, are crucial to the rate of rock weathering. One important example from mechanical weathering is that the frequency of freeze-thaw cycles greatly affects the amount of frost wedging. Temperature and moisture also exert a strong influence on rates of chemical weathering and on the kind and amount of vegetation present. Regions with lush vegetation generally have a thick mantle of soil rich in decayed organic matter from which chemically active fluids such as carbonic and humic acids are derived.

The optimum environment for chemical weathering is a combination of warm temperatures and abundant moisture. In polar regions chemical weathering is ineffective because frigid temperatures keep the available moisture locked up as ice, whereas in arid regions there is insufficient moisture to foster rapid chemical weathering.

Human activities can influence the composition of the atmosphere, which in turn can impact the rate of chemical weathering. One well-known example is acid rain (Figure 5.9).

Differential Weathering

Masses of rock do not weather uniformly. Take a moment to look back at the photo of Shiprock, New Mexico, in Figure 4.22 (p. 88). The durable volcanic neck protrudes high above the surrounding terrain. A glance at the chapter-opening photo shows an additional example of this phenomenon, called **differential weathering.** The results vary in scale from the rough, uneven surface of the marble headstone in Figure 5.8 to the boldly sculpted exposures in Bryce Canyon (Figure 5.10).

Many factors influence the rate of rock weathering. Among the most important are variations in the composition of the rock. More resistant rock protrudes as ridges or pinnacles, or as steeper cliffs on an irregular hillside (see Figure 6.3, p. 125). The number and spacing of joints can also be a significant factor (see Figure 5.7A). Differential weathering and subsequent erosion is responsible for creating many unusual and sometimes spectacular rock formations and landforms.

Figure 5.8 An examination of headstones reveals the rate of chemical weathering on diverse rock types. The granite headstone (left) was erected four years before the marble headstone (right). The inscription date of 1872 on the marble monument is nearly illegible. (Photos by E. J. Tarbuck)

Figure 5.9 Acid rain accelerates the chemical weathering of stone monuments and structures, including this building facade in Leipzig, Germany. As a consequence of burning large quantities of coal and petroleum, more than 40 million tons of sulfur and nitrogen oxides are released into the atmosphere each year in the United States. Through a series of complex chemical reactions, some of these pollutants are converted into acids that then fall to Earth's surface as rain or snow. (Photo by Doug Plummer/Photo Researchers, Inc.)

Figure 5.10 Differential weathering is illustrated by these sculpted rock pinnacles in Bryce National Park, Utah. (Photo by Ray Mathis/The Stock Market)

Soil

Soil covers most land surfaces. Along with air and water, it is one of our most indispensable resources (Figure 5.11). Also like air and water, soil is taken for granted by many of us. The following quote helps put this vital layer in perspective.

> Science, in recent years, has focused more and more on the Earth as a planet, one that for all we know is unique where a thin blanket of air, a thinner film of water, and the thinnest veneer of soil combine to support a web of life of wondrous diversity in continuous change.*

Soil has accurately been called "the bridge between life and the inanimate world." All life—the entire biosphere—owes its existence to a dozen or so elements that must ultimately come from Earth's crust. Once weathering and other processes create soil, plants carry out the intermediary role of assimilating the necessary elements and making them available to animals, including humans.

An Interface in the Earth System

When Earth is viewed as a system, soil is referred to as an interface—a common boundary where different parts of a system interact. This is an appropriate desig-

*Jack Eddy. "A Fragile Seam of Dark Blue Light," in *Proceedings of the Global Change Research Forum*, U.S. Geological Survey Circular 1086, 1993, p. 15.

nation because soil forms where the solid Earth, the atmosphere, the hydrosphere, and the biosphere meet. Soil is a material that develops in response to complex environmental interactions among different parts of the Earth system. Over time, soil gradually evolves to a state of equilibrium or balance with the environment. Soil is dynamic and sensitive to almost every aspect of its surroundings. Thus, when environmental changes occur, such as climate, vegetative cover, and animal (including human) activity, the soil responds. Any such change produces a gradual alteration of soil characteristics until a new balance is reached. Although thinly distributed over the land surface, soil functions as a fundamental interface, providing an excellent example of the integration among many parts of the Earth system.

Figure 5.11 Soil is an essential resource that we often take for granted. Soil is not a living entity, but it contains a great deal of life. Moreover, this complex medium supports nearly all plant life, which, in turn, supports animal life. (Photo by Tom Bean/Corbis/Stock Market)

What Is Soil?

With few exceptions, Earth's land surface is covered by **regolith,** the layer of rock and mineral fragments produced by weathering. Some would call this material soil, but soil is more than an accumulation of weathered debris. **Soil** is a combination of mineral and organic matter, water, and air—that portion of the regolith that supports the growth of plants. Although the proportions of the major components in soil vary, the same four components always are present to some extent (Figure 5.12). About one half of the total volume of a good-quality surface soil is a mixture of disintegrated and decomposed rock (mineral matter) and *humus*, the decayed remains of animal and plant life (organic matter). The remaining half consists of pore spaces among the solid particles where air and water circulate.

Figure 5.12 Composition (by volume) of a soil in good condition for plant growth. Although the percentages vary, each soil is composed of mineral and organic matter, water, and air.

Although the mineral portion of the soil is usually much greater than the organic portion, humus is an essential component. In addition to being an important source of plant nutrients, humus enhances the soil's ability to retain water. Because plants require air and water to live and grow, the portion of the soil consisting of pore spaces that allow for the circulation of these fluids is as vital as the solid soil constituents.

Soil water is far from "pure" water; instead, it is a complex solution containing many soluble nutrients. Soil water not only provides the necessary moisture for the chemical reactions that sustain life, it also supplies plants with nutrients in a form they can use. The pore spaces that are not filled with water contain air. This air is the source of necessary oxygen and carbon dioxide for most microorganisms and plants that live in the soil.

Controls of Soil Formation

Soil is the product of the complex interplay of several factors. The most important of these are parent material, time, climate, plants, animals, and slope. Although all of these factors are interdependent, their roles will be examined separately.

Parent Material

The source of the weathered mineral matter from which soils develop is called the **parent material** and is a major factor influencing a newly forming soil. Gradually it undergoes physical and chemical changes as the processes of soil formation progress. Parent material might be the underlying bedrock or it can be a layer of unconsolidated deposits, as in a stream valley. When the parent material is bedrock, the soils are termed *residual soils*. By contrast, those developed on unconsolidated sediment are called *transported soils* (Figure 5.13). Note that transported soils form *in place* on parent materials that have been carried from elsewhere and deposited by gravity, water, wind, or ice.

The nature of the parent material influences soils in two ways. First, the type of parent material affects the rate of weathering and thus the rate of soil formation. (Consider the weathering rates of granite versus marble.) Also, because unconsolidated deposits are already partly weathered and provide more surface area for chemical weathering, soil development on such material usually progresses more rapidly. Second, the chemical makeup of the parent material affects the soil's fertility. This influences the character of the natural vegetation the soil can support.

At one time the parent material was believed to be the primary factor causing differences among soils. Today, soil scientists realize that other factors, especially climate, are more important. In fact, it has been found that similar soils are often produced from different par-

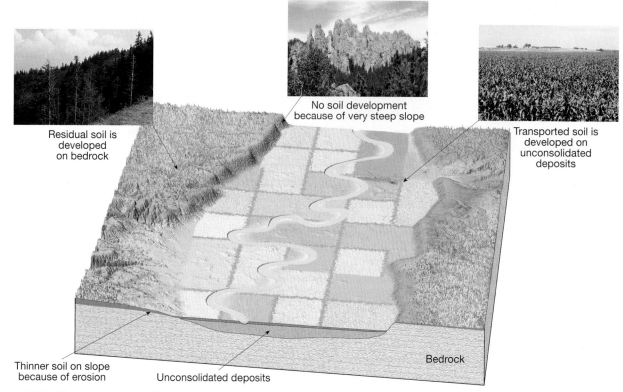

Residual soil is developed on bedrock

No soil development because of very steep slope

Transported soil is developed on unconsolidated deposits

Thinner soil on slope because of erosion

Unconsolidated deposits

Bedrock

Figure 5.13 The parent material for residual soils is the underlying bedrock, whereas transported soils form on unconsolidated deposits. Also note that as slopes become steeper, soil becomes thinner.

ent materials and that dissimilar soils have developed from the same parent material. Such discoveries reinforce the importance of the other soil-forming factors.

Time

Time is an important component of every geological process, and soil formation is no exception. The nature of soil is strongly influenced by the length of time that processes have been operating. If weathering has been going on for a comparatively short time, the parent material determines to a large extent the characteristics of the soil. As weathering processes continue, the influence of parent material on soil is overshadowed by the other soil-forming factors, especially climate. The amount of time required for various soils to evolve cannot be specified, because the soil-forming processes act at varying rates under different circumstances. However, as a rule, the longer a soil has been forming, the thicker it becomes and the less it resembles the parent material.

Climate

Climate is the most influential control of soil formation. Just as temperature and precipitation are the climatic elements that influence people the most, so too are they the elements that exert the strongest impact on soil formation. Variations in temperature and precipitation determine whether chemical or mechanical weathering predominates. They also greatly influence the rate and

depth of weathering. For instance, a hot, wet climate may produce a thick layer of chemically weathered soil in the same amount of time that a cold, dry climate produces a thin mantle of mechanically weathered debris. Also, the amount of precipitation influences the degree to which various materials are removed (leached) from the soil, thereby affecting soil fertility. Finally, climatic conditions are important factors controlling the type of plant and animal life present.

Plants and Animals

The biosphere plays a vital role in soil formation. The types and abundance of organisms present have a strong influence on the physical and chemical properties of a soil. In fact, for well-developed soils in many regions, the significance of natural vegetation in influencing soil type is frequently implied in the description used by soil scientists. Such phrases as *prairie soil, forest soil,* and *tundra soil* are common.

Plants and animals furnish organic matter to the soil. Certain bog soils are composed almost entirely of organic matter, whereas desert soils may contain only a tiny percentage. Although the quantity of organic matter varies substantially among soils, it is the rare soil that completely lacks it.

The primary source of organic matter is plants, although animals and the uncountable microorganisms also contribute. When organic matter decomposes, important nutrients are supplied to plants, as well as to animals and microorganisms living in the soil. Consequently, soil fertility depends in part on the amount of organic matter present. Furthermore, the decay of plant and animal remains causes the formation of various organic acids. These complex acids hasten the weathering process. Organic matter also has a high water-holding ability and thus aids water retention in a soil.

Microorganisms, including fungi, bacteria, and single-celled protozoa, play an active role in the decay of plant and animal remains. The end product is humus, a material that no longer resembles the plants and animals from which it was formed. In addition, certain microorganisms aid soil fertility because they have the ability to convert atmospheric nitrogen into soil nitrogen.

Earthworms and other burrowing animals act to mix the mineral and organic portions of a soil. Earthworms, for example, feed on organic matter and thoroughly mix soils in which they live, often moving and enriching many tons per acre each year. Burrows and holes also aid the passage of water and air through the soil.

Slope

The slope of the land can vary greatly over short distances. Such variations, in turn, can lead to widely varied local soil types. Many of the differences exist because slope has a significant impact on the amount of erosion and the water content of soil.

On steep slopes, soils are often poorly developed. In such situations little water can soak in; as a result, soil moisture may be insufficient for vigorous plant growth. Further, because of accelerated erosion on steep slopes, the soils are thin or nonexistent (Figure 5.13).

In contrast, poorly drained and waterlogged soils in bottomlands have a much different character. Such soils are usually thick and dark. The dark color results from generous organic matter that accumulates because saturated conditions retard the decay of vegetation. The optimum terrain for soil development is a flat-to-undulating upland surface. Here we find good drainage, minimum erosion, and sufficient infiltration of water into the soil.

Slope orientation, or the direction the slope is facing, also is significant. In the midlatitudes of the Northern Hemisphere, a south-facing slope receives a great deal more sunlight than a north-facing slope. In fact, a steep north-facing slope may receive no direct sunlight at all. The difference in the amount of solar radiation received causes substantial differences in soil temperature and moisture, which in turn may influence the nature of the vegetation and the character of the soil.

Although we have dealt separately with each of the soil-forming factors, remember that all work together to form soil. No single factor is responsible for a soil's character. Rather, it is the combined influence of parent material, time, climate, plants, animals, and slope that determines this character.

The Soil Profile

Because soil-forming processes operate from the surface downward, variations in composition, texture, structure, and color gradually evolve at varying depths. These vertical differences, which usually become more pronounced as time passes, divide the soil into zones or layers known as **horizons.** If you were to dig a trench in soil, you would see that its walls are layered. Such a vertical section through all of the soil horizons constitutes the **soil profile** (Figure 5.14).

Figure 5.15 presents an idealized view of a well-developed soil profile in which five horizons are identified. From the surface downward, they are designated as *O, A, E, B,* and *C,* respectively. These five horizons are common to soils in temperate regions. The characteristics and extent of development of horizons vary in different environments. Thus, different localities exhibit soil profiles that can contrast greatly with one another.

The *O* horizon consists largely of organic material. This is in contrast to the layers beneath it that consist mainly of mineral matter. The upper portion of the *O* horizon is primarily plant litter such as loose leaves and other organic debris that are still recognizable. By con-

Figure 5.14 A soil profile is a vertical cross-section from the surface down to the parent material. Well-developed soils show distinct layers called *horizons.* (Photo by E. J. Tarbuck)

Figure 5.15 Idealized soil profile from a humid climate in the middle latitudes.

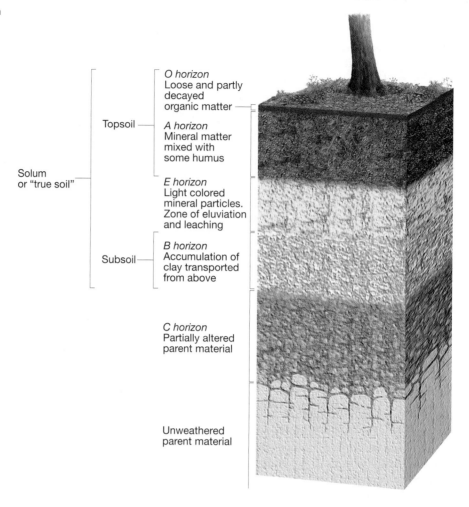

Solum or "true soil"

Topsoil

Subsoil

O horizon
Loose and partly decayed organic matter

A horizon
Mineral matter mixed with some humus

E horizon
Light colored mineral particles. Zone of eluviation and leaching

B horizon
Accumulation of clay transported from above

C horizon
Partially altered parent material

Unweathered parent material

trast, the lower portion of the *O* horizon is made up of partly decomposed organic matter (humus) in which plant structures can no longer be identified. In addition to plants, the *O* horizon is teeming with microscopic life, including bacteria, fungi, algae, and insects. All of these organisms contribute oxygen, carbon dioxide, and organic acids to the developing soil.

Underlying the organic-rich *O* horizon is the *A* horizon. This zone is largely mineral matter, yet biological activity is high and humus is generally present at up to 30 percent in some instances. Together the *O* and *A* horizons make up what is commonly called *topsoil*. Below the *A* horizon, the *E* horizon is a light-colored layer that contains little organic material. As water percolates downward through this zone, finer particles are carried away. This washing out of the fine soil components is termed **eluviation.** Water percolating downward also dissolves soluble inorganic soil components and carries them to deeper zones. This depletion of soluble materials from the upper soil is termed **leaching.**

Immediately below the *E* horizon is the *B* horizon, or *subsoil*. Much of the material removed from the *E* horizon by eluviation is deposited in the *B* horizon, which is often referred to as the *zone of accumulation*. The accumulation of the fine clay particles enhances water

retention in the subsoil. However, in extreme cases clay accumulation can form a very compact and impermeable layer called *hardpan*. The *O, A, E,* and *B* horizons together constitute the **solum,** or true soil. It is in the solum that the soil-forming processes are active and that living roots and other plant and animal life are largely confined.

Below the solum and above the unaltered parent material is the *C* horizon, a layer characterized by partially altered parent material. Whereas the *O, A, E,* and *B* horizons bear little resemblance to the parent material, it is easily identifiable in the *C* horizon. Although this material is undergoing changes that will eventually transform it into soil, it has not yet crossed the threshold that separates regolith from soil.

Soil characteristics and development can vary greatly in different environments. The boundaries between soil horizons may be very distinct, as in Figures 5.14 and 5.15, or the horizons may blend gradually from one to another. A well-developed soil profile indicates that environmental conditions have been relatively stable over an extended time span and that the soil is mature. By contrast, some soils lack horizons altogether.

Such soils are called *immature*, because soil building has been going on for only a short time. Immature soils are also characteristic of steep slopes where erosion continually strips away the soil, preventing full development.

Soil Types

In the following discussion, we will briefly examine some common soil types. As you read, notice that the characteristics of each soil type primarily depend on the prevailing climatic conditions. A summary of the soil characteristics discussed in this section is provided in Table 5.2.

In cold or dry climates, soils are generally very thin and poorly developed. The reasons for this are fairly obvious. Chemical weathering progresses very slowly in such climates, and the scanty plant life yields very little organic matter.

Pedalfer

The term **pedalfer** is derived from the Greek **ped**on, meaning soil, and the chemical symbols **Al** (aluminum) and **Fe** (Iron). Pedalfers are characterized by an accumulation of iron oxides and aluminum-rich clays in the *B* horizon. In midlatitude areas where the annual rainfall exceeds 63 centimeters (25 inches), most of the soluble materials, such as calcium carbonate, are leached from the soil and carried away by underground water. The less-soluble iron oxides and clays are carried from the *E* horizon and deposited in the *B* horizon, giving it a brown to red-brown color. These soils are best developed under forest vegetation where large quantities of decomposing organic matter provide the acid conditions necessary for leaching. In the United States, pedalfers are found east of a line from northwestern Minnesota to south-central Texas.

Pedocal

The term **pedocal** is derived from the Greek **ped**on, meaning soil, and the first three letters of **cal**cite (calcium carbonate). As the name implies, pedocals are characterized by an accumulation of calcium carbonate. This soil type is found in the drier western United States in association with grassland and brush vegetation. Because chemical weathering is less intense in drier areas, pedocals generally contain a smaller percentage of clay materials than do pedalfers.

In the arid and semiarid western states, a calcite-enriched layer called *caliche* may be present in the soils. In these areas, not much rainfall penetrates to great depths. Rather, the rainwater is held by the soil particles near the surface until it evaporates. As a result, the soluble materials, chiefly calcium carbonate, are removed from the uppermost layer and redeposited below, forming the caliche layer.

Laterite

In hot, wet tropical climates, soils called **laterites** may develop. Chemical weathering is intense under such climate conditions, so these soils are usually deeper than soils developing over a similar period in the midlatitudes. Not only does leaching remove the soluble materials such as calcite, but the great quantities of percolating water also remove much of the silica, resulting in the concentration of iron and aluminum oxides in the soil. The iron gives the soil a distinctive orange to red color.

Because bacterial activity is great in the tropics, laterites contain practically no humus. This fact, coupled with the highly leached and bricklike nature of these soils, makes laterites poor for growing crops. Their infertility has been demonstrated repeatedly in tropical

Table 5.2 Summary of soil types.

Climate	Temperate humid (>63 cm rainfall)	Temperate dry (<63 cm rainfall)	Tropical (heavy rainfall)		Extreme arctic or desert
Vegetation	Forest	Grass and brush	Grass and trees		Almost none, so no humus develops
Typical Area	Eastern U.S.	Western U.S.			
Soil Type	Pedalfer	Pedocal	Laterite		
Topsoil	Sandy, light colored; acid	Commonly enriched in calcite; whitish color	Enriched in iron (and aluminum); brick red color	Zones not developed	No real soil forms because there is no organic material. Chemical weathering is very slow.
Subsoil	Enriched in aluminum, iron and clay; brown color	Enriched in calcite; whitish color	All other elements removed by leaching		
Remarks	Extreme development in conifer forests, because abundant humus makes groundwater very acidic. Produces light gray soil because of removal of iron	*Caliche* is name applied to the accumulation of calcite	Apparently bacteria destroy humus, so no acid is available to remove iron		

countries where cultivation has been attempted (see Box 5.1).

Soil Erosion

Soils are just a tiny fraction of all Earth materials, yet they are a vital resource. Because soils are necessary for the growth of rooted plants, they are the very foundation of the human life-support system. Just as human ingenuity can increase the agricultural productivity of soils through fertilization and irrigation, soils can be damaged or destroyed by careless activities. Despite their basic role in providing food, fiber, and other basic materials, soils are among our most abused resources.

Perhaps this neglect and indifference has occurred because a substantial amount of soil seems to remain even where soil erosion is serious. Nevertheless, although the loss of fertile topsoil may not be obvious to the untrained eye, it is a growing problem as human activities expand and disturb more and more of Earth's surface.

How Soil Is Eroded

Soil erosion is a natural process. It is part of the constant recycling of Earth materials that we call the *rock cycle*. Once soil forms, erosional forces, especially water and wind, move soil components from one place to another. Every time it rains, raindrops strike the land with surprising force (Figure 5.16). Each drop acts like a tiny bomb, blasting movable soil particles out of their positions in the soil mass. Then, water flowing across the surface carries away the dislodged soil particles. Because the soil is moved by thin sheets of water, this process is termed *sheet erosion*.

After flowing as a thin, unconfined sheet for a relatively short distance, threads of current typically develop, and tiny channels called *rills* begin to form. Still deeper cuts in the soil, known as *gullies*, are created as rills enlarge (Figure 5.17). When normal farm cultivation cannot eliminate the channels, we know the rills have grown large enough to be called gullies. Although most dislodged soil particles move only a short distance during each rainfall, substantial quantities eventually leave the fields and make their way downslope to a stream. Once in the stream channel, these soil particles, which can now be called *sediment*, are transported downstream and eventually deposited.

Rates of Erosion

We know that soil erosion is the ultimate fate of practically all soils. In the past, erosion occurred at slower rates than it does today because more of the land surface was covered and protected by trees, shrubs, grasses,

and other plants. However, human activities such as farming, logging, and construction, which remove or disrupt the natural vegetation, have greatly accelerated the rate of soil erosion. Without the stabilizing effect of plants, the soil is more easily swept away by the wind or carried downslope by sheet wash.

Natural rates of soil erosion vary greatly from one place to another and depend on soil characteristics as well as such factors as climate, slope, and type of vegetation. Over a broad area, erosion caused by surface runoff may be estimated by determining the sediment loads of the streams that drain the region.

When studies of this kind were made on a global scale, they indicated that prior to the appearance of humans, sediment transport by rivers to the ocean amounted to just over 9 billion metric tons per year (1 metric ton = 1000 kilograms). By contrast, the amount of material currently transported to the sea by rivers is about 24 billion metric tons per year, or more than two and a half times the earlier rate.

It is more difficult to measure the loss of soil due to wind erosion. However, the removal of soil by wind is generally much less significant than erosion by flowing water except during periods of prolonged drought. When dry conditions prevail, strong winds can remove large quantities of soil from unprotected fields. Such was the case in the 1930s in the portions of the Great Plains that came to be called the *Dust Bowl* (see Figure 12.8, p. 255).

In many regions the rate of soil erosion is significantly greater than the rate of soil formation. This means that a renewable resource has become nonrenewable in these places. At present, it is estimated that topsoil is eroding faster than it forms on more than one third of the world's croplands. The result is lower productivity, poorer crop quality, reduced agricultural income, and an ominous future.

Sedimentation and Chemical Pollution

Another problem related to excessive soil erosion involves the deposition of sediment. Each year in the United States hundreds of millions of tons of eroded soil are deposited in lakes, reservoirs, and streams. The detrimental impact of this process can be significant. For example, as more and more sediment is deposited in a reservoir, the capacity of the reservoir is reduced, limiting its usefulness for flood control, water supply, and/or hydroelectric power generation. In addition, sedimentation in streams and other waterways can restrict navigation and lead to costly dredging operations.

In some cases soil particles are contaminated with pesticides used in farming. When these chemicals are

BOX 5.1

Laterites and the Clearing of the Rain Forest

Laterites are thick red soils that form in the wet tropics and subtropics. They are the end product of extreme chemical weathering. Because lush tropical rain forests have lateritic soils, we might assume that the soils are fertile, with great potential for agriculture. However, just the opposite is true: Laterites are among the poorest soils for farming. How can this be?

Because laterites develop under rain-forest conditions of high temperature and heavy rainfall, they are severely leached. Leaching destroys fertility because most plant nutrients are removed by the large volume of downward-percolating water. Therefore, even though the vegetation may be dense and luxuriant, the soil itself contains few available nutrients.

Most nutrients that support the rain forest are locked up in the trees themselves. As vegetation dies and decomposes, the roots of the rain-forest trees quickly absorb the nutrients before they are leached from the soil. The nutrients are continuously recycled as trees die and decompose.

Therefore, when forests are cleared to provide land for farming or to harvest the timber, most of the nutrients are removed as well (Figure 5.A). What remains is a soil that contains little to nourish planted crops.

The clearing of rain forests not only removes plant nutrients but also accelerates erosion. When vegetation is present, its roots anchor the soil and its leaves and branches provide a canopy that protects the ground from the full force of the frequent heavy rains. The removal of vegetation also exposes the ground to strong direct sunlight. When baked by the Sun, laterites can harden to a bricklike consistency and become practically impenetrable to water and crop roots. In only a few years, lateritic soils in a freshly cleared area may no longer be cultivable.

The term *laterite* is derived from the Latin word *latere*, meaning brick, and was first applied to the use of this material for brickmaking in India and Cambodia. Laborers simply excavated

Figure 5.A   Clearing the Amazon rain forest in Surinam. The thick lateric soil is highly leached. (Photo by Wesley Bocxe/Photo Researchers)

the soil, shaped it, and allowed it to harden in the Sun. Ancient but still well-preserved structures built of laterite remain standing today in the wet tropics (Figure 5.B). Such structures have withstood centuries of weathering because all of the original soluble materials were already removed from the soil by chemical weathering. Laterites are therefore virtually insoluble and thus very stable.

In summary, laterites are highly leached soils that are the products of extreme chemical weathering in the warm, wet tropics. Although they may be associated with lush tropical rain forests, these soils are unproductive when vegetation is removed. Moreover, when cleared of plants, laterites are subject to accelerated erosion and can be baked to bricklike hardness by the Sun.

Figure 5.B   This ancient temple at Angor Wat, Cambodia, was built of bricks made of laterite. (Photo by R. Ian Lloyd/The Stock Market)

Figure 5.16 When it is raining, millions of water drops are falling at velocities approaching 10 meters per second (35 kilometers per hour). When water drops strike an exposed surface, soil particles may splash as high as 1 meter into the air and land more than a meter away from the point of raindrop impact. Soil dislodged by splash erosion is more easily moved by sheet erosion. (Photo courtesy of U.S. Department of Agriculture)

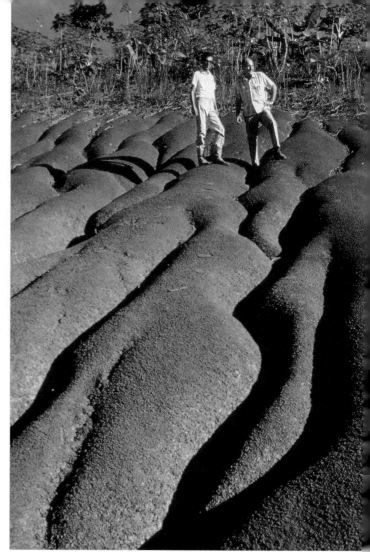

Figure 5.17 Gully erosion in poorly protected soil, southern Colombia. (Photo by Carl Purcell/Photo Researchers, Inc.)

introduced into a lake or reservoir, the quality of the water supply is threatened, and aquatic organisms may be endangered. In addition to pesticides, nutrients found naturally in soils as well as those added by agricultural fertilizers make their way into streams and lakes, where they stimulate the growth of plants. Over a period of time, excessive nutrients accelerate the process by which plant growth leads to the depletion of oxygen and an early death of the lake.

The availability of good soils is critical if the world's rapidly growing population is to be fed. On every continent, unnecessary soil loss is occurring because appropriate conservation measures are not being used. Although it is a recognized fact that soil erosion can never be completely eliminated, soil conservation programs can substantially reduce the loss of this basic resource. Windbreaks (rows of trees), terracing, and plowing along the contours of hills are some of the effective measures, as are special tillage practices and crop rotation.

Weathering Creates Ore Deposits

Weathering creates many important mineral deposits by concentrating minor amounts of metals that are scattered through unweathered rock into economically valuable concentrations. Such a transformation is often termed **secondary enrichment** and takes place in one

of two ways. In one situation, chemical weathering coupled with downward-percolating water removes undesired materials from decomposing rock, leaving the desired elements enriched in the upper zones of the soil. The second way is basically the reverse of the first. That is, the desirable elements that are found in low concentrations near the surface are removed and carried to lower zones, where they are redeposited and become more concentrated.

Bauxite

The formation of *bauxite*, the principal ore of aluminum, is one important example of an ore created as a result of enrichment by weathering processes (Figure 5.18). Although aluminum is the third most abundant element in Earth's crust, economically valuable concentrations of this important metal are not common, because most aluminum is tied up in silicate minerals, from which it is extremely difficult to extract.

Bauxite forms in rainy tropical climates in association with laterites. (In fact, bauxite is sometimes called

Figure 5.18 Bauxite is the principal ore of aluminum and forms as a result of weathering processes under tropical conditions. Its color varies from red or brown to nearly white. (Photo by E. J. Tarbuck)

aluminum laterite.) When aluminum-rich source rocks are subjected to the intense and prolonged chemical weathering of the tropics, most of the common elements, including calcium, sodium, and silicon, are removed by leaching. Because aluminum is extremely insoluble, it becomes concentrated in the soil (as baux-

ite, a hydrated aluminum oxide). Thus, the formation of bauxite depends on climatic conditions in which chemical weathering and leaching are pronounced, plus, of course, the presence of aluminum-rich source rock. Important deposits of nickel and cobalt are also found in laterite soils that develop from igneous rocks rich in other silicate minerals.

Other Deposits

Many copper and silver deposits result when weathering processes concentrate metals that are dispersed through a low-grade primary ore. Usually such enrichment occurs in deposits containing pyrite (FeS_2), the most common and widespread sulfide mineral. Pyrite is important because when it chemically weathers, sulfuric acid forms, which enables percolating waters to dissolve the ore metals. Once dissolved, the metals gradually migrate downward through the primary ore body until they are precipitated. Deposition takes place because of changes that occur in the chemistry of the solution when it reaches the groundwater zone (the zone beneath the surface where all pore spaces are filled with water). In this manner, the small percentage of dispersed metal can be removed from a large volume of rock and redeposited as a higher-grade ore in a smaller volume of rock.

The Chapter in Review

- External processes include (1) *weathering*—the disintegration and decomposition of rock at or near Earth's surface; (2) *mass wasting*—the transfer of rock material downslope under the influence of gravity; and (3) *erosion*—the removal of material by a mobile agent, usually water, wind, or ice. They are called *external processes* because they occur at or near Earth's surface and are powered by energy from the Sun. By contrast, *internal processes*, such as volcanism and mountain building, derive their energy from Earth's interior.

- *Mechanical weathering* is the physical breaking up of rock into smaller pieces. Rocks can be broken into smaller fragments by *frost wedging* (where water works its way into cracks or voids in rock and upon freezing, expands and enlarges the openings), *unloading* (expansion and breaking due to a great reduction in pressure when the overlying rock is eroded away), and *biological activity* (by humans, burrowing animals, plant roots, etc.).

- *Chemical weathering* alters a rock's chemistry, changing it into different substances. Water is by far the most important agent of chemical weathering. Oxygen dissolved in water will *oxidize* iron-rich minerals, while carbon dioxide (CO_2) dissolved in water forms carbonic acid, which attacks and alters rock. The chemical weathering of silicate minerals frequently produces (1) soluble products

containing sodium, calcium, potassium, and magnesium ions, and silica in solution; (2) insoluble iron oxides, including limonite and hematite; and (3) clay minerals.

- The rate at which rock weathers depends on such factors as (1) *particle size*—small pieces generally weather faster than large pieces; (2) *mineral makeup*—calcite readily dissolves in mildly acidic solutions, and silicate minerals that form first from magma are least resistant to chemical weathering; and (3) *climatic factors*, particularly temperature and moisture. Frequently, rocks exposed at Earth's surface do not weather at the same rate. This *differential weathering* of rocks is influenced by such factors as mineral makeup and degree of jointing.

- *Soil* is a combination of mineral and organic matter, water, and air—that portion of the *regolith* (the layer of rock and mineral fragments produced by weathering) that supports the growth of plants. About half of the total volume of a good-quality soil is a mixture of disintegrated and decomposed rock (mineral matter) and *humus* (the decayed remains of animal and plant life); the remaining half consists of pore spaces, where air and water circulate. The most important factors that control soil formation are *parent material, time, climate, plants and animals,* and *slope.*

- Soil-forming processes operate from the surface downward and produce zones or layers in the soil called *horizons.* From the surface downward, the soil horizons are respectively designated as *O* (largely organic matter), *A* (largely mineral matter), *E* (where the fine soil components and soluble materials have been removed by eluviation and leaching), *B* (or *subsoil*, often referred to as the *zone of accumulation*), and *C* (partially altered parent material). Together the *O* and *A* horizons make up what is commonly called the *topsoil.*

- Although there are hundreds of soil types and subtypes worldwide, the three very generic types are (1) *pedalfer* characterized by an accumulation of iron oxides and aluminum-rich clays in the *B* horizon; (2) *pedocal* characterized by an accumulation of calcium carbonate; and (3) *laterite* deep soils that develop in the hot, wet tropics but are poor for growing because they are highly leached.

- Soil erosion is a natural process. It is part of the constant recycling of Earth materials that we call the *rock cycle.* Once in a stream channel, soil particles, which can now be called *sediment,* are transported downstream and eventually deposited. *Rates of soil erosion* vary from one place to another and depend on the soil's characteristics as well as such factors as climate, slope, and type of vegetation.

- Weathering creates ore deposits by concentrating minor amounts of metals into economically valuable deposits. The process, often called *secondary enrichment,* is accomplished by either (1) removing undesirable materials and leaving the desired elements enriched in the upper zones of the soil, or (2) removing and carrying the desirable elements to lower soil zones where they are redeposited and become more concentrated. *Bauxite,* the principal ore of aluminum, is one important ore created as a result of enrichment by weathering processes. In addition, many copper and silver deposits result when weathering processes concentrate metals that were formerly dispersed through low-grade primary ore.

Key Terms

chemical weathering (p. 102)
differential weathering (p. 107)
eluviation (p. 112)
erosion (p. 101)
exfoliation dome (p. 103)
external process (p. 101)

frost wedging (p. 103)
horizon (p. 111)
internal process (p. 101)
laterite (p. 113)
leaching (p. 112)
mass wasting (p. 101)
mechanical weathering (p. 102)

parent material (p. 109)
pedalfer (p. 113)
pedocal (p. 113)
regolith (p. 109)
secondary enrichment (p. 116)
sheeting (p. 103)
soil (p. 109)

soil profile (p. 111)
solum (p. 112)
spheroidal weathering (p. 106)
talus slope (p. 103)
weathering (p. 101)

Questions for Review

1. Describe the role of external processes in the rock cycle.

2. If two identical rocks were weathered, one mechanically and the other chemically, how would the products of weathering for the two rocks differ?

3. How does mechanical weathering add to the effectiveness of chemical weathering?

4. Describe the formation of an exfoliation dome. Give an example of such a feature.

5. Granite and basalt are exposed at the surface in a hot, wet region.
   a. Which type of weathering will predominate?
   b. Which of the rocks will weather most rapidly? Why?

6. Heat speeds up a chemical reaction. Why then does chemical weathering proceed slowly in a hot desert?

7. How is carbonic acid (H_2CO_3) formed in nature? What results when this acid reacts with potassium feldspar?

8. Relate soil to the Earth system.

9. What factors might cause different soils to develop from the same parent material, or similar soils to form from different parent materials?

10. Which control of soil formation is most important? Explain.

11. How can slope affect the development of soil? What is meant by the term *slope orientation?*

12. List the characteristics associated with each of the horizons in a well-developed soil profile. Which of the horizons constitute the solum? Under what circumstances do soils lack horizons?

13. Distinguish between pedalfers and pedocals.

14. What soil type is associated with tropical rain forests? Because this soil supports the growth of lush natural vegetation, is it also excellent for growing crops? Briefly explain. (See Box 5.1.)

15. Is soil erosion a natural process or primarily the result of inappropriate land use by people?

16. List three detrimental effects of soil erosion other than the loss of topsoil from croplands.

17. Name the primary ore of aluminum and describe its formation.

Web Resources

The *Essentials of Geology* Website uses the resources and flexibility of the Internet to aid in your study of the topics in this chapter. Written and developed by geology instructors, this site will help improve your understanding of geology. Visit **www.prenhall.com/lutgens** and click on the cover of *Essentials of Geology, 8/e* to find:

- On-line review quizzes.
- Web-based critical thinking and writing exercises.
- Links to chapter-specific Web resources.
- Internet-wide key term searches.

www.prenhall.com/lutgens

6 Sedimentary Rocks

Focus on Learning

To assist you in learning the important concepts in this chapter, you will find it helpful to focus on the following questions:

- What is a sedimentary rock? About what percentage of Earth's crust is sedimentary?

- How is sediment turned into sedimentary rock?

- What are the two general types of sedimentary rocks, and how does each type form?

- What is the primary basis for distinguishing among various types of detrital sedimentary rocks? What criterion is used to subdivide the chemical rocks?

- What are sedimentary structures? Why are these structures useful to geologists?

- What are the two broad groups of nonmetallic mineral resources?

- Which energy resources are associated with sedimentary rocks?

Sedimentary rocks near Sedona, Arizona. (Photo by Galen Rowell)

Chapter 5 gave you the background needed to understand the origin of sedimentary rocks. Recall that weathering of existing rocks begins the process. Next, gravity and erosional agents (running water, wind, waves, and ice) remove the products of weathering and carry them to a new location where they are deposited. Usually the particles are broken down further during this transport phase. Following deposition, this material, which is now called **sediment,** becomes lithified (turned to rock). In most cases, the sediment is lithified into solid sedimentary rock by the processes of *compaction* and *cementation*.

What Is a Sedimentary Rock?

The products of mechanical and chemical weathering constitute the raw materials for sedimentary rocks. The word *sedimentary* indicates the nature of these rocks, for it is derived from the Latin *sedimentum*, which means "to settle," a reference to solid material settling out of a fluid (water or air). Most, but not all, sediment is deposited in this fashion.

Weathered debris is constantly being swept from bedrock, carried away, and eventually deposited in lakes, river valleys, seas, and countless other places. The particles in a desert sand dune, the mud on the floor of a swamp, the gravel in a stream bed, and even household dust are examples of this never-ending process. Because the weathering of bedrock and the transport and deposition of the weathering products are continuous, sediment is found almost everywhere. As piles of sediment accumulate, the materials near the bottom are compacted. Over long periods, these sediments become cemented together by mineral matter deposited into the spaces between particles, forming solid rock.

Geologists estimate that sedimentary rocks account for only about 5 percent (by volume) of Earth's outer 16 kilometers (10 miles). However, the importance of this group of rocks is far greater than this percentage would imply. If we were to sample the rocks exposed at the surface, we would find that the great majority are sedimentary. Indeed, about 75 percent of all rock outcrops on the continents are sedimentary (Figure 6.1). Therefore, we may think of sedimentary rocks as comprising a relatively thin and somewhat discontinuous layer in the uppermost portion of the crust. This fact is readily understood when we consider that sediment accumulates at Earth's surface.

Because sediments are deposited at Earth's surface, the rock layers that eventually form contain evidence of past events that occurred at the surface. By their very nature, sedimentary rocks contain within them indications of past environments in which their particles were deposited, and in some cases, clues to the mechanisms involved in their transport. Furthermore, it is sedimentary rocks that contain fossils, which are vital tools in the study of the geologic past. Thus, it is largely from this group of rocks that geologists must reconstruct the details of Earth history.

Finally, it should be mentioned that many sedimentary rocks are very important economically. Coal, which is burned to provide a significant portion of U.S. electrical energy, is classified as a sedimentary rock. Our other major energy sources—petroleum and natural gas—are associated with sedimentary rocks. So are major sources of iron, aluminum, manganese, and fertilizer, plus numerous materials essential to the construction industry.

Turning Sediment into Sedimentary Rock: Diagenesis and Lithification

A great deal of change can occur to sediment from the time it is deposited until it becomes a sedimentary rock and is subsequently subjected to the temperatures and pressures that convert it to metamorphic rock. The term **diagenesis** is a collective term for all of the chemical, physical, and biological changes that take place after sediments are deposited and during and after lithification.

Burial promotes diagenesis because as sediments are buried, they are subjected to increasingly higher temperatures and pressures. Diagenesis occurs within the upper few kilometers of Earth's crust at temperatures that are generally less than 150° to 200°C. Beyond this somewhat arbitrary threshold, metamorphism is said to occur.

One example of diagenetic change is *recrystallization*, the development of more stable minerals from less stable ones. It is illustrated by the mineral aragonite, the less stable form of calcium carbonate ($CaCO_3$). Aragonite is secreted by many marine organisms to form shells and other hard parts, such as the skeletal structures produced by corals. In some environments, large quantities of these solid materials accumulate as sediment. As burial takes place, aragonite recrystallizes to the more stable form of calcium carbonate, calcite, the main constituent in the sedimentary rock limestone.

Diagenesis includes **lithification,** the processes by which unconsolidated sediments are transformed into solid sedimentary rocks. Basic lithification processes include compaction and cementation.

The most common physical diagenetic change is **compaction.** As sediment accumulates, the weight of overlying material compresses the deeper sediments. The deeper a sediment is buried, the more it is compacted

Figure 6.1 Sedimentary rocks are exposed at Earth's surface more than igneous and metamorphic rocks. In fact, about 75 percent of all rock outcrops on the continents are sedimentary. Because they contain fossils and other clues about the geologic past, sedimentary rocks are important in the study of Earth history. The sedimentary rocks shown here are in Minnewaska State Park, New York. (Photo by Carr Clifton)

and the firmer it becomes. As the grains are pressed closer and closer, there is considerable reduction in pore space (the open space between particles). For example, when clays are buried beneath several thousand meters of material, the volume of clay may be reduced by as much as 40 percent. As pore space decreases, much of the water that was trapped in the sediments is driven out. Because sands and other coarse sediments are less compressible, compaction is most significant as a lithification process in fine-grained sedimentary rocks.

Cementation is the most important process by which sediments are converted to sedimentary rock. It is a chemical diagenetic change that involves the precipitation of minerals among the individual sediment grains. The cementing materials are carried in solution by water percolating through the open spaces between particles. Through time, the cement precipitates onto the sediment grains, fills the open spaces, and joins the particles. Just as the amount of pore space is reduced during compaction, the addition of cement into a sedimentary deposit reduces its porosity as well.

Calcite, silica, and iron oxide are the most common cements. It is often a relatively simple matter to identify the cementing material. Calcite cement will effervesce with dilute hydrochloric acid. Silica is the hardest cement and thus produces the hardest sedimentary rocks. An orange or dark red color in a sedimentary rock means that iron oxide is present.

Most sedimentary rocks are lithified by means of compaction and cementation. However, some initially form as solid masses of intergrown crystals rather than beginning as accumulations of separate particles that later become solid. Other crystalline sedimentary rocks do not begin that way but are transformed into masses of interlocking crystals sometime after the sediment is deposited.

For example, with time and burial, loose sediment consisting of delicate calcareous skeletal debris may be recrystallized into a relatively dense crystalline limestone. Because crystals grow until they fill all the available space, pore spaces are frequently lacking in crystalline sedimentary rocks. Unless the rocks later de-

122

velop joints and fractures, they will be relatively impermeable to fluids like water and oil.

Types of Sedimentary Rocks

Earth Materials
↳ Sedimentary Rocks

Sediment has two principal sources. First, sediment may be an accumulation of material that originates and is transported as solid particles derived from both mechanical and chemical weathering. Deposits of this type are termed *detrital*, and the sedimentary rocks that they form are called **detrital sedimentary rocks.** The second major source of sediment is soluble material produced largely by chemical weathering. When these dissolved substances are precipitated by either inorganic or organic processes, the material is known as chemical sediment, and the rocks formed from it are called **chemical sedimentary rocks.**

We will now look at each type of sedimentary rock and some examples of each.

Detrital Sedimentary Rocks

Earth Materials
↳ Sedimentary Rocks

Though a wide variety of minerals and rock fragments may be found in detrital rocks, clay minerals and quartz are the chief constituents of most sedimentary rocks in this category. Recall from Chapter 5 that clay minerals are the most abundant product of the chemical weathering of silicate minerals, especially the feldspars. Clays are fine-grained minerals with sheetlike crystalline structures similar to the micas. The other common mineral, quartz, is abundant because it is extremely durable and very resistant to chemical weathering. Thus, when igneous rocks such as granite are attacked by weathering processes, individual quartz grains are freed.

Other common minerals in detrital rocks are feldspars and micas. Because chemical weathering rapidly transforms these minerals into new substances, their presence in sedimentary rocks indicates that erosion and deposition were fast enough to preserve some of the primary minerals from the source rock before they could be decomposed.

Particle size is the primary basis for distinguishing among various detrital sedimentary rocks. Table 6.1 presents the size categories for particles making up detrital rocks. Note that in this context the term *clay* refers only to a particular size and not to the minerals of the same name. Although most clay minerals are

Table 6.1 Particle size classification for detrital rocks.

Size Range (millimeters)	Particle Name	Common Sediment Name	Detrital Rock
>256	Boulder		
64–256	Cobble		Conglomerate
4–64	Pebble	Gravel	or breccia
2–4	Granule		
1/16–2	Sand	Sand	Sandstone
1/256–1/16	Silt		Shale or
<1/256	Clay	Mud	mudstone

of clay size, not all clay-sized sediment consists of clay minerals.

Particle size is not only a convenient method of dividing detrital rocks, but the sizes of the component grains also provide useful information about environments of deposition. Currents of water or air sort the particles by size—the stronger the current, the larger the particle size carried. Gravels, for example, are moved by swiftly flowing rivers as well as by landslides and glaciers. Less energy is required to transport sand; thus, it is common to such features as windblown dunes and some river deposits and beaches. Very little energy is needed to transport clay, so it settles very slowly. Accumulations of these tiny particles are generally associated with the quiet waters of a lake, lagoon, swamp, or certain marine environments.

Common detrital sedimentary rocks, in order of increasing particle size, are shale, sandstone, and conglomerate or breccia. We will now look at each type and how it forms.

Shale

Shale is a sedimentary rock consisting of silt- and clay-sized particles (Figure 6.2). These fine-grained detrital rocks account for well over half of all sedimentary rocks. The particles in these rocks are so small that they cannot be readily identified without great magnification and for this reason make shale more difficult to study and analyze than most other sedimentary rocks.

Much of what can be learned is based on particle size. The tiny grains in shale indicate that deposition occurs as the result of gradual settling from relatively quiet, nonturbulent currents. Such environments include lakes, river floodplains, lagoons, and portions of the deep-ocean basins. Even in these quiet environments, there is usually enough turbulence to keep clay-sized particles suspended almost indefinitely. Consequently, much of the clay is deposited only after the individual particles coalesce to form larger aggregates.

Sometimes the chemical composition of the rock provides additional information. One example is black

Figure 6.2   Shale is a fine-grained detrital rock that is by far the most abundant of all sedimentary rocks. Dark shales containing plant remains are relatively common. (Photo courtesy of E. J. Tarbuck)

shale, which is black because it contains abundant organic matter (carbon). When such a rock is found, it strongly implies that deposition occurred in an oxygen-poor environment such as a swamp, where organic materials do not readily oxidize and decay.

As silt and clay accumulate, they tend to form thin layers commonly referred to as *laminae*. Initially the particles in the laminae are oriented randomly. This disordered arrangement leaves a high percentage of open space (called *pore space*) that is filled with water. However, this situation usually changes with time as additional layers of sediment pile up and compact the sediment below.

During this phase the clay and silt particles take on a more parallel alignment and become tightly packed. This rearrangement of grains reduces the size of the pore spaces and forces out much of the water. Once the grains are pressed closely together, the tiny spaces between particles do not readily permit solutions containing cementing material to circulate. Therefore, shales are often described as being weak because they are poorly cemented and therefore not well lithified. The inability of water to penetrate its microscopic pore spaces explains why shale often forms barriers to the subsurface movement of water and petroleum. Indeed, rock layers that contain groundwater are commonly underlain by shale beds that block further downward movement.* The opposite is true for underground reser-

*The relationship between impermeable beds and the occurrence and movement of groundwater is examined in Chapter 10.

voirs of petroleum. They are often capped by shale beds that effectively prevent oil and gas from escaping to the surface.

It is common to apply the term *shale* to all fine-grained sedimentary rocks, especially in a nontechnical context. However, be aware that there is a more restricted use of the term. In this narrower usage, shale must exhibit the ability to split into thin layers along well-developed, closely spaced planes. This property is termed *fissility*. If the rock breaks into chunks or blocks, the name *mudstone* is applied. Another fine-grained sedimentary rock that, like mudstone, is often grouped with shale but lacks fissillity is *siltstone*. As its name implies, siltstone is composed largely of silt-sized particles and contains less clay-sized material than shale and mudstone.

Although shale is far more common than other sedimentary rocks, it does not usually attract as much notice as other less abundant members of this group. The reason is that shale does not form prominent outcrops as sandstone and limestone often do. Rather, shale crumbles easily and usually forms a cover of soil that hides the unweathered rock below. This is illustrated nicely in the Grand Canyon, where the gentler slopes of weathered shale are quite inconspicuous and overgrown with vegetation, in sharp contrast with the bold cliffs produced by more durable rocks (Figure 6.3).

Although shale beds may not form striking cliffs and prominent outcrops, some deposits have economic value. Certain shales are quarried to obtain raw material for pottery, brick, tile, and china. Moreover, when mixed with limestone, shale is used to make portland cement. In the future, one type of shale, called oil shale, may become a valuable energy resource.

Sandstone

Sandstone is the name given rocks in which sand-sized grains predominate (Figure 6.4). After shale, sandstone is the most abundant sedimentary rock, accounting for approximately 20 percent of the entire group. Sandstones form in a variety of environments and often contain significant clues about their origin, including sorting, particle shape, and composition.

Sorting is the degree of similarity in particle size in a sedimentary rock. For example, if all the grains in a sample of sandstone are about the same size, the sand is considered *well sorted*. Conversely, if the rock contains mixed large and small particles, the sand is said to be *poorly sorted*. By studying the degree of sorting, we can learn much about the depositing current. Deposits of windblown sand are usually better sorted than deposits sorted by wave activity (Figure 6.5). Particles washed by waves are commonly better sorted than materials deposited by streams. Sediment accumulations that exhibit poor sorting usually result when particles are transported for only a relatively short time and then

Figure 6.3 Sedimentary rock layers exposed in the walls of the Grand Canyon, Arizona. Beds of resistant sandstone and limestone produce bold cliffs. By contrast, weaker, poorly cemented shale crumbles and produces a gentler slope of weathered debris in which some vegetation is growing. (Photo by Tom & Susan Bean, Inc.)

rapidly deposited. For example, when a turbulent stream reaches the gentler slopes at the base of a steep mountain, its velocity is quickly reduced, and poorly sorted sands and gravels are deposited.

The shapes of sand grains can also help decipher the history of a sandstone. When streams, winds, or waves move sand and other sedimentary particles, the grains lose their sharp edges and corners and become more rounded as they collide with other particles during transport. Thus, rounded grains likely have been airborne or waterborne. Further, the degree of rounding indicates the distance or time involved in the transportation of sediment by currents of air or water. Highly rounded grains indicate that a great deal of abrasion and hence a great deal of transport has occurred.

Very angular grains, on the other hand, imply two things: that the materials were transported only a short distance before they were deposited, and that some other medium may have transported them. For example, when glaciers move sediment, the particles are usu-

Figure 6.4 Quartz sandstone. After shale, sandstone is the most abundant sedimentary rock. (Photos by E. J. Tarbuck)

5 cm

Close up

125

Figure 6.5 Sorting is the degree of similarity in particle size. The wind-transported sand grains in these dunes are well sorted because they are all practically the same size. Great Sand Dunes National Monument, Colorado. (Photo by David Muench/David Muench Photography, Inc.)

ally made more irregular by the crushing and grinding action of the ice.

In addition to affecting the degree of rounding and the amount of sorting that particles undergo, the length of transport by turbulent air and water currents also influences the mineral composition of a sedimentary deposit. Substantial weathering and long transport lead to the gradual destruction of weaker and less stable minerals, including the feldspars and ferromagnesians. Because quartz is very durable, it is usually the mineral that survives the long trip in a turbulent environment.

The preceding discussion has shown that the origin and history of a sandstone can often be deduced by examining the sorting, roundness, and mineral composition of its constituent grains. Knowing this information allows us to infer that a well-sorted, quartz-rich sandstone consisting of highly rounded grains must be the result of a great deal of transport. Such a rock, in fact, may represent several cycles of weathering, transport, and deposition. We may also conclude that a sandstone containing a significant amount of feldspar and angular grains of ferromagnesian minerals underwent little chemical weathering and transport and was probably deposited close to the source area of the particles.

Owing to its durability, quartz is the predominant mineral in most sandstones. When this is the case, the rock may simply be called *quartz sandstone.* When a sandstone contains appreciable quantities of feldspar, the rock is called *arkose.* In addition to feldspar, arkose usually contains quartz and sparkling bits of mica. The mineral composition of arkose indicates that the grains were derived from granitic source rocks. The particles are generally poorly sorted and angular, which suggests short-distance transport, minimal chemical weathering in a relatively dry climate, and rapid deposition and burial.

A third variety of sandstone is known as *graywacke.* Along with quartz and feldspar, this dark-colored rock contains abundant rock fragments and matrix. *Matrix* refers to the silt- and clay-sized particles found in spaces between larger sand grains. More than 15 percent of graywacke's volume is matrix. The poor sorting and angular grains characteristic of graywacke suggest that the particles were transported only a relatively short distance from their source area and then rapidly deposited. Before the sediment could be reworked and sorted further, it was buried by additional layers of material. Graywacke is frequently associated with submarine deposits made by dense sediment-choked torrents called turbidity currents.

Conglomerate and Breccia

Conglomerate consists largely of gravels (Figure 6.6). As Table 6.1 indicates, these particles may range in size from large boulders to particles as small as garden peas. The particles are commonly large enough to be identified as distinctive rock types; thus, they can be valuable in identifying the source areas of sediments. More often than not, conglomerates are poorly sorted because the openings between the large gravel particles contain sand or mud.

Gravels accumulate in a variety of environments and usually indicate the existence of steep slopes or very turbulent currents. The coarse particles in a conglomerate may reflect the action of energetic mountain streams or result from strong wave activity along a rapidly eroding coast. Some glacial and landslide deposits also contain plentiful gravel.

If the large particles are angular rather than rounded, the rock is called *breccia* (Figure 6.7). Because large particles abrade and become rounded very rapidly during transport, the pebbles and cobbles in a breccia indicate that they did not travel far from their source area before they were deposited. Thus, as with many other sedimentary rocks, conglomerates and breccias contain

Close up

Figure 6.6   Conglomerate is composed primarily of rounded gravel-sized particles. (Photos by E. J. Tarbuck)

Close up

Figure 6.7   When the gravel-sized particles in a detrital rock are angular, the rock is called breccia. (Photos by E. J. Tarbuck)

clues to their history. Their particle sizes reveal the strength of the currents that transported them, whereas the degree of rounding indicates how far the particles traveled. The fragments within a sample identify the source rocks that supplied them.

Chemical Sedimentary Rocks

GEODe Earth Materials
↳ Sedimentary Rocks

In contrast to detrital rocks, which form from the solid products of weathering, chemical sediments derive from material that is carried *in solution* to lakes and seas. This material does not remain dissolved in the water indefinitely, however. Some of it precipitates to form chemical sediments. These become rocks such as limestone, chert, and rock salt.

This precipitation of material occurs in two ways. *Inorganic* processes such as evaporation and chemical activity can produce chemical sediments. *Organic* (life) processes of water-dwelling organisms also form chemical sediments, said to be of **biochemical origin.**

One example of a deposit resulting from inorganic chemical processes is the dripstone that decorates many caves (Figure 6.8). Another is the salt left behind as a body of seawater evaporates. In contrast, many water-dwelling animals and plants extract dissolved mineral matter to form shells and other hard parts. After the organisms die, their skeletons collect by the millions on the floor of a lake or ocean as biochemical sediment (Figure 6.9).

Limestone

Representing about 10 percent of the total volume of all sedimentary rocks, *limestone* is the most abundant chemical sedimentary rock. It is composed chiefly of the mineral calcite ($CaCO_3$) and forms either by inorganic means or as the result of biochemical processes. Regardless of its origin, the mineral composition of all limestone is similar, yet many different types exist. This is true because limestones are produced under a variety of conditions. Those forms having a marine biochemical origin are by far the most common (see Box 6.1).

Coral Reefs.   Corals are one important example of organisms that are capable of creating large quantities of marine limestone. These relatively simple invertebrate animals secrete a calcareous (calcite-rich) external skeleton. Although they are small, corals are capable of creating massive structures called *reefs* (Figure 6.10A).

Figure 6.8 Because many cave deposits are created by the seemingly endless dripping of water over long time spans, they are commonly called *dripstone*. The material being deposited is calcium carbonate ($CaCO_3$) and the rock is a form of limestone called *travertine*. The calcium carbonate is precipitated when some dissolved carbon dioxide escapes from a water drop. (Photo by Clifford Stroud, National Park Service)

Close up

Figure 6.9 This rock, called coquina, consists of shell fragments; therefore, it has a biochemical origin. (Photo by E. J. Tarbuck)

Reefs consist of coral colonies made up of great numbers of individuals that live side by side on a calcite structure secreted by the animals. In addition, calcium carbonate-secreting algae live with the corals and help cement the entire structure into a solid mass. A wide variety of other organisms also live in and near reefs.

Certainly the best-known modern reef is Australia's Great Barrier Reef, 2000 kilometers long, but many lesser reefs also exist. They develop in the shallow, warm waters of the tropics and subtropics equatorward of about 30° latitude. Striking examples exist in the Bahamas and Florida Keys.

Of course, not only modern corals build reefs. Corals have been responsible for producing vast quantities of limestone in the geologic past as well. In the United States, reefs of Silurian age are prominent features in Wisconsin, Illinois, and Indiana. In west Texas and adjacent southeastern New Mexico, a massive reef complex formed during the Permian period is strikingly exposed in Guadalupe Mountains National Park (Figure 6.10B).

Coquina and Chalk. Although most limestone is the product of biological processes, this origin is not always evident, because shells and skeletons may undergo considerable change before becoming lithified into rock. However, one easily identified biochemical limestone is *coquina*, a coarse rock composed of poorly cemented shells and shell fragments (see Figure 6.9). Another less obvious but nevertheless familiar example is *chalk*, a soft, porous rock made up almost entirely of the hard parts of microscopic marine organisms smaller than the head of a pin. Among the most famous chalk deposits are those exposed along the southeast coast of England (Figure 6.11).

Inorganic Limestones. Limestones having an inorganic origin form when chemical changes or high water temperatures increase the concentration of calcium carbonate to the point that it precipitates. *Travertine*, the type of limestone commonly seen in caves, is an example (see Figure 6.8). When travertine is deposited in caves, groundwater is the source of the calcium carbonate. As water droplets become exposed to the air in a cavern, some of the carbon dioxide dissolved in the water escapes, causing calcium carbonate to precipitate.

Another variety of inorganic limestone is *oolitic limestone*. It is a rock composed of small spherical grains called *ooids*. Ooids form in shallow marine waters as tiny seed particles (commonly small shell fragments) are moved back and forth by currents. As the grains are rolled about in the warm water, which is supersaturated with calcium carbonate, they become coated with layer upon layer of the precipitate.

BOX 6.1   The Carbon Cycle and Sedimentary Rocks

To illustrate the movement of material and energy in the Earth system, let us take a brief look at the *carbon cycle* (Figure 6.A). Pure carbon is relatively rare in nature. It is found predominantly in two minerals: diamond and graphite. Most carbon is bonded chemically to other elements to form compounds such as carbon dioxide, calcium carbonate, and the hydrocarbons found in coal and petroleum. Carbon is also the basic building block of life as it readily combines with hydrogen and oxygen to form the fundamental organic compounds that compose living things.

In the atmosphere, carbon is found mainly as carbon dioxide (CO_2). Atmospheric carbon dioxide is significant because it is a greenhouse gas, which means it is an efficient absorber of energy emitted by Earth and thus influences the heating of the atmosphere. Because many of the processes that operate on Earth involve carbon dioxide, this gas is constantly moving into and out of the atmosphere. For example, through the process of photosynthesis, plants absorb carbon dioxide from the atmosphere to produce the essential organic compounds needed for growth. Animals that consume these plants (or consume other animals that eat plants) use these organic compounds as a source of energy and, through the process of respiration, return carbon dioxide to the atmosphere. (Plants also return some CO_2 to the atmosphere via respiration.) Further, when plants die and decay or are burned, this biomass is oxidized, and carbon dioxide is returned to the atmosphere.

Not all dead plant material decays immediately back to carbon dioxide. A small percentage is deposited as sediment. Over long spans of geologic time, considerable biomass is buried with sediment. Under the right conditions,

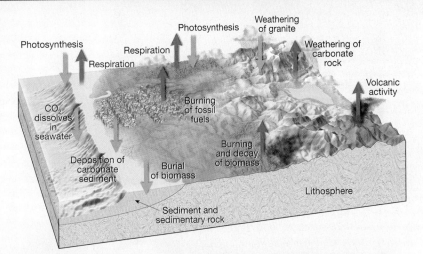

Figure 6.A   Simplified diagram of the carbon cycle, with emphasis on the flow of carbon between the atmosphere and the hydrosphere, lithosphere and biosphere. The colored arrows show whether the flow of carbon is into or out of the atmosphere.

some of these carbon-rich deposits are converted to fossil fuels—coal, petroleum, or natural gas. Eventually some of the fuels are recovered (mined or pumped from a well) and burned to run factories and to fuel our transportation system. One result of fossil-fuel combustion is the release of huge quantities of CO_2 into the atmosphere. Certainly one of the most active parts of the carbon cycle is the movement of CO_2 from the atmosphere to the biosphere and back again.

Carbon also moves from the lithosphere and hydrosphere to the atmosphere and back again. For example, volcanic activity early in Earth's history is thought to be the source of much of the carbon dioxide found in the atmosphere. One way that carbon dioxide makes its way back to the hydrosphere and then to the solid Earth is by first combining with water to form carbonic acid (H_2CO_3), which then attacks the rocks that compose the lithosphere. One product of this chemical weathering of solid rock is the soluble bicarbonate ion ($2HCO_3^-$), which is carried by ground-

water and streams to the ocean. Here water-dwelling organisms extract this dissolved material to produce hard parts of calcium carbonate ($CaCO_3$). When the organisms die, these skeletal remains settle to the ocean floor as biochemical sediment and become sedimentary rock. In fact, the lithosphere is by far Earth's largest depository of carbon, where it is a constituent of a variety of rocks, the most abundant being limestone. Eventually the limestone may be exposed at Earth's surface, where chemical weathering will cause the carbon stored in the rock to be released to the atmosphere as CO_2.

In summary, carbon moves among all four of Earth's major spheres. It is essential to every living thing in the biosphere. In the atmosphere carbon dioxide is an important greenhouse gas. In the hydrosphere, carbon dioxide is dissolved in lakes, rivers, and the ocean. In the lithosphere carbon is contained in carbonate sediments and sedimentary rocks and is stored as organic matter dispersed through sedimentary rocks and as deposits of coal and petroleum.

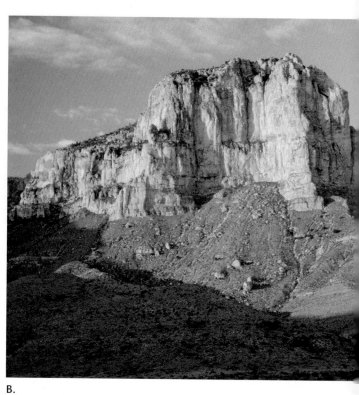

A.

B.

Figure 6.10 A. This modern coral reef is at Bora-Bora in French Polynesia. (Photo by Nancy Sefton/Photo Researchers) **B.** El Capitan Peak, a massive limestone cliff in Guadalupe Mountains National Park, Texas. The rocks here are an exposed portion of a large reef that formed during the Permian period. (Photo by Steve Elmore/The Stock Market)

Dolostone

Closely related to limestone is *dolostone,* a rock composed of the calcium-magnesium carbonate mineral dolomite. Although dolostone can form by direct precipitation from seawater, it is thought that most originates when magnesium in seawater replaces some of the calcium in limestone. The latter hypothesis is reinforced by the fact that there is practically no young

dolostone. Rather, most dolostones are ancient rocks in which there was ample time for magnesium to replace calcium.

Chert

Chert is a name used for a number of very compact and hard rocks made of microcrystalline silica (SiO_2). One

Figure 6.11 The White Chalk Cliffs of Dover. This prominent chalk deposit underlies large portions of southern England as well as parts of northern France. (Photo by ZEFA/UK/Chorley & Handford/The Stock Market)

Figure 6.12 Agate is the banded form of chert. (Photo by Jeff Scovil)

well-known form is *flint,* whose dark color results from the organic matter it contains. *Jasper,* a red variety, gets its bright color from iron oxide. The banded form is usually referred to as *agate* (Figure 6.12).

Chert deposits are commonly found in one of two situations: as irregularly shaped nodules in limestone and as layers of rock. The silica composing many chert nodules may have been deposited directly from water. Such nodules have an inorganic origin. However, it is unlikely that a very large percentage of chert layers was precipitated directly from seawater, because seawater is seldom saturated with silica. Hence, beds of chert are thought to have originated largely as biochemical sediment.

Most water-dwelling organisms that produce hard parts make them of calcium carbonate. But some, such as diatoms and radiolarians, produce glasslike silica skeletons. These tiny organisms are able to extract silica even though seawater contains only tiny quantities of the dissolved material. It is from their remains that most chert beds are believed to originate.

Some bedded cherts occur in association with lava flows and layers of volcanic ash. For these occurrences it is probable that the silica was derived from the decomposition of the volcanic ash and not from biochemical sources. Note that when a hand-specimen of chert is being examined, there are few reliable criteria by which the mode of origin (inorganic versus biochemical) can be determined.

Like glass, most chert has a conchoidal fracture. Its hardness, ease of chipping, and ability to hold a sharp edge made chert a favorite of Native Americans for fashioning points for spears and arrows. Because of chert's durability and extensive use, "arrowheads" are found in many parts of North America.

Evaporites

Very often evaporation is the mechanism triggering deposition of chemical precipitates. Minerals commonly precipitated in this fashion include halite (sodium chloride, NaCl), the chief component of *rock salt,* and gypsum (hydrous calcium sulfate, $CaSO_4 \cdot 2H_2O$), the main ingredient of *rock gypsum.* Both have significant commercial importance. Halite is familiar to everyone as the common salt used in cooking and for seasoning foods. Of course, it has many other uses—from melting ice on roads to making hydrochloric acid—and has been considered important enough that people have sought, traded, and fought over it for much of human history. Gypsum is the basic ingredient of plaster of Paris. This material is used most extensively in the construction industry for wallboard and interior plaster.

In the geologic past, many areas that are now dry land were basins, submerged under shallow arms of a sea that had only narrow connections to the open ocean. Under these conditions, seawater continually moved into the bay to replace water lost by evaporation. Eventually the waters of the bay became saturated and salt deposition began. Such deposits are called **evaporites.**

When a body of seawater evaporates, the minerals that precipitate do so in a sequence determined by their solubility. Less soluble minerals precipitate first, and more soluble minerals precipitate later as salinity increases. For example, gypsum precipitates when about two-thirds to three-quarters of the seawater has evaporated, and halite settles out when nine-tenths of the water has been removed. During the last stages of this process, potassium and magnesium salts precipitate. One of these last-formed salts, the mineral *sylvite,* is mined as a significant source of potassium ("potash") for fertilizer.

On a smaller scale, evaporite deposits can be seen in many desert basins in the western United States. Here, following rains or periods of snowmelt in the mountains, streams flow from the surrounding mountains into an enclosed basin. As the water evaporates, **salt flats** form when dissolved materials are precipitated as a white crust on the ground (Figure 6.13).

Coal

Coal is quite different from other chemical sedimentary rocks. Unlike other rocks in this category, which are calcite- or silica-rich, coal is made mostly of organic matter. Close examination of a piece of coal under a microscope or magnifying glass often reveals plant structures such as leaves, bark, and wood that have been chemically altered but are still identifiable. This supports the conclusion that coal is the end product of large amounts of plant material buried for millions of years.

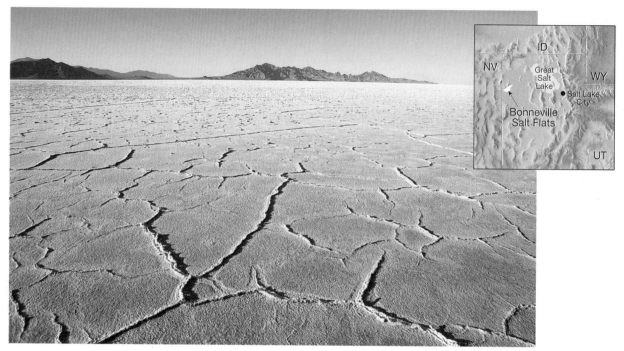

Figure 6.13   The Bonneville salt flats in Utah are a well-known example of evaporite deposits. (Photo by Tom & Susan Bean, Inc.)

The initial stage in coal formation is the accumulation of large quantities of plant remains. However, special conditions are required for such accumulations, because dead plants normally decompose when exposed to the atmosphere. An ideal environment that allows for the buildup of plant material is a swamp. Because stagnant swamp water is oxygen-deficient, complete decay (oxidation) of the plant material is not possible. At various times during Earth's history, such environments have been common. Coal undergoes successive stages of formation. With each successive stage, higher temperatures and pressures drive off impurities and volatiles, as shown in Figure 6.14.

Lignite and bituminous coals are sedimentary rocks, but anthracite is a metamorphic rock. Anthracite forms when sedimentary layers are subjected to the folding and deformation associated with mountain building.

Classification of Sedimentary Rocks

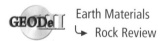 Earth Materials
↳ Rock Review

The classification scheme in Table 6.2 divides sedimentary rocks into two major groups: detrital and chemical. Further, we can see that the main criterion for subdividing the detrital rocks is particle size,

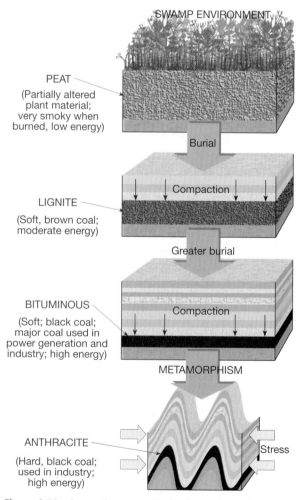

Figure 6.14   Successive stages in the formation of coal.

Table 6.2 Classification of sedimentary rocks.

| | | DETRITAL ROCKS | |
Texture	Sediment Name and Particle Size	Comments	Rock Name
Clastic	Gravel (>2 mm)	Rounded rock fragments	Conglomerate
		Angular rock fragments	Breccia
	Sand (1/16–2 mm)	Quartz predominates	Quartz sandstone
		Quartz with considerable feldspar	Arkose
		Dark color; quartz with considerable feldspar, clay, and rocky fragments	Graywacke
	Mud (<1/16 mm)	Splits into thin layers	Shale
		Breaks into clumps or blocks	Mudstone

| | | CHEMICAL ROCKS | |
Group	Texture	Composition	Rock Name
Inorganic	Clastic or nonclastic	Calcite, $CaCO_3$	Limestone
	Nonclastic	Dolomite, $CaMg(CO_3)_2$	Dolostone
	Nonclastic	Microcrystalline quartz, SiO_2	Chert
	Nonclastic	Halite, $NaCl$	Rock salt
	Nonclastic	Gypsum, $CaSO_4 \cdot 2H_2O$	Rock gypsum
Biochemical	Clastic or nonclastic	Calcite, $CaCO_3$	Limestone
	Nonclastic	Microcrystalline quartz, SiO_2	Chert
	Nonclastic	Altered plant remains	Coal

whereas the primary basis for distinguishing among different rocks in the chemical group is their mineral composition.

As is the case with many (perhaps most) classifications of natural phenomena, the categories presented in Table 6.2 are more rigid than the actual state of nature. In reality, many of the sedimentary rocks classified into the chemical group also contain at least small quantities of detrital sediment. Many limestones, for example, contain varying amounts of mud or sand, giving them a "sandy" or "shaly" quality. Conversely, because practically all detrital rocks are cemented with material that was originally dissolved in water, they too are far from being "pure."

As was the case with the igneous rocks examined in Chapter 3, *texture* is a part of sedimentary rock classification. Two major textures are used in the classification of sedimentary rocks: clastic and nonclastic. The term **clastic** is taken from a Greek word meaning "broken." Rocks that display a clastic texture consist of discrete fragments and particles that are cemented and compacted together. Although cement is present in the spaces between particles, these openings are rarely filled completely. Table 6.2 shows that *all* detrital rocks have a clastic texture. The table also shows that some chemical sedimentary rocks exhibit this texture. For example, coquina, the limestone composed of shells and shell

fragments, is obviously as clastic as conglomerate or sandstone. The same applies for some varieties of oolitic limestone.

Some chemical sedimentary rocks have a **nonclastic** (*crystalline*) texture in which the minerals form a pattern of interlocking crystals. The crystals may be microscopically small or large enough to be visible without magnification. Common examples of rocks with nonclastic textures are evaporites (Figure 6.15). The materials that make up many other nonclastic rocks may actually have originated as detrital deposits. In these instances, the particles probably consisted of shell fragments and other hard parts rich in calcium carbonate or silica. The clastic nature of the grains was subsequently obliterated or obscured because the particles recrystallized when they were consolidated into limestone or chert.

Nonclastic rocks consist of intergrown crystals, and some may resemble igneous rocks, which are also crystalline. The two rock types are usually easy to distinguish because the minerals that make up nonclastic sedimentary rocks are different from those found in most igneous rocks. For example, rock salt, rock gypsum, and some forms of limestone consist of intergrown crystals, but the minerals contained within these rocks (halite, gypsum, and calcite) are seldom associated with igneous rocks.

← 5 cm →

Close up

Figure 6.15 Like other evaporites, this sample of rock salt is said to have a nonclastic texture because it is composed of intergrown crystals. (Photos by E. J. Tarbuck)

Sedimentary Rocks Represent Past Environments

Earth Materials
↳ Sedimentary Rocks

Sedimentary rocks are important in the interpretation of Earth history. By understanding the conditions under which sedimentary rocks form, geologists can often deduce the history of a rock, including information about the origin of its component particles, the method and length of sediment transport, and the nature of the place where the grains eventually came to rest; that is, the environment of deposition. An **environment of deposition** or **sedimentary environment** is simply a geographic setting where sediment is accumulating. Each site is characterized by a particular combination of geologic processes and environmental conditions. Thus, when a series of sedimentary layers are studied, we can see the successive changes in environmental conditions that occurred at a particular place with the passage of time.

At any given time, the geographic setting and environmental conditions of a sedimentary environment

determine the nature of the sediments that accumulate. Therefore, geologists carefully study the sediments in present-day depositional environments because the features they find can also be observed in ancient sedimentary rocks. By applying a thorough knowledge of present-day conditions, geologists attempt to reconstruct the ancient environments and geographical relationships of an area at the time a particular set of sedimentary layers were deposited. Such analyses often lead to the creation of maps depicting the geographic distribution of land and sea, mountains and river valleys, deserts and glaciers, and other environments of deposition. The foregoing description is an excellent example of the application of a fundamental principle of modern geology, namely that "the present is the key to the past."*

Sedimentary environments are commonly placed into one of three categories: continental, marine, or transitional (shoreline). Each category includes many specific subenvironments. Figure 6.16 is an idealized diagram illustrating a number of important sedimentary environments associated with each category. Later, Chapters 9 through 13, as well as portions of Chapter 14, will describe these environments in detail. Each is an area where sediment accumulates and where organisms live and die. Each produces a characteristic sedimentary rock or assemblage that reflects prevailing conditions.

Sedimentary Structures

Earth Materials
↳ Sedimentary Rocks

In addition to variations in grain size, mineral composition, and texture, sediments exhibit a variety of structures. Some, such as graded beds, are created when sediments are accumulating and are a reflection of the transporting medium. Others, such as mud cracks, form after the materials have been deposited and result from processes occurring in the environment. When present, sedimentary structures provide additional information that can be useful in the interpretation of Earth history.

Sedimentary rocks form as layer upon layer of sediment accumulates in various depositional environments. These layers, called **strata,** or **beds,** are probably *the single most common and characteristic feature of sedimentary rocks.* Each stratum is unique. It may be a coarse sandstone, a fossil-rich limestone, a black shale, and so on. When you look at Figure 6.17, or look back through this chapter at Figures 6.1, 6.3, and 6.10B, you will see many such layers, each different from the others. The

*For more on this idea, see "The Birth of Modern Geology" in Chapter 1.

variations in texture, composition, and thickness reflect the different conditions under which each layer was deposited.

The thickness of beds ranges from microscopically thin to tens of meters thick. Separating the strata are **bedding planes,** flat surfaces along which rocks tend to separate or break. Changes in the grain size or in the composition of the sediment being deposited can create bedding planes. Pauses in deposition can also lead to layering because chances are slight that newly deposited material will be exactly the same as previously deposited sediment. Generally, each bedding plane marks the end of one episode of sedimentation and the beginning of another.

Because sediments usually accumulate as particles that settle from a fluid, most strata are originally deposited as horizontal layers. There are circumstances, however, when sediments do not accumulate in horizontal beds. Sometimes when a bed of sedimentary rock is examined, we see layers within it that are inclined to the horizontal. When this occurs, it is called **cross-bedding** and is most characteristic of sand dunes, river deltas, and certain stream channel deposits (Figure 6.18).

Graded beds represent another special type of bedding. In this case the particles within a single sedimentary layer gradually change from coarse at the bottom to fine at the top (Figure 6.19). Graded beds are most characteristic of rapid deposition from water containing sediment of varying sizes. When a current experiences a rapid energy loss, the largest particles settle first, followed by successively smaller grains. The deposition of a graded bed is most often associated with a turbidity current, a mass of sediment-choked water that is denser than clear water and that moves downslope along the bottom of a lake or ocean.*

As geologists examine sedimentary rocks, much can be deduced. A conglomerate, for example, may indicate a high-energy environment, such as a surf zone or rushing stream, where only coarse materials settle out and finer particles are kept suspended (Figure 6.20). If the rock is arkose, it may signify a dry climate where little chemical alteration of feldspar is possible. Carbonaceous shale is a sign of a low-energy, organic-rich environment, such as a swamp or lagoon.

Other features found in some sedimentary rocks also give clues to past environments. Ripple marks are such a feature. **Ripple marks** are small waves of sand that develop on the surface of a sediment layer by the action of moving water or air (Figure 6.21A). The ridges form at right angles to the direction of motion. If the ripple marks were formed by air or water moving in essentially one direction, their form will be asymmetrical.

*More on these currents and graded beds can be found in the section on "Submarine Canyons and Turbidity Currents" in Chapter 14.

These *current ripple marks* will have steeper sides in the downcurrent direction and more gradual slopes on the upcurrent side. Ripple marks produced by a stream flowing across a sandy channel or by wind blowing over a sand dune are two common examples of current ripples. When present in solid rock, they may be used to determine the direction of movement of ancient wind or water currents. Other ripple marks have a symmetrical form. These features, called *oscillation ripple marks,* result from the back-and-forth movement of surface waves in a shallow nearshore environment.

Mud cracks (Figure 6.21B) indicate that the sediment in which they were formed was alternately wet and dry. When exposed to air, wet mud dries out and shrinks, producing cracks. Mud cracks are associated with such environments as tidal flats, shallow lakes, and desert basins.

Fossils, the remains or traces of prehistoric life, are important inclusions in sediment and sedimentary rock (Figure 6.22). They are important tools for interpreting the geologic past. Knowing the nature of the life forms that existed at a particular time helps researchers decipher past environmental conditions. Further, fossils are important time indicators and play a key role in correlating rocks that are of similar ages but are from different places. Fossils will be examined in more detail in Chapter 18.

Nonmetallic Mineral Resources from Sedimentary Rocks

Earth materials that are not used as fuels or processed for the metals they contain are referred to as **nonmetallic mineral resources.** Realize that use of the word "mineral" is very broad in this economic context and is quite different from the geologist's strict definition of mineral found in Chapter 2. Nonmetallic mineral resources are extracted and processed either for the nonmetallic elements they contain or for the physical and chemical properties they possess (Table 6.3). Although these resources have diverse origins, many are sediments or sedimentary rocks.

People often do not realize the importance of nonmetallic minerals, because they see only the products that resulted from their use and not the minerals themselves. That is, many nonmetallics are used up in the process of creating other products. Examples include the fluorite and limestone that are part of the steel-making process, the abrasives required to make a piece of machinery, and the fertilizers needed to grow a food crop.

The quantities of nonmetallic minerals used each year are enormous. Per capita consumption of nonfuel resources in the United States exceeds 11 metric tons, of which 94 percent are nonmetallics (Figure 6.23). Nonmetallic mineral resources are commonly divided into

Figure 6.16 Sedimentary environments are those places where sediment accumulates. Each is characterized by certain physical, chemical, and biological conditions. Because each sediment contains clues about the environment in which it was deposited, sedimentary rocks are important in the interpretation of Earth history. A number of important continental, transitional, and marine sedimentary environments are represented in this idealized diagram. (Photos by E. J. Tarbuck, except alluvial fan, by Martin G. Miller)

Alluvial fans

Glacial deposits

Salt
flat

Playa lake

Stream

Swamp

Floodplain

Delta

Lagoon

Reef

Barrier
island

Figure 6.17 This outcrop of sedimentary strata illustrates the characteristic layering of this group of rocks. (Photo by Tom Till)

two broad groups: *building materials* and *industrial minerals.* Because some substances have many different uses, they are found in both categories. Limestone, perhaps the most versatile and widely used rock of all, is the best example. As a building material, it is used not only as crushed rock and building stone but also in making cement. Moreover, as an industrial mineral, limestone is an ingredient in the manufacture of steel and is used in agriculture to neutralize acidic soils.

Other important building materials include cut stone, aggregate (sand, gravel, and crushed rock), gyp-

sum for plaster and wallboard, clay for tile and bricks, and cement, which is made from limestone and shale. Cement and aggregate go into the making of concrete, a material that is essential to practically all construction.

A wide variety of resources are classified as industrial minerals. In some instances these materials are important because they are sources of specific chemical elements or compounds. Such minerals are used in the manufacture of chemicals and the production of fertilizers. In other cases their importance is related to the physical properties they exhibit. Examples include min-

A.

B.

Figure 6.18 **A.** The cutaway section of this sand dune shows cross-bedding. (Photo by John S. Shelton) **B.** The cross-bedding of this sandstone indicates it was once a sand dune. (Photo by David Muench Photography, Inc.)

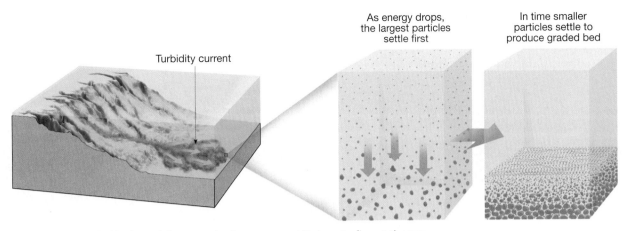

As energy drops, the largest particles settle first

In time smaller particles settle to produce graded bed

Turbidity current

Figure 6.19 Graded beds. Each layer grades from coarse at its base to fine at the top.

erals such as corundum and garnet, which are used as abrasives. Although supplies are generally plentiful, most industrial minerals are not nearly as abundant as building materials.

Moreover, deposits are far more restricted in distribution and extent. As a result, many of these non-metallic resources must be transported considerable distances, which of course adds to their cost. Unlike most building materials, which need a minimum of processing before they are ready to use, many industrial minerals require considerable processing to extract the

Figure 6.20 In a turbulent stream channel, only large particles settle out. Finer sediments remain suspended and continue their downstream journey. (Photo by E. J. Tarbuck)

desired substance at the proper degree of purity for its ultimate use.

Energy Resources from Sedimentary Rocks

Coal, petroleum, and natural gas are the primary fuels of our modern industrial economy. Nearly 90 percent of the energy consumed in the United States today comes from these basic fossil fuels. Although major shortages of oil and gas may not occur for many years, proven reserves are declining. Despite new exploration, even in very remote regions and severe environments, new sources of oil are not keeping pace with consumption.

Unless large, new petroleum reserves are discovered (which is possible but not likely), a greater share of our future needs will eventually have to come from coal and/or from alternative energy sources such as nuclear, solar, wind, tidal, and hydroelectric power. Two fossil-fuel alternatives, tar sands and oil shale, are sometimes mentioned as promising new sources of liquid fuels. In the following sections, we will briefly examine the fuels that have traditionally supplied our energy needs.

Coal

Along with oil and natural gas, coal is commonly called a *fossil fuel*. Such a designation is appropriate, because each time we burn coal we are using energy from the Sun that was stored by plants many millions of years ago. We are indeed burning a "fossil."

Coal has been an important fuel for centuries. In the nineteenth and early twentieth centuries, cheap and plentiful coal powered the Industrial Revolution. By 1900, coal was providing 90 percent of the energy used in the United States. Although still important, coal currently provides only about 25 percent of the energy needs of the United States (Figure 6.24).

Until the 1950s, coal was an important domestic heating fuel as well as a power source for industry. How-

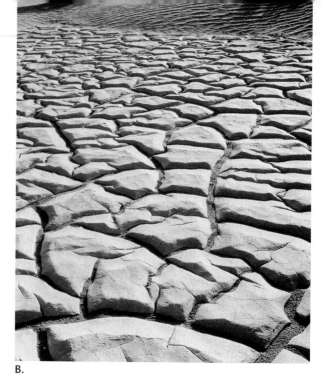

A.

B.

Figure 6.21 **A.** Ripple marks are produced by currents of water or wind. (Photo by Stephen Trimble) **B.** Mud cracks form when wet mud or clay dries out and shrinks. (Photo by Gary Yeowell/Tony Stone Images)

ever, its direct use in the home has been largely replaced by oil, natural gas, and electricity. These fuels are preferred because they are more readily available (delivered via pipes, tanks, or wiring) and cleaner to use.

Nevertheless, coal remains the major fuel used in power plants to generate electricity, and it is therefore indirectly an important source of energy for our homes. More than 70 percent of present-day coal usage is for the generation of electricity. As oil reserves gradually diminish in the years to come, the use of coal will prob-

ably increase. Expanded coal production is possible because the world has enormous reserves and the technology to mine coal efficiently. In the United States, coal fields are widespread and contain supplies that should last for hundreds of years (Figure 6.25).

Table 6.3 Uses of nonmetallic minerals.

Mineral	Uses
Apatite	Phosphorus fertilizers
Asbestos (chrysotile)	Incombustible fibers
Calcite	Aggregate; steelmaking; soil conditioning; chemicals; cement; building stone
Clay minerals (kaolinite)	Ceramics; china
Corundum	Gemstones; abrasives
Diamond	Gemstones; abrasives
Fluorite	Steelmaking; aluminum refining; glass; chemicals
Garnet	Abrasives; gemstones
Graphite	Pencil lead; lubricant; refractories
Gypsum	Plaster of Paris
Halite	Table salt; chemicals; ice control
Muscovite	Insulator in electrical applications
Quartz	Primary ingredient in glass
Sulfur	Chemicals; fertilizer manufacture
Sylvite	Potassium fertilizers
Talc	Powder used in paints, cosmetics, etc.

Figure 6.22 Fossils are important tools for the geologist. They are valuable time indicators and are very useful in determining past environmental conditions. This extinct amphibian is from Permian age rocks in Texas. (Photo by T. A. Wiewandt/DRK Photo)

140

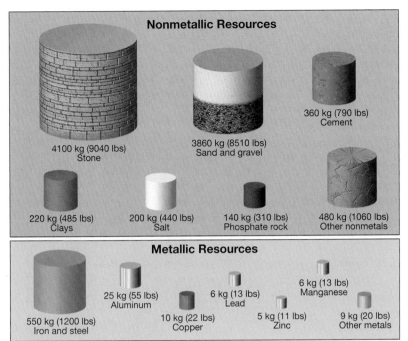

Figure 6.23 The annual per capita consumption of nonmetallic and metallic mineral resources for the United States is nearly 10,000 kilograms! About 94 percent of the materials used are nonmetallic.

Although coal is plentiful, its recovery and its use present a number of problems. Surface mining can turn the countryside into a scarred wasteland if careful (and costly) reclamation is not carried out to restore the land. (Today all U.S. surface mines must reclaim the land.) Although underground mining does not scar the landscape to the same degree, it has been costly in terms of human life and health. Underground mining long ago ceased to be a pick-and-shovel operation and is today a highly mechanized and computerized process. Strong federal safety regulations have made U.S. mining quite safe. However, the hazards of roof falls, gas explosions, and working with heavy equipment remain.

Air pollution is a major problem associated with the burning of coal. Much coal contains significant quantities of sulfur. Despite efforts to remove sulfur before the coal is burned, some remains. When the coal is burned, the sulfur is converted into noxious sulfur-oxide gases. Through a series of complex chemical reactions in the atmosphere, the sulfur oxides are converted to sulfuric acid, which then falls to Earth's surface as rain or snow. This *acid precipitation* can have detrimental ecological effects over widespread areas.

Because none of the problems just mentioned are likely to prevent the increased use of this important and abundant fuel, stronger efforts must be made to correct the problems associated with the mining and use of coal.

Oil and Natural Gas

Petroleum and natural gas obviously are not rocks, but we class them as mineral resources because they *come from* sedimentary rocks. They are found in similar environments and typically occur together. Both consist of various hydrocarbon compounds (compounds consisting of hydrogen and carbon) mixed together. They may also contain small quantities of other elements, such as sulfur, nitrogen, and oxygen. Like coal, petroleum and natural gas are biological products derived from the remains of organisms. However, the environments in which they form are very different, as are the organisms. Coal is formed mostly from plant material that accumulated in a swampy environment above sea level. Oil and gas are derived from the remains of both plants and animals having a marine origin.

Petroleum formation is complex and not completely understood. Nevertheless, we know that it begins with the accumulation of sediment in ocean areas that are rich in plant and animal remains. These accumulations must occur where biological activity is high, such as nearshore areas. However, most marine environments are oxygen-rich, which leads to the decay of organic remains before they can be buried by other

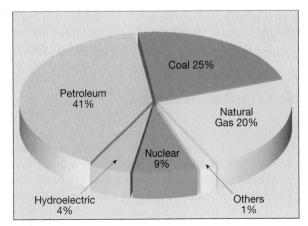

Figure 6.24 Consumption of energy in the United States, 2000. (Source: U.S. Department of Energy)

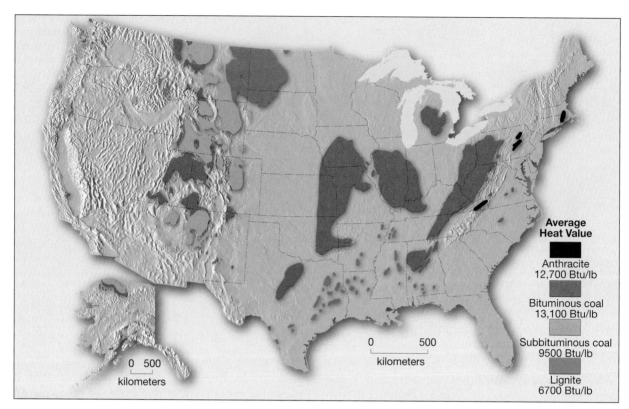

Figure 6.25   Coal fields of the United States. (Courtesy of the Bureau of Mines, U.S. Department of the Interior)

sediments. Therefore, accumulations of oil and gas are not as widespread as the marine environments that support abundant biological activity. This limiting factor notwithstanding, large quantities of organic matter are buried and protected from oxidation in many offshore sedimentary basins. With increasing burial over millions of years, chemical reactions gradually transform some of the original organic matter into the liquid and gaseous hydrocarbons we call petroleum and natural gas.

Unlike the solid organic matter from which they formed, the newly created petroleum and natural gas are *mobile*. These fluids are gradually squeezed from the compacting, mud-rich layers where they originate into adjacent permeable beds such as sandstone, where openings between sediment grains are larger. Because all of this occurs underwater, the rock layers containing the oil and gas are already saturated with water. But oil and gas are less dense than water, so they migrate upward through the waterfilled pore spaces of the enclosing rocks. Unless something acts to halt this upward migration, the fluids will eventually reach the surface. There the volatile components will evaporate.

Sometimes the upward migration is halted. A geologic environment that allows for economically significant amounts of oil and gas to accumulate underground is termed an **oil trap.** Several geologic structures can act as oil traps. All have two basic conditions in common: a porous, permeable **reservoir rock** that will yield petroleum and natural gas in sufficient quantities to make drilling worthwhile; and a **cap rock,** such as shale, that is virtually impermeable to oil and gas. The cap rock keeps the upwardly mobile oil and gas from escaping at the surface.

The Chapter in Review

- *Sedimentary rock* consists of *sediment* that has been *lithified* into solid rock. Sediment has two principal sources: (1) as *detrital material,* which originates and is transported as solid particles from both mechanical and chemical weathering, which, when lithified, forms detrital sedimentary rocks; and (2) from soluble material produced largely by chemical weathering, which, when precipitated, forms *chemical sedimentary rocks.*

- *Diagenesis* refers to all of the physical, chemical, and biological changes that occur after sediments are deposited and during and after the time they are turned into sedimentary rock. Burial promotes diagenesis. Diagenesis includes lithification.

- *Lithification* refers to the processes by which unconsolidated sediments are transformed into solid sedimentary rock. Most sedimentary rocks are lithified by means of *compaction* and/or *cementation*. Compaction occurs when the weight of overlying materials compresses the deeper sediments. Cementation, the most important process by which sediments are converted to sedimentary rock, occurs when soluble cementing materials, such as *calcite, silica,* and *iron oxide,* are precipitated onto sediment grains, fill open spaces, and join the particles. Although most sedimentary rocks are lithified by compaction or cementation, certain chemical rocks, such as the evaporites, initially form as solid masses of intergrown crystals.

- *Particle size* is the primary basis for distinguishing among various detrital sedimentary rocks. The size of the particles in a detrital rock indicates the energy of the medium that transported them. For example, gravels are moved by swiftly flowing rivers, whereas less energy is required to transport sand. Common detrital sedimentary rocks include *shale* (silt- and clay-sized particles), *sandstone,* and *conglomerate* (rounded gravel-sized particles) or *breccia* (angular gravel-sized particles).

- Precipitation of chemical sediments occurs in two ways: (1) by *inorganic processes,* such as evaporation and chemical activity; or (2) by *organic processes* of water-dwelling organisms that produce sediments of *biochemical origin. Limestone,* the most abundant chemical sedimentary rock, consists of the mineral calcite ($CaCO_3$) and forms either by inorganic means or as the result of biochemical processes. Inorganic limestones include *travertine,* which is commonly seen in caves, and *oolitic limestone,* consisting of small spherical grains of calcium carbonate. Other common chemical sedimentary rocks include *dolostone* (composed of the calcium-magnesium carbonate mineral dolomite), *chert* (made of microcrystalline quartz), *evaporites* (such as rock salt and rock gypsum), and *coal* (lignite and bituminous).

- Sedimentary rocks can be divided into two main groups: *detrital* and *chemical.* All detrital rocks have a *clastic texture,* which consists of discrete fragments and particles that are cemented and compacted together. The main criterion for subdividing the detrital rocks is particle size. Common detrital rocks include *conglomerate, sandstone,* and *shale.* The primary basis for distinguishing among different rocks in the chemical group is their mineral composition. Some chemical rocks, such as those deposited when seawater evaporates, have a *nonclastic texture* in which the minerals form a pattern of interlocking crystals. However, in reality, many of the sedimentary rocks classified into the chemical group also contain at least small quantities of detrital sediment. Common chemical rocks include *limestone, rock gypsum,* and *coal* (e.g., lignite and bituminous).

- Sedimentary environments are those places where sediment accumulates. They are grouped into continental, marine, and transitional (shoreline) environments. Each is characterized by certain physical, chemical, and biological conditions. Because sediment contains clues about the environment in which it was deposited, sedimentary rocks are important in the interpretation of Earth's history.

- Layers, called *strata,* or *beds,* are probably the single most characteristic feature of sedimentary rocks. Other features found in some sedimentary rocks, such as *ripple marks, mud cracks, cross-bedding, graded bedding,* and *fossils,* also give clues to past environments.

- Earth materials that are not used as fuels or processed for the metals they contain are referred to as *nonmetallic resources.* Many are sediments or sedimentary rocks. The two broad groups of nonmetallic resources are *building materials* and *industrial minerals.* Limestone, perhaps the most versatile and widely used rock of all, is found in both groups.

- *Coal, petroleum,* and *natural gas,* the *fossil fuels* of our modern economy, are all associated with sedimentary rocks. Coal originates from large quantities of plant remains that accumulate in an oxygen-deficient environment, such as a swamp. More than 70 percent of present-day coal usage is for the generation of electricity. Air pollution from the sulfur-oxide gases that form from burning most types of coal is a major environmental problem. Through a series of complex reactions in the atmosphere, the sulfur oxides produced when coal is burned are converted to sulfuric acid, which falls to Earth's surface in rain or snow.

- Oil and natural gas, which commonly occur together in the pore spaces of some sedimentary rocks, consist of various *hydrocarbon compounds* (compounds made of hydrogen and carbon) mixed together. Although petroleum formation is complex and not completely understood, it is associated with the accumulation of sediment in ocean areas that are rich in plant and animal remains that become buried and isolated in an oxygen-deficient environment. As the mobile petroleum and natural gas form, they migrate and accumulate in adjacent permeable beds such as sandstone. If the upward migration is halted by an impermeable rock layer, referred to as a *cap rock,* a geologic environment develops that allows for economically significant amounts of oil and gas to accumulate underground in what is termed an *oil trap.* The two basic conditions common to all oil traps are (1) a porous, permeable *reservoir rock* that will yield petroleum and/or natural gas in sufficient quantities, and (2) a cap rock.

Key Terms

bedding plane (p. 135)
beds (strata) (p. 134)
biochemical origin (p. 127)
cap rock (p. 142)

cementation (p. 122)
chemical sedimentary rock
    (p. 123)
clastic (p. 133)

compaction (p. 121)
cross-bedding (p. 135)
detrital sedimentary rock
    (p. 123)

diagenesis (p. 121)
environment of deposition
    (p. 134)
evaporite (p. 131)

fossil (p. 135)
graded bed (p. 135)
lithification (p. 121)
mud crack (p. 135)

nonclastic (p. 133)
nonmetallic mineral
   resource (p. 135)
oil trap (p. 142)

reservoir rock (p. 142)
ripple mark (p. 135)
salt flat (p. 131)
sediment (p. 114)

sedimentary environment
   (p. 134)
sorting (p. 124)
strata (beds) (p. 134)

Questions for Review

1. How does the volume of sedimentary rocks in Earth's crust compare with the volume of igneous rocks in the crust? Are sedimentary rocks evenly distributed throughout the crust?

2. What is diagenesis? Give an example.

3. Compaction is most important as a lithification process with which sediment size?

4. List three common cements for sedimentary rocks. How might each be identified?

5. What minerals are most common in detrital sedimentary rocks? Why are these minerals so abundant?

6. What is the primary basis for distinguishing among various detrital sedimentary rocks?

7. Why does shale usually crumble quite easily?

8. Distinguish between conglomerate and breccia.

9. Distinguish between the two categories of chemical sedimentary rocks.

10. What are evaporite deposits? Name a rock that is an evaporite.

11. How is bituminous coal different from lignite? How is anthracite different from bituminous?

12. Each of the following statements describes one or more characteristics of a particular sedimentary rock. For each statement, name the sedimentary rock that is being described.
    a. An evaporite used to make plaster.
    b. A fine-grained detrital rock that breaks into chunks or blocks.

c. Dark-colored sandstone containing angular rock particles as well as clay, quartz, and feldspar.
    d. The most abundant chemical sedimentary rock.
    e. A dark-colored, dense, hard rock made of microcrystalline quartz.
    f. A variety of limestone composed of small spherical grains.

13. What is the primary basis for distinguishing among different chemical sedimentary rocks?

14. Distinguish between clastic and nonclastic textures. What type of texture is common to all detrital sedimentary rocks?

15. What is probably the single most characteristic feature of sedimentary rocks?

16. Distinguish between cross-bedding and graded bedding.

17. Nonmetallic resources are commonly divided into two broad groups. List the two groups and some examples of materials that belong to each.

18. Coal enjoys the advantage of being plentiful. What are some disadvantages associated with the production and use of coal?

19. What is an oil trap? List two conditions common to all oil traps.

Web Resources

The *Essentials of Geology* Website uses the resources and flexibility of the Internet to aid in your study of the topics in this chapter. Written and developed by geology instructors, this site will help improve your understanding of geology. Visit **www.prenhall.com/lutgens** and click on the cover of *Essentials of Geology, 8/e* to find:

- On-line review quizzes.
- Web-based critical thinking and writing exercises.
- Links to chapter-specific Web resources.
- Internet-wide key term searches.

http://www.prenhall.com/lutgens

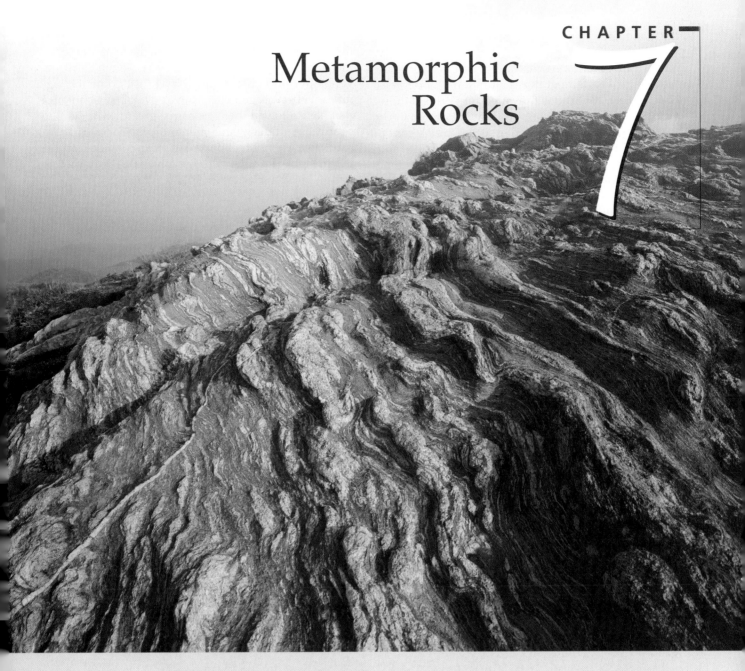

CHAPTER 7

Metamorphic Rocks

Focus on Learning

To assist you in learning the important concepts in this chapter, you will find it helpful to focus on the following questions:

- What are metamorphic rocks, and how do they form?
- In which three geologic settings is metamorphism most likely to occur?
- What are the agents of metamorphism?
- What are the two textural divisions of metamorphic rocks and the conditions associated with the occurrence of each?
- What are the names, textures, and compositions of the common metamorphic rocks?
- How is the intensity, or degree, of metamorphism reflected in the texture and mineralogy of metamorphic rocks?

Metamorphic rocks exposed at Graveyard Ridge, Shining Rock Wilderness, North Carolina. (Photo by David Muench)

The folded and metamorphosed rocks shown in Figure 7.1 were once flat-lying sedimentary strata. Compressional forces of unimaginable magnitude and temperatures hundreds of degrees above surface conditions prevailed for perhaps thousands or millions of years to produce the deformation displayed by these rocks. Under such extreme conditions, solid rock responds by folding, fracturing, and often by flowing. This chapter looks at the tectonic forces that forge metamorphic rocks and how these rocks change in appearance, mineralogy, and sometimes even in overall chemical composition.

Extensive areas of metamorphic rocks are exposed on every continent in the relatively flat regions known as **shields.** These metamorphic regions are found in eastern Canada, Brazil, Africa, India, Australia, and Greenland. Moreover, metamorphic rocks are an important component of many mountain belts, where they make up a large portion of a mountain's crystalline core. Even those portions of the stable continental interiors that are covered by sedimentary rocks are underlain by metamorphic basement rocks. In these settings the metamorphic rocks are highly deformed and intruded by large igneous masses. Indeed, significant parts of Earth's continental crust are composed of metamorphic and associated igneous rocks.

Unlike some igneous and sedimentary processes that take place in surface or near-surface environments, metamorphism invariably occurs deep within Earth, beyond our direct observation. Notwithstanding this significant obstacle, geologists have developed techniques that have allowed them to learn a great deal about the conditions under which metamorphic rocks form. In turn, the study of metamorphic rocks provides important insights into the geologic processes that operate within Earth's crust and upper mantle.

Metamorphism

Recall from the section on the rock cycle in Chapter 1 that metamorphism is the transformation of one rock type into another. Metamorphic rocks are produced from preexisting igneous, sedimentary, or even from other metamorphic rocks. Thus, every metamorphic rock has a **parent rock**—the rock from which it was formed.

Metamorphism is a very appropriate name for this process because it literally means to "change form." Metamorphism takes place *where preexisting rock is subjected to temperatures and pressures unlike those in which it formed*. In response to these new conditions, the rock gradually changes until a state of equilibrium with the

Figure 7.1   Deformed and metamorphosed rocks, Anza-Borrego Desert State Park, California. (Photo by A. P. Trujillo/APT Photos)

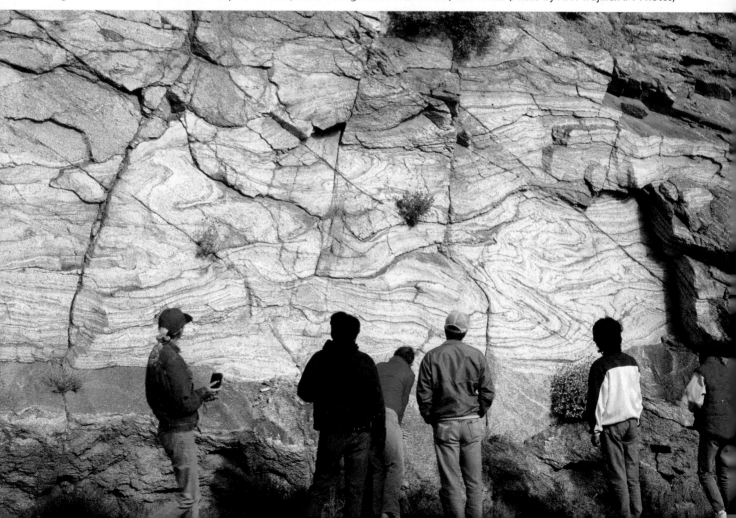

new environment is reached. Most metamorphic changes occur at the elevated temperatures and pressures that exist in the zone beginning a few kilometers below Earth's surface and extending into the upper mantle.

Metamorphism often progresses incrementally, from slight changes (*low-grade metamorphism*) to substantial changes (*high-grade metamorphism*). For example, under low-grade metamorphism, the common sedimentary rock *shale* becomes the more compact metamorphic rock called *slate*. Hand samples of these rocks are sometimes difficult to distinguish, illustrating that the stage at which a sedimentary rock becomes a metamorphic rock is somewhat arbitrary.

In more extreme environments, metamorphism causes a transformation so complete that the identity of the parent rock cannot be determined. In high-grade metamorphism, such features as bedding planes, fossils, and vesicles that may have existed in the parent rock are obliterated. Further, when rocks at depth (where temperatures are high) are subjected to directed pressure, they slowly deform by flowing to generate intricate folds (Figure 7.1). In the most extreme metamorphic environments, the temperatures approach those at which rocks melt. However, *during metamorphism the rock must remain essentially solid*, for if complete melting occurs, we have entered the realm of igneous activity.

Most metamorphism occurs in one of three settings:

1. When rock is intruded by a magma body, *contact* or *thermal metamorphism* may take place. Here, change is driven by a rise in temperature within the host rock surrounding an igneous intrusion.

2. *Hydrothermal metamorphism* involves chemical alterations that occur as hot ion-rich water circulates through fractures in rock. This type of metamorphism is usually associated with igneous activity that provides the heat required to drive chemical reactions and circulate these fluids through rock.

3. During mountain building, great quantities of rock are subjected to directed pressures and high temperatures associated with large-scale deformation called *regional metamorphism*.

Regional metamorphism, which produces the greatest volume of metamorphic rock, occurs at convergent boundaries where lithospheric plates collide (Figure 7.2). Here, large segments of Earth's crust become highly folded, faulted, and metamorphosed. Further, deep burial, together with the emplacement of magma originating in the mantle, are responsible for

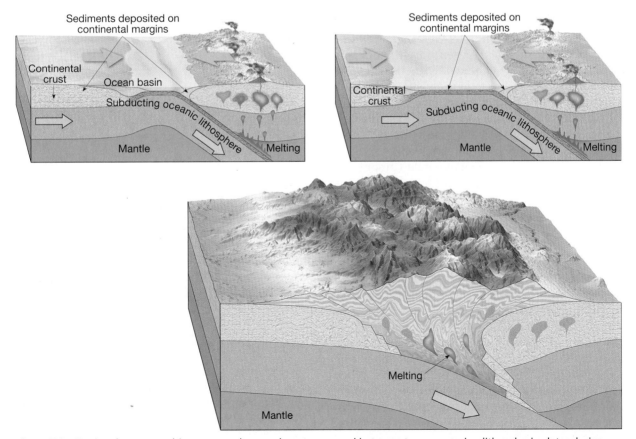

Figure 7.2 Regional metamorphism occurs where rocks are squeezed between two converging lithospheric plates during mountain building.

the elevated temperatures that produce the most intense zones of metamorphism. Rocks deformed by regional metamorphism usually display zones of contact metamorphism, as well as hydrothermal metamorphism.

After considering the agents of metamorphism and some common metamorphic rocks, we will examine these metamorphic environments as well as others.

Agents of Metamorphism

GEODe [Earth Materials
          ↳ Metamorphic Rocks

The agents of metamorphism include *heat, pressure (stress),* and *chemically active fluids*. During metamorphism, rocks are usually subjected to all three metamorphic agents simultaneously. However, the degree of metamorphism and the contribution of each agent vary greatly from one environment to another.

Heat as a Metamorphic Agent

The most important agent of metamorphism is *heat* because it provides the energy to drive chemical reactions that result in the recrystallization of existing minerals and/or the formation of new minerals. Recall from our discussion of igneous rocks that an increase in temperature causes the ions within a mineral to vibrate more rapidly. Even in a crystalline *solid*, where ions are strongly bonded, this continuous vibration allows individual atoms to migrate between sites in the crystal structure. As a result, when minerals are heated sufficiently, the constituent ions tend to arrange themselves into crystalline structures that are more stable in the new high-energy environment. Thus, recrystallization results in the creation of new minerals with stable configurations that have an overall composition roughly equivalent to that of the original minerals. (In some environments ions may migrate into or out of a rock unit, thereby changing its overall chemical composition.)

The heat to metamorphose rocks comes mainly from two sources. First, rocks experience a rise in temperature when they are intruded by magma rising from below—*contact metamorphism*. Here the adjacent host rock is "baked" by the emplaced magma.

Second, rocks that formed at Earth's surface will experience a gradual increase in temperature as they are transported to greater depths. Recall that temperatures increase with depth at a rate known as the *geothermal gradient*. In the upper crust, this increase in temperature averages between 20°C and 30°C per kilometer. When buried to a depth of about 8 kilometers (5 miles), where temperatures are between 150°C and 200°C, clay minerals tend to become unstable and begin to recrystallize into minerals, such as chlorite and muscovite, that are stable in this environment. (Chlorite is a micalike mineral formed by the metamorphism of ferromagnesian silicates.) However, many silicate minerals, particularly those found in crystalline igneous rocks—quartz, for example—remain stable at these temperatures. Thus, metamorphic changes in these minerals occur at much higher temperatures.

Environments where rocks may be carried to great depths include convergent plate boundaries where slabs of sediment-laden oceanic crust are being subducted. In addition, rocks may become deeply buried in large basins where gradual subsidence results in very thick accumulations of sediment. Such locations, exemplified by the Gulf of Mexico, are known to develop metamorphic conditions near the base of the pile.

Pressure (Stress) as a Metamorphic Agent

Pressure, like temperature, also increases with depth as the thickness of the overlying rock increases. Buried rocks are subjected to **confining pressure**, which is analogous to water pressure, where the forces are applied equally in all directions (Figure 7.3A). The deeper you go in the ocean, the greater the confining pressure. The

Figure 7.3   Pressure (stress) as a metamorphic agent. **A.** In a depositional environment, as confining pressure increases, rocks deform by decreasing in volume. **B.** During mountain building, rocks subjected to differential stress are shortened in the direction that pressure is applied, and lengthened in the direction perpendicular to that force.

same is true for rock that is buried. Confining pressure causes the spaces between mineral grains to close, producing a more compact rock having a greater density (Figure 7.3A). Further, at great depths, confining pressure may cause minerals to recrystallize into new minerals that display a more compact crystalline form. Confining pressure does *not*, however, fold and deform rocks, like those shown in Figure 7.1.

In addition to confining pressure, rocks may be subjected to directed pressure where tectonic plates collide to generate mountains (Figure 7.3B). These forces, which are unequal in different directions, are called **differential stress.** Unlike confining pressure, which "squeezes" the rock equally in all directions, differential stress acts mainly along one plane. As shown in Figure 7.3B, rocks subjected to differential stress are shortened in the direction pressure is applied and elongated, or lengthened, in the direction perpendicular to that force. The deformation caused by differential stresses plays a major role in developing metamorphic textures.

In surface environments where temperatures are comparatively low, rocks are *brittle* and tend to fracture when subjected to differential stress. Continued deformation grinds and pulverizes the mineral grains into small fragments. By contrast, in high-temperature environments rocks are *ductile*. When rocks exhibit ductile behavior, their mineral grains tend to flatten and elongate when subjected to differential stress (Figure 7.4). This accounts for their ability to deform by flowing (rather than fracturing) to generate intricate folds.

Chemically Active Fluids

Fluids composed mainly of water and other volatile components, including carbon dioxide, are believed to play an important role in some types of metamorphism. Fluids that surround mineral grains act as catalysts to promote recrystallization by enhancing ion migration. Further, in progressively hotter environments these ion-

Figure 7.4 Metaconglomerate, also called stretched pebble conglomerate. These once nearly spherical pebbles have been heated and flattened into elongated structures. (Photo by E. J. Tarbuck)

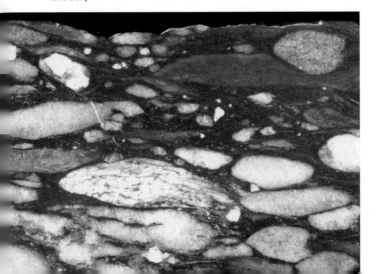

rich fluids become correspondingly more reactive. For example, when two mineral grains are squeezed together, the part of their crystalline structures that touch are the most highly stressed. Ions located at these sites are readily dissolved by the hot fluids and migrate along the surface of the grain to locations where the two grains are not in contact. Thus, hydrothermal fluids aid in the recrystallization of mineral grains by dissolving material from regions of high stress and then precipitating (depositing) this material in areas of low stress. As a result, *minerals tend to recrystallize and grow longer in a direction perpendicular to compressional stresses.*

Where hot fluids circulate freely through rocks, ionic exchange may occur between two adjacent rock layers, or ions may migrate great distances before they are finally deposited. When ions are exchanged between two or more separate rock units, a change in the overall composition results.

What is the source of these chemically active fluids? Water is plentiful in the pore spaces of most sedimentary rocks, as well as in fractures in igneous rocks. In addition, many minerals, such as clays, micas, and amphiboles, are *hydrated* and thus contain water in their crystalline structures. Elevated temperatures associated with low to moderate-grade metamorphism cause the dehydration of these minerals. Once expelled, the water moves along the surfaces of individual grains and is available to facilitate ion transport. However, in high-grade metamorphic environments, where temperatures are extreme, these fluids may be driven from the rocks. Recall that when oceanic crust is subducted to depths exceeding 100 kilometers, water expelled from these slabs migrates into the mantle wedge above, where it is thought to trigger melting.

The Importance of Parent Rock

Most metamorphic rocks have the same overall chemical composition as the parent rock from which they formed, except for the possible loss or acquisition of volatiles such as water (H_2O) and carbon dioxide (CO_2). For example, the metamorphism of shale results in slate, where clay minerals recrystallize to form micas. (The minute crystals of quartz and feldspar found in shale are not altered in the transformation of shale to slate and thus remain intermixed with the micas.) Although the mineralogy changes in the transformation of shale to slate, the overall chemical composition of slate is comparable to that of the rock from which it was derived. Further, when the parent rock has a mafic composition, such as basalt, the metamorphic product will be rich in minerals containing iron and magnesium unless, of course, there has been a substantial loss of these atoms.

In addition, the mineral makeup of the parent rock determines, to a large extent, the degree to which each metamorphic agent will cause change. For example, when magma forces its way into existing rock, high temperatures and associated hot ion-rich fluids tend to alter the host rock. When the host rock is composed of minerals that are comparatively unreactive, such as the quartz grains found in a clean quartz sandstone, very little alteration may take place. However, if the host rock is a "dirty" limestone that contains abundant silica-rich clay, the calcite $(CaCO_3)$ in the limestone may react with silica (SiO_2) in the clays to form wollastonite $(CaSiO_3)$ plus carbon dioxide (CO_2). In this situation the zone of metamorphism may extend for several kilometers from the magma body.

Metamorphic Textures

 Earth Materials
↳ Metamorphic Rocks

Recall that the term **texture** is used to describe the size, shape, and arrangement of grains within a rock. Most igneous and many sedimentary rocks consist of mineral grains that have a random orientation and thus appear the same when viewed from any direction. By contrast, deformed metamorphic rocks that contain platy minerals (micas) and/or elongated minerals (amphiboles), typically display some kind of *preferred orientation* in which the mineral grains exhibit a parallel to subparallel alignment. Like a fistful of pencils, rocks containing elongated minerals that are oriented parallel to each other will appear different when viewed from the side than when viewed head-on. A rock that exhibits a preferred orientation of its minerals is said to possess *foliation*.

Foliation

The term **foliation** refers to any planar (nearly flat) arrangement of mineral grains or structural features within a rock. Although foliation occurs in some sedimentary and even a few types of igneous rocks, it is a fundamental characteristic of regionally metamorphosed rocks—that is, rock units that have been strongly folded and distorted. In metamorphic environments, foliation is ultimately driven by compressional stresses that shorten rock units, causing mineral grains in preexisting rocks to develop parallel, or nearly parallel, alignments. Examples of foliation include the parallel alignment of platy and/or elongated minerals; the parallel alignment of flattened mineral grains and pebbles; compositional banding where the separation of dark and light minerals generate a layered appearance; and slaty cleavage where rocks can be easily split into thin, tabular slabs along parallel surfaces. These diverse types of foliation can form in many different ways, including:

1. Rotation of platy and/or elongated mineral grains into a new orientation.

2. Recrystallization of minerals to form new grains growing in the direction of preferred orientation.

3. Changing the shape of equidimensional grains into elongated shapes that are aligned in a preferred orientation.

The rotation of existing mineral grains is the easiest of these mechanisms to envision. Figure 7.5 illustrates the mechanics by which platy or elongated minerals are rotated. Note that the new alignment is perpendicular to the direction of maximum shortening. Although physical rotation of platy minerals contributes to the development of foliation in low-grade metamorphism, other mechanisms dominate in more extreme environments.

Recall that recrystallization is the creation of new mineral grains out of old ones. During the transformation of shale to slate, minute clay minerals (stable at the surface) recrystallize into microscopic flakes of chlorite and mica (stable at higher temperatures and pressures). This process may occur in places where the crystalline structure of existing minerals is altered. In other settings some of the old grains are dissolved and migrate to a different site, where they precipitate to form new mineral grains. The growth of new mineral grains tends to develop on old crystals of similar structure and grow in the same orientation as the older crystals. In this way the new growth "mimics" the growth of old grains and enhances any preexisting preferred orientation. However, recrystallization that accompanies deformation often results in a *new* preferred orientation. As rock units are folded and generally shortened during metamorphism, elongated and platy minerals tend to recrystallize perpendicular to the direction of maximum stress.

Mechanisms that change the shapes of individual grains are especially important for the development of preferred orientations in rocks that contain minerals such as quartz, calcite, and olivine. When these minerals are stressed, they develop elongated grains that align

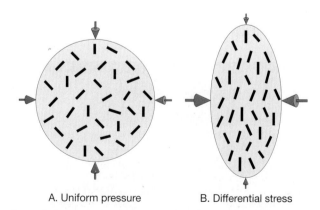

A. Uniform pressure   B. Differential stress

Figure 7.5 Mechanical rotation of platy or elongated mineral grains. **A.** Existing mineral grains keep their random orientation if force is uniformly applied. **B.** As differential stress causes rocks to flatten, mineral grains rotate toward the plane of flattening.

in a direction parallel to maximum flattening (Figure 7.6). This type of deformation occurs in high-temperature environments where ductile deformation is dominant (as opposed to brittle fracturing).

A change in grain shape can occur as units of a mineral's crystalline structure slide relative to one another along discrete planes, thereby distorting the grain as shown in Figure 7.6B. This type of gradual solid-state flow involves slippage that disrupts the crystal lattice and recrystallization that repairs the structure. In addition, the shape of a mineral may change as ions move from a point along the margin of the grain that is highly stressed to a less-stressed position on the same grain (Figure 7.6C). This type of deformation occurs by mass transfer of material from one location to another. As you might expect, this mechanism is aided by chemically active fluids and is a type of recrystallization.

Foliated Textures

Various types of foliation exist, depending largely upon the grade of metamorphism and the mineralogy of the parent rock. We will look at three; *rock* or *slaty cleavage*; *schistosity*; and *gneissic texture*.

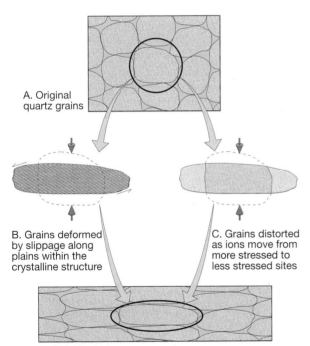

A. Original quartz grains

B. Grains deformed by slippage along plains within the crystalline structure

C. Grains distorted as ions move from more stressed to less stressed sites

D. Flattened rock exhibiting distorted quartz grains

Figure 7.6 Development of preferred orientations of minerals such as quartz, calcite, and olivine. **A.** Ductile deformation (flattening) of these roughly equidimensional mineral grains can occur in one of two ways. **B.** The first mechanism is a solid-state plastic flow that involves intracrystalline gliding of individual units within each grain. **C.** The second mechanism involves dissolving material from areas of high stress and depositing that material in locations of low stress. **D.** Both mechanisms change the shape of the grains, but the volume and composition of each grain remains essentially unchanged.

Rock or Slaty Cleavage. **Rock cleavage** refers to closely spaced planar surfaces along which rocks split into thin, tabular slabs when hit with a hammer. Rock cleavage is developed in various metamorphic rocks but is best displayed in slates that exhibit an excellent splitting property called **slaty cleavage.**

Depending on the metamorphic environment and the composition of the parent rock, rock cleavage develops in a number of different ways. In a low-grade metamorphic environment, rock cleavage is known to develop where beds of shale (and related sedimentary rocks) are strongly folded and metamorphosed to form slate. The process begins as platy grains are kinked and bent—generating microscopic folds having limbs (sides) that are roughly aligned (Figure 7.7). With further deformation, this new alignment is enhanced as old grains break down and recrystallize preferentially in the direction of the newly developed orientation. In this manner the rock develops narrow parallel zones where mica flakes are concentrated. These planar features alternate with zones containing quartz and other mineral grains that do not exhibit a pronounced linear orientation. It is along these very thin zones, where platy minerals display a parallel alignment, that slate splits.

Because slate typically forms during low-grade metamorphism of shale, evidence of the original sedimentary bedding planes is often preserved. However, as shown in Figure 7.7D, the orientation of slate's cleavage usually develops at an oblique angle to the original sedimentary layering. Thus, unlike shale, which splits along bedding planes, slate often splits across them. Other metamorphic rocks, such as schists and gneisses, may also split along planar surfaces and thus exhibit rock cleavage.

Schistosity. Under more extreme temperature-pressure regimes, the minute mica and chlorite grains in slate begin to grow many times larger. When these platy minerals grow large enough to be discernible with the unaided eye and exhibit a planar or layered structure, the rock is said to exhibit a type of foliation called **schistosity.** Rocks having this texture are referred to as *schist.* In addition to platy minerals, schist often contains deformed quartz and feldspar grains that appear as flat, or lens-shaped, grains hidden among the mica grains.

Gneissic Texture. During high-grade metamorphism, ion migrations can result in the segregation of minerals as shown in Figure 7.8. Notice that the dark biotite crystals and light silicate minerals (quartz and feldspar) have separated, giving the rock a banded appearance called **gneissic texture.** A metamorphic rock with this texture is called *gneiss* (pronounced "nice"). Although foliated, gneisses will not usually split as eas-

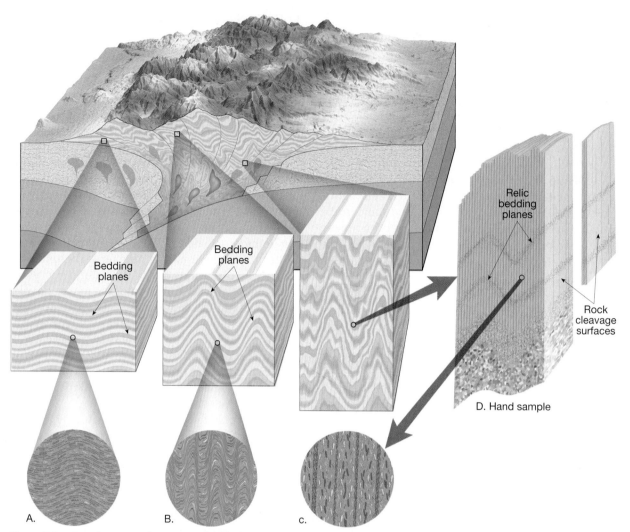

Figure 7.7   Development of one type of rock cleavage. As shale is strongly folded **(A, B)** and metamorphosed to form slate, the developing mica flakes are bent into microfolds. **C.** Further metamorphism results in the recrystallization of mica grains along the limbs of these folds to enhance the foliation. **D.** Hand sample of slate illustrates rock cleavage and its orientation to relic bedding surfaces.

ily as slates and some schists. Gneisses that do cleave tend to break parallel to their foliation and expose mica-rich surfaces that resemble schist.

Other Metamorphic Textures

Not all metamorphic rocks exhibit a foliated texture. Those that *do not* are referred to as **nonfoliated.** Nonfoliated metamorphic rocks typically develop in environments where deformation is minimal and the parent rocks are composed of minerals that exhibit equidimensional crystals, such as quartz or calcite. For example, when a fine-grained limestone (made of calcite) is metamorphosed by the intrusion of a hot magma body, the small calcite grains recrystallize to form larger interlocking crystals. The resulting rock, *marble*, displays large, equidimensional grains that are randomly oriented similar to those of a coarse-grained igneous rock.

Another texture common to metamorphic rocks consists of particularly large grains, called *porphyroblasts*, that are surrounded by a fine-grained matrix of other minerals (Figure 7.9). **Porphyroblastic textures** develop in a wide range of rock types and metamorphic environments when minerals in the parent rock recrystallize to form new minerals. During recrystallization certain metamorphic minerals, including garnet, staurolite, and andalusite, invariably develop *a small number of very large crystals*. By contrast, minerals such as muscovite, biotite, and quartz typically form *a large number of very small grains*. As a result, when metamorphism generates the minerals garnet, biotite, and muscovite in the same setting, the rock will contain large crystals (porphyroblasts) of garnet embedded in a fine-grained matrix consisting of biotite and muscovite (Figure 7.9).

Figure 7.8 This rock displays a gneissic texture. Notice that the dark biotite flakes and light silicate minerals are segregated, giving the rock a banded or layered appearance. (Photo by E. J. Tarbuck)

←—5 cm—→

Close up

Figure 7.9 Garnet-mica schist. The dark red garnet crystals (porphyroblasts) are embedded in a matrix of fine-grained micas. (Photo by E. J. Tarbuck)

Common Metamorphic Rocks

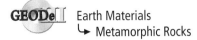

GEODe Earth Materials
  ↳ Metamorphic Rocks

Recall that metamorphism causes many changes in rocks, including increased density, change in grain size, reorientation of mineral grains into a planar arrangement known as foliation, and the transformation of low-temperature minerals into high-temperature minerals. Moreover, the introduction of ions may generate new minerals, some of which are economically important.

The major characteristics of some common metamorphic rocks are summarized in Figure 7.10. Notice that metamorphic rocks can be broadly classified by the type of foliation exhibited and to a lesser extent on the chemical composition of the parent rock.

Foliated Rocks

Slate. *Slate* is a very fine-grained (less than 0.5 millimeter) foliated rock composed of minute mica flakes that are too small to be visible. Thus, slate generally appears dull and closely resembles shale. A noteworthy characteristic of slate is its excellent rock cleavage, or tendency to break into flat slabs. This property traditionally made slate a most useful rock for roof and floor tile, blackboards, and billiard tables (Figure 7.11).

Slate is most often generated by the low-grade metamorphism of shale, mudstone, or siltstone. Less frequently it is produced when volcanic ash is metamorphosed. Slate's color depends on its mineral constituents. Black (carbonaceous) slate contains organic material, red slate gets its color from iron oxide, and green slate usually contains chlorite.

Phyllite. Phyllite represents a gradation in the degree of metamorphism between slate and schist. Its constituent platy minerals are larger than those in slate but not yet large enough to be readily identifiable with the unaided eye. Although phyllite appears similar to slate, it can be easily distinguished from slate by its glossy sheen and wavy surface (Figure 7.12). Phyllite usually exhibits rock cleavage and is composed mainly of very fine crystals of either muscovite, chlorite, or both.

Schist. Schists are medium- to coarse-grained metamorphic rocks in which platy minerals predominate. These flat components commonly include the micas (muscovite and biotite), which display a planar alignment that gives the rock its foliated texture. In addition, schists contain smaller amounts of other minerals, often quartz and feldspar. Schists composed mostly of dark minerals (amphiboles) are known. Like slate, the parent rock of many schists is shale, which

Rock Name	Texture		Grain Size	Comments	Parent Rock
Slate	Foliated (Increasing Metamorphism)		Very fine	Excellent rock cleavage, smooth dull surfaces	Shale, mudstone, or siltstone
Phyllite			Fine	Breaks along wavey surfaces, glossy sheen	Slate
Schist			Medium to Coarse	Micaceous minerals dominate, scaly foliation	Phyllite
Gneiss			Medium to Coarse	Compositional banding due to segregation of minerals	Schist, granite, or volcanic rocks
Migmatite			Medium to Coarse	Banded rock with zones of light-colored crystalline minerals	Gneiss
Mylonite	Weakly Foliated		Fine	When very fine-grained, resembles chert, often breaks into slabs	Any rock type
Metaconglomerate			Coarse-grained	Stretched pebbles with preferred orientation	Quartz-rich conglomerate
Marble	Nonfoliated		Medium to coarse	Interlocking calcite or dolomite grains	Limestone, dolostone
Quartzite			Medium to coarse	Fused quartz grains, massive, very hard	Quartz sandstone
Hornfels			Fine	Usually, dark massive rock with dull luster	Any rock type
Anthracite			Fine	Shiny black rock that may exhibit conchoidal fracture	Bituminous coal
Fault breccia			Medium to very coarse	Broken fragments in a haphazard arrangement	Any rock type

Figure 7.10   Classification of common metamorphic rocks.

Figure 7.11   Because slate breaks into flat slabs, it has a number of uses. Here, it is used to roof this house in Switzerland. (Photo by E. J. Tarbuck)

has undergone medium- to high-grade metamorphism during major mountain-building episodes.

The term *schist* describes the texture of a rock. To indicate the composition, mineral names are used. For example, schists composed primarily of muscovite and biotite are called *mica schists*. Depending upon the degree of metamorphism and composition of the parent rock, mica schists often contain *accessory minerals*, some of which are unique to metamorphic rocks. Some common accessory minerals that occur as porphyroblasts include *garnet, staurolite,* and *sillimanite,* in which case the rock is called *garnet-mica schist, staurolite-mica schist,* and so forth (see Figure 7.9).

Figure 7.12 Phyllite (left) can be distinguished from slate (right) by its glossy sheen and wavy surface. (Photo by E. J. Tarbuck)

In addition, schists may be composed largely of the minerals chlorite or talc, in which case they are called *chlorite schist* and *talc schist*, respectively. Both chlorite and talc schists can form when rocks with a basaltic composition undergo metamorphism. Others contain the mineral *graphite*, which is used as pencil "lead," graphite fibers (used in fishing rods), and lubricant (commonly for locks).

Gneiss. *Gneiss* is the term applied to medium- to coarse-grained banded metamorphic rocks in which granular and elongated (as opposed to platy) minerals predominate. The most common minerals in gneiss are quartz, potassium feldspar, and sodium-rich plagioclase feldspar. Most gneisses also contain lesser amounts of biotite, muscovite, and amphibole that develop a preferred orientation. Some gneisses will split along the layers of platy minerals, but most break in an irregular fashion.

Recall that during high-grade metamorphism, the light and dark components separate, giving gneisses their characteristic banded or layered appearance. Thus, most gneisses consist of alternating bands of white or reddish feldspar-rich zones and layers of dark ferromagnesian minerals (see Figure 7.8). These banded gneisses often exhibit evidence of deformation, including folds and faults (see Figure 7.1).

Most gneisses have a felsic composition and are derived from granite or its aphanitic equivalent, rhyolite. However, some may form from the high-grade metamorphism of shale. In this instance, gneiss represents the last rock in the sequence of shale, slate, phyllite, schist, and gneiss. Like schists, gneisses may also include large crystals of accessory minerals such as garnet and staurolite. Gneisses made up primarily of dark minerals such as those that compose basalt also occur. For example, an amphibole-rich rock that exhibits a gneissic texture is called *amphibolite*.

Nonfoliated Rocks

Marble. *Marble* is a coarse, crystalline metamorphic rock whose parent was limestone or dolostone (Figure 7.13). Pure marble is white and composed essentially of the mineral calcite. Because of its relative softness (hardness of 3), marble is easy to cut and shape. White

← 5 cm →

Photomicrograph (6.5x)

Figure 7.13 Marble, a crystalline rock formed by the metamorphism of limestone. Photomicrograph shows interlocking calcite crystals using polarized light. (Photos by E. J. Tarbuck)

marble is particularly prized as a stone from which to create monuments and statues, such as the famous statue of *David* by Michelangelo. Unfortunately, since marble is composed of calcium carbonate, it is readily attacked by acid rain. Some historic monuments and tombstones already show severe chemical weathering (see Figure 5.9).

The parent rock from which marble forms often contains impurities that tend to color the stone. Thus, marble can be pink, gray, green, or even black and may contain a variety of accessory minerals (chlorite, mica, garnet, and wollastonite). When marble forms from limestone interbedded with shales, it will appear banded. When deformed, these banded marbles may develop highly contorted mica-rich folds that give the rock a rather artistic design. Hence, these decorative marbles have been used as a building stone since prehistoric times.

Quartzite. *Quartzite* is a very hard metamorphic rock formed from quartz sandstone (Figure 7.14).

Figure 7.14 Quartzite is a nonfoliated metamorphic rock formed from quartz sandstone. The photomicrograph shows the interlocking quartz grains typical of quartzite. (Photos by E. J. Tarbuck)

Under moderate- to high-grade metamorphism, the quartz grains in sandstone fuse like chips of glass melting together (inset in Figure 7.14). The recrystallization is so complete that when broken, quartzite will split through the quartz grains rather than along their boundaries. In some instances, sedimentary features such as crossbedding are preserved and give the rock a banded appearance. Pure quartzite is white, but iron oxide may produce reddish or pinkish stains, while dark mineral grains may impart a gray color.

Metamorphic Environments

There are a number of environments in which metamorphism occurs. Most are in the vicinity of plate margins, and many are associated with igneous activity. We will consider the following types of metamorphism: 1) *contact* or *thermal metamorphism;* 2) *hydrothermal metamorphism;* 3) *regional metamorphism;* 4) *burial metamorphism;* 5) *impact metamorphism;* and 6) *metamorphism along faults.*

With the exception of shock metamorphism, there is considerable overlap among the other types. Recall that regional metamorphism occurs where lithospheric plates collide to generate mountains. Here, large segments of Earth's crust are folded and faulted while often being intruded by magma rising from the mantle. Hence, the rocks that are deformed and metamorphosed in a regional setting exhibit metamorphic features common to other types of metamorphism.

Contact or Thermal Metamorphism

Contact or **thermal metamorphism** occurs as the result of a rise in temperature when magma invades a host rock. Here, a zone of alteration called an **aureole** forms in the rock that surrounds the emplaced magma body (Figure 7.15). Small intrusions, such as thin dikes and sills, have aureoles only a few centimeters thick. By contrast, magma bodies that crystallize to form massive batholiths may create zones of metamorphic rock that extend for several kilometers.

In addition to the size of the magma body, the mineral composition of the host rock and the availability of water greatly affect the size of the aureole produced. In chemically active rock such as limestone, the zone of alteration can be 10 kilometers (6 miles) thick. These large aureoles often consist of distinct *zones of metamorphism.* Near the magma body, high-temperature minerals such as garnet may form, whereas farther away low-grade minerals such as chlorite are produced.

Contact metamorphism is most easily recognized when it occurs at the surface or in a near-surface environment, where the temperature contrast between the magma and host rock is large. During contact meta-

Roof pendant

A. Implacement of igneous body and metamorphism

B. Crystallization of pluton

C. Uplift and erosion to expose pluton and metamorphic cap rock

Figure 7.15 Contact metamorphism produces a zone of alteration called an *aureole* around an intrusive igneous body. In the photo, the dark layer, called a *roof pendant*, consists of metamorphosed host rock adjacent to the upper part of the light-colored igneous pluton. The term *roof pendant* implies that the rock was once the roof of a magma chamber. Sierra Nevada, near Bishop, California. The inset photo is a close-up of a contact between an igneous pluton and metamorphosed host rock. (Photos by John S. Shelton)

morphism, clay minerals are baked, as if placed in a kiln, and can generate a very hard, fine-grained rock. Because directed pressure is not a major factor in forming these rocks, they generally are nonfoliated. *Hornfels* is the name applied to the wide variety of rather hard, nonfoliated metamorphic rocks formed during contact metamorphism.

Hydrothermal Metamorphism

When hot, ion-rich fluids circulate through fissures and cracks that develop in rock, a chemical alteration called **hydrothermal metamorphism** occurs. This type of metamorphism is closely associated with igneous activity, since it provides the heat required to circulate these ion-rich solutions. Thus, hydrothermal metamorphism often occurs simultaneously with contact metamorphism in regions where large plutons are emplaced.

As these large magma bodies cool and solidify, the ions that are not incorporated into the crystalline structures of the newly formed silicate minerals, as well as any remaining volatiles (water), are expelled. These ion-rich fluids are called **hydrothermal solutions.** In addition to chemically altering the host rocks, the ions contained in hydrothermal solutions eventually precipitate to form a variety of economically important mineral deposits.

As our understanding of plate tectonics expanded, it became clear that the most widespread occurrence of hydrothermal metamorphism is along the axis of the oceanic ridge system. Here, as plates move apart, upwelling magma from the mantle generates new seafloor. As seawater percolates through the young, hot oceanic crust, it is heated and chemically reacts with the newly formed basaltic rocks. The result is the conversion of ferromagnesian minerals, such as olivine and pyroxene, into hydrated silicates, including serpentine and talc. In addition, large amounts of metals, such as iron, cobalt, nickel, silver, gold and copper, are dissolved from the newly formed crust. These hot, metal-rich fluids eventually rise along fractures and gush from the seafloor at temperatures of about 350°C, generating particle-filled clouds called *black smokers.* Upon mixing with the cold seawater, sulfides and carbonate minerals containing these heavy metals precipitate to form metallic deposits, some of which are economically valuable. This is believed to be the origin of the copper ores mined today on the island of Cyprus.

Regional Metamorphism

The greatest quantity of metamorphic rock is produced during **regional metamorphism** in association with mountain building. During these dynamic events, large segments of Earth's crust are intensely squeezed as lithospheric plates collide (see Figure 7.2). As the rocks are folded and faulted, the crust is shortened and thickened, like a rumpled carpet. This general thickening of the crust results in terrains that are lifted high above sea level.

At the same time that material is elevated to great heights during mountain building, an equal quantity of rock is gradually forced downward. Here in the "roots" of mountains, the most intense metamorphic activity occurs. Deep burial, together with the emplacement of magma from the mantle, are responsible for the elevated temperatures that produce these most intense zones of metamorphism.

Consequently, the cores of many mountain ranges consist of folded and faulted metamorphic rocks, often intermixed with intruded igneous rocks. As these deformed rock masses are uplifted, erosion removes the overlying material to expose the igneous and metamorphic rocks that comprise the central core of the mountain range.

Other Metamorphic Environments

Other types of metamorphism occur that generate comparatively smaller amounts of metamorphic rock in localized concentrations.

Burial Metamorphism. Burial metamorphism occurs in association with very thick accumulations of sedimentary strata in a subsiding basin. Here, low-grade metamorphic conditions may be attained within the lowest layers. Confining pressure and geothermal heat drive the recrystallization of the constituent minerals to change the texture and/or mineralogy of the rock without appreciable deformation.

The depth required for burial metamorphism varies from one location to another, depending on the prevailing geothermal gradient. Low-grade metamorphism often begins at depths of about 8 kilometers (5 miles), where temperatures range between 100°C and 200°C. However, in areas that exhibit steep geothermal gradients, such as near the Salton Sea in California and in northern New Zealand, drilling operations have collected metamorphic minerals from a depth of only a few kilometers.

Metamorphism Along Fault Zones. Near the surface, rock behaves like a brittle solid. Consequently, movement along a fault zone fractures and pulverizes rock. The result is a loosely coherent rock called *fault breccia,* which is composed of broken and crushed rock fragments (Figure 7.16). Displacements along California's San Andreas Fault have created a zone of fault breccia and related rock types over 1000 kilometers long and up to 3 kilometers wide.

Figure 7.16 Fault breccia consisting of large angular fragments. This outcrop, located in Titus Canyon, Death Valley, California, was produced along a fault zone. (Photo by A. P. Trujillo/APT Photos)

Much of the intense deformation associated with fault zones occurs at great depth and thus at high temperatures. In this environment preexisting minerals deform by ductile flow. As large slabs of rock move in opposite directions, the minerals in the fault zone between them tend to form elongated grains that give the rock a foliated or lineated appearance. Rocks formed in these zones of intense ductile deformation are termed *mylonites*.

Impact Metamorphism.   **Impact** (or *shock*) **metamorphism** occurs when high-speed projectiles called *meteorites* (fragments of comets or asteroids) strike Earth's surface. Upon impact the energy of the rapidly moving meteorite is transformed into heat energy and shock waves that pass through the surrounding rocks. The result is pulverized, shattered, and sometimes melted rock. The products of these impacts, called *impactiles*, include mixtures of fused fragmented rock plus glass-rich ejecta that resemble volcanic bombs (Box 7.1). In some cases, a very dense form of quartz (*coesite*) and minute *diamonds* are found. These high-pressure minerals provide convincing evidence that pressures and temperatures at least as great as those

existing in the upper mantle must have been attained at least briefly at Earth's surface.

Metamorphic Zones

In areas affected by metamorphism, there usually exist systematic variations in the mineralogy and often the textures of the rocks. These differences are clearly related to variations in the degree of metamorphism.

Textural Variations

If we begin with a clay-rich sedimentary rock such as shale, a gradual increase in metamorphic intensity is accompanied by a general coarsening of the grain size. Thus, we observe shale changing to a fine-grained slate, which then forms phyllite and through continued recrystallization generates a coarse-grained schist (Figure 7.17). Under more intense conditions a gneissic texture that exhibits layers of dark and light minerals may develop. This systematic transition in metamorphic textures can be observed as we approach the Appalachian Mountains from the west. Beds of shale, which once extended over large areas of the eastern United States, still

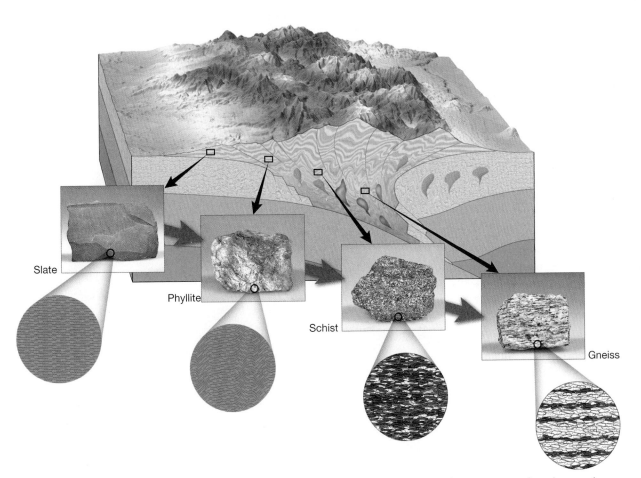

Slate

Phyllite

Schist

Gneiss

Figure 7.17   Idealized illustration of progressive regional metamorphism. From left to right, we progress from low-grade metamorphism (slate) to high-grade metamorphism (gneiss). (Photos by E. J. Tarbuck)

BOX 7.1

Impact Metamorphism and Tektites

It is now clear that comets and asteroids have collided with Earth far more frequently than once thought. The evidence: More than a hundred giant impact structures called *astroblemes* have been identified to date. Many had been thought to result from some poorly understood volcanic process. Most astroblemes, like those found in Indiana and Ohio, are too old to look like an impact crater, but a notable exception is the very fresh-looking Meteor Crater in Arizona (Figure 7.A).

One earmark of astroblemes is impact or shock metamorphism. When high-velocity projectiles (comets, asteroids) impact Earth's surface, pressures reach millions of atmospheres and temperatures exceed 2000°C momentarily. The result is pulverized, shattered, and melted rock. Where impact craters are relatively fresh, shock-melted ejecta and rock fragments ring the impact site. Although most material is deposited close to its source, some ejecta can travel great distances. One example is *tektites,* beads of silica-rich glass, some of which have been aerodynamically shaped like teardrops during flight (Figure 7.B). Most tektites are no more than a few centimeters across and are jet-black to dark green or yellowish. In Australia, millions of tektites are strewn over an area seven times the size of Texas. Several such tektite groupings have been identified worldwide, one stretching nearly halfway around the globe.

No tektite falls have been observed, so their origin is not known with certainty. Because tektites are much higher in silica than volcanic glass (obsidian), a volcanic origin is unlikely. Most researchers agree that tektites are the result of impacts of large projectiles.

One hypothesis suggests an extraterrestrial origin for tektites. Asteroids may have struck the Moon with such force that ejecta "splashed" outward fast enough to escape the Moon's

Figure 7.A   Meteor Crater, located west of Winslow, Arizona. (Photo by Michael Collier)

gravity. Others argue that tektites are terrestrial, but an objection is that some groupings, such as Australia's, lack an identifiable impact crater. However, the object that produced the Australian tektites might have struck the continental shelf, leaving the remnant crater out of sight below sea level. Evidence supporting a terrestrial origin includes tektite falls in western Africa that appear to be the same age as a crater in the same region.

Figure 7.B   Tektites recovered from Nullarbor Plain, Australia. (Photo by Brian Mason/Smithsonian Institution)

occur as nearly flat-lying strata in Ohio. However, in the broadly folded Appalachians of central Pennsylvania, the rocks that once formed flat-lying beds are folded and display a preferred orientation of platy mineral grains as exhibited by well-developed slaty cleav-

age. As we move farther eastward into the intensely deformed crystalline Appalachians, we find large outcrops of schists. The most intense zones of metamorphism are found in Vermont and New Hampshire, where gneissic rocks outcrop.

Figure 7.18 The typical transition in mineralogy that results from progressive metamorphism of shale.

Index Minerals and Metamorphic Grade

In addition to textural changes, we encounter corresponding changes in mineralogy as we shift from regions of low-grade metamorphism to regions of high-grade metamorphism. An idealized transition in

mineralogy that results from the regional metamorphism of shale is shown in Figure 7.18. The first new mineral to form as shale changes to slate is chlorite. At higher temperatures flakes of muscovite and biotite begin to dominate. Under more extreme conditions, metamorphic rocks may contain garnet and staurolite crystals. At temperatures approaching the melting point

Figure 7.19 Zones of metamorphic intensities in New England.

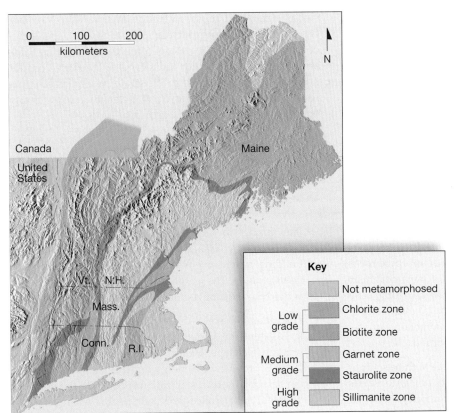

Figure 7.20 Migmatite. The lightest-colored layers are igneous rock composed of quartz and feldspar, while the darker layers have a metamorphic origin. (Photo by Harlan H. Roepke)

of rock, sillimanite forms. Sillimanite is a high-temperature metamorphic mineral used to make refractory porcelains such as those used in spark plugs.

Through the study of metamorphic rocks in their natural settings (called *field studies*) and through experimental studies, researchers have learned that certain minerals are good indicators of the metamorphic environment in which they formed. Using these **index minerals,** geologists distinguish among different zones of regional metamorphism. For example, the mineral chlorite begins to form when temperatures are relatively low, less than 200°C (Figure 7.19). Thus, rocks that contain chlorite (usually slates) are referred to as *low-grade*. By contrast, the mineral sillimanite only forms in very extreme environments where temperatures exceed 600°C, and rocks containing it are considered *high-grade*. By mapping the occurrences of index minerals, geologists are in effect mapping zones of varying metamorphic grade. *Grade* is a term used in a relative sense to refer to the conditions of temperature (or sometimes pressure) to which a rock has been subjected.

Migmatites.   In the most extreme environments, even the highest-grade metamorphic rocks undergo change. For example, gneissic rocks may be heated sufficiently to cause melting to begin. However, recall from our discussion of igneous rocks that different minerals melt at different temperatures. The light-colored silicates, usually quartz and potassium feldspar, have the lowest melting temperatures and begin to melt first, whereas the mafic silicates, such as amphibole and biotite, remain solid. When this partially melted rock cools, the light bands will be composed of igneous, or igneous-appearing components, while the dark bands will consist of unmelted metamorphic material. Rocks of this type are called **migmatites** (Figure 7.20). The light-colored bands in migmatites often form tortuous folds and may contain tabular inclusions of the dark components. Migmatites serve to illustrate the fact that some rocks are transitional and do not clearly belong to any one of the three basic rock groups.

The Chapter in Review

- *Metamorphism* is the transformation of one rock type into another. *Metamorphic rocks* form from preexisting rocks (either igneous, sedimentary, or other metamorphic rocks) that have been altered by the agents of metamorphism, which include *heat, pressure (stress)*, and *chemically active fluids*. During metamorphism the material essentially remains solid. The changes that occur in metamorphosed rocks are textural as well as mineralogical.

- Metamorphism most often occurs in one of three settings: (1) when rock is in contact with magma, *contact* or *thermal metamorphism* occurs; (2) where hot, ion-rich water circu-

lates through rock, chemical alteration occurs by a process called *hydrothermal metamorphism*; or (3) during mountain building, where extensive areas of rock undergo *regional metamorphism*. The greatest volume of metamorphic rock is produced during regional metamorphism.

- The mineral makeup of the parent rock determines, to a large extent, the degree to which each metamorphic agent will cause change. Heat is the most important agent because it provides the energy to drive chemical reactions that result in the recrystallization of minerals. Pressure, like temperature, also increases with depth. When subjected to *confining pressure*, minerals may recrystallize into more compact forms. During mountain building, rocks are subjected to *differential stress*, which tends to shorten them in the direction pressure is applied and lengthen them in the direction perpendicular to that force. At depth, rocks are warm and *ductile*, which accounts for their ability to deform by flowing when subjected to differential stresses. Chemically active fluids, most commonly water containing ions in solution, also enhance the metamorphic process by dissolving minerals and aiding the migration and precipitation of this material at other sites.

- *The grade of metamorphism is reflected in the texture and mineralogy of metamorphic rocks.* During regional metamorphism, rocks typically display a *preferred orientation* called *foliation* in which their platy and elongated minerals are aligned. Foliation develops as platy or elongated minerals are rotated into parallel alignment, recrystallize to form new grains that exhibit a preferred orientation, or are plastically deformed into flattened grains that exhibit a planar alignment. *Rock cleavage* is a type of foliation in which rocks split cleanly into thin slabs along surfaces where platy minerals are aligned. *Schistosity* is a type of foliation defined by the parallel alignment of medium- to coarse-grained platy minerals. During high-grade metamorphism, ion migrations can cause minerals to segregate into distinct layers or bands. Metamorphic rocks with a banded texture are called *gneiss*. Metamorphic rocks composed of only one mineral forming equidimensional crystals often appear *nonfoliated*. *Marble* (metamorphosed limestone) is often nonfoliated. Further, metamorphism can cause the transformation of low-temperature minerals into high-temperature minerals and, through the introduction of ions from *hydrothermal solutions*, generate new miner-

als, some of which form economically important metallic ore deposits.

- Common foliated metamorphic rocks include *slate, phyllite,* various types of *schists* (e.g., garnet-mica schist), and *gneiss.* Nonfoliated rocks include *marble* (parent rock—limestone) and *quartzite* (most often formed from quartz sandstone).

- The three geologic environments in which metamorphism commonly occurs are (1) *contact* or *thermal metamorphism,* (2) *hydrothermal metamorphism* and (3) *regional metamorphism.* Contact metamorphism occurs when rocks are in contact with an igneous body, resulting in the formation of zones of alteration around the magma called *aureols.* Most contact metamorphic rocks are fine-grained, dense, tough rocks of various chemical compositions. Because directional pressure is not a major factor, these rocks are not generally foliated. Hydrothermal metamorphism occurs where hot, ion-rich fluids circulate through rock and cause chemical alteration of the constituent minerals. Most hydrothermal alteration occurs along the oceanic ridge system where seawater migrates through hot oceanic crust and chemically alters newly formed basaltic rocks. Metallic ions that are removed from the crust are eventually carried to the floor of the ocean where they precipitate from black smokers to form metallic deposits, some of which may be economically important. Regional metamorphism takes place at considerable depths over an extensive area and is associated with the process of mountain building. A gradation in the degree of change usually exists in association with regional metamorphism, in which the intensity of metamorphism (low- to high-grade) is reflected in the texture and mineralogy of the rocks. In the most extreme metamorphic environments, rocks called *migmatites* fall into a transition zone *somewhere between* "true" igneous rocks and "true" metamorphic rocks.

Key Terms

aureole (p. 156)
burial metamorphism
   (p. 158)
confining pressure (p. 148)
contact metamorphism
   (p. 156)
differential stress (p. 149)
foliation (p. 150)

gneissic texture (p. 151)
hydrothermal
   metamorphism (p. 157)
hydrothermal solution
   (p. 158)
impact metamorphism
   (p. 159)
index mineral (p. 162)

metamorphism (p. 146)
migmatite (p. 162)
nonfoliated (p. 152)
parent rock (p. 146)
porphyroblastic texture
   (p. 152)
regional metamorphism
   (p. 158)

rock cleavage (p. 151)
schistosity (p. 151)
shield (p. 146)
slaty cleavage (p. 151)
texture (p. 150)
thermal metamorphism
   (p. 156)

Questions for Review

1. What is metamorphism? What are the agents that change rocks?

2. Why is heat considered the most important agent of metamorphism?

3. How is confining pressure different than differential stress?

4. What role do chemically active fluids play in metamorphism?

5. In what two ways can the parent rock affect the metamorphic process?

6. What is foliation? Distinguish between *slaty cleavage, schistosity,* and *gneissic* textures.

7. Briefly describe the three mechanisms by which minerals develop a preferred orientation.

8. List some changes that might occur to a rock in response to metamorphic processes.

9. Slate and phyllite resemble each other. How might you distinguish one from the other?

10. Each of the following statements describes one or more characteristics of a particular metamorphic rock. For each statement, name the metamorphic rock that is being described.
   a. calcite-rich and nonfoliated
   b. loosely coherent rock composed of broken fragments that formed along a fault zone
   c. represents a grade of metamorphism between slate and schist
   d. very fine-grained and foliated; excellent rock cleavage
   e. foliated and composed predominately of platy minerals
   f. composed of alternating bands of light and dark silicate minerals
   g. hard, nonfoliated rock resulting from contact metamorphism

11. Distinguish between contact metamorphism and regional metamorphism. Which creates the greatest quantity of metamorphic rock?

12. Where does most hydrothermal metamorphism occur?

13. Describe burial metamorphism.

14. How do geologists use index minerals?

15. Briefly describe the textural changes that occur in the transformation of slate to phyllite to schist and then to gneiss.

16. How are gneisses and migmatites related?

Web Resources

 The *Essentials of Geology* Website uses the resources and flexibility of the Internet to aid in your study of the topics in this chapter. Written and developed by geology instructors, this site will help improve your understanding of geology. Visit **http://www.prenhall.com/lutgens** and click on the cover of *Essentials of Geology, 8/e* to find:

- On-line review quizzes.
- Web-based critical thinking and writing exercises.
- Links to chapter-specific Web resources.
- Internet-wide key term searches.

http://www.prenhall.com/lutgens

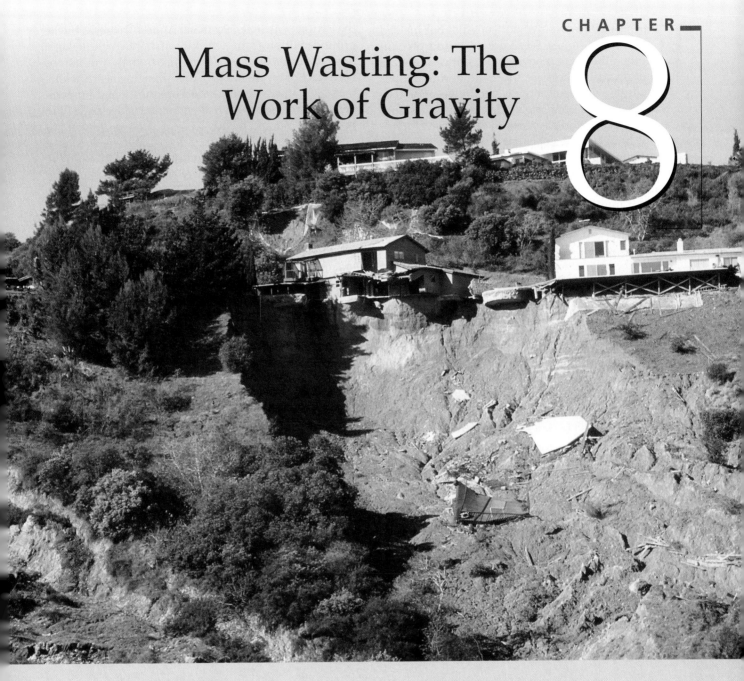

Mass Wasting: The Work of Gravity

Focus on Learning

To assist you in learning the important concepts in this chapter, you will find it helpful to focus on the following questions:

- What is the process of mass wasting?
- What role does mass wasting play in the development of valleys?
- What are the controls and triggers of mass wasting?
- What criteria are used to divide and describe the various types of mass wasting?
- What are the general characteristics of slump, rockslide, debris flow, earthflow, and creep?

This large landslide in Los Angeles's Stone Canyon, left some homes in precarious situations. (Photo by Tim McHugh/Photo Researchers, Inc.)

Earth's surface is never perfectly flat but instead consists of slopes. Some are steep and precipitous; others are moderate or gentle. Some are long and gradual; others are short and abrupt. Slopes can be mantled with soil and covered by vegetation or consist of barren rock and rubble. Taken together, slopes are the most common elements in our physical landscape. Although most slopes may appear to be stable and unchanging, the force of gravity causes material to move downslope. At one extreme, the movement may be gradual and practically imperceptible. At the other extreme, it may consist of a roaring debris flow or a thundering rock avalanche. Landslides are a worldwide natural hazard. When these hazardous processes lead to loss of life and property, they become natural disasters.

A Landslide Disaster in Peru

Occasionally, news media report the terrifying and often grim details of landslides. For example, on May 31, 1970, a gigantic rock avalanche buried more than 20,000 people in Yungay and Ranrahirca, Peru. There was little warning of the impending disaster; it began and ended in just a matter of a few minutes. The avalanche started about 14 kilometers (9 miles) from Yungay, near the summit of the 6700-meter (22,000-foot) Nevados Huascaran,

the loftiest peak in the Peruvian Andes. Triggered by the ground motion from a strong offshore earthquake, a huge mass of rock and ice broke free from the precipitous north face of the mountain. After plunging nearly a kilometer, the material pulverized on impact and immediately began rushing down the mountainside, made fluid by trapped air and melted ice.

The falling debris ripped loose millions of tons of additional debris as it roared downhill. Although the material followed a previously eroded gorge, a portion of the debris jumped a 200 to 300-meter (650- to 1000-foot) bedrock ridge that had protected Yungay from similar events in the past and buried the entire city. After destroying another town in its path, Ranrahirca, the mass of debris finally reached the bottom of the valley. There, its momentum carried it across the Rio Santa and tens of meters up the valley wall on the opposite side.

Rapid forms of *mass wasting* are significant geologic hazards. Each year landslides in the United States are responsible for 25 to 50 deaths and produce more than $1.5 billion in losses. Worldwide, landslides cause an average of 600 deaths annually. Sometimes extreme events occur. Such was the case in November 1998, when debris flows were triggered by heavy rains from Hurricane Mitch (Figure 8.1). Thousands perished in Nicaragua and Honduras. Thousands more lost their

Figure 8.1 Heavy rains from Hurricane Mitch in November 1998 set off devastating debris flows in parts of Central America. The lack of adequate anchoring vegetation and the fact that many people lived in vulnerable sites added to the tragedy. (Photo by Noel Quidu/Liaison Agency, Inc.)

homes. The economies of these two small countries were devastated.

As with many geologic hazards, the tragic rock avalanche in Peru was triggered by a natural event—in this case, an earthquake. In fact, most mass-wasting events, whether spectacular or subtle, are the result of circumstances that are completely independent of human activities. In places where mass wasting is a recognized threat, steps can often be taken to control downslope movements or limit the damages that such movements can cause. If the potential for mass wasting goes unrecognized or is ignored, the results can be costly and dangerous. It should also be pointed out that, although most downslope movements occur whether people are present or not, many occurrences each year are aggravated or even triggered by human actions.

Mass Wasting and Landform Development

Landslides are spectacular examples of a common geologic process called mass wasting. **Mass wasting** refers to the downslope movement of rock, regolith, and soil under the direct influence of gravity. It is distinct from the erosional processes that are examined in subsequent chapters because mass wasting does not require a transporting medium such as water, wind, or glacial ice.

The Role of Mass Wasting

In the evolution of most landforms, mass wasting is the step that follows weathering. By itself weathering does not produce significant landforms. Rather, landforms develop as the products of weathering are removed from the places where they originate. Once weathering weakens and breaks rock apart, mass wasting transfers the debris downslope, where a stream, acting as a conveyor belt, usually carries it away. Although there may be many intermediate stops along the way, the sediment is eventually transported to its ultimate destination: the sea.

The combined effects of mass wasting and running water produce stream valleys, which are the most common and conspicuous of Earth's landforms. If streams alone were responsible for creating the valleys in which they flow, valleys would be very narrow features. However, the fact that most river valleys are much wider than they are deep is a strong indication of the significance of mass-wasting processes in supplying material to streams. This is illustrated by the Grand Canyon (Figure 8.2). The walls of the canyon extend far from the Colorado River, owing to the transfer of weathered debris downslope to the river and its tributaries by mass-wasting processes. In this manner, streams and mass wasting combine to modify and sculpture the surface. Of course, glaciers, groundwater, waves, and wind are also important agents in shaping landforms and developing landscapes.

Slopes Change through Time

It is clear that if mass wasting is to occur, there must be slopes down which rock, soil, and regolith can move. It is Earth's mountain building and volcanic processes that produce these slopes through sporadic changes in the elevations of landmasses and the ocean floor. If dynamic internal processes did not continually produce regions having higher elevations, the system that moves debris to lower elevations would gradually slow and eventually cease.

Most rapid and spectacular mass-wasting events occur in areas of rugged, geologically young mountains. Newly formed mountains are rapidly eroded by rivers and glaciers into regions characterized by steep and un-

Figure 8.2 The walls of the Grand Canyon extend far from the channel of the Colorado River. This results primarily from the transfer of weathered debris downslope to the river and its tributaries by mass-wasting processes. (Photo by Tom and Susan Bean, Inc.)

stable slopes. It is in such settings that massive destructive landslides, such as the Yungay disaster, occur. As mountain building subsides, mass wasting and erosional processes lower the land. Through time, steep and rugged mountain slopes give way to gentler, more subdued terrain. Thus, as a landscape ages, massive and rapid mass-wasting processes give way to smaller, less dramatic downslope movements.

Controls and Triggers of Mass Wasting

Gravity is the controlling force of mass wasting, but several factors play an important role in overcoming inertia and creating downslope movements. Long before a landslide occurs, various processes work to weaken slope material, gradually making it more and more susceptible to the pull of gravity. During this span, the slope remains stable but gets closer and closer to being unstable. Eventually, the strength of the slope is weakened to the point that something causes it to cross the threshold from stability to instability. Such an event that initiates downslope movement is called a *trigger*. Remember that the trigger is not the sole cause of the mass-wasting event, but just the last of many causes. Among the common factors that trigger mass-wasting processes are saturation of material with water, oversteepening of slopes, removal of anchoring vegetation, and ground vibrations from earthquakes.

The Role of Water

Mass wasting is sometimes triggered when heavy rains or periods of snowmelt saturate surface materials (see Figure 8.1). When the pores in sediment become filled with water, the cohesion among particles is destroyed, allowing them to slide past one another with relative ease. For example, when sand is slightly moist, it sticks together quite well. However, if enough water is added to fill the openings between the grains, the sand will ooze out in all directions. Thus, saturation reduces the internal resistance of materials, which are then easily set in motion by the force of gravity. When clay is wetted, it becomes very slick—another example of the lubricating effect of water. Water also adds considerable weight to a mass of material. The added weight in itself may be enough to cause the material to slide or flow downslope.

Oversteepened Slopes

Oversteepening of slopes is another trigger of many mass movements. There are many situations in nature where oversteepening takes place. A stream undercut-

ting a valley wall and waves pounding against the base of a cliff are but two familiar examples. Furthermore, through their activities, people often create oversteepened and unstable slopes that become prime sites for mass wasting.

Unconsolidated, granular particles (sand-sized or coarser) assume a stable slope called the **angle of repose.** This is the steepest angle at which material remains stable. Depending on the size and shape of the particles, the angle varies from 25 to 40 degrees. The larger, more angular particles maintain the steepest slopes. If the angle is increased, the rock debris will adjust by moving downslope.

Oversteepening is not just important because it triggers movements of unconsolidated granular materials. Oversteepening also produces unstable slopes and mass movements in cohesive soils, regolith, and bedrock. The response will not be immediate, as with loose, granular material, but sooner or later one or more mass-wasting processes will eliminate the oversteepening and restore stability to the slope.

Removal of Vegetation

Plants protect against erosion and contribute to the stability of slopes because their root systems bind soil and regolith together. Where plants are lacking, mass wasting is enhanced, especially if slopes are steep and water is plentiful. When anchoring vegetation is removed by forest fires or by people (for timber, farming, or development), surface materials frequently move downslope.

An unusual example illustrating the anchoring effect of plants occurred several decades ago on steep slopes near Menton, France. Farmers replaced olive trees, which have deep roots, with a more profitable but shallow-rooted crop: carnations. When the less stable slope failed, the landslide took 11 lives.

Earthquakes as Triggers

Conditions that favor mass wasting may exist in an area for a long time without movement occurring. An additional factor is sometimes necessary to trigger the movement. Among the more important and dramatic triggers are earthquakes. An earthquake and its aftershocks can dislodge enormous volumes of rock and unconsolidated material. The event in the Peruvian Andes described at the beginning of this chapter is one tragic example.

Landslides Triggered by the Northridge Earthquake. In January 1994 an earthquake struck the Los Angeles region of southern California. Named for its epicenter in the town of Northridge, the 6.7-magnitude event produced estimated losses of $20 billion. Some of these losses were the result of thousands of landslides over an area of 10,000 square kilometers

Figure 8.3   Various forms of mass wasting can be triggered by earthquakes. This home in Pacific Palisades, California, was destroyed by a landslide triggered by the January 1994 Northridge earthquake. In some cases, damages from earthquake-induced mass wasting are greater than damages caused directly by an earthquake's ground vibrations (Photo by Chromo Sohm/Corbis/The Stock Market)

(3900 square miles) that were set in motion by the quake (Figure 8.3). Most were shallow rock falls and slides, but some were much larger and filled canyon bottoms with jumbles of soil, rock, and plant debris. The debris in canyon bottoms created a secondary threat because it can be mobilized during rainstorms to produce debris flows. Such flows are common and often disastrous in southern California.

The mass-wasting processes triggered by the Northridge earthquake destroyed dozens of homes and caused extensive damage to roads, pipelines, and well machinery in oil fields. In some places more than 75 percent of slope areas were denuded by landslides, making them vulnerable to subsequent mass wasting triggered by heavy rains.

Liquification.   Intense ground shaking during earthquakes can cause water-saturated surface materials to lose their strength and behave as fluidlike masses that flow. This process, called *liquefaction*, was a major cause of property damage in Anchorage, Alaska, during the massive 1964 Good Friday earthquake described in Chapter 15.

Landslides Without Triggers?

Do rapid mass-wasting events always require some sort of trigger such as heavy rains or earthquakes? The answer is no. Such events sometimes occur without being triggered. For example, on the afternoon of May 9, 1999, a landslide killed 10 hikers and injured many others at Sacred Falls State Park near Hauula on the north shore of Oahu, Hawaii. The tragic event occurred when a mass of rock from a canyon wall plunged 150 meters (500 feet) down a nearly vertical slope to the valley floor. Because of safety concerns, the park was closed so that

landslide specialists from the U.S. Geological Survey could investigate the site. Their study concluded that the landslide occurred *without being triggered* by any discernible external conditions.

Many rapid mass-wasting events occur without a discernible trigger. Slope materials gradually weaken over time under the influence of long-term weathering, infiltration of water, and other physical processes. Eventually, if the strength falls below what is necessary to maintain slope stability, a landslide will occur. The timing of such events is random, and thus accurate prediction is impossible.

Classification of Mass-Wasting Processes

There is a broad array of different processes that geologists call mass wasting. Four processes are illustrated in Figure 8.4. Generally, the different types are classified based on the type of material involved, the kind of motion displayed, and the velocity of the movement.

Type of Material

The classification of mass-wasting processes on the basis of the material involved in the movement depends upon whether the descending mass began as unconsolidated material or as bedrock. If soil and regolith dominate, terms such as debris, mud, or earth are used in the description. In contrast, when a mass of bedrock breaks loose and moves downslope, the term rock may be part of the description.

Type of Motion

In addition to characterizing the type of material involved in a mass-wasting event, the way in which the material moves may also be important. Generally, the kind of motion is described as either a fall, a slide, or a flow.

Fall.   When the movement involves the freefall of detached individual pieces of any size, it is termed a **fall.** Fall is a common form of movement on slopes that are so steep that loose material cannot remain on the surface. The rock may fall directly to the base of the slope or move in a series of leaps and bounds over other rocks along the way. Many falls result when freeze and thaw cycles and/or the action of plant roots loosen rock to the point that gravity takes over. Although signs along bedrock cuts on highways warn of falling rock, few of us have actually witnessed such an event. However, as Figure 8.5 illustrates, they do indeed occur. In fact, this is the primary way in which **talus slopes** are built and maintained (see Figure 5.3, p. 103). Sometimes falls may trigger other forms of downslope movement. For example, recall that the Yungay disas-

A. Slump

B. Rockside

C. Debris flow

D. Earthflow

Figure 8.4 The four processes illustrated here are all considered to be relatively rapid forms of mass wasting. Because material in slumps **A.** and rockslides **B.** move along well-defined surfaces, they are said to move by sliding. By contrast, when material moves downslope as a viscous fluid, the movement is described as a flow. Debris flow **C.** and earthflow **D.** advance downslope in this manner.

ter described at the beginning of the chapter was initiated by a mass of freefalling material that broke from the nearly vertical summit of Nevado Huascaran.

Slide. Many mass-wasting processes are described as **slides.** Slides occur whenever material remains fairly coherent and moves along a well-defined surface. Sometimes the surface is a joint, a fault, or a bedding plane that is approximately parallel to the slope. However, in the case of the movement called *slump*, the descending material moves en masse along a curved surface of rupture.

Figure 8.5 Rockfall blocking Montana Highway 2 east of Pipestone Pass, May 1998. (AP/Wide World Photo)

A note of clarification is appropriate at this point. Sometimes the word *slide* is used as a synonym for the word *landslide*. It should be pointed out that although many people, including geologists, use the term, the word *landslide* has no specific definition in geology. Rather, it should be considered as a popular nontechnical term used to describe all perceptible forms of mass wasting, including those in which sliding does not occur.

Flow.   The third type of movement common to mass-wasting processes is termed **flow.** Flow occurs when material moves downslope as a viscous fluid. Most flows are saturated with water and typically move as lobes or tongues.

Rate of Movement

The event described at the beginning of this chapter clearly involved rapid movement. The rock and debris moved downslope at speeds well in excess of 200 kilometers (125 miles) per hour. This most rapid type of mass movement is termed a **rock avalanche.** Many researchers believe that rock avalanches, such as the one that produced the scene in Figure 8.6, must literally float on air as they move downslope. That is, high velocities result when air becomes trapped and compressed beneath the falling mass of debris, allowing it to move as a buoyant, flexible sheet across the surface.

Most mass movements, however, do not move with the speed of a rock avalanche. In fact, a great deal of mass wasting is imperceptibly slow. One process that we will examine later, termed *creep*, results in particle movements that are usually measured in millimeters or centimeters per year. Thus, as you can see, rates of movement can be spectacularly sudden or exceptionally gradual. Although various types of mass wasting are often classified as either rapid or slow, such a distinction is highly subjective because there is a wide range of rates between the two extremes. Even the velocity of a single process at a particular site can vary considerably.

Slump

Slump refers to the downward sliding of a mass of rock or unconsolidated material moving as a unit along a curved surface (Figure 8.4A). Usually the slumped material does not travel spectacularly fast nor very far. This is a common form of mass wasting, especially in thick accumulations of cohesive materials such as clay. The rupture surface is characteristically spoon-shaped and concave upward or outward. As the movement occurs, a crescent-shaped scarp is created at the head, and the block's upper surface is sometimes tilted backward. Although slump may involve a single mass, it often consists of multiple blocks. Sometimes water collects between the base of the scarp and the top of the tilted block. As this water percolates downward along the surface of rupture, it may promote further instability and additional movement.

Slump commonly occurs because a slope has been oversteepened. The material on the upper portion of a slope is held in place by the material at the bottom of the slope. As this anchoring material at the base is removed, the material above is made unstable and reacts to the pull of gravity. One relatively common example is a valley wall that becomes oversteepened by a meandering river. Another is a coastal cliff that has been undercut by wave action at its base.

Slumping may also occur when a slope is overloaded, causing internal stress on the material below. This type of slump often occurs where weak, clay-rich material underlies layers of stronger, more resistant rock such as sandstone. The seepage of water through the

Figure 8.6   This 4-kilometer-long tongue of rubble was deposited atop Alaska's Sherman Glacier by a rock avalanche. The event was triggered by a tremendous earthquake in March 1964. (Photo by Austin Post, U.S. Geological Survey)

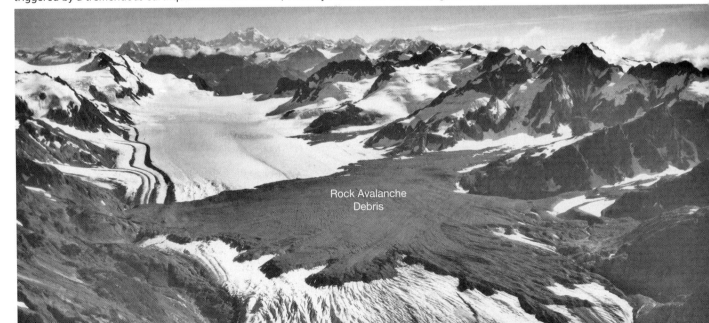

upper layers reduces the strength of the clay below, causing slope failure.

Rockslide

Rockslides frequently occur in high mountain areas such as the Andes, Alps, and Canadian Rockies. They are sudden and rapid movements that happen when detached segments of bedrock break loose and slide downslope (Figure 8.4B). As the moving mass thunders along the surface, it breaks into many smaller pieces. Such events are among the fastest and most destructive mass movements.

Rockslides usually take place where there is an inclined surface of weakness. Such surfaces tend to form where strata are tilted or where joints and fractures exist parallel to the slope. When rock in such a setting is undercut at the base of the slope, it loses support and eventually gives way. Sometimes the rockslide is triggered when rain or melting snow lubricates the underlying surface to the point that friction is no longer sufficient to hold the rock unit in place. As a result, rockslides tend to be more common during the spring, when heavy rains and melting snow are most prevalent.

As was noted earlier, earthquakes can trigger rockslides and other mass movements. The 1811 earthquake at New Madrid, Missouri, for example, caused slides in an area of more than 13,000 square kilometers (5000 square miles) along the Mississippi River valley. A more recent example occurred on August 17, 1959, when a severe earthquake west of Yellowstone National Park triggered a massive slide in the canyon of the Madison River in southwestern Montana (Figure 8.7). In a mat-

ter of moments an estimated 27 million cubic meters of rock, soil, and trees slid into the canyon. The debris dammed the river and buried a campground and highway. More than 20 unsuspecting campers perished.

Not far from the site of the Madison Canyon slide, the classic Gros Ventre rockslide occurred 34 years earlier. The Gros Ventre River flows west from the northernmost part of the Wind River Range in northwestern Wyoming, through the Grand Teton National Park, and eventually empties into the Snake River. On June 23, 1925, a massive rockslide took place in its valley, just east of the small town of Kelly. In the span of just a few minutes a great mass of sandstone, shale, and soil crashed down the south side of the valley, carrying with it a dense pine forest. The volume of debris, estimated at 38 million cubic meters (50 million cubic yards), created a 70-meter-high dam on the Gros Ventre River. Because the river was completely blocked, a lake was created. It filled so quickly that a house that had been 18 meters (60 feet) above the river was floated off its foundation 18 hours after the slide. In 1927, the lake overflowed the dam, partially draining the lake and resulting in a devastating flood downstream.

Why did the Gros Ventre rockslide take place? Figure 8.8 shows a diagrammatic cross-sectional view of the geology of the valley. Notice the following points: (1) The sedimentary strata in this area dip (tilt) 15 to 21 degrees; (2) underlying the bed of sandstone is a relatively thin layer of clay; and (3) at the bottom of the valley, the river had cut through much of the sandstone layer. During the spring of 1925, water from heavy rains and melting snow seeped through the sandstone, saturating the clay below. Because much of the sandstone

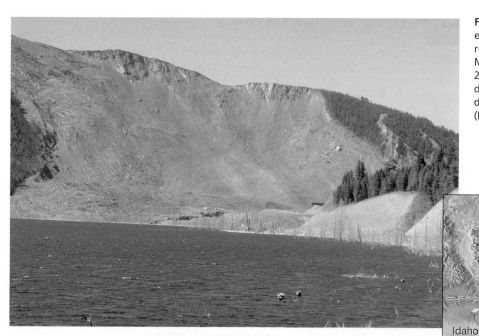

Figure 8.7   On August 17, 1959, an earthquake triggered a massive rockslide in the canyon of Montana's Madison River. An estimated 27 million cubic meters of debris slid down the canyon wall, forming a dam that created Earthquake Lake. (Photo by John Montagne)

Figure 8.8 Part A shows a cross-sectional view of the Gros Ventre rockslide. The slide occurred when the tilted and undercut sandstone bed could no longer maintain its position atop the saturated bed of clay. As the photo in part B illustrates, even though the Gros Ventre rockslide occurred in 1925, the scar left on the side of Sheep Mountain is still a prominent feature. (Part A after W.C. Alden, "Landslide and Flood at Gros Ventre, Wyoming," Transactions (AIME) 76 (1928); 348. Part B photo by Stephen Trimble)

layer had been cut through by the Gros Ventre River, the layer had virtually no support at the bottom of the slope. Eventually the sandstone could no longer hold its position on the wetted clay, and gravity pulled the mass down the side of the valley. The circumstances at this location were such that the event was inevitable.

Debris Flow

Debris flow is a relatively rapid type of mass wasting that involves a flow of soil and regolith containing a large amount of water (Figure 8.4C, p. 170). Debris flows, which are also called *mudflows*, are most characteristic of semiarid mountainous regions and are also common on the slopes of some volcanoes. Because of their fluid properties, debris flows frequently follow canyons and stream channels. In populated areas, debris flows can pose a significant hazard to life and property (see Box 8.1).

Debris Flows in Semiarid Regions

When a cloudburst or rapidly melting mountain snows create a sudden flood in a semiarid region, large quantities of soil and regolith are washed into nearby stream channels because there is usually little vegetation to anchor the surface material. The end product is a flowing tongue of well-mixed mud, soil, rock, and water. Its consistency can range from that of wet concrete to a soupy mixture not much thicker than muddy water. The rate of flow therefore depends not only on the slope but also on the water content. When dense, debris flows are capable of carrying or pushing large boulders, trees, and even houses with relative ease.

Debris flows pose a serious hazard to development in relatively dry mountainous areas such as southern California. Here construction of homes on canyon hillsides and the removal of native vegetation by brush fires and other means have increased the frequency of these destructive events (Figure 8.9). Moreover, when a debris flow reaches the end of a steep, narrow canyon, it spreads out, covering the area beyond the mouth of the canyon with a mixture of wet debris. This material contributes to the buildup of fanlike deposits at canyon mouths.* The fans are relatively easy to build on, often have nice views, and are close to the mountains; in fact, like the nearby canyons, many have become preferred sites for development. Because debris flows occur only

*These structures are called *alluvial fans* and will be discussed in greater detail in Chapters 9 and 12

BOX 8.1

Real-Time Monitoring of Active Landslides*

Landslides threaten lives and property in all 50 states. To reduce the risk from active landslides, the U.S. Geological Survey (USGS) develops and uses real-time landslide monitoring systems. Monitoring can detect early indications of rapid catastrophic movement.

During heavy rains in January 1997, thousands of tons of rock and debris gave way in the Sierra Nevada, destroying three homes, blocking a major northern California highway (U.S. 50), and briefly damming the adjacent American River (Figure 8.A). Waters dammed by the landslide quickly flooded two vehicles traveling on the highway, leading to dramatic rescues. Reopening the highway cost $4.5 million, and indirect economic losses from the highway closure were estimated to exceed $50 million.

Several large, active landslides in the same area continue to threaten U.S. Highway 50. Although these slides often move slowly, some have the po-

tential to collapse catastrophically. To help reduce the risk posed by these rapid movements, the USGS, in cooperation with other agencies, established continuous real-time monitoring of the landslide activity using systems developed for monitoring active volcanoes in remote areas. Data from a variety of sensors installed at the landslides are transmitted by radio to USGS computers (Figure 8.B).

Why is it necessary to collect real-time landslide data remotely? The immediate detection of landslide activity provided by real-time systems can be crucial in saving human lives and protecting property. Monitoring can determine the speed of landslide movement and can also detect early indications of catastrophic movement. Rapid movement at some landslides, for example, is preceded by gradual acceleration. By using real-time data from such slides, geologists can anticipate possible catastrophic movement. USGS landslide monitoring focuses on detecting (1) pre-

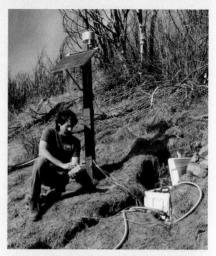

Figure 8.B Testing a solar-powered radio telemetry system for remote transmission of real-time landslide data. At remotely monitored landslide areas, ground movement and groundwater pressures are measured every second. The amount of downslope movement is recorded by extensometers that can detect stretching or shortening of the ground. Ground vibrations are monitored by geophones buried within the slides. Groundwater conditions within slides are monitored by porewater-pressure sensors, and on-site rain gauges record precipitation. (Photo by Mark Reid, U.S. Geological Survey)

cipitation and groundwater conditions that could destabilize a hillslope, (2) acceleration of slide movement, and (3) ground vibrations associated with movement.

Currently, five landslides along U.S. Highway 50 are monitored in real time using 58 instruments. This network, operated in cooperation with the California Department of Transportation, provides early notification of landslide activity and information useful in the design of remedial measures to halt these slides. The USGS operates other remote real-time landslide monitoring sites elsewhere in California and in Washington, New Mexico, and Colorado.

Figure 8.A Home destroyed by a sudden catastrophic landslide in 1997 along U.S. Highway 50, about 25 miles east of Placerville, California. (Photo by Mark Reid, U.S. Geological Survey)

*Based on material prepared by the U.S. Geological Survey.

sporadically, the public is often unaware of the potential hazard of such sites.

Lahars

Debris flows composed mostly of volcanic materials on the flanks of volcanoes are called **lahars.** The word originated in Indonesia, a volcanic region that has experienced many of these often destructive events. Historically, lahars have been one of the deadliest volcano hazards. They can occur either during an eruption or when a volcano is quiet. They take place when highly unstable layers of ash and debris become saturated with water and flow down steep volcanic slopes, generally following existing stream channels. Heavy rainfalls often trigger these flows. Others are initiated when large volumes of ice and snow are suddenly melted by heat flowing to the surface from within the volcano or by the hot gases and near-molten debris emitted during a violent eruption.

When Mount St. Helens erupted in May 1980, several lahars were created. The flows and accompanying floods raced down the valleys of the north and south forks of the Toutle River at speeds that were often in excess of 30 kilometers (20 miles) per hour. Fortunately, the affected area was not densely settled. Nevertheless, more than 200 homes were destroyed or severely damaged. Most bridges met a similar fate.

In November 1985, lahars were produced during the eruption of Nevado del Ruiz, a 5300-meter (17,400-foot) volcano in the Andes Mountains of Colombia. The eruption melted much of the snow and ice that capped the uppermost 600 meters (2000 feet) of the peak, producing torrents of hot, thick mud, ash, and debris. The lahars moved outward from the volcano, following the valleys of three rain-swollen rivers that radiate from the peak. The flow that moved down the valley of the Lagunilla River was the most destructive. It devastated the town of Armero, 48 kilometers (30 miles) from the mountain. Most of the more than 25,000 deaths caused by the event occurred in this once thriving agricultural community.

Death and property damage due to the lahars also occurred in 13 other villages within the 180-square-kilometer (70-square-mile) disaster area. Although a great deal of pyroclastic material was explosively ejected from Nevado del Ruiz, it was the lahars triggered by this eruption that made this such a devastating natural disaster. In fact, it was the worst volcanic disaster since 28,000 people died following the 1902 eruption of Mount Pelée on the Caribbean island of Martinique.*

Earthflow

We have seen that debris flows are frequently confined to channels in semiarid regions. In contrast, **earthflows** most often form on hillsides in humid areas during times of heavy precipitation or snowmelt (see Figure 8.4D, p. 170). When water saturates the soil and regolith on a hillside, the material may break away, leaving a scar on the slope and forming a tongue- or teardrop-shaped mass that flows downslope (Figure 8.10).

The materials most commonly involved are rich in clay and silt and contain only small proportions of sand and coarser particles. Earthflows range in size from bodies a few meters long, a few meters wide, and less than a meter deep to masses more than a kilometer long, several hundred meters wide, and more than 10 meters deep.

*A discussion of the Mount Pelée eruption as well as additional material on lahars can be found in Chapter 4.

Figure 8.10 This small, tongue-shaped earthflow occurred on a newly formed slope along a recently constructed highway. It formed in clay-rich material following a period of heavy rain. Notice the small slump at the head of the earthflow. (Photo by E. J. Tarbuck)

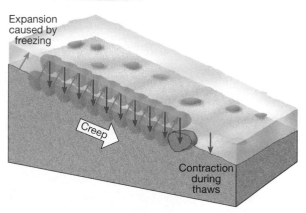

Figure 8.11 The repeated expansion and contraction of surface material causes a net downslope migration of rock particles—a process called *creep*.

BOX 8.2

The Sensitive Permafrost Landscape

Many of the mass-wasting events described in this chapter had sudden and disastrous impacts on people. When the activities of people cause ice contained in permanently frozen ground to melt, the impact is more gradual and less deadly. Nevertheless, because permafrost regions are sensitive and fragile landscapes, the scars resulting from poorly planned actions can remain for generations.

Permanently frozen ground, known as *permafrost*, occurs where summers are too cool to melt more than a shallow surface layer. Deeper ground remains frozen year-round. Permafrost covers more than 80 percent of Alaska and

Figure 8.D This building, located south of Fairbanks, Alaska, subsided because of thawing permafrost. Notice that the right side, which was heated, settled much more than the unheated porch on the left.

about 50 percent of Canada, as well as a substantial portion of northern Siberia.

When people disturb the surface by removing the insulating vegetation mat or by constructing roads and buildings, the delicate thermal balance is disturbed, and the permafrost can thaw (Figure 8.C). Thawing produces unstable ground that may slide, slump, subside, and undergo severe frost heaving.

As Figure 8.D illustrates, when a heated structure is built directly on permafrost that contains a high proportion of ice, thawing creates soggy material into which a building can sink. One solution is to place buildings and other structures on piles, like stilts. Such piles allow subfreezing air to circulate between the floor of the building and the soil and thereby keep the ground frozen.

When oil was discovered on Alaska's North Slope, many people were

concerned about the building of a pipeline linking the oil fields of Prudhoe Bay to the ice-free port of Valdez 1300 kilometers to the south. There was serious concern that such a massive project might damage the sensitive permafrost environment. Many also worried about possible oil spills.

Because oil must be heated to about 60°C to flow properly, special engineering procedures had to be developed to isolate this heat from the permafrost. Methods included insulating the pipe, elevating portions of the pipeline above ground level, and even placing cooling devices in the ground to keep it frozen. The Alaska pipeline is clearly one of the most complex and costly projects ever built in the Arctic tundra. Detailed studies and careful engineering helped minimize adverse effects resulting from the disturbance of frozen ground.

Figure 8.C When a rail line was built across this permafrost landscape in Alaska, the ground subsided. (Photo by Lynn A. Yehle/O. J. Ferrians, Jr./U.S. Geological Survey)

Because earthflows are quite viscous, they generally move at slower rates than the more fluid debris flows described in the preceding section. They are characterized by a slow and persistent movement and may remain active for periods ranging from days to years. Depending on the steepness of the slope and the material's consistency, measured velocities range from less than 1 millimeter per day up to several meters per day. Over the time span that earthflows are active, movement is typically faster during wet periods than during drier times. In addition to occurring as isolated hillside phenomena, earthflows commonly take place in association with large slumps. In this situation, they may be seen as tonguelike flows at the base of the slump block.

Slow Movements

Movements such as rockslides, rock avalanches, and lahars are certainly the most spectacular and catastrophic forms of mass wasting. These dangerous events deserve intensive study to enable more effective prediction, timely warnings, and better controls to save lives. However, because of their large size and spectacular nature, they give us a false impression of their importance as a mass-wasting process. Indeed, sudden movements are responsible for moving less material than the slower and far more subtle action of creep. Whereas rapid types of mass wasting are characteristic of mountains and steep hillsides, creep takes place on both steep and gentle slopes and is thus much more widespread.

Creep

Creep is a type of mass wasting that involves the gradual downhill movement of soil and regolith. One factor that contributes to creep is the alternate expansion and contraction of surface material caused by freezing and thawing or wetting and drying. As shown in Figure 8.11, freezing or wetting lifts particles at right angles to the slope, and thawing or drying allows the particles to fall back to a slightly lower level. Each cycle therefore moves the material a tiny distance downslope. Creep is aided by anything that disturbs the soil. For example, raindrop impact and disturbance by plant roots and burrowing animals may contribute.

Creep is also promoted when the ground becomes saturated with water. Following a heavy rain or snowmelt, a waterlogged soil may lose its internal cohesion, allowing gravity to pull the material downslope.

Figure 8.12 Solifluction lobes northeast of Fairbanks, Alaska. Solifluction occurs in permafrost regions when the active layer thaws in summer. (Photo by James E. Patterson)

Because creep is imperceptibly slow, the process cannot be observed in action. What can be observed, however, are the effects of creep. Creep causes fences and utility poles to tilt and retaining walls to be displaced.

Solifluction

When soil is saturated with water, the soggy mass may flow downslope at a rate of a few millimeters or a few centimeters per day or per year. Such a process is called **solifluction** (literally "soil flow"). It is a type of mass wasting that is common wherever water cannot escape from the saturated surface layer by infiltrating to deeper levels. A dense clay hardpan in soil or an impermeable bedrock layer can promote solifluction.

Solifluction is a form of mass wasting that is common in regions underlain by **permafrost.** Permafrost refers to the permanently frozen ground that occurs in association with Earth's harsh tundra and ice-cap climates (see Box 8.2). Solifluction occurs in a zone above the permafrost called the *active layer*, which thaws to a depth of about a meter during the brief high-latitude summer and then refreezes in winter. During the summer season, water is unable to percolate into the impervious permafrost layer below. As a result, the active layer becomes saturated and slowly flows. The process can occur on slopes as gentle as 2 to 3 degrees. Where there is a well-developed mat of vegetation, a solifluction sheet may move in a series of well-defined lobes or as a series of partially overriding folds (Figure 8.12).

The Chapter in Review

- *Mass wasting* refers to the downslope movement of rock, regolith, and soil under the direct influence of gravity. In the evolution of most landforms, mass wasting is the step that follows weathering. The combined effects of mass wasting and erosion by running water produce stream valleys.

- *Gravity is the controlling force of mass wasting.* Other factors that influence or trigger downslope movements are saturation of the material with water, oversteepening of slopes beyond the *angle of repose*, removal of vegetation, and ground shaking by earthquakes.

- The various processes included under the name of mass wasting are divided and described on the basis of (1) the type of material involved (debris, mud, earth, or rock); (2) the type of motion (fall, slide, or flow); and (3) the rate of movement (rapid or slow).

- The more rapid forms of mass wasting include *slump*, the downward sliding of a mass of rock or unconsolidated material moving as a unit along a curved surface; *rockslide*, blocks of bedrock breaking loose and sliding downslope; *debris flow*, a relatively rapid flow of soil and regolith con-

taining a large amount of water; and *earthflow*, an unconfined flow of saturated, clay-rich soil that most often occurs on a hillside in a humid area following heavy precipitation or snowmelt.

- The slowest forms of mass wasting include *creep*, the gradual downhill movement of soil and regolith, and *solifluction*, the gradual flow of a saturated surface layer that is underlain by an impermeable zone. Common sites for solifluction include regions underlain by *permafrost* (permanently frozen ground associated with tundra and ice-cap climates).

Key Terms

angle of repose (p. 168)	fall (p. 169)	permafrost (p. 177)	slump (p. 171)
creep (p. 177)	flow (p. 171)	rock avalanche (p. 171)	solifluction (p. 177)
debris flow (p. 173)	lahar (p. 172)	rockslide (p. 172)	talus slope (p. 169)
earthflow (p. 175)	mass wasting (p. 170)	slide (p. 170)	

Questions for Review

1. Describe how mass-wasting processes contribute to the development of stream valleys.

2. What is the controlling force of mass wasting?

3. How does water affect mass wasting?

4. Describe the significance of the angle of repose.

5. How might the removal of vegetation by fire or logging promote mass wasting?

6. How are earthquakes linked to landslides?

7. Distinguish among fall, slide, and flow.

8. Why can rock avalanches move at such great speeds?

9. Slump and rockslide both move by sliding. In what ways do these processes differ?

10. What factors led to the massive rockslide at Gros Ventre, Wyoming?

11. Compare and contrast debris flow and earthflow.

12. Describe the mass wasting that occurred at Mount St. Helens during its active period in 1980 and at Nevado del Ruiz in 1985.

13. Describe the mechanisms that contribute to the slow downslope movement called *creep*.

14. What is permafrost? What areas in the Northern Hemisphere are affected? (See Box 8.2.)

15. During what season does solifluction occur in permafrost regions?

Web Resources

The *Essentials of Geology* Website uses the resources and flexibility of the Internet to aid in your study of the topics in this chapter. Written and developed by geology instructors, this site will help improve your understanding of geology. Visit **www.prenhall.com/lutgens** and click on the cover of *Essentials of Geology, 8/e* to find:

- On-line quizzes.
- Web-based critical thinking and writing exercises.
- Links to chapter-specific Web resources.
- Internet-wide key term searches.

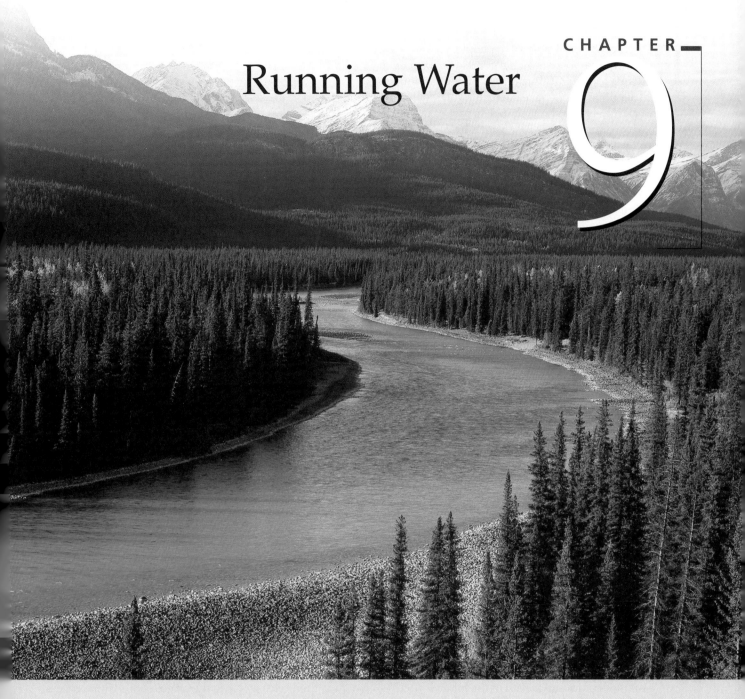

Running Water

Focus on Learning

To assist you in learning the important concepts in this chapter, you will find it helpful to focus on the following questions:

- What is the hydrologic cycle? What is the source of energy that powers the cycle?
- What are the factors that determine the velocity of water in a stream?
- In what way is base level related to a stream's ability to erode?
- The work of a stream includes what three processes?
- What are the two general types of stream valleys and some features associated with each?
- What are the common drainage patterns produced by streams?
- Why do floods occur? What are some basic flood-control strategies?

Athabasca River, Jasper National Park, Alberta, Canada. (Photo by Carr Clifton Photography)

Rivers are very important to people. We use them as highways for moving goods, as sources of water for irrigation, and as an energy source. Their fertile floodplains have been cultivated since the dawn of civilization. When viewed as part of the Earth system, rivers and streams represent a basic link in the constant cycling of the planet's water. Moreover, running water is the dominant agent of landscape alteration, eroding more terrain and transporting more sediment than any other process. Because so many people live near rivers, floods are among the most destructive of all geologic hazards. Despite huge investments in levees and dams, rivers cannot always be controlled.

Earth as a System: The Hydrologic Cycle

External Processes
↳ Hydrologic Cycle

All the rivers run into the sea; yet the sea is not full; unto the place from whence the rivers come, thither they return again.
(Ecclesiastes 1:7)

As the perceptive writer of Ecclesiastes indicated, water is continually on the move, from the ocean to the land and back again in an endless cycle. Water is everywhere on Earth—in the oceans, glaciers, rivers, lakes, air, soil, and in living tissue. All of these reservoirs constitute Earth's *hydrosphere*. In all, the water content of the hydrosphere is about 1.36 billion cubic kilometers (326 million cubic miles).

The vast bulk of it, about 97.2 percent, is stored in the global oceans. Ice sheets and glaciers account for another 2.15 percent, leaving only 0.65 percent to be divided among lakes, streams, subsurface water, and the atmosphere (Figure 9.1). Although the percentages of Earth's total water found in each of the latter sources is but a small fraction of the total inventory, the absolute quantities are great.

The water found in each of the reservoirs depicted in Figure 9.1 does not remain in these places indefinitely. Water can readily change from one state of matter (solid, liquid, or gas) to another at the temperatures and pressures that occur at Earth's surface. Therefore, water is constantly moving among the hydrosphere, the atmosphere, the solid Earth, and the biosphere. This unending circulation of Earth's water supply is called the **hydrologic cycle.** The cycle shows us many critical interrelationships among different parts of the Earth system.

The hydrologic cycle is a gigantic worldwide system powered by energy from the Sun in which the atmosphere provides the vital link between the oceans

Figure 9.1 Distribution of Earth's water.

and continents (Figure 9.2). Water evaporates into the atmosphere from the ocean and to a much lesser extent from the continents. Winds transport this moisture-laden air often great distances until conditions cause the moisture to condense into clouds and precipitation to fall. The precipitation that falls into the ocean has completed its cycle and is ready to begin another. The water that falls on land, however, must make its way back to the ocean.

What happens to precipitation once it has fallen on land? A portion of the water soaks into the ground (called **infiltration**), slowly moving downward, then laterally, finally seeping into lakes, streams, or directly into the ocean. When the rate of rainfall exceeds Earth's ability to absorb it, the surplus water flows over the surface into lakes and streams, a process called **runoff.** Much of the water that infiltrates or runs off eventually returns to the atmosphere because of evaporation from the soil, lakes, and streams. Also, some of the water that infiltrates the ground surface is absorbed by plants, which then release it into the atmosphere. This process is called **transpiration.**

Each year a field of crops may transpire the equivalent of a water layer 60 centimeters (2 feet) deep over the entire field. The same area of trees may pump twice this amount into the atmosphere. Because we cannot clearly distinguish between the amount of water that is evaporated and the amount that is transpired by plants, the term *evapotranspiration* is often used for the combined effect.

When precipitation falls in very cold areas—at high elevations or high latitudes—the water may not immediately soak in, run off, or evaporate. Instead, it may become part of a snowfield or a glacier. In this way, glaciers store large quantities of water on land. If present-day glaciers were to melt and release all their water, sea level would rise by several dozen meters. This would

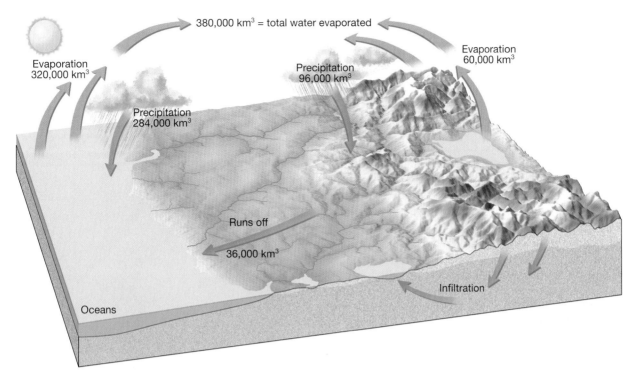

Figure 9.2   Earth's water balance. Each year, solar energy evaporates about 320,000 cubic kilometers of water from the oceans, while evaporation from the land (including lakes and streams) contributes 60,000 cubic kilometers of water. Of this total of 380,000 cubic kilometers of water, about 284,000 cubic kilometers fall back to the ocean, and the remaining 96,000 cubic kilometers fall on the land surface. Of that 96,000 cubic kilometers, only 60,000 cubic kilometers of water evaporate from the land, leaving 36,000 cubic kilometers of water to erode the land during the journey back to the oceans.

submerge many heavily populated coastal areas. As we shall see in Chapter 11, over the past 2 million years huge ice sheets have formed and melted on several occasions, each time changing the balance of the hydrologic cycle.

Figure 9.2 also shows Earth's overall *water balance*, or the volume of water that passes through each part of the cycle annually. The amount of water vapor in the air at any one time is just a tiny fraction of Earth's total water supply. But the *absolute* quantities that are cycled through the atmosphere over a one-year period are immense—some 380,000 cubic kilometers. Enough to cover Earth's entire surface to a depth of about 1 meter (39 inches). Estimates show that over North America almost six times more water is carried by moving currents of air than is transported by all the continent's rivers.

It is important to know that the hydrologic cycle is *balanced*. Because the total amount of water vapor in the atmosphere remains about the same, the average annual precipitation over Earth must be equal to the quantity of water evaporated. However, for all of the continents taken together, precipitation exceeds evaporation. Conversely, over the oceans, evaporation exceeds precipitation. Since the level of the world ocean is not dropping, the system must be in balance. In Figure 9.2, the 36,000 cubic kilometers of water that annually runs off from the land to the ocean causes enormous erosion.

In fact, this immense volume of moving water is *the single most important agent sculpturing Earth's land surface*.

To summarize, the hydrologic cycle represents the continuous movement of water from the oceans to the atmosphere, from the atmosphere to the land, and from the land back to the sea. The wearing down of Earth's land surface is largely attributable to the last of these steps and is the primary focus of the remainder of this chapter.

Running Water

Although we have always depended on running water, its source eluded us for centuries. Not until the sixteenth century did we realize that streams were supplied by surface runoff and underground water, which ultimately had their sources as rain and snow.

Runoff initially flows in broad, thin sheets across the ground, appropriately termed **sheet flow.** The amount of water that runs off in this manner rather than sinking into the ground depends upon the **infiltration capacity** of the soil. Infiltration capacity is controlled by many factors, including (1) the intensity and duration of the rainfall, (2) the prior wetted condition of the soil, (3) the soil texture, (4) the slope of the land, and (5) the nature of the vegetative cover. When the soil becomes saturated, sheet flow commences as a layer only a few millimeters thick. After flowing as a thin, unconfined sheet for only a short distance, threads of current typically develop and tiny channels called **rills** begin to form and carry the water to a stream.

To some, the term *stream* implies relative size. That is to say, streams are thought of as being larger than creeks or brooks but smaller than rivers. In geology, however, this is not the case. Here the word **stream** is used to denote channelized flow of any size, from the smallest trickle to the mightiest river. It should be pointed out, however, that although the terms *river* and *stream* are used interchangeably, the term *river* is often preferred when describing a main stream into which several tributaries flow.

The remainder of this chapter will concentrate on that part of the hydrologic cycle in which the water moves in stream channels. The discussion will deal primarily with streams in humid regions. Streams are also important in arid landscapes, but we will examine that in Chapter 12, Deserts and Wind.

Streamflow

External Processes
↳ Hydrologic Cycle

Water makes its way to the sea under the influence of gravity. The time required for the journey depends on the velocity of the stream. Velocity is the distance that water travels in a unit of time. Some sluggish streams flow at less than 1 kilometer per hour, whereas a few rapid ones may exceed 30 kilometers per hour. Velocities are determined at gaging stations where measurements are taken at several locations across the stream channel and then averaged. This is done because the rate of water movement is not uniform within a stream channel. When the channel is straight, the highest velocities occur in the center just below the surface (Figure 9.3). It is here that friction is least. Minimum velocities occur along the sides and bottom (bed) of the channel where friction is always greatest. When a stream channel is crooked or curved, the fastest flow is not in the center. Rather, the zone of maximum velocity shifts toward the outside of each bend. As we shall see later, this shift plays an important part in eroding the stream's channel on that side.

The ability of a stream to erode and transport materials depends on its velocity. Even slight changes in velocity can lead to significant changes in the load of sediment that water can transport. Several factors determine the velocity of a stream, including (1) gradient; (2) shape, size, and roughness of the channel; and (3) discharge.

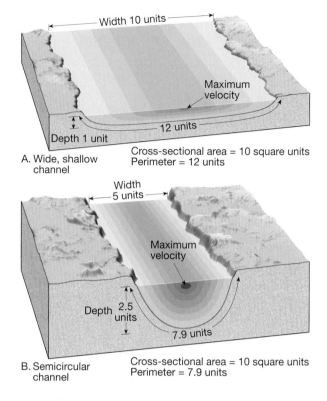

A. Wide, shallow channel

Cross-sectional area = 10 square units
Perimeter = 12 units

B. Semicircular channel

Cross-sectional area = 10 square units
Perimeter = 7.9 units

C. Gaging station

Figure 9.3   Influence of channel shape on velocity. **A.** The stream in this wide, shallow channel moves more slowly than does water in the semicircular channel because of greater frictional drag. **B.** The cross-sectional area of this semicircular channel is the same as the one in part A, but it has less water in contact with its channel and therefore less frictional drag. Thus, water will flow more rapidly in channel B, all other factors being equal. **C.** Continuous records of stage and discharge are collected by the U.S. Geological Survey at more than 7000 gaging stations in the United States. Average velocities are determined by using measurements from several spots across the stream. This station is on the Rio Grande south of Taos, New Mexico. (Photo by E. J. Tarbuck)

Gradient and Channel Characteristics

Certainly one of the most obvious factors controlling stream velocity is the **gradient,** or slope, of a stream channel. Gradient is typically expressed as the vertical drop of a stream over a fixed distance. Gradients vary considerably from one stream to another, as well as along the course of a given stream.

Portions of the lower Mississippi River, for example, have very low gradients of 10 centimeters per kilometer and less. By way of contrast, some mountain stream channels decrease in elevation at a rate of more than 40 meters per kilometer, 400 times more abruptly than the lower Mississippi. The steeper the gradient, the more energy available for streamflow. If two streams were identical in every respect except gradient, the stream with the higher gradient would obviously have the greater velocity.

The *cross-sectional shape* of a channel determines the amount of water in contact with the channel and hence affects the frictional drag. The most efficient channel is one with the least perimeter for its cross-sectional area. Figure 9.3 compares two channel shapes. Although the cross-sectional area of both is identical, the semicircular shape has less water in contact with the channel and therefore less frictional drag. As a result, if all other factors are equal, the water will flow more rapidly in the semicircular channel.

The size and roughness of the channel also affect the amount of friction. An increase in the size of a channel reduces the ratio of perimeter to cross-sectional area and therefore increases the efficiency of flow. The effect of roughness is obvious. A smooth channel promotes a more uniform flow, whereas an irregular channel filled with boulders creates enough turbulence to significantly retard the stream's forward motion.

Discharge

The **discharge** of a stream is the amount of water flowing past a certain point in a given unit of time. This is usually measured in cubic meters per second or cubic feet per second. Discharge is determined by multiplying a stream's cross-sectional area by its velocity:

Discharge (m^3/second) =
     Channel width (meters)
     × Channel depth (meters)
     × Velocity (meters/second)

The largest river in North America, the Mississippi, discharges an average of 17,300 cubic meters per second. Although this is a huge quantity of water, it is nevertheless dwarfed by the mighty Amazon, the world's largest river. Draining an area that is nearly three-quarters the size of the conterminous United States and that averages about 200 centimeters of rain annually, the Amazon discharges 12 times more water than the Mississippi. In fact, it has been estimated that the flow of the Amazon accounts for about 15 percent of all the fresh water discharged into the ocean by all of the world's rivers. Just one day's discharge would supply the water needs of New York City for nine years!

The discharges of most rivers are far from constant. This is true because of such variables as rainfall and snowmelt. When discharge changes, the factors noted earlier must also change. When discharge increases, the width or depth of the channel must increase or the water must flow faster, or some combination of these factors must change. Indeed, measurements show that when the amount of water in a stream increases, the width, depth, and velocity all increase in an orderly fashion (Figure 9.4). To handle the additional water, the stream will increase the size of its channel by widening and deepening it. As you saw earlier, when the size of the channel increases, a proportionally smaller amount of water is in contact with the bed and banks of the channel. This means that friction, which acts to retard

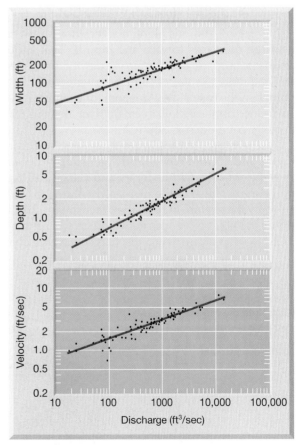

Figure 9.4 Relationship of width, depth, and velocity to discharge of the Powder River at Locate, Montana. As discharge increases, width, depth, and velocity all increase in an orderly fashion. (From L. B. Leopold and Thomas Maddock, Jr., *U.S. Geological Survey Professional Paper 252*, 1953)

the flow, is reduced. The less friction, the more swiftly the water will flow.

Changes from Upstream to Downstream

External Processes
↳ Running Water

One useful way of studying a stream is to examine its **longitudinal profile.** Such a profile is simply a cross-sectional view of a stream from its source area (called the **head** or **headwaters**) to its **mouth,** the point downstream where the river empties into another water body. By examining Figure 9.5, you can see that the most obvious feature of a typical longitudinal profile is a constantly decreasing gradient from the head to the mouth. Although many local irregularities may exist, the overall profile is a smooth concave-upward curve.

The longitudinal profile shows that the gradient decreases downstream. To see how other factors change in a downstream direction, observations and measurements must be made. When data are collected from successive gaging stations along a river, they show that discharge increases toward the mouth. This should come as no surprise because, as we move downstream, more and more tributaries contribute water to the main channel. In the case of the Amazon, for example, about a thousand tributaries join the main river along its 6500-kilometer course across South America.

Furthermore, in most humid regions, additional water is added from the groundwater supply. Because this is the case, the width, depth, and velocity all must change in response to the increased volume of water carried by the stream. Indeed, the downstream changes in these variables have been shown to vary in a manner similar to what occurs when discharge increases at one place; that is, width, depth, and velocity all increase systematically.

The observed increase in average velocity that occurs downstream contradicts our intuitive impressions concerning wild, turbulent mountain streams and wide, placid rivers in the lowlands. The mountain stream has much higher instantaneous and turbulent velocities, but the water moves vertically, laterally, and actually upstream in some cases (Figure 9.6). Thus, the average flow velocity may be lower than in a wide, placid river that is just rolling along very efficiently with far less turbulence.

In the headwaters region where the gradient is steepest, the water must flow in a relatively small and often boulder-strewn channel. The small channel and rough bed create great drag and inhibit movement by sending water in all directions with almost as much backward motion as forward motion. However, as one progresses downstream, the material on the bed of the stream becomes much smaller, offering less resistance to flow, and both the width and the depth of the channel increase to accommodate the greater discharge. These factors, especially the wider and deeper channel, permit the water to flow more freely and hence more rapidly.

In summary, we have seen that an inverse relationship exists between gradient and discharge. Where the gradient is high, the discharge is small, and where the discharge is great, the gradient is small. Stated another way, a stream can maintain a higher velocity near its mouth even though it has a lower gradient than upstream because of the greater discharge, larger channel, and smoother bed.

Base Level and Graded Streams

In 1875, John Wesley Powell, the pioneering geologist who first explored the Grand Canyon and later headed the U.S. Geological Survey, introduced the concept that there is a downward limit to stream erosion, which he

Figure 9.5 A longitudinal profile is a cross-section along the length of a stream. Note the concave-upward curve of the profile, with a steeper gradient upstream and a gentler gradient downstream.

Figure 9.6 Rapids are common in mountain streams where the gradient is steep and the channel is rough and irregular. (Photo by Bruce Gaylord/Visuals Unlimited)

called *base level*. Although the idea is relatively straightforward, it is nevertheless a key concept in the study of stream activity. **Base level** is defined as the lowest elevation to which a stream can erode its channel. Essentially this is the level at which the mouth of a stream enters the ocean, a lake, or another stream. Base level accounts for the fact that most stream profiles have low gradients near their mouths, because the streams are approaching the elevation below which they cannot erode their beds.

Two general types of base level are recognized. Sea level is considered the **ultimate base level,** because it usually represents the lowest level to which a stream can erode the land. **Local** or **temporary base levels** include lakes, resistant layers of rock, and main streams that act as base levels for their tributaries. All have the capacity to limit a stream at a certain level.

For example, when a stream enters a lake, its velocity quickly approaches zero and its ability to erode ceases. Thus, the lake prevents the stream from eroding below its level at any point upstream from the lake. However, because the outlet of the lake can cut downward and drain the lake, the lake is only a temporary hindrance to the stream's ability to downcut its channel. In a similar manner, the layer of resistant rock at the lip of the waterfall in Figure 9.7 acts as a temporary base level. Until the ledge of hard rock is eliminated, it will limit the amount of downcutting upstream.

Any change in base level will cause a corresponding readjustment of stream activities. When a dam is built along a stream course, the reservoir that forms behind it raises the base level of the stream (Figure 9.8). Upstream from the dam the stream gradient is reduced, lowering its velocity and, hence, its sediment-transporting ability. The stream, now unable to transport all of its load, will deposit material, thereby building up its channel. This process continues until the

A.

B.

C.

Figure 9.7 A resistant layer of rock can act as a local (temporary) base level. Because the durable layer is eroded more slowly, it limits the amount of downcutting upstream.

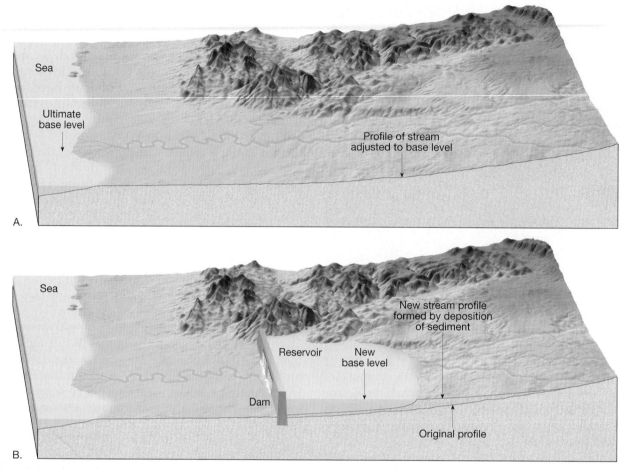

Figure 9.8   When a dam is built and a reservoir forms, the stream's base level is raised. This reduces the stream's velocity and leads to deposition and a reduction of the gradient upstream from the reservoir.

stream again has a gradient sufficient to carry its load. The profile of the new channel would be similar to the old, except that it would be somewhat higher.

The opposite situation is illustrated in Figure 9.9. A stream in a broad, flat-floored valley that is near base level often develops a course with many bends and turns called *meanders* (Figure 9.9A). What would happen to such a stream if the base level dropped or if the land were uplifted? The river, now considerably above base level, would have excess energy and would downcut its channel to establish a balance with its new base level. The result could be *incised meanders*, that is, a winding river in a steep, narrow valley (Figure 9.9B).

The observation that streams adjust their profiles to changes in base level led to the concept of a graded stream. A **graded stream** has the correct slope and other channel characteristics necessary to maintain just the velocity required to transport the material supplied to it. On the average, a graded system is neither eroding nor depositing material but is simply transporting it. Once a stream has reached this state of equilibrium, it becomes a self-regulating system in which a change in one characteristic causes an adjustment in the others to counteract the effect. Referring again to our example of a stream adjusting to a lowering of its base level, the

stream would not be graded while it was cutting its new channel but would achieve this state after downcutting had ceased.

Work of Streams

The work of streams includes erosion, transportation, and deposition. These activities go on simultaneously in stream channels, even though they are presented individually here.

Erosion

Erosion is the removal of rock and soil. Much of the material carried by streams reaches them through underground water, overland flow, and mass wasting. Streams also create some of their load by eroding their own channels. If a channel is composed of bedrock, most of the erosion is accomplished by the abrasive action of water armed with sediment, a process analogous to sandblasting. Pebbles caught in swirling eddies act like cutting tools and bore circular "potholes" into the channel floor. In channels composed of loose material, considerable lifting and removal can be accomplished by the impact of water alone.

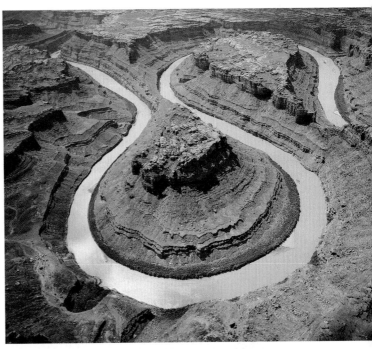

A.

B.

Figure 9.9   A river in a broad, flat-floored valley near base level often exhibits a meandering channel. **B.** Incised meanders of the Colorado River in Canyonlands National Park, Utah. Here, as the Colorado Plateau was gradually uplifted, the meandering river adjusted to being higher above base level by downcutting. (Photos by Michael Collier)

Transportation

Streams are Earth's most important erosional agent. Not only do they have the ability to downcut their channels, but streams also have the capacity to transport the enormous quantities of sediment produced by weathering. Although erosion by running water in a stream's channel contributes significant amounts of material for transport, by far the greatest quantity of sediment carried by a stream is derived from the products of weathering. Weathering produces tremendous amounts of material that are delivered to the stream by sheet flow, mass wasting, and groundwater.

Streams transport their load of sediment in three ways: (1) in solution **(dissolved load)**; (2) in suspension **(suspended load)**; and (3) scooting or rolling along the bottom **(bed load)**.

Dissolved Load.   The greatest portion of the *dissolved load* transported by most streams is supplied by groundwater. As water percolates through the ground, it first acquires soluble soil compounds. As the water seeps deeper through cracks and pores in the bedrock below, it may dissolve additional mineral matter. Eventually much of this mineral-rich water finds its way into streams.

The velocity of streamflow has essentially no effect on a stream's ability to carry its dissolved load. Once material is in solution, it goes wherever the stream goes, regardless of velocity. Precipitation (deposition) occurs only when the chemistry of the water changes.

The quantity of material carried in solution is highly variable and depends on climate and the geologic setting. Usually the dissolved load is expressed as parts of dissolved material per million parts of water (parts per million, or ppm). Although some rivers may have a dissolved load of 1000 ppm or more, the average figure for the world's rivers is estimated at 115 to 120 ppm. Almost 4 billion metric tons of dissolved mineral matter are supplied to the oceans each year by streams.

Suspended Load.   Most streams (but not all) carry the largest part of their load in *suspension*. Indeed, the visible cloud of sediment suspended in the water is the most obvious portion of a stream's load. Usually only sand, silt, and clay can be carried this way, but during a flood larger particles are transported as well. Also, during a flood the total quantity of material carried in suspension increases dramatically, as can be verified by anyone whose home has been a site for the deposition of this material.

Bed Load.   A portion of a stream's load of solid material consists of sediment too large to be carried in suspension. These coarser particles move along the bottom of the stream and constitute the *bed load*. In terms of the erosional work accomplished by a downcutting stream, the grinding action of the bed load is of great importance.

The particles making up the bed load move along the bottom by rolling, sliding, and saltation. Sediment moving by **saltation** appears to jump or skip along the stream bed. This occurs as particles are propelled upward by collisions or lifted by the current and then carried downstream a short distance until gravity pulls them back to the bed of the stream. Particles that are too large or heavy to move by saltation either roll or slide along the bottom, depending on their shapes.

Unlike the suspended and dissolved loads, which are constantly in motion, the bed load is in motion only intermittently, when the force of the water is sufficient to move the larger particles. The bed load usually does not exceed 10 percent of a stream's total load, although it may constitute up to 50 percent of the total load of a few streams. For example, consider the distribution of the 750 million tons of material carried to the Gulf of Mexico by the Mississippi River each year. Of this total, it is estimated that approximately 67 percent is carried in suspension, 26 percent in solution, and the remaining 7 percent as bed load. Estimates of a stream's bed load should be viewed cautiously, however, because this fraction of the load is very difficult to measure accurately. Not only is the bed load more inaccessible than the suspended and dissolved loads, but it moves primarily during periods of flooding when the bottom of a stream channel is most difficult to study.

Capacity and Competence. A stream's ability to carry solid particles is typically described using two criteria. First, the maximum load of solid particles that a stream can transport is termed its **capacity.** The greater the amount of water flowing in a stream (discharge), the greater the stream's capacity for hauling sediment. Second, the **competence** of a stream indicates the maximum particle size that a stream can transport. The stream's velocity determines its competence; the stronger the flow, the larger the particles it can carry in suspension and as bed load.

It is a general rule that the competence of a stream increases as the square of its velocity. Thus, if the velocity of a stream doubles, the impact force of the water increases four times; if the velocity triples, the force increases nine times, and so forth. Hence, the large boulders that are often visible during a low-water stage and that seem immovable can, in fact, be transported during flood stage because of the stream's increased competence.

By now it should be clear why the greatest erosion and transportation of sediment occur during floods (Figure 9.10). The increase in discharge results in greater capacity; the increased velocity produces greater competence. With rising velocity the water becomes more turbulent, and larger and larger particles are set in motion. In the course of just a few days, or perhaps just a few hours, a stream in flood stage can erode and transport more sediment than it does during months of normal flow (see Box 9.1).

Figure 9.10 The suspended load is clearly visible because it gives this flooding river a "muddy" appearance. During floods both capacity and competency increase. Therefore the greatest erosion and sediment transport occur during these high-water periods. Here we see the sediment-filled floodwaters of Kenya's Mara River. (Photo by Tim Davis/Tony Stone Images)

Deposition

Whenever a stream slows down, the situation reverses. As its velocity decreases, its competence is reduced and sediment begins to drop out, largest particles first. Each particle size has a *critical settling velocity*. As streamflow drops below the critical settling velocity of a certain particle size, sediment in that category begins to settle out. Thus, stream transport provides a mechanism by which solid particles of various sizes are separated. This process, called **sorting,** explains why particles of similar size are deposited together.

The well-sorted material typically deposited by a stream is called **alluvium,** the general term for any stream-deposited sediment. Many different depositional features are composed of alluvium. Some occur within stream channels, some occur on the valley floor adjacent to the channel, and some exist at the mouth of the stream.

Channel Deposits. As a river transports sediment toward the sea, some material may be deposited within the channel. Channel deposits are most often composed of sand and gravel, the coarser components of a stream's load, and are commonly referred to as **bars.** Such features, however, are only temporary, for the material will be picked up again by the running water and be transported farther downstream. Eventually most of the material will be carried to its ultimate destination: the ocean.

Sand and gravel bars can form in a number of situations. For example, they are common where streams flow in a series of bends, called *meanders.* As a stream flows around a bend, the velocity of the

BOX 9.1

The 1997 Red River Floods

The Red River of the North originates near the point where South Dakota, North Dakota, and Minnesota meet. It flows northward, finally emptying into Manitoba's Lake Winnepeg (Figure 9.A). Along most of its course in the United States, the river serves as the North Dakota–Minnesota boundary.

In April 1997, Grand Forks, North Dakota, and other communities along

Figure 9.B   Floodwaters surround Hendrum, Minnesota, during the record-breaking 1997 floods. (Photo by Dan Koeck/Liaison Agency, Inc.)

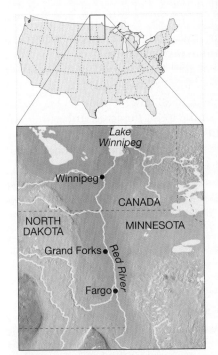

Figure 9.A   The Red River flows north into Manitoba's Lake Winnipeg.

the Red River experienced disastrous flooding—the worst ever recorded in the region (Figure 9.B). The flood was preceded by an especially snowy winter. As April began, the winter snows were melting and flooding seemed imminent. Then on April 5 and 6 a final winter storm pounded the northern Great Plains. The blizzard rebuilt once shrinking snowdrifts back to heights of 6 meters (20 feet) in some places. However, the drifts did not last long because rapidly rising temperatures melted the snow in a matter of days, causing rapid runoff into the Red River.

The discharge at Grand Forks jumped from about 70 cubic meters (2500 cubic feet) per second on April 1 to nearly 4200 cubic meters (150,000 cubic feet) per second three weeks later. The floodwaters at Grand Forks crested at 17.7 meters (58.1 feet), almost 8 meters (more than 26 feet) above flood stage! This exceeded the previous record flood stage set in 1897 by 1.2 meters (4 feet). Roughly 4.5 million acres of land were underwater. A large portion of downtown Grand Forks was destroyed by water and fire, and more than 24,000 homes were either partially or totally destroyed by floodwaters. Only 27 homes in East Grand Forks, Minnesota, survived unscathed. The losses in Grand Forks and East Grand Forks were estimated to exceed $3.5 billion.

---

water on the outside increases, leading to erosion at that site. At the same time, the water on the inside of the meander slows, which causes some of the sediment load to settle out. As these deposits occur on the inside "point" of the bend, they are called **point bars.** More accurately, these deposits would be better described as "crescent-shaped accumulations of sand and gravel."

Sometimes a stream deposits materials on the floor of its channel. If these accumulations become thick enough to choke the channel, they force the stream to split and follow several paths. What results is a complex network of converging and diverging channels that thread their way among the bars. Because such chan-

nels have an interwoven appearance, the stream is said to be **braided.**

Floodplain Deposits.   As its name implies, a *floodplain* is that part of a valley that is inundated during a flood. Most streams are flanked by floodplains. Although some are impressive features that are many kilometers across, others have modest widths of just a few meters. If we were to sample the alluvium that covers a floodplain, we would find that some of it consists of coarse sands and gravels that were originally deposited as point bars by meanders shifting laterally across the valley floor. Other sediments would be composed of fine sands, silts, and clays that were spread across the floodplain whenever water overflowed the channel during flood stage.

Rivers that occupy valleys with broad, flat floors sometimes create landforms called **natural levees** that flank the stream channel. Natural levees are built by successive floods over many years. When a stream overflows its banks onto the floodplain, the water flows over the surface as a broad sheet. Because such a flow pattern significantly reduces the water's velocity and turbulence, the coarser portion of the suspended load is deposited in strips bordering the channel. As the water spreads out over the floodplain, a lesser amount of finer sediment is laid down over the valley floor. This uneven distribution of material produces the gentle, almost imperceptible slope of the natural levee (Figure 9.11).

The natural levees of the lower Mississippi River rise 6 meters above the lower portions of the valley floor. The area behind the levee is characteristically poorly drained for the obvious reason that water cannot flow up over the levee and into the river. Marshes called **back swamps** often result. When a tributary stream enters a valley having substantial natural levees, it may not be able to make its way into the main channel. As a consequence, the tributary may flow through the back swamp parallel to the main river for many kilometers before crossing the natural levee and joining the main river. Such streams are called **yazoo tributaries,** after the Yazoo River, which parallels the lower Mississippi River for more than 300 kilometers.

Deltas. When a stream enters the relatively still waters of an ocean or lake, its velocity drops abruptly, and the resulting deposits form a **delta** (Figure 9.12). As the delta grows outward, the stream's gradient continually lessens. This circumstance eventually causes the channel to become choked with sediment from the slowing water. As a consequence, the river seeks a shorter, higher-gradient route to base level, as illustrated in Figure 9.12B. This illustration shows the main channel dividing into several smaller ones called **distributaries.** Most deltas are characterized by these shifting channels that act in an opposite way to that of tributaries. Rather than carrying water into the main channel, distributaries carry water away from the main channel. After numerous shifts of the channel, a simple delta may grow into a rough triangular shape like the Greek letter delta (Δ), for which it is named. Note, however, that many deltas do not exhibit the idealized shape. Differences in the configurations of shorelines and variations in the nature and strength of wave activity result in many shapes.

Although deltas that form in the ocean generally exhibit the same basic form as the simple lake-deposited feature just described, most large marine deltas are far more complex than the delta depicted in Figure 9.12. Many of the world's great rivers have created massive deltas, each with its own peculiarities and none as simple as the one illustrated in Figure 9.12.

Many large rivers have deltas extending over thousands of square kilometers. The delta of the Mississippi River is one example. It resulted from the accumulation of huge quantities of sediment derived from the vast region drained by the river and its tributaries. Today, New

Figure 9.11 Natural levees are gently sloping structures that are created by repeated floods. Because the ground next to the stream channel is higher than the adjacent floodplain, back swamps and yazoo tributaries may develop.

Figure 9.12   A. Structure of a simple delta that forms in the relatively quiet waters of a lake. The finer silts and clays will settle out some distance from the mouth into nearly horizontal layers called *bottomset beds*. Prior to the accumulation of bottomset beds, *foreset beds* begin to form. These beds are composed of coarse sediment, which is dropped almost immediately upon entering a lake or ocean, forming sloping layers. The foreset beds are usually covered by thin, horizontal *topset beds* deposited during floodstage. **B.** Growth of a simple delta. As a stream extends its channel, the gradient is reduced. Frequently, during flood stage the river is diverted to a higher-gradient route, forming a new distributary. Old abandoned distributaries gradually fill with sediment and vegetation. (After Ward's Natural Science Establishment, Inc., Rochester, N.Y.)

Orleans rests where there was ocean less than 5000 years ago. Figure 9.13 shows that portion of the Mississippi delta that has been built over the past 5000 to 6000 years. As the figure illustrates, the delta is actually a series of seven coalescing subdeltas. Each formed when the river left its existing channel in favor of a shorter, more direct path to the Gulf of Mexico. The individual subdeltas interfinger and partially cover one another to produce a very complex structure. It is also apparent from Figure 9.13 that after each portion was abandoned, coastal erosion (largely wave activity) modified the features. The present subdelta, called a *bird-foot* delta because of the configuration of its distributaries, has been built by the Mississippi in the last 500 years.

Stream Valleys

 External Processes
↳ Running Water

Stream valleys can be divided into two general types. Narrow V-shaped valleys, and wide valleys with flat floors exist as the ideal forms, with many gradations between.

Most stream valleys are much broader at the top than at the bottom. This would not be the case if the only agents responsible for eroding valleys were the streams flowing through them. The sides of most valleys are shaped by a combination of weathering, overland flow, and mass wasting. In some arid regions, where downcutting is rapid and weathering is slow, and in places where rock is particularly resistant, narrow valleys may not be V-shaped but rather may have nearly vertical walls.

Narrow Valleys

The Yellowstone River provides an excellent example of a narrow valley (Figure 9.14). A narrow V-shaped valley indicates that the primary work of the stream has been downcutting toward base level. The most prominent features of a narrow valley are *rapids* and *waterfalls*. Both occur where the stream profile drops rapidly, a situation usually caused by variations in the erodibility of the bedrock into which a stream channel is cutting. Resistant beds create rapids by acting as a temporary base level upstream while allowing downcutting to continue downstream. Once erosion has eliminated the resistant rock, the stream profile smooths out again.

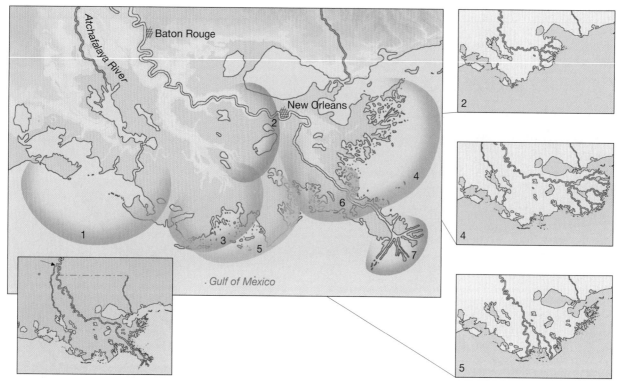

Figure 9.13   During the past 5000 to 6000 years, the Mississippi River has built a series of seven coalescing subdeltas. The numbers indicate the order in which the subdeltas were deposited. The present bird-foot delta (number 7) represents the activity of the past 500 years. Without ongoing human efforts, the present course will shift and follow the path of the Atchafalaya River. The inset on left shows the point where the Mississippi may someday break through (arrow) and the shorter path it would take to the Gulf of Mexico. (After C. R. Kolb and J. R. Van Lopik)

Waterfalls are places where the stream profile makes a vertical drop.

Wide Valleys

Once a stream has cut its channel closer to base level, downward erosion becomes less dominant. At this point more of the stream's energy is directed from side to side. The result is a widening of the valley as the river cuts away at one bank and then at the other (Figure 9.15).

Floodplains.   The side-to-side cutting of a stream eventually produces a flat valley floor, or **floodplain.** It is appropriately named because the river is confined to its channel, except during flood stage, when it overflows its banks and inundates the floodplain.

When a river erodes laterally, creating a floodplain as just described, it is called an *erosional floodplain.* Floodplains can also be depositional in nature. *Depositional floodplains* are produced by a major fluctuation in conditions, such as a change in base level. The floodplain in California's Yosemite Valley is one such feature, and it was produced when a glacier gouged the former stream valley deeper by about 300 meters (1000 feet). After the glacial ice melted, the stream readjusted to its former base level by refilling the valley with alluvium.

Meanders.   Streams that flow on floodplains often move in sweeping bends called **meanders** (Figure 9.15). Once a bend in a channel begins to form, it grows larger. Erosion occurs on the outside of the meander where velocity and turbulence are greatest. Most often, the outside bank is undermined, especially during periods of high water. As the bank becomes oversteepened, it fails by slumping into the channel. Because the outside of a meander is a zone of active erosion, it is often referred to as the **cut bank** (Figure 9.16). Much of the debris detached by the stream at the cut bank moves downstream and is soon deposited as point bars in zones of decreased velocity on the insides of meanders. In this manner meanders migrate laterally, while keeping the same cross-sectional area, by eroding the outside of the bends and depositing on the inside (see Figure 9.16). Growth ceases when the meander reaches a critical size that is determined by the size of the stream. The larger the stream, the larger its meanders can be.

Because of the slope of the channel, erosion is more effective on the downstream side of a meander. Therefore, in addition to migrating laterally, the bends also gradually migrate down the valley. Sometimes the downstream migration of a meander is slowed when it reaches a more resistant portion of the floodplain. This allows the next meander upstream to "catch up." Gradually the neck of land between the meanders is narrowed. When

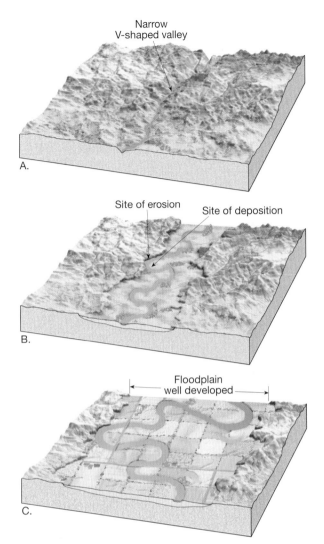

Figure 9.15 Stream eroding its floodplain.

Figure 9.14 V-shaped valley of the Yellowstone River. The rapids and waterfalls indicate that the river is vigorously downcutting. (Photo by Art Wolfe, Inc.)

they get close enough, the river erodes through the narrow neck of land to the next loop (Figure 9.17). The new, shorter channel segment is called a **cutoff,** and because of its shape, the abandoned bend is called an **oxbow lake.**

Drainage Basins and Patterns

External Processes
↳ Running Water

Every stream, no matter how large or small, has a **drainage basin.** This is the land area that contributes water to the stream. The drainage basin of one stream is separated from the drainage basin of another by an imaginary line called a **divide.** Divides range in scale from a ridge separating two small gullies on a hillside to *continental divides* that split continents into enormous drainage basins. The Mississippi River has the largest

drainage basin in North America (Figure 9.18). Extending between the Rocky Mountains in the West and the Appalachian Mountains in the East, the Mississippi River and its tributaries collect water from more than 3.2 million square kilometers (1.2 million square miles) of the continent.

Drainage systems are networks of streams that together form distinctive patterns. The nature of a drainage pattern can vary greatly from one type of terrain to another, primarily in response to the kinds of rock on which the streams developed or the structural pattern of faults and folds.

Certainly the most commonly encountered drainage pattern is the **dendritic pattern** (Figure 9.19A). This pattern of irregularly branching tributary streams resembles the branching pattern of a deciduous tree. In fact, the word *dendritic* means "treelike." The dendritic pattern forms where the underlying material is relatively uniform. Because the surface material is essentially uniform in its resistance to erosion, it does not control the pattern of streamflow. Rather, the pattern is determined chiefly by the direction of slope of the land.

Figure 9.16 When a stream meanders, its zone of maximum speed shifts toward the outer bank. A point bar is deposited when the water on the inside of a meander slows. By eroding its outer bank and depositing material on the inside of the bend, a stream is able to shift its channel.

Figure 9.17 Formation of a cutoff and oxbow lake.

When streams diverge from a central area like spokes from the hub of a wheel, the pattern is said to be **radial** (Figure 9.19B). This pattern typically develops on isolated volcanic cones and domal uplifts.

Figure 9.19C illustrates a **rectangular** "pattern" in which many right-angle bends can be seen. This pattern develops when the bedrock is crisscrossed by a series of joints and/or faults. Because these structures are eroded more easily than unbroken rock, their geometric pattern guides the directions of valleys.

Figure 9.19D illustrates a **trellis** "drainage pattern," a rectangular pattern in which tributary streams are nearly parallel to one another and have the appearance of a garden trellis. This pattern forms in areas underlain by alternating bands of resistant and less resistant rock.

Floods and Flood Control

When the discharge of a stream becomes so great that it exceeds the capacity of its channel, it overflows its banks as a **flood.** Floods are the most common and most destructive of all geologic hazards. They are, nevertheless, simply part of the *natural* behavior of streams.

Causes of Floods

Rivers flood because of the weather. Rapid melting of snow in the spring, and/or major storms that bring heavy rains over a large region, cause most floods. The extensive 1997 flood along the Red River of the North (described in Box 9.1) is a recent example of an event triggered by rapid snowmelt. Exceptional rains caused the devastating floods in the upper Mississippi River Valley during the summer of 1993 (Figure 9.20).

Another type of flood, called a *flash flood,* can occur with little warning and can be deadly because it produces a rapid rise in water levels and can have a devastating flow velocity (see Box 9.2). Several factors influence flash flooding. Among them are rainfall intensity and duration, surface conditions, and the topog-

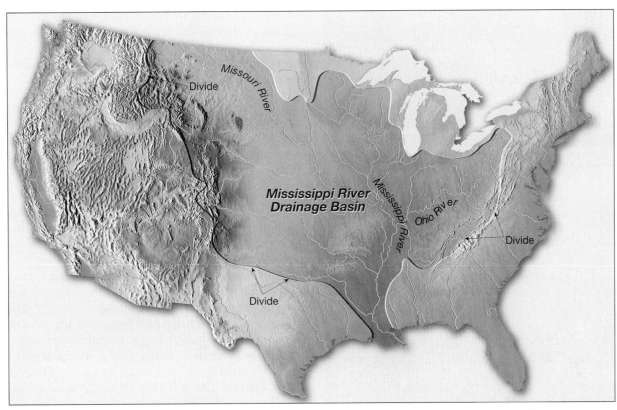

Figure 9.18   The drainage basin of the Mississippi River, North America's largest river, covers more than 3 million square kilometers. *Divides* are the boundaries that separate drainage basins from each other. Drainage basins and divides exist for all streams.

Figure 9.19   Drainage patterns. **A.** Dendritic. **B.** Radial. **C.** Rectangular. **D.** Trellis.

Figure 9.20 Satellite views of the Missouri River flowing into the Mississippi River. St. Louis is just south of their confluence. The upper image shows the rivers during a drought that occurred in summer 1988. The lower image depicts the peak of the record-breaking 1993 flood. Exceptional rains produced the wettest spring and early summer of the twentieth century in the upper Mississippi River basin. In all, nearly 14 million acres were inundated, displacing at least 50,000 people. (Courtesy of Spaceimaging.com)

raphy. Urban areas are susceptible to flash floods because a high percentage of the surface area is composed of impervious roofs, streets, and parking lots where runoff is very rapid. Mountainous areas are susceptible because steep slopes can funnel runoff into narrow canyons.

Human interference with the stream system can worsen or even cause floods. A prime example is the failure of a dam or an artificial levee. Dams and artificial levees are built for flood protection. They are designed to contain floods of a certain magnitude. If a larger flood occurs, the dam or levee is overtopped. If the dam or levee fails or is washed out, the water be-

hind it is released, causing a flash flood. The bursting of a dam in 1889 on the Little Conemaugh River caused the devastating Johnstown, Pennsylvania, flood that took some 3000 lives. A second dam failure occurred there again in 1977 and caused 77 fatalities.

Flood Control

Several strategies have been devised to eliminate or lessen the catastrophic effects of floods. Engineering efforts include the construction of artificial levees, the building of flood-control dams, and river channelization.

Artificial Levees. *Artificial levees* are earthen mounds built on the banks of a river to increase the volume of water the channel can hold. These most common of stream-containment structures have been used since ancient times and continue to be used today.

Artificial levees are usually easy to distinguish from natural levees because their slopes are much steeper. When a river is confined by levees during periods of high water, it frequently deposits material in its channel as the discharge diminishes. This is sediment that otherwise would have been dropped on the floodplain. Thus, each time there is a high flow, deposits are left on the river bed, and the bottom of the channel is built up. With the buildup of the bed, less water is required to overflow the original levee. As a result, the height of the levee may have to be raised periodically to protect the floodplain. Moreover, many artificial levees are not built to withstand periods of extreme flooding. For example, levee failures were numerous in the Midwest during the summer of 1993, when the upper Mississippi and many of its tributaries experienced record floods (Figure 9.21).

Flood-Control Dams. *Flood-control dams* are built to store floodwater and then let it out slowly. This lowers the flood crest by spreading it out over a longer time span. Ever since the 1920s, thousands of dams have been built on nearly every major river in the United States. Many dams have significant nonflood-related functions, such as providing water for irrigated agriculture and for hydroelectric power generation. Many reservoirs are also major regional recreational facilities.

Although dams may reduce flooding and provide other benefits, building these structures also has significant costs and consequences. For example, reservoirs created by dams may cover fertile farmland, useful forests, historic sites, and scenic valleys. Of course, dams trap sediment. Therefore, deltas and floodplains downstream erode because they are no longer replenished with silt during floods. Large dams can also cause significant ecological damage to river environments that took thousands of years to establish.

Building a dam is not a permanent solution to flooding. Sedimentation behind a dam means that the

BOX 9.2   Flash Floods

Tornadoes and hurricanes are nature's most awesome storms. Yet surprisingly, these dreaded events are not responsible for the greatest number of storm-related deaths. That distinction is reserved for flash floods. For a recent 30-year period, the number of storm-related deaths in the United States from flooding averaged 135 per year. By contrast, tornado fatalities averaged 73 annually, and hurricanes 25.

Flash floods are local floods of great volume and short duration. The rapidly rising surge of water usually occurs with little advance warning and can destroy roads, bridges, homes, and other substantial structures (Figure 9.C). Discharges quickly reach a maximum and diminish almost as rapidly. Flood flows often contain large quantities of sediment and debris as they sweep channels clean.

Frequently flash floods result from the torrential rains associated with a slow-moving severe thunderstorm or take place when a series of thunderstorms repeatedly pass over the same location. Occasionally floating debris or ice can accumulate at a natural or artificial obstruction and restrict the flow of water. When such temporary dams fail, torrents of water can be released as a flash flood.

Flash floods can take place in almost any area of the country. They are particularly common in mountainous terrain, where steep slopes can quickly channel runoff into narrow valleys. The hazard is most acute when the soil is already nearly saturated from earlier rains or consists of impermeable materials. A disaster in Shadydale, Ohio, demonstrates what can happen when even moderately heavy rains fall on saturated ground with steep slopes.

On the evening of 14 June 1990, 26 people lost their lives as rains es-

Figure 9.C   The disastrous nature of flash floods is illustrated by the Big Thompson River flood of July 31, 1976, in Colorado. During a four-hour period more than 30 centimeters (12 inches) of rain fell on portions of the small area drained by the river. This amounted to nearly three-quarters of the average yearly total. The flash flood in the narrow canyon lasted only a few hours but cost 139 people their lives. Damages were estimated at $39 million. (U.S. Geological Survey, Denver)

timated to be in the range of 3 to 5 inches fell on saturated soil, which generated flood waves in streams that reached tens of feet in height, destroying near-bank residences and businesses. Preceding months of above-normal rainfall had generated soil moisture contents of near saturation. As a result, moderate amounts of rainfall caused large amounts of surface and near-surface runoff. Steep valleys with practically vertical walls channeled the floods, creating very fast, high, and steep wave crests.*

Why do so many people perish in flash floods? Aside from the factor of

surprise (many are caught sleeping), people do not appreciate the power of moving water. A glance at Figure 9.C helps illustrate the force of a flood wave. Just 15 centimeters (6 inches) of fast-moving floodwater can knock a person down. Most automobiles will float and be swept away in only 0.6 meter (2 feet) of water. *More than half of all U.S. flash-flood fatalities are auto related!* Clearly, people should never attempt to drive over a flooded road. The depth of water is not always obvious. Also, the road bed may have been washed out under water. Present-day flash floods are calamities with potential for very high death tolls and huge property losses. Although efforts are being made to improve observations and warnings, flash floods remain elusive natural killers.

*"Prediction and Mitigation of Flash Floods: A Policy Statement of the American Meteorological Society." *Bulletin of the American Meteorological Society*, vol. 74, no.8 (Aug. 1993), p. 1586.

---

volume of its reservoir will gradually diminish, reducing the effectiveness of this flood-control measure.

Channelization.   *Channelization* involves altering a stream channel to speed the flow of water to prevent it from reaching flood height. This may simply involve clearing a channel of obstructions or dredging a channel to make it wider and deeper.

A more radical alteration involves straightening a channel by creating *artificial cutoffs.* The idea is that by shortening the stream, the gradient and hence the velocity are increased. By increasing velocity, the larger discharge associated with flooding can be dispersed more rapidly.

Since the early 1930s, the U.S. Army Corps of Engineers has created many artificial cutoffs on the Mis-

Figure 9.21 Water rushes through a break in an artificial levee in Monroe County, Illinois. During the record-breaking 1993 Midwest floods, many artificial levees could not withstand the force of the floodwaters. Sections of many weakened structures were overtopped or simply collapsed. (Photo by James A. Finley/AP/Wide World Photos)

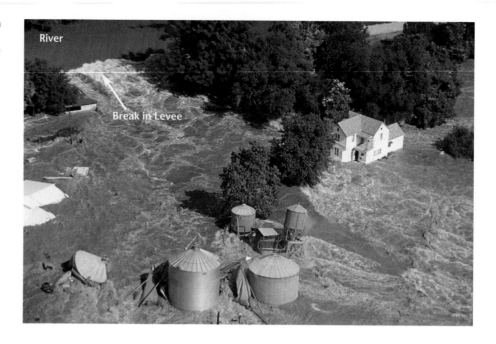

sissippi for the purpose of increasing the efficiency of the channel and reducing the threat of flooding. In all, the river has been shortened more than 240 kilometers (150 miles). The program has been somewhat successful in reducing the height of the river in flood stage. However, because the river's tendency toward meandering still exists, preventing the river from returning to its previous course has been difficult.

A Nonstructural Approach.   All of the flood-control measures described so far have involved structural so-lutions aimed at "controlling" a river. These solutions are expensive and often give people residing on the floodplain a false sense of security.

Today many scientists and engineers advocate a nonstructural approach to flood control. They suggest that an alternative to artificial levees, dams, and channelization is sound floodplain management. By identifying high-risk areas, appropriate zoning regulations can be implemented to minimize development and promote more appropriate land use.

The Chapter in Review

- The *hydrologic cycle* describes the continuous interchange of water among the oceans, atmosphere, and continents. Powered by energy from the Sun, it is a global system in which the atmosphere provides the link between the oceans and continents. The processes involved in the hydrologic cycle include *precipitation, evaporation, infiltration* (the movement of water into rocks or soil through cracks and pore spaces), *runoff* (water that flows over the land), and *transpiration* (the release of water vapor to the atmosphere by plants). *Running water is the single most important agent sculpturing Earth's land surface.*

- The amount of water that runs off the land rather than sinking into the ground depends upon the *infiltration capacity* of the soil. Initially, runoff flows as broad, thin sheets across the ground, appropriately termed *sheet flow*. After a short distance, threads of current typically develop, and tiny channels called *rills* begin to form.

- The factors that determine a stream's *velocity* are *gradient* (slope of the stream channel), *cross-sectional shape, size* and *roughness* of the channel, and the stream's *discharge* (amount of water passing a given point per unit of time, frequently measured in cubic feet per second). Most often, the gradient and roughness of a stream decrease downstream, while width, depth, discharge, and velocity increase.

- The two general types of *base level* (the lowest point to which a stream may erode its channel) are (1) *ultimate base level* (sea level) and (2) *temporary*, or *local base level*. Any change in base level will cause a stream to adjust and establish a new balance. Lowering base level will cause a stream to downcut, whereas raising base level results in deposition of material in the channel.

- The work of a stream includes *erosion* (the incorporation of material), *transportation* (as *dissolved load, suspended load*, and *bed load*), and, whenever a stream's velocity decreases, *deposition*. A stream's ability to carry solid particles is typically described using two criteria: *capacity* (the maximum load of solid particles a stream can transport) and *competence* (the maximum particle size a stream can transport). The depositional features of a stream include *deltas* and *natural levees*.

- Although many gradations exist, the two general types of stream valleys are (1) *narrow V-shaped valleys* and (2) *wide valleys with flat floors*. Because the dominant activity is downcutting toward base level, narrow valleys often contain *waterfalls* and *rapids*. When a stream has cut its channel closer to base level, its energy is directed from side to side, and erosion produces a flat valley floor, or *floodplain*. Streams that flow upon floodplains often move in sweeping bends called *meanders*. Widespread meandering may result in shorter channel segments, called *cutoffs*, and/or abandoned bends, called *oxbow lakes*.

- The land area that contributes water to a stream is called a *drainage basin*. Drainage basins are separated by an imaginary line termed a *divide*. Common *drainage patterns* (the form of a network of streams) produced by a main channel and its tributaries include (1) *dendritic*, (2) *radial*, (3) *rectangular*, and (4) *trellis*.

- *Floods* are triggered by heavy rains and/or snowmelt. Sometimes human interference can worsen or even cause floods. Flood-control measures include the building of *artificial levees* and dams, as well as *channelization*, which could involve creating *artificial cutoffs*. Many scientists and engineers advocate a nonstructural approach to flood control that involves more appropriate land use.

Key Terms

alluvium (p. 188)
back swamp (p. 190)
bar (p. 188)
base level (p. 185)
bed load (p. 187)
braided stream (p. 189)
capacity (p. 188)
competence (p. 188)
cut bank (p. 192)
cutoff (p. 192)
delta (p. 190)
dendritic pattern (p. 193)
discharge (p. 183)

dissolved load (p. 187)
distributary (p. 190)
divide (p. 193)
drainage basin (p. 193)
flood (p. 194)
floodplain (p. 192)
graded stream (p. 186)
gradient (p. 183)
head (headwaters) (p. 184)
hydrologic cycle (p. 184)
infiltration (p. 180)
infiltration capacity (p. 181)

local (temporary) base
    level (p. 185)
longitudinal profile (p. 184)
meander (p. 192)
mouth (p. 184)
natural levee (p. 190)
oxbow lake (p. 193)
point bar (p. 189)
radial pattern (p. 194)
rectangular pattern (p. 184)
rill (p. 192)
runoff (p. 180)

saltation (p. 187)
sheet flow (p. 181)
sorting (p. 188)
stream (p. 182)
suspended load (p. 187)
temporary (local) base
    level (p. 185)
transpiration (p. 180)
trellis drainage pattern
    (p. 194)
ultimate base level (p. 185)
yazoo tributary (p. 190)

Questions for Review

1. Describe the movement of water through the hydrologic cycle. Once precipitation has fallen on land, what paths are available to it?

2. Over the oceans, evaporation exceeds precipitation, yet sea level does not drop. Can you explain why?

3. List several factors that influence infiltration capacity.

4. A stream starts out 2000 meters above sea level and travels 250 kilometers to the ocean. What is its average gradient in meters per kilometer?

5. Suppose that the stream mentioned in Question 4 developed extensive meanders so that its course was lengthened to 500 kilometers. Calculate its new gradient. How does meandering affect gradient?

6. When the discharge of a stream increases, what happens to the stream's velocity?

7. How do channel width, channel depth, velocity, and discharge typically change from the point where a stream begins to the point where it ends? Briefly explain why these changes take place.

8. Define *base level*. Name the main river in your area. For what streams does it act as base level? What is the base level for the Mississippi River?

9. In what three ways does a stream transport its load?

10. If you collect a jar of water from a stream, what part of its load will settle to the bottom of the jar? What portion will remain in the water? What part of a stream's load would probably not be present in your sample?

11. Distinguish between *competence* and *capacity*.

12. Why does a river flowing across a delta eventually change course?

13. Briefly describe the formation of a natural levee. How is this feature related to back swamps and yazoo tributaries?

14. Each of the following statements refers to a particular drainage pattern. Identify the pattern.
    a. Streams diverging from a central high area, such as a dome.
    b. Branching treelike pattern.
    c. A pattern that develops when bedrock is crisscrossed by joints and faults.

15. How is a flash flood different from other floods?

16. List and briefly describe three basic flood-control strategies. What are some drawbacks of each?

Web Resources

 The *Essentials of Geology* Website uses the resources and flexibility of the Internet to aid in your study of the topics in this chapter. Written and developed by geology instructors, this site will help improve your understanding of geology. Visit **http://www.prenhall.com/lutgens** and click on the cover of *Essentials of Geology, 8/e* to find:

- On-line review quizzes.
- Web-based critical thinking and writing exercises.
- Links to chapter-specific Web resources.
- Internet-wide key term searches.

http://www.prenhall.com/lutgens

Groundwater

Focus on Learning

To assist you in learning the important concepts in this chapter, you will find it helpful to focus on the following questions:

• What is the importance of groundwater as a resource and as a geological agent?

• What is groundwater, and how does it move?

• How do springs, geysers, wells, and artesian wells form?

• What are some environmental problems associated with groundwater?

• What features are produced by the geologic work of groundwater?

Old Faithful Geyser in Wyoming's Yellowstone National Park. (Photo by Marc Muench)

Worldwide, wells and springs provide water for cities, crops, livestock, and industry. In the United States, groundwater is the source of about 40 percent of the water used for all purposes (except hydroelectric power generation and powerplant cooling). Groundwater is the drinking water for more than 50 percent of the population, is 40 percent of the water used for irrigation, and provides more than 25 percent of industry's needs. In some areas, however, overuse of this basic resource has resulted in water shortage, streamflow depletion, land subsidence, contamination by saltwater, increased pumping cost, and groundwater pollution.

Importance of Underground Water

External Processes
↳ Groundwater

Groundwater is one of our most important and widely available resources, yet people's perceptions of the subsurface environment from which it comes are often unclear and incorrect. The reason is that the groundwater environment is largely hidden from view except in caves and mines, and the impressions people gain from these subsurface openings are misleading. Observations on the land surface give an impression that Earth is solid. This view remains when we enter a cave and see water flowing in a channel that appears to have been cut into solid rock.

Because of such observations, many people believe that groundwater occurs only in underground rivers. In reality, most of the subsurface environment is not solid at all. It includes countless tiny *pore spaces* between grains of soil and sediment, plus narrow joints and fractures in bedrock. Together, these spaces add up to an immense volume. It is in these small openings that groundwater collects and moves.

Considering the entire hydrosphere, or all of Earth's water, only about six-tenths of 1 percent occurs underground. Nevertheless, this small percentage, stored in the rocks and sediments beneath Earth's surface, is a vast quantity. When the oceans are excluded and only sources of freshwater are considered, the significance of groundwater becomes more apparent.

Table 10.1 contains estimates of the distribution of fresh water in the hydrosphere. Clearly, the largest volume occurs as glacial ice. Second in rank is groundwater, with slightly more than 14 percent of the total. However, when ice is excluded and just liquid water is considered, more than 94 percent of all fresh water is groundwater. Without question, *groundwater represents the largest reservoir of fresh water that is readily available to*

humans. Its value in terms of economics and human well-being is incalculable.

Geologically, groundwater is important as an erosional agent. The dissolving action of groundwater slowly removes rock, allowing surface depressions known as sinkholes to form as well as creating subterranean caverns (Figure 10.1). Groundwater is also an equalizer of streamflow. Much of the water that flows in rivers is not direct runoff from rain and snowmelt. Rather, a large percentage of precipitation soaks in and then moves slowly underground to stream channels. Groundwater is thus a form of storage that sustains streams during periods when rain does not fall. When we see water flowing in a river during a dry period, it is water from rain that fell at some earlier time and was stored underground.

Distribution of Underground Water

External Processes
↳ Groundwater

When rain falls, some of the water runs off, some evaporates, and the remainder soaks into the ground. This last path is the primary source of practically all underground water. The amount of water that takes each of these paths, however, varies greatly both in time and space. Influential factors include steepness of slope, nature of surface material, intensity of rainfall, and type and amount of vegetation. Heavy rains falling on steep slopes underlain by impervious materials will obviously result in a high percentage of the water running off. Conversely, if rain falls steadily and gently on more gradual slopes composed of materials that are easily penetrated by the water, a much larger percentage of water soaks into the ground.

Some of the water that soaks in does not travel far, because it is held by molecular attraction as a surface film on soil particles. This near-surface zone is called the *belt of soil moisture*. It is crisscrossed by roots, voids left by decayed roots, and animal and worm burrows that enhance the infiltration of rainwater into the soil. Soil water is used by plants in life functions and transpiration. Some water also evaporates directly back into the atmosphere.

Water that is not held as soil moisture will percolate downward until it reaches a zone where all of the open spaces in sediment and rock are completely filled with water. This is the **zone of saturation.** Water within it is called **groundwater.** The upper limit of this zone is known as the **water table.** The area above the water table where the soil, sediment, and rock are not saturated is called the **zone of aeration** (Figure 10.2). Although a considerable amount of water can be present in the zone of

Table 10.1 Fresh water of the hydrosphere.

Parts of the Hydrosphere	Volume of Fresh Water (km³)	Share of Total Volume of Fresh Water (percent)	Rate of Water Exchange
Ice sheets and glaciers	24,000,000	84.945	8000 years
Groundwater	4,000,000	14.158	280 years
Lakes and reservoirs	155,000	0.549	7 years
Soil moisture	83,000	0.294	1 year
Water vapor in the atmosphere	14,000	0.049	9.9 days
River water	1,200	0.004	11.3 days
Total	28,253,200	100.00	

Source: U.S. Geological Survey Water Supply Paper 2220, 1987.

aeration, this water cannot be pumped by wells because it clings too tightly to rock and soil particles. By contrast, below the water table, the water pressure is great enough to allow water to enter wells, thus permitting groundwater to be withdrawn for use. We will examine wells more closely later in the chapter.

The Water Table

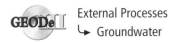

External Processes
↳ Groundwater

The water table, the upper limit of the zone of saturation, is a very significant feature of the groundwater system. The water-table level is important in predicting the productivity of wells, explaining the changes in the flow of springs and streams, and accounting for fluctuations in the levels of lakes.

Variations in the Water Table

The depth of the water table is highly variable and can range from zero, when it is at the surface, to hundreds of meters in some places. An important characteristic of the water table is that its configuration varies seasonally and from year to year because the addition of water to the groundwater system is closely related to the quantity, distribution, and timing of precipitation. Except where the water table is at the surface, we cannot observe it directly. Nevertheless, its elevation can be mapped and studied in detail where wells are nu-

Figure 10.1 The dissolving action of groundwater creates caves. Three Fingers Cave, Lincoln County, New Mexico. (Photo by Tom Stack and Associates)

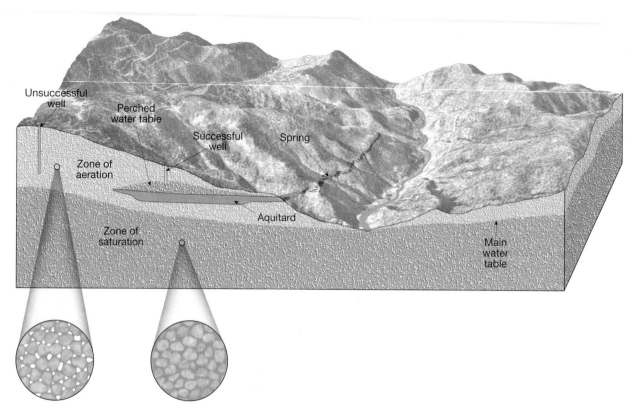

Figure 10.2   This diagram illustrates the relative positions of many features associated with subsurface water.

merous, because the water level in wells coincides with the water table (Figure 10.3). Such maps reveal that the water table is rarely level, as we might expect a table to be. Instead, its shape is usually a subdued replica of the surface topography, reaching its highest elevations beneath hills and then descending toward valleys (Figure 10.2). Where a wetland (swamp) is encountered, the water table is right at the surface. Lakes and streams generally occupy areas low enough that the water table is above the land surface.

Several factors contribute to the irregular surface of the water table. One important influence is the fact that groundwater moves very slowly and at varying rates under different conditions. Because of this, water tends to "pile up" beneath high areas between stream valleys. If rainfall were to cease completely, these water table "hills" would slowly subside and gradually approach the level of the valleys. However, new supplies of rainwater are usually added frequently enough to prevent this. Nevertheless, in times of extended drought, the water table may drop enough to dry up shallow wells. Other causes for the uneven water table are variations in rainfall and permeability from place to place.

Interaction Between Groundwater and Streams

The interaction between the groundwater system and streams is a basic link in the hydrologic cycle. It can take place in one of three ways. Streams may gain water from the inflow of groundwater through the streambed. Such streams are called **gaining streams** (Figure 10.4A). For this to occur, the elevation of the water table must be higher than the level of the surface of the stream. Streams may lose water to the groundwater system by outflow through the streambed. The term **losing stream** is applied to this situation (Figure 10.4B). When this happens, the elevation of the water table must be lower than the surface of the stream. The third possibility is a combination of the first two—a stream gains in some sections and loses in others.

Losing streams can be connected to the groundwater system by a continuous saturated zone, or they can be disconnected from the groundwater system by an unsaturated zone. Compare parts B and C in Figure 10.4. When the stream is disconnected, the water table may have a discernible bulge beneath the stream if the rate of water movement through the streambed and zone of aeration is greater than the rate of groundwater movement away from the bulge.

In some settings, a stream might always be a gaining stream or always be a losing stream. However, in many situations flow direction can vary a great deal along a stream; some sections receive groundwater and other sections lose water to the groundwater system. Moreover, the direction of flow can change over a short time span as the result of storms adding water near the streambank or when temporary flood peaks move down the channel.

A.

B.

EXPLANATION

⬤   Location of well and altitude of water table above
     sea level, in feet

_120_   Water table contour shows altitude of water table,
          contour interval 10 feet

◄ – – – – Ground-water flow line

Figure 10.3   Preparing a map of the water table. The water
level in wells coincides with the water table. **A.** First, the
locations of wells and the elevation of the water table above
sea level are plotted on a map. **B.** These data points are used
to guide the drawing of water-table contour lines at regular
intervals. On this sample map the interval is 10 feet.
Groundwater flow lines can be added to show water movement
in the upper portion of the zone of saturation. Groundwater
tends to move approximately perpendicular to the contours and
down the slope of the water table. (After U.S Geological Survey)

A. Gaining stream

B. Losing stream (connected)

C. Losing stream (disconnected)

Figure 10.4   Interaction between the groundwater system
and streams. **A.** Gaining streams receive water from the
groundwater system. **B.** Losing streams lose water to the
groundwater system. **C.** When losing streams are separated
from the groundwater system by the zone of aeration, a bulge
may form in the water table. (After U.S. Geological Survey)

Groundwater contributes to streams in most geologic and climatic settings. Even where streams are primarily losing water to the groundwater system, certain
sections may receive groundwater inflow during some
seasons. In one study of 54 streams in all parts of the
United States, the analysis indicated that 52 percent of
the streamflow was contributed by groundwater. The
groundwater contribution ranged from a low of 14 percent to a maximum of 90 percent.

How Groundwater Moves

We noted the common misconception that groundwater occurs in underground rivers that resemble surface
streams. Although subsurface streams do exist, they are
not common. Rather, as you learned in the preceding
sections, groundwater exists in the pore spaces and fractures in rock and sediment. Thus, contrary to any impressions of rapid flow that an underground river might
evoke, the movement of most groundwater is exceedingly slow, from pore to pore. By exceedingly slow, we

mean anywhere from millimeters per year to perhaps a
kilometer per year, depending on conditions.

The energy that makes groundwater move is provided by the force of gravity. In response to gravity,
water moves from areas where the water table is high to
zones where the water table is lower. This means that
water gravitates toward a stream channel, lake, or
spring as shown in Figure 10.2. Although some water
takes the most direct path down the slope of the water
table, much of the water follows long, curving paths toward the zone of discharge.

Figure 10.5 shows how water percolates into a
stream from all possible directions. Some paths actually
turn upward, apparently against the force of gravity,
and enter through the bottom of the channel. This is easily explained: The deeper you go into the zone of saturation, the greater the water pressure. Thus, the looping
curves followed by water in the saturated zone may be
thought of as a compromise between the downward
pull of gravity and the tendency of water to move to-

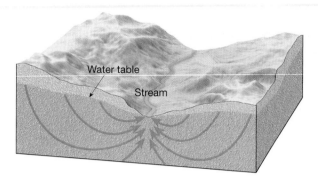

Figure 10.5 Arrows indicate groundwater movement through uniformly permeable material. The looping curves may be thought of as a compromise between the downward pull of gravity and the tendency of water to move toward areas of reduced pressure.

ward areas of reduced pressure. As a result, water at any given height is under greater pressure beneath a hill than beneath a stream channel, and the water tends to migrate toward points of lower pressure.

Factors Influencing the Storage and Movement of Groundwater

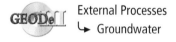
External Processes
↳ Groundwater

The nature of subsurface materials strongly influences the rate of groundwater movement and the amount of groundwater that can be stored. Two factors are especially important—porosity and permeability.

Porosity

Water soaks into the ground because bedrock, sediment, and soil contain countless voids, or openings. These openings are similar to those of a sponge and are often called *pore spaces*. The quantity of groundwater that can be stored depends on the **porosity** of the material, which is the percentage of the total volume of rock or sediment that consists of pore spaces. Voids most often are spaces between sedimentary particles, but also common are joints, faults, cavities formed by the dissolving of soluble rocks such as limestone, and vesicles (voids left by gases escaping from lava).

Variations in porosity can be great. Sediment is commonly quite porous, and open spaces may occupy 10 to 50 percent of the sediment's total volume. Pore space depends on the size and shape of the grains, how they are packed together, the degree of sorting, and in sedimentary rocks, the amount of cementing material.

For example, clay may have a porosity as high as 50 percent, whereas some gravels may have only 20 percent voids.

Where sediments of various sizes are mixed, the porosity is reduced because the finer particles tend to fill the openings among the larger grains. Most igneous and metamorphic rocks, as well as some sedimentary rocks, are composed of tightly interlocking crystals, so the voids between the grains may be negligible. In these rocks, fractures must provide the voids.

Permeability, Aquitards, and Aquifers

Porosity alone cannot measure a material's capacity to yield groundwater. Rock or sediment might be very porous yet still not allow water to move through it. The pores must be *connected* to allow water flow, and they must be *large enough* to allow flow. Thus, the **permeability** of a material, its ability to *transmit* a fluid, is also very important.

Groundwater moves by twisting and turning through interconnected small openings. The smaller the pore spaces, the slower the water moves. For example, clay's ability to store water is great, owing to its high porosity, but its pore spaces are so small that water is unable to move through it. Thus, clay's porosity is high but its permeability is poor.

Impermeable layers that hinder or prevent water movement are termed **aquitards.** Clay is a good example. In contrast, larger particles, such as sand or gravel, have larger pore spaces. Therefore, the water moves with relative ease. Permeable rock strata or sediments that transmit groundwater freely are called **aquifers.** Sands and gravels are common examples.

In summary, you have seen that porosity is not always a reliable guide to the amount of groundwater that can be produced, and permeability is significant in determining the rate of groundwater movement and the quantity of water that might be pumped from a well.

Springs

External Processes
↳ Groundwater

Springs have aroused the curiosity and wonder of people for thousands of years. The fact that springs were, and to some people still are, rather mysterious phenomena is not difficult to understand, for here is water flowing freely from the ground in all kinds of weather in seemingly inexhaustible supply but with no obvious source.

Not until the middle of the 1600s did the French physicist Pierre Perrault invalidate the age-old assumption that precipitation could not adequately account for the amount of water emanating from springs and flowing in rivers. Over several years Perrault computed the quantity of water that fell on France's Seine River basin. He then calculated the mean annual runoff by measuring the river's discharge. After allowing for the loss of water by evaporation, he showed that there *was* sufficient water remaining to feed the springs. Thanks to Perrault's pioneering efforts and the measurements by many afterward, we now know that the source of springs is water from the zone of saturation and that the ultimate source of this water is precipitation.

Whenever the water table intersects Earth's surface, a natural outflow of groundwater results, which we call a **spring.** Springs such as the one pictured in Figure 10.6 form when an aquitard blocks the downward movement of groundwater and forces it to move laterally. Where the permeable bed outcrops, a spring results. Another situation leading to the formation of a spring is illustrated in Figure 10.2. Here an aquitard is situated above the main water table. As water percolates downward, a portion of it is intercepted by the aquitard, thereby creating a localized zone of saturation called a **perched water table.**

Figure 10.6 Spring in Arizona's Marble Canyon. (Photo by Michael Collier)

Many geological situations lead to the formation of springs because subsurface conditions vary greatly from place to place. Even in areas underlain by impermeable crystalline rocks, permeable zones may exist in the form of fractures or solution channels. If these openings fill with water and intersect the ground surface along a slope, a spring will result.

Wells

External Processes
  ↳ Groundwater

The most common device used by people for removing groundwater is the **well,** a hole bored into the zone of saturation (see Figure 10.2). Wells serve as small reservoirs into which groundwater migrates and from which it can be pumped to the surface. The use of wells dates back many centuries and continues to be an important method of obtaining water today. By far the single greatest use of this water in the United States is irrigation for agriculture. More than 65 percent of the groundwater used each year is for this purpose. Industrial uses rank a distant second, followed by the amount used by homes in cities and rural areas.

The water-table level may fluctuate considerably during the course of a year, dropping during dry seasons and rising following periods of rain. Therefore, to ensure a continuous supply of water, a well must penetrate below the water table. Often when water is withdrawn from a well, the water table around the well is lowered. This effect, termed **drawdown,** decreases with increasing distance from the well. The result is a depression in the water table, roughly conical in shape, known as a **cone of depression** (Figure 10.7). Because the cone of depression increases the slope of the water table near the well, groundwater will flow more rapidly toward the opening. For most small domestic wells, the cone of depression is negligible. However, when wells are heavily pumped for irrigation or industrial purposes, the withdrawal of water can be great enough to create a very wide and steep cone of depression. This may substantially lower the water table in an area and cause nearby shallow wells to become "high and dry." Figure 10.7 illustrates this situation.

Digging a successful well is a familiar problem for people in areas where groundwater is the primary source of supply. One well may be successful at a depth of 10 meters (33 feet), whereas a neighbor may have to go twice as deep to find an adequate supply. Still others may be forced to go deeper or try a different site altogether. When subsurface materials are heterogeneous, the amount of water a well is capable of providing may vary a great deal over short distances. For example,

Figure 10.7 A cone of depression in the water table often forms around a pumping well. If heavy pumping lowers the water table, the shallow wells may be left dry.

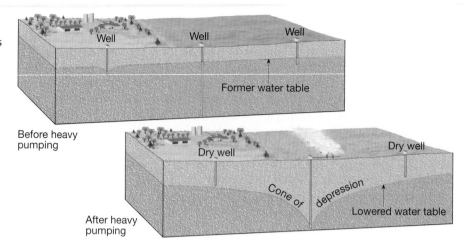

Before heavy pumping

After heavy pumping

when two nearby wells are drilled to the same level and only one is successful, it may be caused by the presence of a perched water table beneath one of them. Such a case is shown in Figure 10.2. Massive igneous and metamorphic rocks provide a second example. These crystalline rocks are usually not very permeable except where they are cut by many intersecting joints and fractures. Therefore, when a well drilled into such rock does not intersect an adequate network of fractures, it is likely to be unproductive.

Artesian Wells

External Processes
↳ Groundwater

In most wells, water cannot rise on its own. If water is first encountered at 30 meters depth, it remains at that level, fluctuating perhaps a meter or two with seasonal wet and dry periods. However, in some wells, water rises, sometimes overflowing at the surface. Such wells are abundant in the *Artois* region of northern France, and so we call these self-rising wells *artesian*.

The term **artesian** is applied to any situation in which groundwater rises in a well above the level where it was initially encountered. For such a situation to occur, two conditions must exist (Figure 10.8): (1) Water must be confined to an aquifer that is inclined so that one end is exposed at the surface, where it can receive water; and (2) impermeable layers (aquitards), both above and below the aquifer, must be present to prevent the water from escaping. When such a layer is tapped, the pressure created by the weight of the water above will force the water to rise. If there were no friction, the water in the well would rise to the level of the water at the top of the aquifer. However, friction reduces the height of this pressure surface. The greater the distance from the recharge area (area where water enters the inclined aquifer), the greater the friction and the less the rise of water.

In Figure 10.8, Well 1 is a *nonflowing artesian well*, because at this location the pressure surface is below ground level. When the pressure surface is above the ground and a well is drilled into the aquifer, a *flowing artesian well* is created (Well 2, Figure 10.8). Not all artesian systems are wells. *Artesian springs* also exist. Here groundwater reaches the surface by rising through a natural fracture rather than through an artificially produced hole.

Artesian systems act as conduits, transmitting water from remote areas of recharge great distances to the points of discharge. In this manner, water that fell in central Wisconsin years ago is now taken from the ground and used by communities many kilometers away in Illinois. In South Dakota, such a system brings water from the Black Hills in the west, eastward across the state.

On a different scale, city water systems can be considered as examples of artificial artesian systems. The water tower, into which water is pumped, can be considered the area of recharge, the pipes the confined aquifer, and the faucets in homes the flowing artesian wells.

Environmental Problems Associated with Groundwater

As with many of our valuable natural resources, groundwater is being exploited at an increasing rate. In some areas, overuse threatens the groundwater supply. In other places, groundwater withdrawal has caused the ground and everything resting upon it to sink. Still other localities are concerned with the possible contamination of their groundwater supply.

Treating Groundwater as a Nonrenewable Resource

Many natural systems tend to establish a condition of equilibrium. The groundwater system is no exception. The water table's height reflects a balance between the

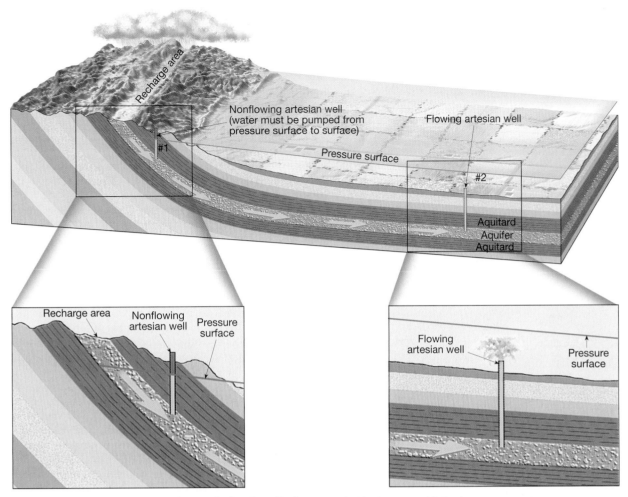

Figure 10.8 Artesian systems occur when an inclined aquifer is surrounded by impermeable beds.

rate of water added by precipitation and the rate of water removed by discharge and withdrawal. Any imbalance will either raise or lower the water table. A long-term drop in the water table can occur if there is either a decrease in recharge due to a prolonged drought, or an increase in groundwater discharge or withdrawal.

For many, groundwater appears to be an endlessly renewable resource, because it is continually replenished by rainfall and melting snow. But in some regions groundwater has been and continues to be treated as a *nonrenewable resource*. Where this occurs, the water available to recharge the aquifer falls significantly short of the amount being withdrawn.

The High Plains provides one example. Here an extensive agricultural economy is largely dependent on irrigation (Figure 10.9). In some parts of the region where intense irrigation has been practiced for an extended period, depletion of groundwater has been severe. Under these circumstances, it can be said that the groundwater is literally being "mined." Even if pumping were to cease immediately, it could take hundreds or thousands of years for the groundwater to be fully replenished. Box 10.1 explores this issue more fully.

Land Subsidence Caused by Groundwater Withdrawal

As you shall see later in this chapter, surface subsidence can result from natural processes related to groundwater. However, the ground may also sink when water

Figure 10.9 In some agricultural regions, water is pumped from the ground faster than it is replenished. In such instances, groundwater is being treated as a nonrenewable resource. (Photo by Bill Kamin/Visuals Unlimited)

BOX 10.1

The Ogallala Aquifer—How Long Will the Water Last?

The High Plains extend from the western Dakotas south to Texas. Despite being a land of little rain, this is one of the most important agricultural regions in the United States. The reason is a vast endowment of groundwater that makes irrigation possible throughout most of the region's 450,000 square kilometers (175,000 square miles). The source of most of this water is the Ogallala Formation, the largest aquifer in the United States (Figure 10.A).

Figure 10.A   The Ogallala Formation underlies about 450,000 square kilometers (175,000 square miles) of the High Plains, making it the largest aquifer in the United States.

Geologically, the Ogallala Formation is young, consisting of a number of sandy and gravelly rock layers of late Tertiary and Quaternary age (see geologic time scale Figure 1.4, p. 7). The sediments were derived from the erosion of the Rocky Mountains and carried eastward by sluggish streams. Erosion has removed much of the formation from eastern Colorado, severing the Ogallala's connection to the Rockies.

The Ogallala Formation averages 60 meters (200 feet) thick, and in places it is as much as 180 meters (600 feet) thick. Groundwater in the aquifer originally came downslope from the Rockies as well as from surface precipitation that soaked into the ground over thousands of years. Because of its high porosity and great size, the Ogallala accumulated huge amounts of groundwater—enough fresh water to fill Lake Huron! Today, with the connection between the aquifer and the Rockies severed, all of the Ogallala's recharge must come from the region's meager rainfall.

The Ogallala aquifer was first used for irrigation in the late 1800s, but its use was limited by the capacity of the pumps available at the time. In the 1920s, with the development of large-capacity irrigation pumps, High Plains farmers, particularly in Texas, began tapping the Ogallala for irrigation. Then in the 1950s, improved technology brought large-scale exploitation of the aquifer. Today, nearly 170,000 wells are being used to irrigate more than 65,000 square kilometers (16 million acres) of land.

With the increase in irrigation has come a drastic drop in the Ogallala's water table, especially in the southern High Plains. Declines in the water table of 3 to 15 meters (10 to 50 feet) are common, and in places the water table is now 60 meters (200 feet) below its level prior to irrigation.

Beginning in the 1980s, irrigated acreage has declined in the High Plains. One important reason has been higher energy costs. As water levels have dropped, the costs of pumping groundwater to the surface have increased.

Although the water table decline has slowed in parts of the southern High Plains, substantial pumping continues, often in excess of recharge. The future of irrigated farming in this region is clearly in jeopardy.

The southern High Plains represent an area of the United States that will return sooner or later to dry-land farming. The transition will come sooner and with fewer ecological and economic crises if the agricultural industry is weaned gradually from its dependence on groundwater irrigation. If nothing is done until all the accessible water in the Ogallala aquifer has been removed, the transition will be ecologically dangerous and economically dreadful.*

*National Research Council. *Solid-Earth Sciences and Society*. Washington, DC: National Academy Press, 1993, p. 148.

---

is pumped from wells faster than natural recharge processes can replace it. This effect is particularly pronounced in areas underlain by thick layers of loose sediments. As water is withdrawn, the weight of the overburden packs the sediment grains more tightly together and the ground subsides.

Many areas can be used to illustrate such land subsidence caused by excessive pumping of groundwater from relatively loose sediment. A classic example in the United States occurred in the San Joaquin Valley of California (Figure 10.10). This important agricultural region relies heavily on irrigation. Land subsidence due to groundwater withdrawal began in the valley in the mid-1920s and locally exceeded 8 meters (28 feet) by 1970. Then, because of the importation of surface water

and a decrease in groundwater pumping, water levels in the aquifer recovered and subsidence ceased.

However, during a drought from 1976 to 1977, heavy groundwater pumping led to renewed subsidence. This time, water levels dropped at a much faster rate than during the previous period because of the reduced storage capacity caused by earlier compaction of material in the aquifer. In all, more than 13,400 square kilometers (5200 square miles) of irrigable land, half the entire valley, were affected by subsidence. Damage to structures, including highways, bridges, water lines, and wells, was extensive. Many other cases of land subsidence due to groundwater pumping exist in the United States, including Las Vegas, Nevada; New Orleans and Baton Rouge, Louisiana; and the Houston–Galveston area of

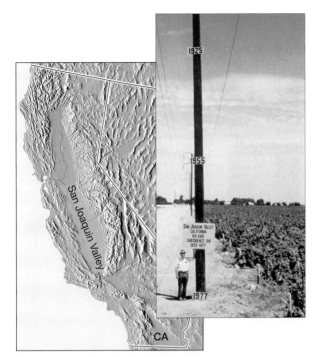

Figure 10.10 The shaded area on the map shows California's San Joaquin Valley. The marks on the utility pole in the photo indicate the level of the surrounding land in preceding years. Between 1925 and 1975 this part of the San Joaquin Valley subsided almost 9 meters because of the withdrawal of groundwater and the resulting compaction of sediments. (Photo courtesy of U.S. Geological Survey)

Texas. In the low-lying coastal area between Houston and Galveston, land subsidence ranges from 1.5 to 3 meters (5 to 9 feet). The result is that about 78 square kilometers (30 square miles) are permanently flooded.

Groundwater Contamination

The pollution of groundwater is a serious matter, particularly in areas where aquifers provide a large part of the water supply. One common source of groundwater pollution is sewage. Its sources include an ever-increasing number of septic tanks, as well as inadequate or broken sewer systems and farm wastes.

If sewage water that is contaminated with bacteria enters the groundwater system, it may become purified through natural processes. The harmful bacteria may be mechanically filtered by the sediment through which the water percolates, destroyed by chemical oxidation, and/or assimilated by other organisms. For purification to occur, however, the aquifer must be of the correct composition. For example, extremely permeable aquifers (such as highly fractured crystalline rock, coarse gravel, or cavernous limestone) have such large openings that contaminated groundwater might travel long distances without being cleansed. In this case, the water flows too rapidly and is not in contact with the surrounding material long enough for purification to occur. This is the problem at Well 1 in Figure 10.11.

In contrast, when the aquifer is composed of sand or permeable sandstone, the water can sometimes be purified after traveling only a few dozen meters through it. The openings between sand grains are large enough to permit water movement, yet the movement of the water is slow enough to allow ample time for its purification (Well 2, Figure 10.11).

Other sources and types of contamination also threaten groundwater supplies (Figure 10.12). These include widely used substances such as highway salt, fertilizers that are spread across the land surface, and pesticides. In addition, a wide array of chemicals and industrial materials may leak from pipelines, storage tanks, landfills, and holding ponds. Some of these pollutants are classified as *hazardous,* meaning that they are either flammable, corrosive, explosive, or toxic. As rainwater oozes through the refuse, it may dissolve a variety of potential contaminants. If the leached material reaches the water table, it will mix with the groundwater and contaminate the supply. Similar problems may result from leakage of shallow excavations called holding ponds into which a variety of liquid wastes are disposed.

Because groundwater movement is usually slow, polluted water may go undetected for a long time. In fact, most contamination is discovered only after drinking water has been affected and people become ill. By this time, the volume of polluted water may be very large, and even if the source of contamination is removed immediately, the problem is not solved. Although the sources of groundwater contamination are numerous, the solutions are relatively few.

Once the source of the problem has been identified and eliminated, the most common practice is simply to abandon the water supply and allow the pollutants to be flushed away gradually. This is the least costly and easiest solution, but the aquifer must remain unused for many years. To accelerate this process, polluted water is sometimes pumped out and treated. Following removal of the tainted water, the aquifer is allowed to recharge naturally, or in some cases, the treated water or other fresh water is pumped back in. This process is costly, time-consuming, and it may be risky because there is no way to be certain that all of the contamination has been removed. Clearly, the most effective solution to groundwater contamination is prevention.

Hot Springs and Geysers

By definition, the water in **hot springs** is 6–9°C (11–16°F) warmer than the mean annual air temperature for the localities where they occur. In the United States alone, there are well over 1000 such springs.

Temperatures in deep mines and oil wells usually rise with increasing depth, an average of about 2°C per 100 meters (1°F per 100 feet). Therefore, when groundwater circulates at great depths, it becomes heated. If it

Figure 10.11 **A.** Although the contaminated water has traveled more than 100 meters before reaching Well 1, the water moves too rapidly through the cavernous limestone to be purified. **B.** As the discharge from the septic tank percolates through the permeable sandstone, it is purified in a relatively short distance.

rises to the surface, the water may emerge as a hot spring. The water of some hot springs in the eastern United States is heated in this manner. The great majority (over 95 percent) of the hot springs (and geysers) in the United States are found in the West. The reason for such a distribution is that the source of heat for most hot springs is cooling igneous rock, and it is in the West that igneous activity has been most recent.

Geysers are intermittent hot springs or fountains where columns of water are ejected with great force at various intervals, often rising 30 to 60 meters (100 to 200 feet) into the air. After the jet of water ceases, a column

Figure 10.12 Sometimes agricultural chemicals **A.** and materials leached from landfills **B.** find their way into the groundwater. These are two of the potential sources of groundwater contamination. (Photo A by Roy Morsch/The Stock Market; Photo B by F. Rossotto/ The Stock Market)

A.

B.

C.

Figure 10.13 A wintertime eruption of Old Faithful, one of the world's most famous geysers. It emits as much as 45,000 liters (almost 12,000 gallons) of hot water and steam about once each hour. (Photo by Marc Muench, David Muench Photography, Inc.)

of steam rushes out, usually with a thunderous roar. Perhaps the most famous geyser in the world is Old Faithful in Yellowstone National Park, which erupts about once each hour (Figure 10.13). The great abundance, diversity, and spectacular nature of Yellowstone's geysers and other thermal features undoubtedly were the primary reason for its becoming the first national park in the United States. Geysers are also found in other parts of the world, notably New Zealand and Iceland. In fact, the Icelandic word *geysa*, "to gush," gives us the name *geyser*.

Geysers occur where extensive underground chambers exist within hot igneous rocks. How they operate is shown in Figure 10.14. As relatively cool groundwater enters the chambers, it is heated by the surrounding rock. At the bottom of the chambers, the water is under great pressure because of the weight of the overlying water. This great pressure prevents the water from boiling at the normal surface temperature of 100°C (212°F).

For example, water at the bottom of a 300-meter (1000-foot) water-filled chamber must reach nearly 230°C (450°F) to boil. The heating causes the water to expand, with the result that some is forced out at the surface. This loss of water reduces the pressure on the remaining water in the chamber, which lowers the boiling point. A portion of the water deep within the chamber quickly turns to steam, and the geyser erupts (Figure 10.14). Follow-

Figure 10.14 Idealized diagrams of a geyser. A geyser can form if the heat is not distributed by convection. **A.** In this figure, the water near the bottom is heated to near its boiling point. The boiling point is higher there than at the surface because the weight of the water above increases the pressure. **B.** The water higher in the geyser system is also heated; therefore, it expands and flows out at the top, reducing the pressure on the water at the bottom. **C.** At the reduced pressure on the bottom, boiling occurs. Some of the bottom water flashes into steam, and the expanding steam causes an eruption.

ing the eruption, cool groundwater again seeps into the chamber, and the cycle begins anew.

When groundwater from hot springs and geysers flows out at the surface, material in solution is often precipitated, producing an accumulation of chemical sedimentary rock. The material deposited at any given place commonly reflects the chemical makeup of the rock through which the water circulated. When the water contains dissolved silica, a material called *siliceous sinter* or *geyserite* is deposited around the spring. When the water contains dissolved calcium carbonate, a form of limestone called *travertine* or *calcareous tufa* is deposited. The latter term is used if the material is spongy and porous. In addition to containing dissolved silica and calcium carbonate, some hot springs contain sulfur, which gives water a poor taste and unpleasant odor. Undoubtedly, Rotten Egg Spring, Nevada, is such a situation.

Geothermal Energy

Geothermal energy is harnessed by tapping natural underground reservoirs of steam and hot water. These occur where subsurface temperatures are high, owing to relatively recent volcanic activity. Geothermal energy is put to use in two ways: The steam and hot water are used for heating and to generate electricity.

Iceland is a large volcanic island with many active volcanoes (Figure 10.15). In Iceland's capital, Reykjavik, hot water is pumped into buildings throughout the city for space heating. (Reykjavik literally means "bay of steam.") It also warms greenhouses, where fruits and vegetables are grown year-round. In the United States, localities in several western states use hot water from geothermal sources for space heating.

As for generating electricity geothermally, the Italians were first to do so in 1904, so the idea is not new. By the turn of the twenty-first century, more than 250 geothermal power plants in 22 countries were producing more than 8000 megawatts (million watts). These plants provide power to more than 60 million people. The leading producers of geothermal power are listed in Table 10.2.

The first commercial geothermal power plant in the United States was built in 1960 at The Geysers, north of San Francisco (Figure 10.16). The Geysers remains the world's largest geothermal power plant, generating about 1700 megawatts or nearly 60 percent of U.S. geothermal power. In addition to The Geysers, geothermal development is occurring elsewhere in the western United States, including Nevada, Utah, and the Imperial Valley in southern California. The U.S. geothermal generating capacity of more than 2800 megawatts is enough to power about 3.5 million homes. This is an amount of electricity comparable to burning about 60 million barrels of oil each year.

Figure 10.15 Iceland straddles the Mid-Atlantic Ridge. This divergent plate boundary is the site of numerous active volcanoes and geothermal systems. Because the entire country consists of geologically young volcanic rocks, warm water can be encountered in holes drilled almost anywhere. More than 45 percent of Iceland's energy comes from geothermal sources. The photo shows a power station in southwestern Iceland. The steam is used to generate electricity. Hot (83°C) water from the plant is sent via an insulated pipeline to Reykjavik for space heating. (Photo by Simon Fraser/Science Photo Library)

Table 10.2 Worldwide Geothermal Power Production, 2000.

Producing Country	Megawatts
United States	2850
Philippines	1848
Italy	768.5
Mexico	743
Indonesia	589.5
Japan	530
New Zealand	345
Iceland	140
Costa Rica	120
El Salvador	105
all others	178
Total	8217

Source: Geothermal Education Office.

What geologic factors favor a geothermal reservoir of commercial value?

1. *A potent source of heat*, such as a large magma chamber deep enough to ensure adequate pressure and slow cooling, yet not so deep that the natural water circulation is inhibited. Such magma chambers are most likely in regions of recent volcanic activity.
2. *Large and porous reservoirs with channels connected to the heat source*, near which water can circulate and then be stored in the reservoir.
3. *A cap of low permeability rocks* that inhibits the flow of water and heat to the surface. A deep, well-insulated reservoir contains much more stored energy than does a similar but uninsulated reservoir.

Figure 10.16 The Geysers, near the city of Santa Rosa in northern California, is the world's largest electricity-generating geothermal development. Most of the steam wells are about 3000 meters deep. (Photo courtesy of Pacific Gas & Electric Company)

We must recognize that geothermal power is not inexhaustible. When hot fluids are pumped from volcanically heated reservoirs, water cannot be replaced and then heated sufficiently to recharge the reservoir. Experience shows that steam and hot water from individual wells usually lasts no more than 10 to 15 years, so more wells must be drilled to maintain power production. Eventually the field is depleted.

As with other alternative methods of power production, geothermal sources are not expected to provide a high percentage of the world's growing energy needs. Nevertheless, in regions where its potential can be developed, its use will no doubt continue to grow.

The Geologic Work of Groundwater

Groundwater dissolves rock. This fact is the key to understanding how caverns and sinkholes form. Because soluble rocks, especially limestone, underlie millions of square kilometers of Earth's surface, it is here that groundwater carries on its important role as an erosional agent. Limestone is nearly insoluble in pure water, but it is quite easily dissolved by water containing small quantities of carbonic acid, and most groundwater contains this acid. It forms because rainwater readily dissolves carbon dioxide from the air and from decaying plants. Therefore, when groundwater comes in contact with limestone, the carbonic acid reacts with calcite (calcium carbonate) in the rocks to form calcium bicarbonate, a soluble material that is then carried away in solution.

Caverns

The most spectacular results of groundwater's erosional handiwork are limestone **caverns.** In the United States alone, about 17,000 caves have been discovered, and new ones are being found every year. Although most are modest, some have spectacular dimensions. Mammoth Cave in Kentucky and Carlsbad Caverns in southeastern New Mexico are famous examples. The Mammoth Cave system is the most extensive in the world, with more than 540 kilometers (340 miles) of interconnected passages. The dimensions at Carlsbad Caverns are impressive in a different way. Here we find the largest and perhaps most spectacular single chamber. The Big Room at Carlsbad Caverns has an area equivalent to 14 football fields and enough height to accommodate the U.S. Capitol Building.

Most caverns are created at or just below the water table in the zone of saturation. Here acidic groundwater follows lines of weakness in the rock, such as joints and bedding planes. As time passes, the dissolving process slowly creates cavities and gradually enlarges them into caverns (Figure 10.17). The material dissolved

Figure 10.17 Development of a karst landscape. **A.** During early stages, groundwater percolates through limestone along joints and bedding planes. Solution activity creates and enlarges caverns at and below the water table. **B.** In this view, sinkholes are well developed and surface streams are funneled below ground. **C.** With the passage of time, caverns grow larger and the number and size of sinkholes increase. Collapse of caverns and coalescence of sinkholes form larger, flat-floored depressions. Eventually, solution activity may remove most of the limestone from the area, leaving only isolated remnants.

Figure 10.18 Speleothems are of many types, including stalactites, stalagmites, and columns. Chinese Theater, Carlsbad Caverns National Park. (Photo by David Muench Photography)

by groundwater is carried away and discharged into streams and transported to the ocean.

In many caves, development has occurred at several levels, with the current cavern-forming activity occurring at the lowest elevation. This situation reflects the close relationship between the formation of major subterranean passages and the river valleys into which they drain. As streams cut their valleys deeper, the water table drops as the elevation of the river drops.

Certainly the features that arouse the greatest curiosity for most cavern visitors are the stone formations that give some caverns a wonderland appearance. These are not *erosional* features like the cavern itself, but are depositional features created by the seemingly endless dripping of water over great spans of time. The calcium carbonate that is left behind produces the limestone we call *travertine*. These cave deposits, however, are also commonly called *dripstone*, an obvious reference to their mode of origin. Although the formation of caverns takes place in the zone of saturation, the deposition of dripstone is not possible until the caverns are above the water table in the zone of aeration. As soon as the chamber is filled with air, the stage is set for the decoration phase of cavern building to begin.

The various dripstone features found in caverns are collectively called **speleothems,** no two of which are exactly alike (Figure 10.18). Perhaps the most familiar speleothems are **stalactites.** These icicle-like pendants hang from the ceiling of the cavern and form where water seeps through cracks above. When the water reaches the air in the cave, some of the carbon dioxide in solution escapes from the drop and calcium carbonate precipitates. Deposition occurs as a ring around the edge of the water drop. As drop after drop follows, each leaves an infinitesimal trace of calcite behind, and a hollow limestone tube is created. Water then moves through the tube, remains suspended momentarily at the end, contributes a tiny ring of calcite, and falls to the cavern floor.

The stalactite just described is appropriately called a *soda straw* (Figure 10.19). Often the hollow tube of the soda straw becomes plugged or its supply of water increases. In either case, the water is forced to flow, and hence deposit, along the outside of the tube. As deposition continues, the stalactite takes on the more common conical shape.

Figure 10.19 "Live" soda straw stalactites. Lehman Caves, Great Basin National Park, Nevada. (Photo by Tom & Susan Bean, Inc.)

Speleothems that form on the floor of a cavern and reach upward toward the ceiling are called **stalagmites.** The water supplying the calcite for stalagmite growth falls from the ceiling and splatters over the surface. As a result, stalagmites do not have a central tube, and they are usually more massive in appearance and more rounded on their upper ends than stalactites. Given enough time, a downward-growing stalactite and an upward-growing stalagmite may join to form a *column.*

Karst Topography

Many areas of the world have landscapes that to a large extent have been shaped by the dissolving power of groundwater. Such areas are said to exhibit **karst topography,** named for the Krs Plateau, located along the northeastern shore of the Adriatic Sea in the border area between Slovenia (formerly a part of Yugoslavia) and Italy where such topography is strikingly developed (Figure 10.17). In the United States, karst landscapes occur in many areas that are underlain by limestone, including portions of Kentucky, Tennessee, Alabama, southern Indiana, and central and northern Florida. Generally, arid and semiarid areas are too dry to develop karst topography. When solution features exist in such regions, they are likely to be remnants of a time when rainier conditions prevailed.

Karst areas typically have irregular terrain punctuated with many depressions, called **sinkholes** or **sinks.** In the limestone areas of Florida, Kentucky, and south-ern Indiana, there are literally tens of thousands of these depressions varying in depth from just a meter or two to a maximum of more than 50 meters (Figure 10.20).

Sinkholes commonly form in two ways. Some develop gradually over many years without any physical disturbance to the rock. In these situations, the limestone immediately below the soil is dissolved by downward-seeping rainwater that is freshly charged with carbon dioxide. With time, the bedrock surface is lowered and the fractures into which the water seeps are enlarged. As the fractures grow in size, soil subsides into the widening voids, from which it is removed by groundwater flowing in the passages below. These depressions are usually shallow and have gentle slopes.

By contrast, sinkholes can also form abruptly and without warning when the roof of a cavern collapses under its own weight. Typically, the depressions created in this manner are steep-sided and deep. When they form in populous areas, they may represent a serious geologic hazard. Such a situation is clearly the case in Figure 10.20B.

In addition to a surface pockmarked by sinkholes, karst regions characteristically show a striking lack of surface drainage (streams). Following a rainfall, the runoff is quickly funneled below ground through the sinks. It then flows through caverns until it finally reaches the water table. Where streams do exist at the surface, their paths are usually short. The names of such streams often give a clue to their fate. In the Mammoth Cave area of Kentucky, for example, there is Sinking

A.

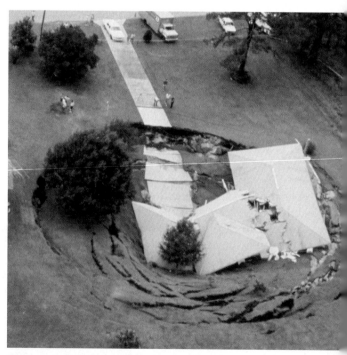

B.

Figure 10.20 **A.** This high-altitude infrared image shows an area of karst topography in central Florida. The numerous lakes occupy sinkholes. (Courtesy of USDA-ASCS) **B.** This small sinkhole formed suddenly in 1991 when the roof of a cavern collapsed, destroying this home in Frostproof, Florida. (Photo by *St. Petersburg Times*/Liaison Agency, Inc.)

Creek, Little Sinking Creek, and Sinking Branch. Other sinkholes become plugged with clay and debris to create small lakes or ponds.

Some regions of karst development exhibit landscapes that look very different from the sinkhole-studded terrain depicted in Figure 10.17. One striking example is an extensive region in southern China that is described as exhibiting *tower karst*. As the image in Figure 10.21 shows, the term *tower* is appropriate because the landscape consists of a maze of isolated steep-sided hills that rise abruptly from the ground. Each is riddled with interconnected caves and passageways. This type of karst topography forms in wet tropical and subtropical regions having thick beds of highly jointed limestone. Here groundwater has dissolved large volumes of limestone leaving only these residual towers. Karst development is more rapid in tropical climates due to the abundant rainfall and the greater availability of carbon dioxide from the decay of lush tropical vegetation. The extra carbon dioxide in the soil means there is more carbonic acid for dissolving limestone. Other tropical areas of advanced karst development include portions of Puerto Rico, western Cuba, and northern Vietnam.

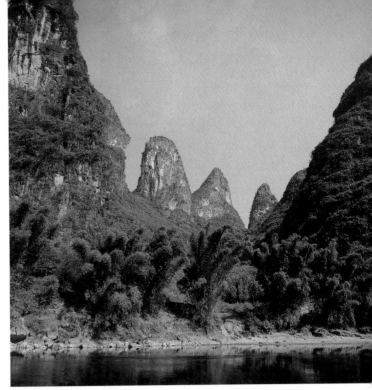

Figure 10.21 One of the best-known and most distinctive regions of tower karst development is the Guilin District of southeastern China. (Photo by Jose Fuste Raga/Corbis/The Stock Market)

The Chapter in Review

- As a resource, *groundwater* represents the largest reservoir of fresh water that is readily available to humans. Geologically, the dissolving action of groundwater produces *caves* and *sinkholes*. Groundwater is also an equalizer of streamflow.

- Groundwater is water that completely fills the pore spaces in sediment and rock in the subsurface *zone of saturation*. The upper limit of this zone is the *water table*. The *zone of aeration* is above the water table where the soil, sediment, and rock are not saturated.

- The interaction between streams and groundwater takes place in one of three ways: Streams gain water from the inflow of groundwater (*gaining stream*), they lose water through the streambed to the groundwater system (*losing stream*), or they do both, gaining in some sections and losing in others.

- The quantity of water that can be stored in a material depends upon its *porosity* (the volume of open spaces). The *permeability* (the ability to transmit a fluid through interconnected pore spaces) of a material is a very important factor controlling the movement of groundwater.

- Materials with very small pore spaces (such as clay) hinder or prevent groundwater movement and are called *aquitards*. *Aquifers* consist of materials with larger pore spaces (such as sand) that are permeable and transmit groundwater freely.

- Groundwater moves in looping curves that are a compromise between the downward pull of gravity and the tendency of water to move toward areas of reduced pressure.

- *Springs* occur whenever the water table intersects the land surface and a natural flow of groundwater results. *Wells*, openings bored into the zone of saturation, withdraw groundwater and create roughly conical depressions in the water table known as *cones of depression*. *Artesian wells* occur when water rises above the level at which it was initially encountered.

- When groundwater circulates at great depths, it becomes heated. If it rises, the water may emerge as a *hot spring*. Geysers occur when groundwater is heated in underground chambers, expands, and some water quickly changes to steam, causing the geyser to erupt. The source of heat for most hot springs and geysers is hot igneous rock. *Geothermal energy* is harnessed by tapping natural underground reservoirs of steam and hot water.

- Some of the current environmental problems involving groundwater include (1) *overuse* by intense irrigation, (2) *land subsidence* caused by groundwater withdrawal, and (3) contamination by pollutants.

- Most *caverns* form in limestone at or below the water table when acidic groundwater dissolves rock along lines of weakness, such as joints and bedding planes. The various *dripstone* features found in caverns are collectively called *speleothems*. Landscapes that to a large extent have been shaped by the dissolving power of groundwater exhibit *karst topography*, an irregular terrain punctuated with many depressions, called *sinkholes* or *sinks*.

219

Key Terms

aquifer (p. 206)
aquitard (p. 206)
artesian (p. 208)
cavern (p. 215)
cone of depression (p. 207)
drawdown (p. 207)
gaining stream (p. 204)
geothermal energy (p. 214)

geyser (p. 212)
groundwater (p. 202)
hot spring (p. 211)
karst topography (p. 218)
losing stream (p. 204)
perched water table
    (p. 207)

permeability (p. 206)
porosity (p. 206)
sinkhole (sink) (p. 218)
speleothem (p. 217)
spring (p. 207)
stalactite (p. 217)
stalagmite (p. 218)

water table (p. 202)
well (p. 207)
zone of aeration (p. 202)
zone of saturation (p. 202)

Questions for Review

1. What percentage of fresh water is groundwater? If glacial ice is excluded and only liquid fresh water is considered, about what percentage is groundwater?

2. Geologically, groundwater is important as an erosional agent. Name another significant geological role of groundwater.

3. Define groundwater, and relate it to the water table.

4. Contrast a gaining stream and a losing stream.

5. Distinguish between porosity and permeability.

6. What is the difference between an aquitard and an aquifer?

7. Under what circumstances can a material have a high porosity but not be a good aquifer?

8. When an aquitard is situated above the main water table, a localized saturated zone may be created. What term is applied to such a situation?

9. What is meant by the term *artesian*? List two conditions that must be present for artesian wells to exist.

10. What problem is associated with the pumping of groundwater for irrigation in the southern part of the High Plains? (See Box 10.1.)

11. Briefly explain what happened in the San Joaquin Valley of California as the result of excessive groundwater withdrawal.

12. Which aquifer would be most effective in purifying polluted groundwater: coarse gravel, sand, or cavernous limestone?

13. What is meant when a groundwater pollutant is classified as hazardous?

14. What is the source of heat for most hot springs and geysers? How is this reflected in the distribution of these features?

15. Is geothermal power considered an inexhaustible energy source?

16. Name two common speleothems and distinguish between them.

17. Areas whose landscapes are largely a reflection of the erosional work of groundwater are said to exhibit what kind of topography?

18. Describe two ways in which sinkholes are created.

Web Resources

The *Essentials of Geology* Website uses the resources and flexibility of the Internet to aid in your study of the topics in this chapter. Written and developed by geology instructors, this site will help improve your understanding of geology. Visit **http://www.prenhall.com/lutgens** and click on the cover of *Essentials of Geology, 8/e* to find:

- On-line review quizzes.
- Web-based critical thinking and writing exercises.
- Links to chapter-specific Web resources.
- Internet-wide key term searches.

http://www.prenhall.com/lutgens

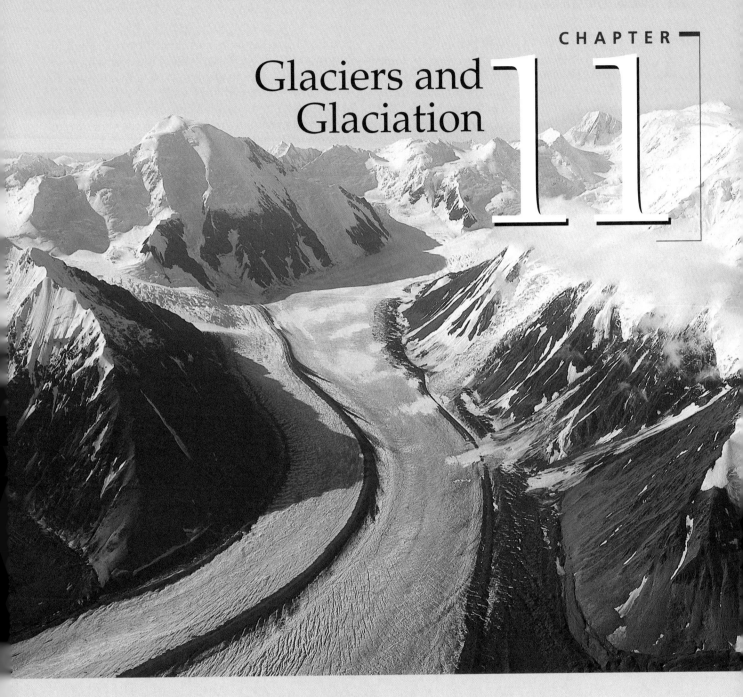

CHAPTER

Glaciers and Glaciation

11

Focus on Learning

To assist you in learning the important concepts in this chapter, you will find it helpful to focus on the following questions:

- What is a glacier? What are the different types of glaciers? Where are glaciers located?
- How do glaciers move, and what are the various processes of glacial erosion?
- What are the features created by glacial erosion and deposition? What materials make up the depositional features?
- What is the evidence for the Ice Age? What are some indirect effects of Ice Age glaciers?
- What are some proposals that attempt to explain the causes of glacial ages?

Peter Glacier is a valley glacier in Alaska's Denali National Park. (Photo by Michael Collier)

221

Today, glaciers cover nearly 10 percent of Earth's land surface. However, in the recent geologic past, ice sheets were three times more extensive, covering vast areas with ice thousands of meters thick. Many regions still bear the mark of these glaciers. The basic character of such diverse places as the Alps, Cape Cod, and Yosemite Valley was fashioned by now vanished masses of glacial ice. Moreover, Long Island, the Great Lakes, and the fiords of Norway and Alaska all owe their existence to glaciers. Glaciers, of course, are not just a phenomenon of the geologic past. As you will see, they are still sculpting and depositing debris in many regions today (Figure 11.1).

Glaciers: A Part of Two Basic Cycles

External Processes
↳ Glaciers

Glaciers are a part of two fundamental cycles in the Earth system—the hydrologic cycle and the rock cycle. Earlier you learned that the water of the hydrosphere is constantly cycled through the atmosphere, biosphere, and solid Earth. Time and time again the same water is evaporated from the oceans into the atmosphere, pre-

cipitated upon the land, and carried by rivers and underground flow back to the sea. However, when precipitation falls at high elevations or high latitudes, the water may not immediately make its way toward the sea. Instead, it may become part of a glacier. Although the ice will eventually melt, allowing the water to continue its path to the sea, water can be stored as glacial ice for many tens, hundreds, or even thousands of years (see Box 11.1).

A **glacier** is a thick ice mass that forms over hundreds or thousands of years. It originates on land from the accumulation, compaction, and recrystallization of snow. A glacier appears to be motionless, but it is not—glaciers move very slowly. Like running water, groundwater, wind, and waves, glaciers are dynamic erosional agents that accumulate, transport, and deposit sediment. As such, glaciers are among the processes that perform a very basic function in the rock cycle. Although glaciers are found in many parts of the world today, most are located in remote areas, either near Earth's poles or in high mountains.

Valley (Alpine) Glaciers

Literally thousands of relatively small glaciers exist in lofty mountain areas, where they usually follow valleys that were originally occupied by streams. Unlike the rivers that previously flowed in these valleys, the glac-

Figure 11.1 Barry Glacier, a valley glacier in the Chugach Mountains, Prince William Sound, Alaska. (Photo by Carr Clifton)

iers advance slowly, perhaps only a few centimeters per day. Because of their setting, these moving ice masses are termed **valley glaciers** or **alpine glaciers** (Figure 11.1). Each glacier is a stream of ice, bounded by precipitous rock walls, that flows downvalley from a snow accumulation center near its head. Like rivers, valley glaciers can be long or short, wide or narrow, single or with branching tributaries. Generally, the widths of alpine glaciers are small compared to their lengths. Some extend for just a fraction of a kilometer, whereas others go on for many tens of kilometers. The west branch of the Hubbard Glacier, for example, runs through 112 kilometers (nearly 70 miles) of mountainous terrain in Alaska and the Yukon Territory.

Ice Sheets

In contrast to valley glaciers, **ice sheets** exist on a much larger scale. Although many ice sheets have existed in the past, just two achieve this status at present (Figure 11.2). In the Northern Hemisphere, Greenland is covered by an imposing ice sheet that occupies 1.7 million square kilometers (0.7 million square miles), or about 80 percent of this large island. Averaging nearly 1500 meters (5000 feet) thick, in places the ice extends 3000 meters (10,000 feet) above the island's bedrock floor.

In the Southern Hemisphere, the huge Antarctic Ice Sheet attains a maximum thickness of nearly 4300 meters (14,000 feet) and covers an area of more than 13.9 million square kilometers (5.4 million square miles). Because of the proportions of these huge features, they are often called continental ice sheets. Indeed, the combined areas of present-day *continental ice sheets* represent almost 10 percent of Earth's land area.

These enormous masses flow out in all directions from one or more snow-accumulation centers and completely obscure all but the highest areas of underlying terrain. Even sharp variations in the topography beneath the glacier usually appear as relatively subdued undulations on the surface of the ice. Such topographic differences, however, do affect the behavior of the ice sheets, especially near their margins, by guiding flow in certain directions and creating zones of faster and slower movement.

Along portions of the Antarctic coast, glacial ice flows into bays, creating features called **ice shelves.** They are large, relatively flat masses of floating ice that extend seaward from the coast but remain attached to the land along one or more sides. The shelves are thickest on their landward sides, and they become thinner seaward. They are sustained by ice from the adjacent ice sheet as well as being nourished by snowfall and the freezing of seawater to their bases. Antarctica's ice shelves extend over approximately 1.4 million square kilometers (0.6 million square miles). The Ross and Filchner ice shelves are the largest, with the Ross Ice

Shelf alone covering an area approximately the size of Texas (Figure 11.2).

Other Types of Glaciers

In addition to valley glaciers and ice sheets, other types of glaciers are also identified. Covering some uplands and plateaus are masses of glacial ice called **ice caps.**

Figure 11.2 The only present-day continental ice sheets are those covering Greenland and Antarctica. Their combined areas represent almost 10 percent of Earth's land area. Greenland's ice sheet occupies 1.7 million square kilometers, or about 80 percent of the island. The area of the Antarctic Ice Sheet is almost 14 million square kilometers. Ice shelves occupy an additional 1.4 million square kilometers adjacent to the Antarctic Ice Sheet.

What If the Ice Melted?

How much water is stored as glacial ice? Estimates by the U.S. Geological Survey indicate that only slightly more than 2 percent of the world's water is accounted for by glaciers. But this small figure may be misleading when the actual amounts of water are considered. The total volume of all valley glaciers is about 210,000 cubic kilometers, comparable to the combined volume of the world's largest saline and freshwater lakes. Furthermore, 80 percent of the world's ice and nearly two-thirds of Earth's fresh water are represented by the Antarctic Ice Sheet, which covers an area almost one and a half times that of the United States. If this ice melted, sea level would rise an estimated 60 to 70 meters, and the ocean would inundate many densely populated coastal areas (Figure 11.A).

The hydrologic importance of the continent and its ice can be illustrated in another way. If the Antarctic Ice Sheet were to melt at a uniform rate, it could feed (1) the Mississippi River for more than 50,000 years, (2) all the rivers in the United States for about 17,000 years,

Figure 11.A This map of a portion of North America shows the present-day coastline compared to the coastline that existed during the last Ice Age maximum (18,000 years ago) and the coastline that would exist if present ice sheets in Greenland and Antarctica melted. (After R. H. Dott, Jr., and R. L. Battan, *Evolution of the Earth*, New York: McGraw-Hill, 1971. Reprinted by permission of the publisher.)

(3) the Amazon River for approximately 5000 years, or (4) all the rivers of the world for about 750 years.

As the foregoing discussion illustrates, the quantity of ice on Earth today is truly immense. However, present glaciers occupy only about one-third the area they did in the very recent geologic past.

Like ice sheets, ice caps completely bury the underlying landscape but are much smaller than the continental-scale features. Ice caps occur in many places, including Iceland and several of the large islands in the Arctic Ocean.

Often ice caps and ice sheets feed **outlet glaciers**. These tongues of ice flow down valleys extending outward from the margins of these larger ice masses. The tongues are essentially valley glaciers that are avenues for ice movement from an ice cap or ice sheet through mountainous terrain to the sea. Where they encounter the ocean, some outlet glaciers spread out as floating ice shelves. Often large numbers of icebergs are produced.

Piedmont glaciers occupy broad lowlands at the bases of steep mountains and form when one or more valley glaciers emerge from the confining walls of mountain valleys. Here the advancing ice spreads out to form a broad sheet. The size of individual piedmont glaciers varies greatly. Among the largest is the broad Malaspina Glacier along the coast of southern Alaska. It covers more than 5000 square kilometers (2000 square miles) of the flat coastal plain at the foot of the lofty St. Elias Range.

How Glaciers Move

External Processes
↳ Glaciers

The movement of glacial ice is generally referred to as *flow*. The fact that glacial movement is described in this way seems paradoxical—how can a solid flow? The way in which ice flows is complex and is of two basic types. The first of these, *plastic flow*, involves movement *within* the ice. Ice behaves as a brittle solid until the pressure upon it is equivalent to the weight of about 50 meters (165 feet) of ice. Once that load is surpassed, ice behaves as a plastic material, and flow begins. A second and often equally important mechanism of glacial movement consists of the entire ice mass slipping along the ground. The lowest portions of most glaciers are thought to move by this sliding process.

The upper 50 meters or so of a glacier is not under sufficient pressure to exhibit plastic flow. Rather, the ice in this uppermost zone is brittle and is appropriately referred to as the *zone of fracture*. The ice in this zone is carried along "piggyback" style by the ice below. When

Figure 11.3 Hikers challenged by a large crevasse on Peru's Mt. Huascaran. Crevasses form in the brittle ice of the zone of fracture. They do not continue down into the zone of flow. (Photo by Tom & Susan Bean, Inc.)

the glacier moves over irregular terrain, the zone of fracture is subjected to tension, resulting in cracks called **crevasses** (Figure 11.3). These gaping cracks, which often make travel across glaciers dangerous, may extend to depths of 50 meters (165 feet). Beyond this depth, plastic flow seals them off.

Rates of Glacial Movement

Unlike streamflow, glacial movement is not obvious. If we could watch an alpine glacier move, we would see that like the water in a river, all of the ice in the valley does not move downvalley at an equal rate. Just as friction with the bedrock floor slows the movement of the ice at the bottom of the glacier, the drag created by the valley walls leads to the flow being greatest in the center of the glacier.

This was first demonstrated by experiments during the nineteenth century, in which markers were carefully placed in a straight line across the top of a valley glacier. Periodically, the positions of the stakes were recorded, revealing the type of movement just described (Figure 11.4).

How rapidly does glacial ice move? Average velocities vary considerably from one glacier to another. Some move so slowly that trees and other vegetation may become well established in the debris that has accumulated on the glacier's surface, whereas others may move at rates of up to several meters per day. For example, Byrd Glacier, an outlet glacier in Antarctica that was the subject of a 10-year study using satellite images, moved at an average rate of 750 to 800 meters per year (about 2 meters per day). Other glaciers in the study advanced at one-fourth that rate.

The advance of some glaciers is characterized by periods of extremely rapid movements called *surges*. Glaciers that exhibit such movement may flow along in an apparently normal manner, then speed up for a rel-

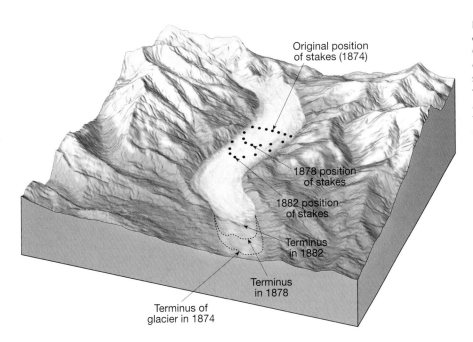

Original position of stakes (1874)

1878 position of stakes

1882 position of stakes

Terminus in 1882

Terminus in 1878

Terminus of glacier in 1874

Figure 11.4 Ice movement and changes in the terminus at Rhone Glacier, Switzerland. In this classic study of a valley glacier, the movement of stakes clearly showed that ice along the sides of the glacier moves slowest. Also notice that even though the ice front was retreating, the ice within the glacier was advancing.

atively short time before returning to the normal rate again. The flow rates during surges are as much as 100 times the normal rate.

Budget of a Glacier

Snow is the raw material from which glacial ice originates; therefore, glaciers form in areas where more snow falls in winter than melts during the summer. Glaciers are constantly gaining and losing ice. Snow accumulation and ice formation occur in the **zone of accumulation.** Its outer limits are defined by the *snowline.* The elevation of the snowline varies greatly. In polar regions, it may be sea level, whereas in tropical areas, the snowline exists only high in mountain areas, often at altitudes exceeding 4500 meters (15,000 feet). Above the snowline, in the zone of accumulation, the addition of snow thickens the glacier and promotes movement. Below the snowline is the **zone of wastage.** Here there is a net loss to the glacier as all of the snow from the previous winter melts, as does some of the glacial ice (Figure 11.5).

In addition to melting, glaciers also waste as large pieces of ice break off the front of the glacier in a process called **calving.** Calving creates *icebergs* in places where the glacier has reached the sea or a lake (Figure 11.6).

Because icebergs are just slightly less dense than seawater, they float very low in the water, with more than 80 percent of their mass submerged. Along the margins of Antarctica's ice shelves, calving is the primary means by which these masses lose ice. The relatively flat icebergs produced here can be several kilometers across and 600 meters thick. By comparison, thousands of irregularly shaped icebergs are produced by outlet glaciers flowing from the margins of the Greenland Ice Sheet. Many drift southward and find their way into the North Atlantic, where they can pose a hazard to navigation.

Whether the margin of a glacier is advancing, retreating, or remaining stationary depends on the budget of the glacier. The **glacial budget** is the balance, or lack of balance, between accumulation at the upper end of the glacier and loss at the lower end. This loss is termed **ablation.** If ice accumulation exceeds ablation, the glacial front advances until the two factors balance. When this happens, the terminus of the glacier is stationary.

If a warming trend increases ablation and/or if a drop in snowfall decreases accumulation, the ice front will retreat. As the terminus of the glacier retreats, the extent of the zone of wastage diminishes. Therefore, in time a new balance will be reached between accumula-

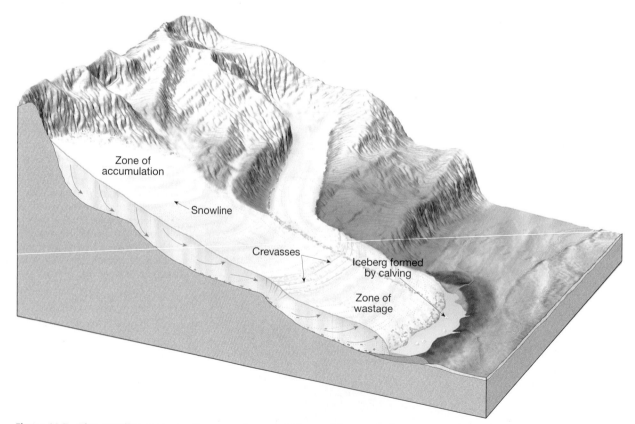

Figure 11.5 The snowline separates the zone of accumulation and the zone of wastage. Above the snowline, more snow falls each winter than melts each summer. Below the snowline, the snow from the previous winter completely melts as does some of the underlying ice. Whether the margin of a glacier advances, retreats, or remains stationary depends on the balance between accumulation and wastage (ablation). When a glacier moves across irregular terrain, *crevasses* form in the brittle upper portion.

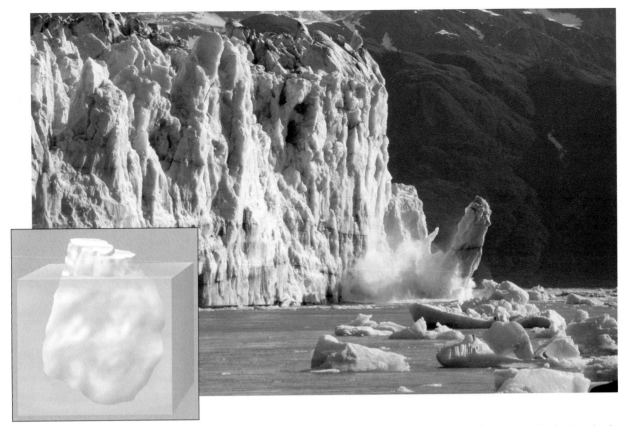

Figure 11.6 Icebergs are created when large pieces calve from the front of a glacier after it reaches a water body. Here ice is calving from the terminus of Alaska's Hubbard Glacier in Wrangell–St. Elias National Park, Alaska. Only about 20 percent of an iceberg protrudes above the water line. (Photo by Tom & Susan Bean, Inc.)

tion and wastage, and the ice front will again become stationary.

Whether the margin of a glacier is advancing, re-treating, or stationary, the ice within the glacier continues to flow forward. In the case of a receding glacier, the ice still flows forward, but not rapidly enough to offset ablation. This point is illustrated well in Figure 11.4. As the line of stakes within the Rhone Glacier continued to move downvalley, the terminus of the glacier slowly retreated upvalley.

Glacial Erosion

Glaciers are capable of great erosion. For anyone who has observed the terminus of an alpine glacier, the evidence of its erosive force is clear. You can witness first-hand the release of rock fragments of various sizes from the ice as it melts. All signs lead to the conclusion that the ice has scraped, scoured, and torn rock from the floor and walls of the valley and carried it downvalley. Indeed, as a medium of sediment transport, ice has no equal.

Once rock debris is acquired by the glacier, the enormous competency of ice will not allow the debris to settle out like the load carried by a stream or by the wind. Consequently, glaciers can transport huge blocks that no other erosional agent could possibly budge (Figure 11.7). Although today's glaciers are of limited importance as erosional agents, many landscapes that were modified by the widespread glaciers of the most recent Ice Age still reflect to a high degree the work of ice.

Glaciers erode the land primarily in two ways—*plucking* and *abrasion*. First, as a glacier flows over a fractured bedrock surface, it loosens and lifts blocks of rock and incorporates them into the ice. This process, known as **plucking,** occurs when meltwater penetrates the cracks and joints of bedrock beneath a glacier and freezes. As the water expands, it exerts tremendous leverage that pries the rock loose. In this manner sediment of all sizes, ranging from particles as fine as flour to blocks as big as houses, becomes part of the glacier's load.

The second major erosional process is **abrasion** (Figure 11.8). As the ice and its load of rock fragments slide over bedrock, they function like sandpaper to smooth and polish the surface below. The pulverized rock produced by the glacial gristmill is appropriately called **rock flour.** So much rock flour may be produced that meltwater streams flowing out of a glacier often have the grayish appearance of skim milk

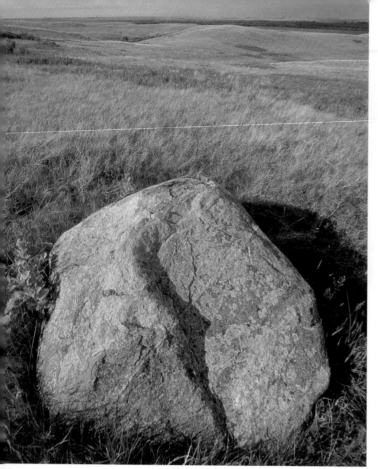

Figure 11.7 A large, glacially transported boulder near Hensler, North Dakota. Such boulders are called glacial erratics. (Photo by Tom Bean/DRK Photo)

Figure 11.8 Glacial abrasion created the scratches and grooves in this bedrock. Glacier Bay National Park, Alaska. (Photo by Carr Clifton)

and offer visible evidence of the grinding power of ice.

When the ice at the bottom of a glacier contains large fragments of rock, long scratches and grooves called **glacial striations** may be gouged into the bedrock (Figure 11.8). These linear grooves provide clues to the direction of ice flow. By mapping the striations over large areas, patterns of glacial flow can often be reconstructed.

In contrast, not all abrasive action produces striations. The rock surfaces over which the glacier moves may also become highly polished by the ice and its load of finer particles. The broad expanses of smoothly polished granite in Yosemite National Park provide an excellent example.

As is the case with other agents of erosion, the rate of glacial erosion is highly variable. This differential erosion by ice is largely controlled by four factors: (1) rate of glacial movement; (2) thickness of the ice; (3) shape, abundance, and hardness of the rock fragments contained in the ice at the base of the glacier; and (4) the erodibility of the surface beneath the glacier. Variations in any or all of these factors from time to time and/or from place to place mean that the features, effects, and degree of landscape modification in glaciated regions can vary greatly.

228

Landforms Created by Glacial Erosion

 External Processes
↳ Glaciers

The erosional effects of valley glaciers and ice sheets are quite different. A visitor to a glaciated mountain region is likely to see a sharp and angular topography. The reason is that as alpine glaciers move downvalley, they tend to accentuate the irregularities of the mountain landscape by creating steeper canyon walls and making bold peaks even more jagged. By contrast, continental ice sheets generally override the terrain and hence subdue rather than accentuate the irregularities they encounter. Although the erosional potential of ice sheets is enormous, landforms carved by these huge ice masses usually do not inspire the same wonderment and awe as do the erosional features created by valley glaciers. Much of the rugged mountain scenery so celebrated for its majestic beauty is the product of erosion by alpine glaciers. Figure 11.9 shows a hypothetical mountain area before, during, and after glaciation.

Figure 11.9   Erosional landforms created by alpine glaciers. The unglaciated landscape in part A is modified by valley glaciers in part B. After the ice recedes, in part C, the terrain looks very different than before glaciation.

A. Unglaciated topography

B. Region during period of maximum glaciation

C. Glaciated topography

Glaciated Valleys

A hike up a glaciated valley reveals a number of striking ice-created features. The valley itself is often a dramatic sight. Unlike streams, which create their own valleys, glaciers take the path of least resistance by following the course of existing stream valleys. Prior to glaciation, mountain valleys are characteristically narrow and V-shaped because streams are well above base level and are therefore downcutting. However, during glaciation these narrow valleys undergo a transformation as the glacier widens and deepens them, creating a U-shaped **glacial trough** (Figure 11.9C and Figure 11.10). In addition to producing a broader and deeper valley, the glacier also straightens the valley. As ice flows around sharp curves, its great erosional force removes the spurs of land that extend into the valley.

The intensity of glacial erosion depends in part upon the thickness of the ice. Consequently, main glaciers, also called *trunk glaciers*, cut their valleys deeper than do their smaller tributary glaciers. Thus, when the glaciers eventually recede, the valleys of tributary glaciers are left standing above the main glacial trough and are termed **hanging valleys** (Figure 11.9C). Rivers flowing through hanging valleys can produce spectacular waterfalls, such as those in Yosemite National Park (Figure 11.11).

Figure 11.10 Prior to glaciation, a mountain valley is typically narrow and V-shaped. During glaciation, an alpine glacier widens, deepens, and straightens the valley, creating the U-shaped glacial trough seen here. The string of lakes is called pater noster lakes. This valley is in Glacier National Park, Montana. (Photo by John Montagne)

As hikers walk up a glacial trough, they may pass a series of bedrock depressions on the valley floor that were probably formed by plucking and scoured by the abrasive force of the ice. If these depressions are filled with water, they are called **pater noster lakes** (Figure 11.10). The Latin name means "our Father" and is a reference to a string of rosary beads.

Figure 11.11 Bridalveil Falls in Yosemite National Park cascades from a hanging valley into the glacial trough below. (Photo by E. J. Tarbuck)

At the head of a glacial valley is a very characteristic and often imposing feature associated with an alpine glacier called a **cirque.** These bowl-shaped depressions have precipitous walls on three sides but are open on the downvalley side. The cirque is the focal point of the glacier's growth, because it is the area of snow accumulation and ice formation. They begin as irregularities in the mountainside that are subsequently enlarged by the frost wedging and plucking that occur along the sides and bottom of the glacier. After the glacier has melted away, the cirque basin is often occupied by a small lake (Figure 11.12).

Before leaving the topic of glacial troughs and their associated features, one more rather well-known feature should be discussed—fiords. **Fiords** are deep, often spectacular, steep-sided inlets of the sea that are present at high latitudes where mountains are adjacent to the ocean (Figure 11.13). They are drowned glacial troughs that became submerged as the ice left the valley and sea level rose following the Ice Age.

The depths of fiords can exceed 1000 meters (3300 feet). However, the great depths of these flooded troughs is only partly explained by the post–Ice Age rise in sea level. Unlike the situation governing the downward erosional work of rivers, sea level does not act as base level for glaciers. As a consequence, glaciers are capable of eroding their beds far below the surface of the sea. For example, a 300-meter-thick alpine glacier can carve its valley floor more than 250 meters below sea level before downward erosion ceases and the ice begins to float. Norway, British Columbia, Greenland, New Zealand, Chile, and Alaska all have coastlines characterized by fiords.

Arêtes and Horns

A visit to the Alps, the Northern Rockies, or many other scenic mountain landscapes carved by valley glaciers reveals not only glacial troughs, cirques, pater noster

Figure 11.12 A small lake occupying a cirque in the Tombstone Mountains, Yukon Territory, Canada. A *cirque* is a bowl-shaped depression that forms at the head of a glacial valley and represents a focal point of ice formation. (Photo by Stephen J. Krasemann)

lakes, and the other related features just discussed. You also are likely to see sinuous, sharp-edged ridges called **arêtes** (French for *knife-edge*) and sharp, pyramid-like peaks called **horns** projecting above the surroundings. Both features can originate from the same basic process, the enlargement of cirques produced by plucking and frost action (Figure 11.9B, C). In the case of the spires of rock called horns, cirques around a single high mountain are responsible. As the cirques enlarge and converge, an isolated horn is produced. The most famous example is the Matterhorn in the Swiss Alps (Figure 11.14).

Arêtes can be formed in a similar manner, except that the cirques are not clustered around a point but rather exist on opposite sides of a divide. As the cirques grow, the divide separating them is reduced to a narrow knifelike partition. An arête, however, may also be created in another way. When two glaciers occupy parallel valleys, an arête can form when the divide separating the moving tongues of ice is progressively narrowed as the glaciers scour and widen their adjacent valleys.

Roches Moutonnées

In many glaciated landscapes, but most frequently where continental ice sheets have modified the terrain, the ice carves small streamlined hills from protruding bedrock knobs. Such an asymmetrical knob of bedrock is called a **roche moutonnée** (French for *sheep rock*). They

Figure 11.13 Like other fiords, this one at Tracy Arm, Alaska, is a drowned glacial trough. (Photo by Tom & Susan Bean, Inc.)

231

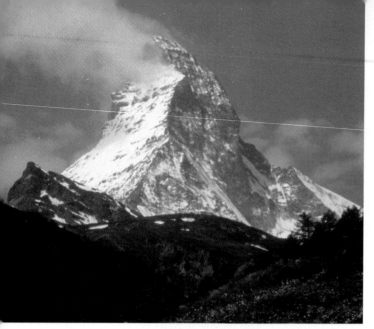

Figure 11.14 Horns are sharp, pyramid-like peaks that are fashioned by alpine glaciers. This example is the famous Matterhorn in the Swiss Alps. (Photo by E. J. Tarbuck)

are formed when glacial abrasion smoothes the gentle slope facing the oncoming ice sheet and plucking steepens the opposite side as the ice rides over the knob (Figure 11.15). Roches moutonnées indicate the direction of glacial flow, because the gentler slope is generally on the side from which the ice advanced.

Glacial Deposits

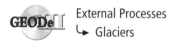

External Processes
↳ Glaciers

Glaciers pick up and transport a huge load of debris as they slowly advance across the land. Ultimately these materials are deposited when the ice melts. In regions where glacial sediment is deposited, it can play a truly

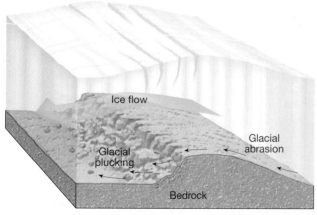

Figure 11.15 Roche moutonnée. The gentle slope was abraded and the steep side was plucked. The ice moved from right to left.

significant role in forming the physical landscape. For example, in many areas once covered by the ice sheets of the recent Ice Age, the bedrock is rarely exposed because glacial deposits that are tens or even hundreds of meters thick completely mantle the terrain. The general effect of these deposits is to reduce the local relief and thus level the topography. Indeed, rural country scenes that are familiar to many of us—rocky pastures in New England, wheat fields in the Dakotas, rolling farmland in the Midwest—result directly from glacial deposition.

Types of Glacial Drift

Long before the theory of an extensive Ice Age was proposed, much of the soil and rock debris covering portions of Europe was recognized as coming from elsewhere. At the time, these foreign materials were believed to have been "drifted" into their present positions by floating ice during an ancient flood. As a consequence, the term *drift* was applied to this sediment. Although rooted in a concept that was not correct, this term was so well established by the time the true glacial origin of the debris became widely recognized that it remained in the basic glacial vocabulary. Today, **glacial drift** is an all-embracing term for sediments of glacial origin, no matter how, where, or in what form they were deposited.

Glacial drift is divided into two distinct types: (1) materials deposited directly by the glacier, which are known as *till*, and (2) sediments laid down by glacial meltwater, called *stratified drift*. Here is the difference: **Till** is deposited as glacial ice melts and drops its load of rock fragments. Unlike moving water and wind, ice cannot sort the sediment it carries; therefore, deposits of till are characteristically unsorted mixtures of many particle sizes (Figure 11.16). A close examination of this sediment often shows that many of the pieces are scratched and polished as a result of being dragged along by the glacier. **Stratified drift** is sorted according to the size and weight of the fragments. Because ice is not capable of such sorting activity, these sediments are not deposited directly by the glacier. Rather, they reflect the sorting action of glacial meltwater.

Some deposits of stratified drift are made by streams issuing directly from the glacier. Other stratified deposits involve sediment that was originally laid down as till and later picked up, transported, and redeposited by meltwater beyond the margin of the ice. Accumulations of stratified drift often consist largely of sand and gravel, because the meltwater is not capable of moving larger material and because the finer rock flour remains suspended and is commonly carried far from the glacier. An indication that stratified drift consists primarily of sand and gravel can be seen in many areas where these deposits are actively mined as aggregate for road work and other construction projects.

Close up
of cobble

Figure 11.16 Glacial till is an unsorted mixture of many different sediment sizes. A close examination often reveals cobbles that have been scratched as they were dragged along by the glacier. (Photos by E. J. Tarbuck)

When boulders are found in the till or lying free on the surface, they are called **glacial erratics** if they are different from the bedrock below (see Figure 11.7). Of course, this means that they must have been derived from a source outside the area where they are found. Although the locality of origin for most erratics is unknown, the origin of some can be determined. In many cases, boulders were transported as far as 500 kilometers from their source area and, in a few instances, more than 1000 kilometers. Therefore, by studying glacial erratics as well as the mineral composition of the till, geologists can sometimes trace the path of a lobe of ice. In portions of New England, as well as other areas, erratics can be seen dotting pastures and farm fields. In some places, these rocks were cleared from fields and piled to make fences and walls (Figure 11.17).

Figure 11.17 Land cleared of glacial erratics, which were then piled atop one another to build this stone wall near West Bend, Wisconsin. (Photo by Tom Bean)

Moraines, Outwash Plains, and Kettles

Perhaps the most widespread features created by glacial deposition are *moraines*, which are simply layers or ridges of till. Several types of moraines are identified; some are common only to mountain valleys, and others are associated with areas affected by either ice sheets or valley glaciers. Lateral and medial moraines fall in the first category, whereas end moraines and ground moraines are in the second.

Lateral and Medial Moraines. The sides of a valley glacier accumulate large quantities of debris from the valley walls. When the glacier wastes away, these materials are left as ridges, called **lateral moraines,** along the sides of the valley (Figure 11.18A). **Medial moraines** are formed when two valley glaciers coalesce to form a single ice stream. The till that was once carried along the edges of each glacier joins to form a single dark stripe of debris within the newly enlarged glacier. The creation of these dark stripes within the ice stream is one obvious proof that glacial ice moves, because the medial moraine could not form if the ice did not flow downvalley (Figure 11.18B). It is common to see several medial moraines within a large alpine glacier, because a streak will form whenever a tributary glacier joins the main valley.

End and Ground Moraines. An **end moraine** is a ridge of till that forms at the terminus of a glacier. These relatively common landforms are deposited when a state of equilibrium is attained between ablation and ice accumulation. That is, the end moraine forms when the ice is melting and evaporating near the end of the glacier at a rate equal to the forward advance of the glacier from its region of nourishment. Although the terminus of the glacier is now stationary, the ice continues to flow forward, delivering a continuous supply of sediment in the same manner a conveyor belt delivers goods to the end of a production

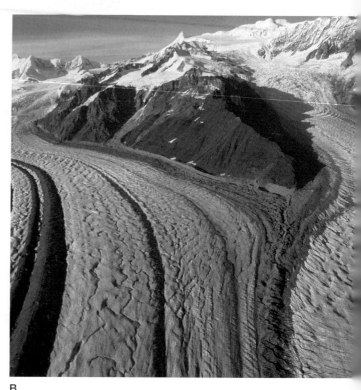

A.

B.

Figure 11.18 **A.** Well-developed lateral moraines deposited by the shrinking Athabaska Glacier in Canada's Jasper National Park. (Photo by David Barnes/The Stock Market) **B.** Medial moraines form when the lateral moraines of merging glaciers join. Saint Elias National Park, Alaska. (Photo by Tom Bean)

line. As the ice melts, the till is dropped and the end moraine grows. The longer the ice front remains stable, the larger the ridge of till will become.

Eventually the time comes when ablation exceeds nourishment. At this point, the front of the glacier begins to recede in the direction from which it originally advanced. However, as the ice front retreats, the conveyor-belt action of the glacier continues to provide fresh supplies of sediment to the terminus. In this manner a large quantity of till is deposited as the ice melts away, creating a rock-strewn, undulating plain. This gently rolling layer of till deposited as the ice front recedes is termed **ground moraine**. Ground moraine has a leveling effect, filling in low spots and clogging old stream channels, often leading to a derangement of the existing drainage system. In areas where this layer of till is still relatively fresh, such as the northern Great Lakes region, poorly drained swampy lands are quite common.

Periodically, a glacier will retreat to a point where ablation and nourishment once again balance. When this happens, the ice front stabilizes and a new end moraine forms.

Figure 11.20 represents a hypothetical area during glaciation and following the retreat of ice sheets. It shows the moraines that were described in this section, as well as the depositional features that are discussed in the sections that follow. This figure depicts landscape features similar to what might be encountered if you were traveling in the upper Midwest or New England. As you read upcoming sections dealing with other glacial deposits, you will be referred to this figure several times.

End moraines deposited by the most recent stage of Ice Age glaciation are prominent features in many parts of the Midwest and Northeast. In Wisconsin, the wooded, hilly terrain of the Kettle Moraine near Milwaukee is a particularly picturesque example. A well-known example in the Northeast is Long Island. This linear strip of glacial sediment that extends northeastward from New York City is part of an end moraine complex that stretches from eastern Pennsylvania to Cape Cod, Massachusetts (Figure 11.19).

Outwash Plains and Valley Trains. At the same time that an end moraine is forming, water from the melting glacier cascades over the till, sweeping some of it out in front of the growing ridge of unsorted debris. Meltwater generally emerges from the ice in rapidly moving streams that are often choked with suspended material and carry a substantial bed load as well. As the water leaves the glacier, it moves onto the relatively flat surface beyond and rapidly loses velocity. As a consequence, much of its bed load is dropped and the meltwater begins weaving a complex pattern of braided channels (Figure 11.20). In this way, a broad, ramplike surface composed of stratified drift is built adjacent to the downstream edge of most end mor-

Figure 11.19   End moraines make up substantial parts of Long Island, Cape Cod, Martha's Vineyard, and Nantucket. Although portions are submerged, the Ronkonkoma moraine extends through central Long Island, Martha's Vineyard, and Nantucket. It was deposited about 20,000 years ago. The Harbor Hill moraine, which formed about 14,000 years ago, extends along the north shore of Long Island, through southern Rhode Island and Cape Cod.

aines. When the feature is formed in association with an ice sheet, it is termed an **outwash plain,** and when largely confined to a mountain valley, it is usually referred to as a **valley train.**

Outwash plains and valley trains often are pockmarked with basins or depressions known as **kettles** (Figure 11.20). Kettles also occur in deposits of till. Kettles are formed when blocks of stagnant ice become wholly or partly buried in drift and eventually melt, leaving pits in the glacial sediment. Although most kettles do not exceed 2 kilometers in diameter, some with diameters exceeding 10 kilometers occur in Minnesota. Likewise, the typical depth of most kettles is less than 10 meters, although the vertical dimensions of some approach 50 meters. In many cases water eventually fills the depression and forms a pond or lake. One well-known example is Walden Pond near Concord, Massachusetts. It is here that Henry David Thoreau lived alone for two years in the 1840s and about which he wrote his famous book *Walden,* or *Life in the Woods.*

Drumlins, Eskers, and Kames

Moraines are not the only landforms deposited by glaciers. Some landscapes are characterized by numerous elongate parallel hills made of till. Other areas exhibit conical hills and relatively narrow winding ridges composed largely of stratified drift.

Drumlins.   **Drumlins** are streamlined asymmetrical hills composed of till (Figure 11.20). They range in height from 15 to 60 meters (50 to 200 feet) and average 0.4 to 0.8 kilometer (0.25 to 0.50 mile) in length. The steep side of the hill faces the direction from which the ice advanced, whereas the gentler slope points in the direction the ice moved. Drumlins are not found singly, but rather occur in clusters, called *drumlin fields.* One such cluster, east of Rochester, New York, is estimated to contain about 10,000 drumlins. Their streamlined shape indicates that they were molded in the zone of flow within an active glacier. It is thought that drumlins originate when glaciers advance over previously deposited drift and reshape the material.

Eskers and Kames.   In some areas that were once occupied by glaciers, sinuous ridges composed largely of sand and gravel can be found. Known as **eskers,** these ridges are deposited by meltwater rivers flowing within, on top of, and beneath a mass of motionless, stagnant glacial ice (Figure 11.20). Many sediment sizes are carried by the torrents of meltwater in the ice-banked channels, but only the coarser material can settle out of the turbulent stream. In some areas they are mined for sand and gravel, and for this reason, eskers are disappearing in some localities.

Kames are steep-sided hills that, like eskers, are composed of sand and gravel (Figure 11.20). Kames originate when glacial meltwater washes sediment into openings and depressions in the stagnant wasting terminus of a glacier. When the ice eventually melts away, the stratified drift is left behind as mounds or hills.

Glaciers of the Ice Age

At various points in the preceding pages we mentioned the Ice Age, a time when ice sheets and alpine glaciers were far more extensive than they are today. There was a time when the most popular explanation for what we now know to be glacial deposits was that the material had been drifted in by means of icebergs or perhaps simply swept across the landscape by a catastrophic flood. However, during the nineteenth century, field investigations by many scientists provided convincing proof that an extensive Ice Age was responsible for these deposits and for many other features.

By the beginning of the twentieth century, geologists had largely determined the extent of Ice Age glaciation. Further, they discovered that many glaciated regions had not one layer of drift but several. Close examination of these older deposits showed well-developed zones of chemical weathering and soil formation as well as the remains of plants that require warm temperatures. The evidence was clear: There had not been just one glacial advance but many, each separated by extended periods when climates were as warm or warmer than at present. The Ice Age had not simply

Figure 11.20 This hypothetical area illustrates many common depositional landforms. The outermost end moraine marks the limit of glacial advance and is called the *terminal end moraine*. End moraines that form as the ice front occasionally becomes stationary during retreat are called *recessional end moraines*. (Drumlin photo courtesy of Ward's Natural Science Establishment; Kame, Esker, and Kettle photos by Richard P. Jacobs/JLM Visuals)

Drumlin field

Terminal end moraine

Recessional end moraine

Outwash plain

Kettle lake

Ground moraine

Bedrock

been a time when the ice advanced over the land, lingered for a while, and then receded. Rather, the period was a very complex event characterized by a number of advances and withdrawals of glacial ice.

The glacial record on land is punctuated by many erosional gaps. This makes it difficult to reconstruct the episodes of the Ice Age clearly. But sediment on the ocean floor provides an uninterrupted record of climate cycles for this period. Studies of cores drilled from these seafloor sediments show that glacial/interglacial cycles have occurred about every 100,000 years. About 20 such cycles of cooling and warming were identified for the span we call the Ice Age.

During the glacial age, ice left its imprint on almost 30 percent of Earth's land area, including about 10 million square kilometers of North America, 5 million square kilometers of Europe, and 4 million square kilometers of Siberia (Figure 11.21). The amount of glacial ice in the Northern Hemisphere was roughly twice that of the Southern Hemisphere. The primary reason is that the Southern Hemisphere has little land in the middle latitudes, and therefore the southern polar ice could not spread far beyond the margins of Antarctica. By contrast, North America and Eurasia provided great expanses of land for the spread of ice sheets.

Today we know that the Ice Age began between 2 million and 3 million years ago. This means that most of the major glacial episodes occurred during a division of the geologic time scale called the **Pleistocene epoch.** Although the Pleistocene is commonly used as a synonym for the Ice Age, this epoch does not encompass it all. The Antarctic Ice Sheet, for example, formed at least 14 million years ago and, in fact, might be much older.

Some Indirect Effects of Ice Age Glaciers

In addition to the massive erosional and depositional work carried on by Pleistocene glaciers, the ice sheets had other, sometimes profound, effects on the landscape. For example, as the ice advanced and retreated, animals and plants were forced to migrate. This led to stresses that some organisms could not tolerate. Furthermore, many present-day stream courses bear little resemblance to their preglacial routes. This effect is described in Box 11.2. Other rivers that today carry only a trickle of water but nevertheless occupy broad channels are testimony to the fact that they once carried torrents of glacial meltwater.

In areas that were centers of ice accumulation, such as Scandinavia and northern Canada, the land has been slowly rising for the past several thousand years. The land had downwarped under the tremendous weight

Figure 11.21 Maximum extent of glaciation in the Northern Hemisphere during the Ice Age.

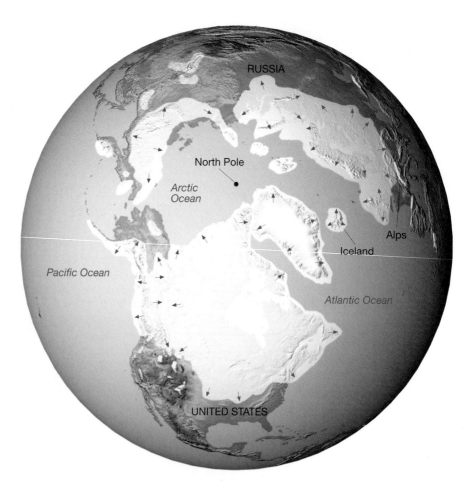

BOX 11.2   Rivers Before and After the Ice Age

Figure 11.B shows the familiar present-day pattern of rivers in the central United States, with the Missouri, Ohio, and Illinois rivers as major tributaries to the Mississippi. Figure 11.C depicts drainage systems in this region prior to the Ice Age. The pattern is *very* different from the present. This remarkable transformation of river systems resulted from the advance and retreat of the ice sheets.

Notice that prior to the Ice Age a significant part of the Missouri River drained north toward Hudson Bay. Moreover, the Mississippi River did not follow the present Iowa–Illinois boundary but rather flowed across west-central Illinois, where the lower Illinois River flows today. The preglacial Ohio River barely reached to the present-day state of Ohio, and the rivers that today feed the Ohio in western Pennsylvania flowed north and drained into the North Atlantic. The Great Lakes were created by glacial erosion during the Ice Age. Prior to the Pleistocene, the basins occupied by these huge lakes were lowlands with rivers that ran eastward to the Gulf of St. Lawrence.

The large Teays River was a significant feature prior to the Ice Age (Figure 11.C). It flowed from West Virginia across Ohio, Indiana, and Illinois, where it discharged into the Mississippi River not far from present-day Peoria. This river valley, which would have rivaled the Mississippi in size, was completely obliterated during the Pleistocene, buried by glacial deposits hundreds of feet thick. Today the sands and gravels in the buried Teays valley make it an important aquifer.

Clearly, if we are to understand the present pattern of rivers in the central United States (and many other places as well), we must be aware of glacial history.

Figure 11.B   This map shows the Great Lakes and the familiar present-day pattern of rivers in the central United States. Pleistocene ice sheets played a major role in creating this pattern.

Figure 11.C   Reconstruction of drainage systems in the central United States prior to the Ice Age. The pattern was very different from today, and there were no Great Lakes.

of 3-kilometer-thick masses of ice. Following the removal of this immense load, the crust has been adjusting by gradually rebounding upward ever since.

A far-reaching effect of the Ice Age was the worldwide change in sea level that accompanied each advance and retreat of the ice sheets. The snow that nourishes glaciers ultimately comes from moisture evaporated from the oceans. Therefore, when the ice sheets increased in size, sea level fell and the shoreline shifted seaward (see Figure 11.A in Box 11.1). Estimates suggest that sea level was as much as 100 meters (330 feet) lower than today. Thus, land that is presently flooded by the oceans was dry. The Atlantic Coast of the United States lay more than 100 kilometers (60 miles) to the east of New York City. Moreover, France and Britain were joined where the English Channel is today, Alaska and Siberia were connected across the Bering Strait, and Southeast Asia was tied by dry land to the islands of Indonesia.

The formation and growth of ice sheets was an obvious response to significant changes in climate. But the existence of the glaciers themselves triggered climatic changes in the regions beyond their margins. In arid and semiarid areas on all continents, temperatures were lower, which meant evaporation rates were also lower. At the same time, precipitation was moderate. This cooler, wetter climate resulted in the formation of many lakes called **pluvial lakes** (from the Latin term *pluvia* meaning "rain"). In North America, pluvial lakes were concentrated in the vast Basin and Range region of Nevada and Utah (Figure 11.22). Although most are now gone, a few remnants remain, the largest being Utah's Great Salt Lake.

Figure 11.22   Pluvial lakes of the Western United States. By far the largest of the pluvial lakes in the vast Basin and Range region of Nevada and Utah was Lake Bonneville. With maximum depths exceeding 300 meters and an area of 50,000 square kilometers, Lake Bonneville was nearly the same size as present-day Lake Michigan. The Great Salt Lake is a remnant of this huge pluvial lake. (After R. F. Flint, *Glacial and Quaternary Geology*, New York: John Wiley & Sons)

Causes of Glaciation

A great deal is known about glaciers and glaciation. Much has been learned about glacier formation and movement, the extent of glaciers past and present, and the features created by glaciers, both erosional and depositional. However, a widely accepted theory for the causes of glacial ages has not yet been established.

Although widespread glaciation has been rare in Earth's history, the Pleistocene Ice Age is not the only glacial period for which a record exists. Two Precambrian glacial episodes have been identified in the geologic record, the first approximately 2 billion years ago and the second about 600 million years ago. Further, a well-documented record of an earlier glacial age is found in late Paleozoic rocks that are about 250 million years old and which exist on several landmasses.

Any theory that attempts to explain the causes of glacial ages must successfully answer two basic questions: (1) *What causes the onset of glacial conditions?* For continental ice sheets to have formed, average temperatures must have been somewhat lower than at present and perhaps substantially lower than throughout much

of geologic time. Thus, a successful theory would have to account for the cooling that finally leads to glacial conditions. (2) *What caused the alternation of glacial and interglacial stages that have been documented for the Pleistocene epoch?* The first question deals with long-term trends in temperature on a scale of millions of years, but this second question relates to much shorter-term changes.

Although the literature of science contains many hypotheses relating to the possible causes of glacial periods, we will discuss only a few major ideas to summarize current thought.

Plate Tectonics

Probably the most attractive proposal for explaining the fact that extensive glaciations have occurred only a few times in the geologic past comes from the theory of plate tectonics.* Because glaciers can form only on land, we

*A brief overview of the theory appears in Chapter 1, and a more extensive discussion is presented in Chapter 16.

know that landmasses must exist somewhere in the higher latitudes before an ice age can commence. Many scientists suggest that ice ages have occurred only when Earth's shifting crustal plates have carried the continents from tropical latitudes to more poleward positions.

Glacial features in present-day Africa, Australia, South America, and India indicate that these regions, which are now tropical or subtropical, experienced an Ice Age near the end of the Paleozoic era, about 250 million years ago. However, there is no evidence that ice sheets existed during this same period in what are today the higher latitudes of North America and Eurasia. For many years this puzzled scientists. Was the climate in these relatively tropical latitudes once like it is today in Greenland and Antarctica? Why did glaciers not form in North America and Eurasia? Until the plate tectonics theory was formulated, there had been no reasonable explanation.

Today scientists understand that the areas containing these ancient glacial features were joined together as a single supercontinent (Pangaea) located at latitudes far to the south of their present positions. Later this landmass broke apart, and its pieces, each moving on a different plate, migrated toward their present locations (Figure 11.23). Now we know that during the geologic past, plate movements accounted for many dramatic climatic changes as landmasses shifted in relation to one another and moved to different latitudinal positions.

Changes in oceanic circulation also must have occurred, altering the transport of heat and moisture and consequently the climate as well. Because the rate of plate movement is very slow—a few centimeters per year—appreciable changes in the positions of the continents occur only over great spans of geologic time. Thus, climatic changes brought about by shifting plates are extremely gradual and happen on a scale of millions of years.

Variations in Earth's Orbit

Because climatic changes brought about by moving plates are extremely gradual, the plate tectonics theory cannot be used to explain the alternation between glacial and interglacial climates that occurred during the Pleistocene epoch. Therefore, we must look to some other triggering mechanism that may cause climatic change on a scale of thousands rather than millions of years. Today many scientists strongly suspect that the climatic oscillations that characterized the Pleistocene may be linked to variations in Earth's orbit. This hypothesis was first developed and strongly advocated by the Yugoslavian scientist Milutin Milankovitch and is based on

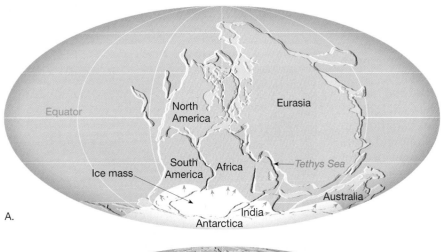

Figure 11.23   A. The supercontinent Pangaea showing the area covered by glacial ice about 300 million years ago. **B.** The continents as they are today. The white areas indicate where evidence of the old ice sheets exists.

the premise that variations in incoming solar radiation are a principal factor controlling Earth's climate.

Milankovitch formulated a comprehensive mathematical model based on the following elements (Figure 11.24):

1. Variations in the shape (*eccentricity*) of Earth's orbit about the Sun;

2. Changes in *obliquity*; that is, changes in the angle that the axis makes with the plane of Earth's orbit; and

3. The wobbling of Earth's axis, called *precession*.

Using these factors, Milankovitch calculated variations in the receipt of solar energy and the corresponding surface temperature of Earth back into time in an attempt to correlate these changes with the climatic fluctuations of the Pleistocene. In explaining climatic changes that result from these three variables, note that they cause little or no variation in the total solar energy reaching the ground. Instead, their impact is felt because they change the degree of contrast between the seasons. Somewhat milder winters in the middle to high latitudes means greater snowfall totals, whereas cooler summers would bring a reduction in snowmelt.

Among the studies that have added credibility to the astronomical hypothesis of Milankovitch is one in which deep-sea sediments containing certain climatically sensitive microorganisms were analyzed to establish a chronology of temperature changes going back nearly 500,000 years.* This time scale of climatic change was then compared to astronomical calculations of eccentricity, obliquity, and precession to determine whether a correlation did indeed exist.

Although the study was very involved and mathematically complex, the conclusions were straightforward. The researchers found that major variations in climate over the past several hundred thousand years were closely associated with changes in the geometry of Earth's orbit; that is, cycles of climatic change were shown to correspond closely with the periods of obliquity, precession, and orbital eccentricity. More specifically, the authors stated: "It is concluded that changes in the earth's orbital geometry are the fundamental cause of the succession of Quaternary ice ages."[†]

Let us briefly summarize the ideas that were just described. The theory of plate tectonics provides us with an explanation for the widely spaced and nonperiodic onset of glacial conditions at various times in the geologic past, whereas the astronomical model proposed by Milankovitch and supported by the work of J. D.

A.

B.

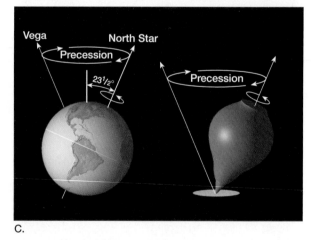

C.

Figure 11.24 Orbital variations. **A.** The shape of Earth's orbit changes during a cycle that spans about 100,000 years. It gradually changes from nearly circular to one that is more elliptical and then back again. This diagram greatly exaggerates the amount of change. **B.** Today the axis of rotation is tilted about 23.5° to the plane of Earth's orbit. During a cycle of 41,000 years, this angle varies from 21.5° to 24.5°. **C.** Precession. Earth's axis wobbles like that of a spinning top. Consequently, the axis points to different spots in the sky during a cycle of about 26,000 years.

*J. D. Hays, John Imbrie, and N. J. Shackelton, Variations in the Earth's Orbit: Pacemaker of the Ice Ages, *Science* 194 (1976): 1121–32.

[†]J. D. Hays et al., ibid., p. 1131. The term *quaternary* refers to the period on the geologic time scale that encompasses the last 1.6 million years.

Hays and his colleagues furnishes an explanation for the alternating glacial and interglacial episodes of the Pleistocene.

In conclusion, we emphasize that the ideas just discussed do not represent the only possible explana-tions for glacial ages. Although interesting and attrac-tive, these proposals are certainly not without critics, nor are they the only possibilities currently under study. Other factors may be, and probably are, involved.

The Chapter in Review

- A *glacier* is a thick mass of ice originating on the land from the compaction and recrystallization of snow, and it shows evidence of past or present flow. Today, *valley* or *alpine glaciers* are found in mountain areas where they usually follow valleys that were originally occupied by streams. *Ice sheets* exist on a much larger scale, covering most of Greenland and Antarctica.

- Near the surface of a glacier, in the *zone of fracture*, ice is brittle. However, below about 50 meters, pressure is great, causing ice to *flow* like a *plastic material*. A second impor-tant mechanism of glacial movement consists of the entire ice mass *slipping* along the ground.

- The average velocity of glacial movement is generally quite slow, but it varies considerably from one glacier to another. The advance of some glaciers is characterized by periods of extremely rapid movements called *surges*.

- Glaciers form in areas where more snow falls in winter than melts during summer. Snow accumulation and ice formation occur in the *zone of accumulation*. Its outer lim-its are defined by the *snowline*. Beyond the snowline is the *zone of wastage*, where there is a net loss to the glacier. The *glacial budget* is the balance, or lack of balance, between ac-cumulation at the upper end of the glacier, and loss, called *ablation*, at the lower end.

- Glaciers erode land by *plucking* (lifting pieces of bedrock out of place) and *abrasion* (grinding and scraping of a rock surface). Erosional features produced by valley glaciers include *glacial troughs, hanging valleys, pater noster lakes, fiords, cirques, arêtes, horns,* and *roches moutonnées*.

- Any sediment of glacial origin is called *drift*. The two dis-tinct types of glacial drift are (1) *till*, which is unsorted sed-iment deposited directly by the ice; and (2) *stratified drift*, which is relatively well-sorted sediment laid down by glacial meltwater.

- The most widespread features created by glacial deposi-tion are layers or ridges of till, called *moraines*. Associated with valley glaciers are *lateral moraines*, formed along the sides of the valley, and *medial moraines*, formed between two valley glaciers that have joined. *End moraines*, which mark the former position of the front of a glacier, and *ground moraine*, an undulating layer of till deposited as the ice front retreats, are common to both valley glaciers and ice sheets. An *outwash plain* is often associated with the end moraine of an ice sheet. A *valley train* may form when the glacier is confined to a valley. Other depositional fea-tures include *drumlins* (streamlined asymmetrical hills composed of till), *eskers* (sinuous ridges composed largely of sand and gravel deposited by streams flowing in tun-nels beneath the ice, near the terminus of a glacier), and *kames* (steep-sided hills composed of sand and gravel).

- The *Ice Age*, which began 2 million to 3 million years ago, was a very complex period characterized by a number of advances and withdrawals of glacial ice. Most of the major glacial episodes occurred during a division of the geologic time scale called the *Pleistocene epoch*. Perhaps the most convincing evidence for the occurrence of several glacial advances during the Ice Age is the widespread existence of *multiple layers of drift* and an uninterrupted record of climate cycles preserved in *seafloor sediments*.

- In addition to massive erosional and depositional work, other effects of Ice Age glaciers included the *migration* of organisms, *changes in stream courses, adjustment of the crust* by rebounding after the removal of the immense load of ice, and *climate changes* caused by the existence of the glac-iers themselves. In the sea, the most far-reaching effect of the Ice Age was the *worldwide change* in *sea level* that ac-companied each advance and retreat of the ice sheets.

- Any theory that attempts to explain the causes of glacial ages must answer two basic questions: (1) What causes the onset of glacial conditions? and (2) What caused the alternating glacial and interglacial stages that have been documented for the Pleistocene epoch? Two of the many hypotheses for the cause of glacial ages involve (1) plate tectonics and (2) variations in Earth's orbit.

Key Terms

ablation (p. 226)	crevasse (p. 225)	glacial drift (p. 232)	hanging valley (p. 229)
abrasion (p. 227)	drumlin (p. 235)	glacial erratic (p. 233)	horn (p. 231)
alpine glacier (p. 223)	end moraine (p. 233)	glacial striations (p. 228)	ice cap (p. 224)
arête (p. 231)	esker (p. 235)	glacial trough (p. 229)	ice sheet (p. 223)
calving (p. 226)	fiord (p. 230)	glacier (p. 222)	ice shelf (p. 223)
cirque (p. 230)	glacial budget (p. 226)	ground moraine (p. 234)	kame (p. 235)

kettle (p. 235)

lateral moraine (p. 233)

medial moraine (p. 233)

outlet glacier (p. 224)

outwash plain (p. 235)

pater noster lakes (p. 230)

piedmont glacier (p. 224)

Pleistocene epoch (p. 238)

plucking (p. 227)

pluvial lake (p. 240)

roche moutonnée (p. 231)

rock flour (p. 227)

stratified drift (p. 232)

till (p. 232)

valley glacier (p. 223)

valley train (p. 235)

zone of accumulation
(p. 226)

zone of wastage (p. 226)

Questions for Review

1. What is a glacier? What percentage of Earth's land area do glaciers cover? Where are glaciers found today?

2. Describe how glaciers fit into the hydrologic cycle. What role do they play in the rock cycle?

3. Each of the following statements refers to a particular type of glacier. Name the type of glacier.
   a. The term *continental* is often used to describe this type of glacier.
   b. This type of glacier is also called an *alpine glacier*.
   c. This is a glacier formed when one or more valley glaciers spreads out at the base of a steep mountain front.
   d. Greenland is the only example in the Northern Hemisphere.
   e. This is a stream of ice leading from the margin of an ice sheet through the mountains to the sea.

4. Describe the two components of glacial flow. At what rates do glaciers move? In a valley glacier, does all of the ice move at the same rate? Explain.

5. Why do crevasses form in the upper portion of a glacier but not below a depth of about 50 meters?

6. Under what circumstances will the front of a glacier advance? Retreat? Remain stationary?

7. Describe two basic processes of glacial erosion.

8. How does a glaciated mountain valley differ in appearance from a mountain valley that was not glaciated?

9. List and describe the erosional features you might expect to see in an area where valley glaciers exist or have recently existed.

10. What is glacial drift? What is the difference between till and stratified drift? What general effect do glacial deposits have on the landscape?

11. List and briefly describe the four basic moraine types. What do all moraines have in common?

12. List and briefly describe four depositional features other than moraines.

13. Examine the photo of the drumlin in Figure 11.20. From what direction (right or left) did the ice sheet advance in this area?

14. How does a kettle form?

15. About what percentage of Earth's land surface was covered at some time by Pleistocene glaciers? How does this compare to the area presently covered by ice sheets and glaciers? (Check your answer with Question 1.)

16. During the Pleistocene epoch the extent of glacial ice in the Northern Hemisphere was about twice as great as in the Southern Hemisphere. Briefly explain why this was the case.

17. List three indirect effects of Ice Age glaciers.

18. How might plate tectonics help us understand the cause of ice ages? Can plate tectonics explain the alternation between glacial and interglacial climates during the Pleistocene?

Web Resources

The *Essentials of Geology* Website uses the resources and flexibility of the Internet to aid in your study of the topics in this chapter. Written and developed by geology instructors, this site will help improve your understanding of geology. Visit **www.prenhall.com/lutgens** and click on the cover of *Essentials of Geology, 8e* to find:

- On-line review quizzes.
- Web-based critical thinking and writing exercises.
- Links to chapter-specific Web resources.
- Internet-wide key term searches.

http://www.prenhall.com/lutgens

Deserts and Wind

Focus on Learning

To assist you in learning the important concepts in this chapter, you will find it helpful to focus on the following questions:

- What are the causes of deserts in the lower and middle latitudes?
- What are the roles of weathering, water, and wind in arid and semiarid climates?
- How have many of the landscapes in the dry Basin and Range region of the United States evolved?
- How does wind erode?
- What are some depositional features produced by wind?

Giant sand dune in Namibia (southwest Africa). Note the trees for scale. (Photo by Michael Fogden/ DRK Photo) climate has a strong influence on the nature and intensity of Earth's external processes. This was clearly demonstrated in the preceding chapter on glaciers. Another excellent example of the strong link between climate and geology is seen when we examine the development of arid landscapes. The word *desert* literally means *deserted* or *unoccupied*. For many dry regions this is a very appropriate description, although where water is available in deserts, plants and animals thrive. Nevertheless, the world's dry regions are among the least familiar land areas on Earth outside the polar realm (see Box 12.1).

Desert landscapes frequently appear stark. Their profiles are not softened by a carpet of soil and abundant plant life. Instead, barren rocky outcrops with steep, angular slopes are common. At some places the rocks are tinted orange and red. At others they are gray and brown and streaked with black. For many visitors, desert scenery exhibits a striking beauty; to others the terrain seems bleak. No matter which feeling is elicited, it is clear that deserts are very different from the more humid places where most people live.

As you shall see, arid regions are not dominated by a single geologic process. Rather, the effects of tectonic forces, running water, and wind are all apparent. Because these processes combine in different ways from place to place, the appearance of desert landscapes varies a great deal as well (Figure 12.1).

Distribution and Causes of Dry Lands

External Processes
↳ Deserts

We all recognize that deserts are dry places, but just what is meant by the term *dry?* That is, how much rain defines the boundary between humid and dry regions? Sometimes it is arbitrarily defined by a single rainfall figure—for example, 25 centimeters per year of precipitation. However, the concept of *dryness* is very relative; it refers to *any situation in which water deficiency exists.* Hence, climatologists define **dry climate** as one in which yearly precipitation is less than the potential loss of water by evaporation.

Dryness, then, is related not only to annual rainfall totals but is also a function of evaporation, which in turn closely depends upon temperature. As temperatures climb, potential evaporation also increases. Twenty-five centimeters of rain may support only a sparse vegetative cover in Nevada, whereas the same amount of precipitation falling in northern Scandinavia is sufficient to support forests.

The dry regions of the world encompass about 42 million square kilometers, a surprising 30 percent of

Figure 12.1   A scene in southern Utah near the San Juan River. The appearance of desert landscapes varies a great deal from place to place. (Photo copyright © by Carr Clifton. All rights reserved.)

Earth's land surface. No other climatic group covers so large a land area. Within these water-deficient regions, two climatic types are commonly recognized: **desert,** which is *arid*, and **steppe,** which is *semiarid*. The two share many features; their differences are primarily a matter of degree. The steppe is a marginal and more humid variant of the desert and is a transition zone that surrounds the desert and separates it from bordering humid climates. The world map showing the distribu-tion of desert and steppe regions reveals that dry lands are concentrated in the subtropics and in the midlatitudes (Figure 12.2).

Low-Latitude Deserts

The heart of the low-latitude dry climates lies in the vicinities of the Tropics of Cancer and Capricorn. Figure 12.2 shows a virtually unbroken desert environment stretching for more than 9300 kilometers (nearly 5800

BOX 12.1

Common Misconceptions About Deserts

*D*eserts are hot, life-less, sand-covered landscapes shaped largely by the force of wind.* The preceding statement summarizes the images of arid regions that many people hold, especially those living in more humid places. Is it an accurate view? The answer is no. Although there are clearly elements of reality in such an impression, it is a generalization that contains a number of misconceptions (Figure 12.A).

One common fallacy about deserts is that they are lifeless or almost lifeless. Although reduced in amount and different in character, plant and animal life are usually present. Desert plants all have one characteristic in common: They have developed adaptations that make them highly tolerant of drought. Many have waxy leaves, stems, or branches or a thickened cuticle (outermost protective layer) to reduce water loss. Others have very small leaves or no leaves at all. Also, the roots of some species often extend to great depths to tap the moisture found there, whereas others produce a shallow but widespread root system that enables them to absorb great amounts of moisture quickly from the infrequent desert downpours. Thus, although widely dispersed and providing little ground cover, plants of many kinds flourish in the desert.

A second widely held belief about the world's dry lands is that they are always hot. This fact seems to be reinforced by commonly quoted temperature statistics. The highest accepted temperature record for the United States as well as the entire Western Hemisphere is 57°C (134°F). This long-standing record was set at Death Valley, California, on July 10, 1913. The nearly 59°C (137°F) reading in Azizia, Libya, in North Africa's Sahara Desert on September 13, 1922, is the world record. Despite these remarkably high figures, cold temperatures are also experienced in desert regions. For example, the average daily minimum in January at Phoenix, Arizona, is 1.7°C (35°F), just barely above freezing. At Ulan Bator in Mongolia's Gobi Desert, the average *high* temperature in January is only −19°C (−2°F)! Dry climates are found from the tropics to the high middle latitudes. Consequently, although tropical deserts lack a cold season, deserts in the middle latitudes do experience seasonal temperature changes.

The last two commonly held misconceptions about the world's deserts are more geologic than climatic. One mistaken assumption is that they consist of mile after mile of drifting sand. It is true that sand accumulations do exist in some areas and may be striking features, but they represent only a small percentage of the total desert area. For example, in the Sahara, the world's largest desert, accumulations of sand cover only one-tenth of its area. The sandiest of all deserts is the Arabian, one-third of which consists of sand.

The final mistaken assumption is the seemingly logical idea that wind is the most important agent of erosion in deserts. Although wind is relatively more significant in dry areas than anywhere else, most erosional landforms in deserts are created by running water. When the rains come, they usually take the form of thunderstorms. Because the heavy rain associated with these storms cannot all soak in, rapid runoff results. Without a thick vegetative cover to protect the ground, erosion is great.

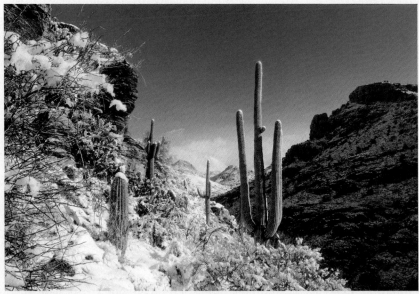

Figure 12.A   Snow blankets the rocky ground of the Sonoran Desert near Tuscon, Arizona. As this scene demonstrates, deserts are not necessarily always hot, lifeless, dune-covered expanses. (Photo by T. Wiewandt/DRK Photo)

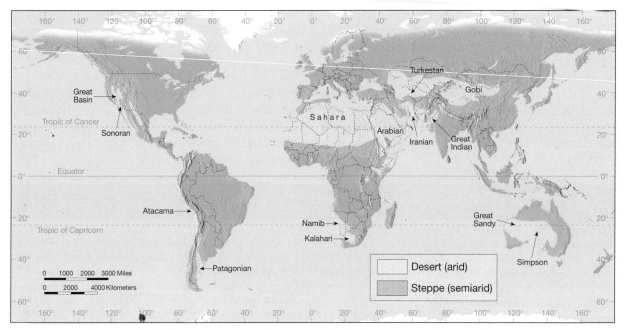

Figure 12.2 Arid and semiarid climates cover about 30 percent of Earth's land surface. No other climate group covers so large an area.

miles) from the Atlantic coast of North Africa to the dry lands of northwestern India. In addition to this single great expanse, the Northern Hemisphere contains another, much smaller area of subtropical desert and steppe in northern Mexico and the southwestern United States.

In the Southern Hemisphere, dry climates dominate Australia. Almost 40 percent of the continent is desert, and much of the remainder is steppe. In addition, arid and semiarid areas occur in southern Africa and make a limited appearance in coastal Chile and Peru.

What causes these bands of low-latitude desert? The answer is the global distribution of air pressure and winds. Zones of high air pressure called *subtropical highs* coincide with dry regions in lower latitudes. These pressure systems are characterized by subsiding air currents. When air sinks, it is compressed and warmed. Such conditions are just the opposite of what is needed to produce clouds and precipitation. Consequently, these regions are known for their clear skies, sunshine, and ongoing drought.

Middle-Latitude Deserts

Unlike their low-latitude counterparts, middle-latitude deserts and steppes are not controlled by the subsiding air masses associated with high pressure. Instead, these dry lands exist principally because they are sheltered in the deep interiors of large landmasses. They are far removed from the ocean, which is the ultimate source of moisture for cloud formation and precipitation. One well-known example is the Gobi Desert of central Asia, shown on the map north of India (Figure 12.2).

The presence of high mountains across the paths of prevailing winds further separates these areas from water-bearing maritime air masses. Also, the mountains force the air to lose much of its water. The mechanism is simple: As prevailing winds meet mountain barriers, the air is forced to ascend. When air rises, it expands and cools, a process that can produce clouds and precipitation. The windward sides of mountains, therefore, often have high precipitation.

By contrast, the leeward sides of mountains are usually much drier (Figure 12.3). This situation exists because air reaching the leeward side has lost much of its moisture, and if the air descends, it is compressed and warmed, making cloud formation even less likely. The dry region that results is often referred to as a **rainshadow desert.**

Because most midlatitude deserts occupy sites on the leeward sides of mountains, they can also be classified as rainshadow deserts. In North America, the Coast Ranges, Sierra Nevada, and Cascades are the foremost mountain barriers to moisture from the Pacific. In Asia, the great Himalayan chain prevents the summertime monsoon flow of moist Indian Ocean air from reaching the interior.

Because the Southern Hemisphere lacks extensive land areas in the middle latitudes, only a small area of desert and steppe occurs in this latitude range, existing primarily near the southern tip of South America in the rainshadow of the towering Andes.

Middle-latitude deserts provide an example of how tectonic processes affect climate. Rainshadow deserts exist by virtue of the mountains produced when plates

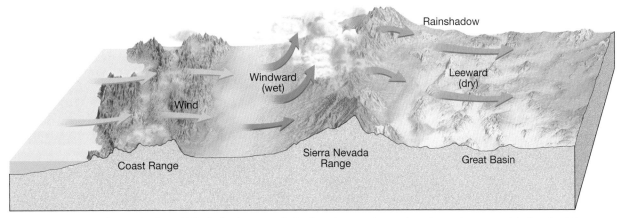

Figure 12.3 Many deserts in the middle latitudes are rainshadow deserts. As moving air meets a mountain barrier, it is forced to rise. Clouds and precipitation on the windward side often result. Air descending the leeward side is much drier. The mountains effectively cut the leeward side off from the sources of moisture, producing a rainshadow desert. The Great Basin desert is a rainshadow desert that covers nearly all of Nevada and portions of adjacent states.

collide. Without such mountain-building episodes, wetter climates would prevail where many dry regions exist today.

Geologic Processes in Arid Climates

 External Processes
    ↳ Deserts

The angular hills, the sheer canyon walls, and the desert surface of pebbles or sand contrast sharply with the rounded hills and curving slopes of more humid places. Indeed, to a visitor from a humid region, a desert landscape may seem to have been shaped by forces altogether different from those operating in well-watered areas. However, although the contrasts might be striking, they do not reflect different processes. They merely disclose the differing effects of the same processes that operate under contrasting climatic conditions.

Weathering

In humid regions, relatively well-developed soils support an almost continuous cover of vegetation. Here the slopes and rock edges are rounded, reflecting the strong influence of chemical weathering in a humid climate. By contrast, much of the weathered debris in deserts consists of unaltered rock and mineral fragments—the results of mechanical weathering processes. In dry lands, rock weathering of any type is greatly reduced because of the lack of moisture and the scarcity of organic acids from decaying plants. However, chemical weathering is not completely lacking in deserts. Over long spans of time, clays and thin soils do form, and

many iron-bearing silicate minerals oxidize, producing the rust-colored stain found tinting some desert landscapes.

The Role of Water

Permanent streams are normal in humid regions, but practically all desert stream beds are dry most of the time (Figure 12.4A). Deserts have **ephemeral streams,** which means that they carry water only in response to specific episodes of rainfall. A typical ephemeral stream might flow only a few days or perhaps just a few hours during the year. In some years, the channel might carry no water at all.

This fact is obvious even to the casual traveler who notices numerous bridges with no streams beneath them or numerous dips in the road where dry channels cross. However, when the rare heavy showers do come, so much rain falls in such a short time that all of it cannot soak in. Because desert vegetative cover is sparse, runoff is largely unhindered and consequently rapid, often creating flash floods along valley floors (Figure 12.4B). These floods are quite unlike floods in humid regions. A flood on a river like the Mississippi may take several days to reach its crest and then subside. But desert floods arrive suddenly and subside quickly. Because much of the surface material in a desert is not anchored by vegetation, the amount of erosional work that occurs during a single short-lived rain event is impressive.

In the dry western United States, different names are used for ephemeral streams, including *wash* and *arroyo*. In other parts of the world, a dry desert stream might be called a *wadi* (Arabia and North Africa), a *donga* (South America), or a *nullah* (India).

Humid regions are notable for their integrated drainage systems. But in arid regions streams usually

A.

B.

Figure 12.4 **A.** Most of the time, desert stream channels are dry. **B.** An ephemeral stream shortly after a heavy shower. Although such floods are short-lived, large amounts of erosion occur. (Photos by E. J. Tarbuck)

lack an extensive system of tributaries. In fact, a basic characteristic of desert streams is that they are small and die out before reaching the sea. Because the water table is usually far below the surface, few desert streams can draw upon it as streams do in humid regions (see Figure 10.4, p. 205). Without a steady supply of water, the combination of evaporation and infiltration soon depletes the stream.

The few permanent streams that do cross arid regions, such as the Colorado and Nile rivers, originate *outside* the desert, often in well-watered mountains. Here the water supply must be great to compensate for the losses occurring as the stream crosses the desert (see Box 12.2). For example, after the Nile leaves its headwaters in the lakes and mountains of central Africa, it traverses almost 3000 kilometers of the Sahara *without a single tributary*. By contrast, in humid regions the discharge of a river increases as it flows downstream because tributaries and groundwater contribute additional water along the way.

It should be emphasized that *running water, although infrequent, nevertheless does most of the erosional work in deserts*. This is contrary to a common belief that wind is the most important erosional agent sculpturing desert landscapes. Although wind erosion is indeed more significant in dry areas than elsewhere, most desert landforms are carved by running water. As you will see shortly, the main role of wind is in the transportation and deposition of sediment, which creates and shapes the ridges and mounds we call *dunes*.

Basin and Range: The Evolution of a Desert Landscape

 External Processes
↳ Deserts

Because arid regions typically lack permanent streams, they are characterized as having **interior drainage.** This means that they have a discontinuous pattern of intermittent streams that do not flow out of the desert to the ocean. In the United States, the dry Basin and Range region provides an excellent example. The region includes southern Oregon, all of Nevada, western Utah, southeastern California, southern Arizona, and southern New Mexico. The name Basin and Range is an apt description for this almost 800,000-square-kilometer (more than 300,000-square-mile) area as it is characterized by more than 200 relatively small mountain ranges that rise 900 to 1500 meters (3000 to 5000 feet) above the basins that separate them.

In this region, as in others like it around the world, most erosion occurs without reference to the ocean (ultimate base level), because the interior drainage never reaches the sea. Even where permanent streams flow to the ocean, few tributaries exist, and thus only a narrow strip of land adjacent to the stream has sea level as its ultimate level of land reduction.

The block models in Figure 12.5 depict how the landscape has evolved in the Basin and Range region.

BOX 12.2

The Shrinking Aral Sea

The Aral Sea lies on the border between Uzbekistan and Kazakhstan in central Asia (Figure 12.B). The setting is the Turkestan desert, a middle-latitude desert in the rainshadow of Afghanistan's high mountains. In this region of interior drainage, two large rivers, the Amu Darya and the Syr Darya, carry water from the mountains of northern Afghanistan across the desert to the Aral Sea. Water leaves the sea by evaporation. Thus, the size of the water body depends upon the balance between river inflow and evaporation.

In 1960 the Aral Sea was one of the world's largest inland water bodies, with an area of about 67,000 square kilometers (26,000 square miles). Only the Caspian Sea, Lake Superior, and Lake Victoria were larger. By the year 2000 the area of the Aral Sea was less than 50 percent of its 1960 size, and its volume was reduced by 80 percent. The shrinking of this water body is depicted in Figure 12.C. By about 2010 all that will remain will be three shallow remnants.

What caused the Aral Sea to dry up over the past 40 years? The answer is that the flow of water from the mountains that supplied the sea was significantly reduced and then all but eliminated. The waters of the Amu Darya and Syr Darya were diverted to supply a major expansion of irrigated agriculture in this dry realm.

Figure 12.C   The shrinking Aral Sea. By the year 2010 all that will remain are three small remnants.

The intensive irrigation greatly increased agricultural productivity, but not without a significant cost. Some of the effects are listed below.

- The once thriving fishing industry is dead, and the 24 species of fish that once lived in the Aral Sea are no longer there. The shoreline is now tens of kilometers from the towns that were once fishing centers (Figure 12.D).
- The shrinking sea has exposed millions of acres of former seabed to Sun and wind. The former seabed is the source of an increasing number of dust storms that blow salts and toxic herbicides and pesticides into nearby towns causing detrimental health effects.

Figure 12.D   In the town of Jamboul, Kazakhstan, boats now lie in the sand because the Aral Sea has dried up. (Photo by Ergun Cagatay/Liaison Agency, Inc.)

- The deltas of the two major rivers have lost their wetlands, and wildlife has disappeared.
- The region has lost a moderating climatic influence. Without the large water body, there are greater extremes of temperature and reduced local precipitation.

If enough fresh water were to once again flow into the Aral Sea, could the situation be reversed? Prospects appear grim. Experts estimate that restoring the Aral Sea to about twice its present size would require stopping all irrigation from the two major rivers for 50 years. This could not be done without ruining the economies of the countries that rely on that water.*

The decline of the Aral Sea is a major environmental disaster that sadly is of human making.

Figure 12.B   The Aral Sea lies east of the Caspian Sea in the Turkestan desert. Two rivers, the Amu Darya and Syr Darya, bring water from the mountains to the south.

*For more on this, see "Coming to Grips with the Aral Sea's Grim Legacy," in *Science*, vol. 284, 2 April 1999, pp. 30–31.

Figure 12.5 Stages of landscape evolution in a mountainous desert such as the Basin and Range region of the American West. As erosion of the mountains and deposition in the basins continue, relief diminishes. **A.** Early stage. **B.** Middle stage. **C.** Late stage.

During and following uplift of the mountains, running water begins carving the elevated mass and depositing large quantities of debris in the basin. In this early stage, relief is greatest, and as erosion lowers the mountains and sediment fills the basins, elevation differences diminish (Figure 12.5A).

When the occasional torrents of water produced by sporadic rains move down the mountain canyons, they are heavily loaded with sediment. Emerging from the confines of the canyon, the runoff spreads over the gentler slopes at the base of the mountains and quickly loses velocity. Consequently, most of its load is dumped within a short distance. The result is a cone of debris known as an **alluvial fan** at the mouth of a canyon (Figure 12.6). Over the years, a fan enlarges, eventually coalescing with fans from adjacent canyons to produce an apron of sediment (*bajada*) along the mountain front (Figure 12.5B).

On the rare occasions of abundant rainfall or snowmelt in the mountains, streams may flow across the alluvial fans to the center of the basin, converting the basin floor into a shallow **playa lake.** Playa lakes

last only a few days or weeks, before evaporation and infiltration remove the water. The dry, flat lake bed that remains is termed a *playa.*

Playas occasionally become encrusted with salts left behind by evaporation. These precipitated salts may be unusual. A case in point is the sodium borate (better known as borax) mined from ancient playa lake deposits in Death Valley, California.

With the ongoing erosion of the mountain mass and the accompanying sedimentation, the local relief continues to diminish. Eventually nearly the entire mountain mass is gone. Thus, by the late stages of erosion, the mountain areas are reduced to a few large bedrock knobs (called *inselbergs*) projecting above the sediment-filled basin (Figure 12.5C).

Each of the stages of landscape evolution in an arid climate depicted in Figure 12.5 can be observed in the Basin and Range region. Recently uplifted mountains in an early stage of erosion are found in southern Oregon and northern Nevada. Death Valley, California, and southern Nevada fit into the more advanced middle

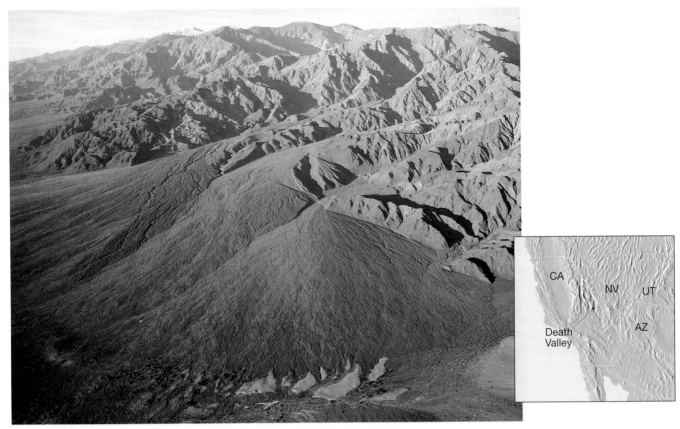

Figure 12.6   Aerial view of alluvial fans in Death Valley, California. The size of the fan depends on the size of the drainage basin. As the fans grow, they eventually coalesce to form a bajada. (Photo by Michael Collier)

stage, whereas the late stage, with its inselbergs, can be seen in southern Arizona.

Transportation of Sediment by Wind

Moving air, like moving water, is turbulent and able to pick up loose debris and transport it to other locations. Just as in a stream, the velocity of wind increases with height above the surface. Also like a stream, wind transports fine particles in suspension while heavier ones are carried as bed load. However, the transport of sediment by wind differs from that by running water in two significant ways. First, wind's lower density compared to water renders it less capable of picking up and transporting coarse materials. Second, because wind is not confined to channels, it can spread sediment over large areas, as well as high into the atmosphere.

Bed Load

The **bed load** carried by wind consists of sand grains. Observations in the field and experiments using wind tunnels indicate that windblown sand moves by skipping and bouncing along the surface—a process termed **saltation.** The term is not a reference to salt, but instead derives from the Latin word meaning "to jump."

The movement of sand grains begins when wind reaches a velocity sufficient to overcome the inertia of the resting particles. At first the sand rolls along the surface. When a moving sand grain strikes another grain, one or both of them may jump into the air. Once in the air, the grains are carried forward by the wind until gravity pulls them back toward the surface. When the sand hits the surface, it either bounces back into the air or dislodges other grains, which then jump upward. In this manner, a chain reaction is established, filling the air near the ground with saltating sand grains in a short period of time (Figure 12.7).

Bouncing sand grains never travel far from the surface. Even when winds are very strong, the height of the saltating sand seldom exceeds a meter and usually is no greater than a half meter.

Suspended Load

Unlike sand, finer dust particles can be swept high into the atmosphere by the wind. Because dust is often composed of rather flat particles that have large surface areas compared to their weight, it is relatively easy for turbulent air to counterbalance the pull of gravity and keep these fine particles airborne for hours or even days. Although both silt and clay can be carried in suspension, silt commonly makes up the bulk of the **suspended load**

Figure 12.7 A cloud of saltating sand grains moving up the gentle slope of a dune. (Photo by Stephen Trimble)

because the reduced level of chemical weathering in deserts produces only small amounts of clay.

Fine particles are easily carried by the wind, but they are not so easily picked up to begin with. The reason is that the wind velocity is practically zero within a very thin layer close to the ground. Thus, the wind cannot lift the sediment by itself. Instead, the dust must be ejected or spattered into the moving air currents by bouncing sand grains or other disturbances. This idea is illustrated nicely by a dry unpaved country road on a windy day. Left undisturbed, little dust is raised by the wind. However, as a car or truck moves over the road, the layer of silt is kicked up, creating a thick cloud of dust.

Although the suspended load is usually deposited relatively near its source, high winds are capable of carrying large quantities of dust great distances. In the 1930s, silt that was picked up in Kansas was transported to New England and beyond into the North Atlantic. Similarly, dust blown from the Sahara has been traced as far as the West Indies.

Wind Erosion

Compared to running water and glaciers, wind is a relatively insignificant erosional agent. Recall that even in deserts, most erosion is performed by intermittent running water, not by the wind. Wind erosion is more effective in arid lands than in humid areas because in humid regions moisture binds particles together and vegetation anchors the soil. For wind to be effective, dryness and scanty vegetation are important prerequisites.

When such circumstances exist, wind may pick up, transport, and deposit great quantities of fine sediment. During the 1930s, parts of the Great Plains experienced great dust storms (Figure 12.8A). The plowing under of the natural vegetative cover for farming, followed by severe drought, made the land ripe for wind erosion and led to the area being labeled the Dust Bowl.

Deflation, Blowouts, and Desert Pavement

One way that wind erodes is by **deflation,** the lifting and removal of loose material. Although the effects of deflation are sometimes difficult to notice because the entire surface is being lowered at the same time, they can be significant. In portions of the 1930s Dust Bowl, vast areas of land were lowered by as much as a meter in only a few years (Figure 12.8B).

The most noticeable results of deflation in some places are shallow depressions called **blowouts.** In the Great Plains region, from Texas north to Montana, thousands of blowouts are visible on the landscape. They range from small dimples less than a meter deep and 3 meters wide to depressions that approach 50 meters in depth and several kilometers across. The factor that controls the depths of these basins (that is, acts as base level) is the local water table. When blowouts are lowered to the water table, damp ground and vegetation prevent further deflation.

In portions of many deserts, the surface is a layer of coarse pebbles and gravels too large to be moved by the wind. Such a layer, called **desert pavement,** is created as the wind lowers the surface by removing fine material until eventually only a continuous cover of coarse sediment remains (Figure 12.9). Once desert pavement becomes established, a process that may take hundreds of years, the surface is protected from further deflation if left undisturbed. However, as the layer is only one or two stones thick, disruption by vehicles or animals can dislodge the pavement and expose the fine-grained material below. If this happens, the surface is once again subject to deflation.

Wind Abrasion

Like glaciers and streams, wind also erodes by **abrasion.** In dry regions and along some beaches, windblown sand cuts and polishes exposed rock surfaces. It sometimes creates interestingly shaped stones called **ventifacts** (Figure 12.10). The side of the stone exposed to the prevailing wind is abraded, leaving it polished, pitted, and with sharp edges. If the wind is not consistently from one direction or if the pebble becomes reoriented, it may have several faceted surfaces.

Abrasion is often credited for accomplishments beyond its actual capabilities. Such features as balanced rocks that stand high atop narrow pedestals and intricate detailing on tall pinnacles are not the results of abrasion. Because sand seldom travels more than a

A.   **B.**

Figure 12.8   A. Dust blackens the sky on May 21, 1937, near Elkhart, Kansas. It was because of storms like this that portions of the Great Plains were called the "Dust Bowl" in the 1930s. (Photo reproduced from the collection of the Library of Congress) **B.** This photo was taken north of Granville, North Dakota, in July 1936, during a prolonged drought. Strong winds removed the soil that was not anchored by vegetation. The mounds are 1.2 meters (4 feet) high and show the level of the land prior to deflation. (Photo courtesy of the State Historical Society of North Dakota)

meter above the surface, the wind's sandblasting effect is obviously quite limited in vertical extent.

Wind Deposits

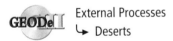

External Processes
↳ Deserts

Although wind is relatively unimportant in producing *erosional* landforms, significant *depositional* landforms are created by the wind in some regions. Accumulations of windblown sediment are particularly conspicuous in the world's dry lands and along many sandy coasts. Wind deposits are of two distinctive types: (1) mounds and ridges of sand from the wind's bed load, which we call *dunes*, and (2) extensive blankets of silt, called *loess*, that once were carried in suspension.

Sand Deposits

As is the case with running water, wind drops its load of sediment when its velocity falls and the energy available for transport diminishes. Thus, sand begins to accumulate wherever an obstruction across the path of the wind slows its movement. Unlike many deposits of silt, which form blanketlike layers over large areas, winds commonly deposit sand in mounds or ridges called **dunes.** The chapter opening photo is a spectacular example.

As moving air encounters an object, such as a clump of vegetation or a rock, the wind sweeps around and over it, leaving a shadow of slower moving air behind the obstacle as well as a smaller zone of quieter air just in front of the obstacle. Some of the saltating sand grains moving with the wind come to rest in these wind shadows. As the accumulation of sand continues, it becomes a more imposing barrier to the wind and thus a more ef-

Figure 12.9   A. Formation of desert pavement. As these cross-sections illustrate, coarse particles gradually become concentrated into a tightly packed layer as deflation lowers the surface by removing sand and silt. **B.** If left undisturbed, desert pavement such as this in Arizona's Sonoran Desert will protect the surface from further deflation. (Photo by David Muench)

Figure 12.10 Ventifacts are rocks that are polished and shaped by sandblasting. (Photo by Stephen Trimble)

ficient trap for even more sand. If there is a sufficient supply of sand and the wind blows steadily for a long enough time, the mound of sand grows into a dune.

Many dunes have an asymmetrical profile with the leeward (sheltered) slope being steep and the windward slope more gently inclined. Sand moves up the gentle slope on the windward side by saltation. Just beyond the crest of the dune, where the wind velocity is reduced, the sand accumulates. As more sand collects, the slope steepens, and eventually some of it slides or slumps under the pull of gravity. In this way, the leeward slope of the dune, called the **slip face,** maintains an angle of about 34 degrees, the angle of repose for loose dry sand (Figure 12.11). (Recall from Chapter 8

that the angle of repose is the steepest angle at which loose material remains stable.) Continued sand accumulation, coupled with periodic slides down the slip face, results in the slow migration of the dune in the direction of air movement.

As sand is deposited on the slip face, layers form that are inclined in the direction the wind is blowing. These sloping layers are called **cross beds.** When the dunes are eventually buried under other layers of sediment and become part of the sedimentary rock record, their asymmetrical shape is destroyed, but the cross beds remain as testimony to their origin. Cross-bedding is especially well developed in the Navajo Sandstone, a prominent layer in parts of northern Arizona and southern Utah (Figure 12.12).

Types of Sand Dunes

Dunes are not just random heaps of windblown sediment. Rather, they are accumulations that usually assume surprisingly consistent patterns (Figure 12.13). A broad assortment of dune forms exists, generally simplified to a few major types for discussion. Of course, gradations exist among different forms as well as irregularly shaped dunes that do not fit easily into any category. Several factors influence the form and size that dunes ultimately assume. These include wind direction and velocity, availability of sand, and the amount of vegetation present. Six basic dune types are shown in Figure 12.13, with arrows indicating wind directions.

Barchan Dunes. Solitary sand dunes shaped like crescents and with their tips pointing downwind are

Figure 12.11 Sand sliding down the steep slip face of a dune in White Sands National Monument, New Mexico. (Photo by Michael Collier)

Figure 12.12   As parts **A** and **B** illustrate, dunes commonly have an asymmetrical shape. The steeper leeward side is called the slip face. Sand grains deposited on the slip face at the angle of repose create the cross-bedding of the dunes. **C.** A complex pattern develops in response to changes in wind direction. Also notice that when dunes are buried and become part of the sedimentary record, the cross-bedded structure is preserved. **D.** Cross beds are an obvious characteristic of the Navajo Sandstone in Zion National Park, Utah. (Photo by Martin G. Miller)

called **barchan dunes** (Figure 12.13A). These dunes form where supplies of sand are limited and the surface is relatively flat, hard, and lacking vegetation. They migrate slowly with the wind at a rate of up to 15 meters per year. Their size is usually modest, with the largest barchans reaching heights of about 30 meters while the maximum spread between their horns approaches 300 meters. When the wind direction is nearly constant, the crescent form of these dunes is nearly symmetrical. However, when the wind direction is not perfectly fixed, one tip becomes larger than the other.

Transverse Dunes.   In regions where the prevailing winds are steady, sand is plentiful, and vegetation is sparse or absent, the dunes form a series of long ridges that are separated by troughs and oriented at right angles to the prevailing wind. Because of this orientation, they are termed **transverse dunes** (Figure 12.13B). Typically, many coastal dunes are of this type. In addition, transverse dunes are common in many arid regions where the extensive surface of wavy sand is sometimes called a *sand sea*. In some parts of the Sahara and Arabian deserts, transverse dunes reach heights of 200 meters, are 1 to 3 kilometers across, and can extend for distances of 100 kilometers or more.

There is a relatively common dune form that is intermediate between isolated barchans and extensive waves of transverse dunes. Such dunes, called **barchanoid dunes,** form scalloped rows of sand oriented at right angles to the wind (Figure 12.13C). The rows resemble a series of barchans that have been positioned side by side. Visitors exploring the gypsum dunes at White Sands National Monument, New Mexico, will recognize this form.

Longitudinal Dunes.   **Longitudinal dunes** are long ridges of sand that form more or less parallel to the prevailing wind and where sand supplies are moderate (Figure 12.13D). Apparently the prevailing wind direction must vary somewhat but still remain in the same quadrant of the compass. Although the smaller types are only 3 or 4 meters high and several dozens of meters long, in some large deserts longitudinal dunes can reach great size. For example, in portions of North Africa, Arabia, and central Australia, these dunes can approach a height of 100 meters and extend for distances of more than 100 kilometers (62 miles).

Parabolic Dunes.   Unlike the other dunes that have been described thus far, **parabolic dunes** form where vegetation partially covers the sand. The shape of these dunes resembles the shape of barchans except that their tips point into the wind rather than downwind (Figure 12.13E). Parabolic dunes often form along coasts where there are strong onshore winds and abundant sand. If the sand's sparse vegetative cover is disturbed at some spot, deflation creates a blowout. Sand is then transported out of the depression and deposited as a curved rim that grows higher as deflation enlarges the blowout.

Star Dunes.   Confined largely to parts of the Sahara and Arabian deserts, **star dunes** are isolated hills of sand that exhibit a complex form (Figure 12.13F). Their name is derived from the fact that the bases of these dunes resemble multipointed stars. Usually three or four sharp-crested ridges diverge from a central high point that in some cases may approach a height of

Figure 12.13 Sand dune types. **A**. Barchan dunes. **B**. Transverse dunes. **C**. Barchanoid dunes. **D**. Longitudinal dunes. **E**. Parabolic dunes. **F**. Star dunes.

90 meters. As their form suggests, star dunes develop where wind directions are variable.

Loess (Silt) Deposits

In some parts of the world the surface topography is mantled with deposits of windblown silt, called **loess.** Over thousands of years dust storms deposited this material. When loess is breached by streams or road cuts, it tends to maintain vertical cliffs and lacks any visible layers, as you can see in Figure 12.14.

The distribution of loess worldwide indicates that there are two principal sources for this sediment: deserts and glacial deposits. The thickest and most extensive deposits of loess on Earth occur in western and

northern China. They were blown there from the extensive desert basins of Central Asia. Accumulations of 30 meters are common, and thicknesses of more than 100 meters have been measured. It is this fine, buff-colored sediment that gives the Yellow River (Hwang Ho) its name.

In the United States, deposits of loess are significant in many areas, including South Dakota, Nebraska, Iowa, Missouri, and Illinois, as well as portions of the Columbia Plateau in the Pacific Northwest. The correlation between the distribution of loess and important farming regions in the Midwest and eastern Washington State is not just a coincidence, because soils derived from this wind-deposited sediment are among the most fertile in the world.

Figure 12.14   This vertical loess bluff near the Mississippi River in southern Illinois is about 3 meters high. (Photo by James E. Patterson)

Unlike the deposits in China, which originated in deserts, the loess in the United States and Europe is an indirect product of glaciation. Its source is deposits of stratified drift. During the retreat of the ice sheets, many river valleys were choked with sediment deposited by meltwater. Strong westerly winds sweeping across the barren floodplains picked up the finer sediment and dropped it as a blanket on the eastern sides of the valleys. Such an origin is confirmed by the fact that loess deposits are thickest and coarsest on the lee side of such major glacial drainage outlets as the Mississippi and Illinois rivers and rapidly thin with increasing distance from the valleys.

The Chapter in Review

- The *concept of dryness is relative*; it refers to any situation in which a water deficiency exists. Dry regions encompass about 30 percent of Earth's land surface. Two climatic types are commonly recognized: *desert*, which is arid, and *steppe* (a marginal and more humid variant of desert), which is semiarid. *Low-latitude deserts* coincide with the zones of subtropical highs in lower latitudes. In contrast, *middle-latitude deserts* exist principally because of their positions in the deep interiors of large landmasses far removed from the ocean.

- The same geologic processes that operate in humid regions also operate in deserts, but under contrasting climatic conditions. In dry lands *rock weathering of any type is greatly reduced* because of the lack of moisture and the scarcity of organic acids from decaying plants. Much of the weathered debris in deserts is the result of *mechanical weathering*. Practically all desert streams are dry most of the time and are said to be *ephemeral*. Stream courses in deserts are seldom well integrated and lack an extensive system of tributaries. Nevertheless, *running water is responsible for most of the erosional work in a desert*. Although wind erosion is more significant in dry areas than elsewhere, the main role of wind in a desert is in the transportation and deposition of sediment.

- Because arid regions typically lack permanent streams, they are characterized as having *interior drainage*. Many of the landscapes of the Basin and Range region of the western and southwestern United States are the result of streams eroding uplifted mountain blocks and depositing the sediment in interior basins. *Alluvial fans, playas,* and *playa lakes* are features often associated with these landscapes. In the late stages of erosion, the mountain areas are reduced to a few large bedrock knobs, called *inselbergs*, projecting above the sediment-filled basin.

- The transport of sediment by wind differs from that by running water in two ways. First, wind has a low density compared to water; thus, it is not capable of picking up and transporting coarse materials. Second, because wind is not confined to channels, it can spread sediment over large areas. The *bed load* of wind consists of sand grains skipping and bouncing along the surface in a process termed *saltation*. Fine dust particles are capable of being carried by the wind great distances as *suspended load*.

- Compared to running water and glaciers, wind is a relatively insignificant erosional agent. *Deflation*, the lifting and removal of loose material, often produces shallow depressions called *blowouts*. In portions of many deserts, the surface is a layer of coarse pebbles and gravels, called *desert pavement* too large to be moved by the wind. Wind also erodes by *abrasion*, sometimes creating interestingly shaped stones termed *ventifacts*. Because sand seldom travels more than a meter above the surface, the effect of abrasion is obviously limited in vertical extent.

- Wind deposits are of two distinct types: (1) *mounds and ridges of sand*, called *dunes*, that are formed from sediment carried as part of the wind's bed load; and (2) extensive *blankets of silt*, called *loess*, that once were carried by wind in *suspension*. The profiles of many dunes show an asymmetrical shape, with the leeward (sheltered) slope being steep and the windward slope more gently inclined. The *types of sand dunes* include (1) *barchan dunes*, (2) *transverse dunes*, (3) *barchanoid dunes*, (4) *longitudinal dunes*, (5) *parabolic dunes*, and (6) *star dunes*. The thickest and most extensive deposits of loess occur in western and northern China. Unlike the deposits in China, which originated in deserts, the loess in the United States and Europe is an indirect product of glaciation.

Key Terms

abrasion (p. 254)
alluvial fan (p. 252)
barchan dune (p. 257)
barchanoid dune (p. 257)
bed load (p. 253)
blowout (p. 254)
cross beds (p. 256)

deflation (p. 254)
desert (p. 247)
desert pavement (p. 254)
dry climate (p. 247)
dune (p. 245)
ephemeral stream (p. 249)
interior drainage (p. 250)

loess (p. 258)
longitudinal dune (p. 257)
parabolic dune (p. 257)
playa lake (p. 252)
rainshadow desert (p. 248)
saltation (p. 253)
slip face (p. 256)

star dune (p. 257)
steppe (p. 247)
suspended load (p. 253)
transverse dune (p. 257)
ventifact (p. 254)

Questions for Review

1. List four common misconceptions about deserts. (See Box 12.1.)

2. How extensive are the desert and steppe regions of Earth?

3. What is the primary cause of subtropical deserts? Of middle-latitude deserts?

4. In which hemisphere (Northern or Southern) are middle-latitude deserts most common?

5. Why is the amount of precipitation that is used to define whether a place has a dry or humid climate a variable figure?

6. Why is rock weathering less in deserts than in humid climates?

7. What is the most important erosional agent in deserts?

8. Describe the features and characteristics associated with each of the stages in the evolution of a mountainous desert. Where in the United States can these stages be observed?

9. Why is sea level (ultimate base level) not a significant factor influencing erosion in desert regions?

10. Describe the way in which wind transports sand. During very strong winds, how high above the surface can sand be carried?

11. Why is wind erosion relatively more important in arid regions than in humid areas?

12. What factor limits the depths of blowouts?

13. How do sand dunes migrate?

14. Indicate which type of dune is associated with each of the statements below.
    a. Dunes whose tips point into the wind.
    b. Long sand ridges oriented at right angles to the wind.
    c. Dunes that often form along coasts where strong winds create a blowout.
    d. Solitary dunes whose tips point downward.
    e. Long sand ridges that are oriented more or less parallel to the prevailing wind.
    f. An isolated dune consisting of three or four sharp-crested ridges diverging from a central high point.
    g. Scalloped rows of sand oriented at right angles to the wind.

15. What is loess? Where are deposits of loess found? What are the origins of this sediment?

Web Resources

The *Essentials of Geology* Website uses the resources and flexibility of the Internet to aid in your study of the topics in this chapter. Written and developed by geology instructors, this site will help improve your understanding of geology. Visit **www.prenhall.com/lutgens** and click on the cover of *Essentials of Geology, 8e* to find:

- On-line review quizzes.
- Web-based critical thinking and writing exercises.
- Links to chapter-specific Web resources.
- Internet-wide key term searches.

http://www.prenhall.com/lutgens

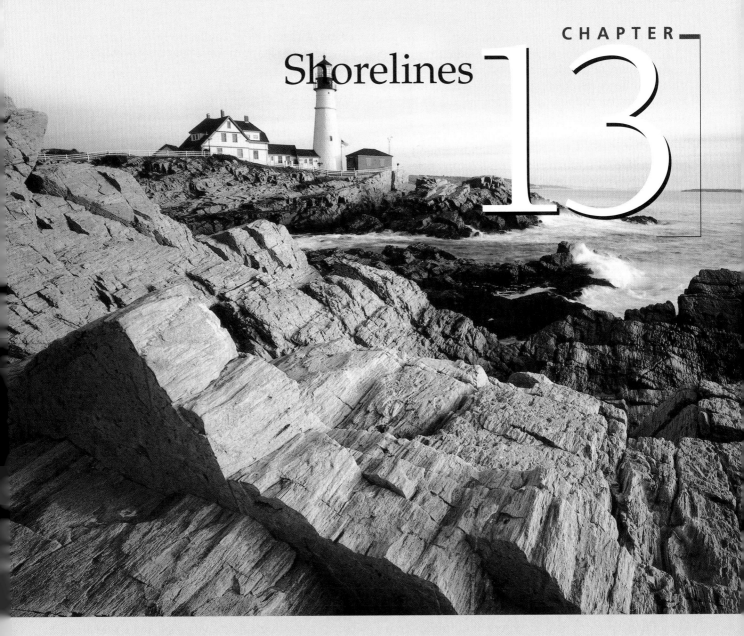

Shorelines

Focus on Learning

To assist you in learning the important concepts in this chapter, you will find it helpful to focus on the following questions:

- Why is the shoreline considered a dynamic interface?
- What factors influence the height, length, and period of a wave?
- What are the two types of wind-generated waves and the motion of water particles within each?
- How do waves erode?
- What are some typical features produced by wave erosion and from sediment deposited by beach drift and longshore currents?
- What are the local factors that influence shoreline erosion, and what are some basic responses to shoreline erosion problems?
- How do emergent and submergent coasts differ in their formation and characteristic features?
- How are tides produced?

The rocky and rugged coast of Maine. Portland Head Lighthouse. (Photo by David Muench)

The restless waters of the ocean are constantly in motion. Winds generate surface currents, the gravity of the Moon and Sun produces tides, and density differences create deep-ocean circulation. Further, waves carry the energy from storms to distant shores, where their impact erodes the land.

Shorelines are dynamic environments. Their topography, geologic makeup, and climate vary greatly from place to place. Continental and oceanic processes converge along coasts to create landscapes that frequently undergo rapid change. When it comes to the deposition of sediment, they are transition zones between marine and continental environments.

The Shoreline: A Dynamic Interface

Nowhere is the restless nature of the ocean's water more noticeable than along the shore—the dynamic interface among air, land, and sea. An *interface* is a common boundary where different parts of a system interact. This is certainly an appropriate designation for the coastal zone. Here we can see the rhythmic rise and fall of tides and observe waves constantly rolling in and breaking. Sometimes the waves are low and gentle. At other times, they pound the shore with awesome fury.

Although it may not be readily apparent, the shoreline is constantly being shaped and modified by the moving ocean waters. For example, along Cape Cod, Massachusetts, wave activity is eroding cliffs of poorly consolidated glacial sediments so aggressively that the cliffs are retreating inland at up to 1 meter per year (Figure 13.1A). By contrast, at Point Reyes, California, the far more durable bedrock cliffs are less susceptible to wave attack and are therefore retreating much more slowly (Figure 13.1B). Along both coasts, wave activity is moving sediment near the shore and building narrow sandbars that protrude into and across some bays.

The nature of present-day shorelines is not just the result of the relentless attack on the land by the sea. Indeed, the shore has a complex character that results from multiple geologic processes. For example, practically all coastal areas were affected by the worldwide rise in sea level that accompanied the melting of ice sheets at the close of the Pleistocene epoch. As the sea encroached landward, the shoreline retreated, becoming superimposed upon existing landscapes that had resulted from such diverse processes as stream erosion, glaciation, volcanic activity, and the forces of mountain building.

Today, the coastal zone is experiencing intensive human activity. Unfortunately, people often treat the shoreline as if it were a stable platform on which structures can safely be built. This attitude inevitably leads to conflicts between people and nature. As we shall see, many coastal landforms, especially beaches and barrier islands, are relatively fragile, short-lived geological features that are inappropriate sites for development.

Waves

GEODe II
External Processes
↳ Coastal Processes

Wind-generated waves provide most of the energy that shapes and modifies shorelines. Where the land and sea meet, waves that may have traveled unimpeded for hundreds or thousands of kilometers suddenly encounter a barrier that will not allow them to advance further. Stated another way, the shore is where a practically irresistible force confronts an almost immovable object. The conflict that results is never-ending and sometimes dramatic.

Characteristics of Waves

It all begins when the wind blows over the ocean. Friction exists between the moving air and the water surface, so the wind drags the water, transferring energy to it. Thus, the undulations of the water surface, called *waves*, derive their energy and motion entirely from the wind. If a breeze of less than 3 kilometers (2 miles) per hour starts to blow across still water, small wavelets appear almost instantly. When the breeze dies, the ripples disappear as suddenly as they formed. However, if the wind exceeds 3 kilometers per hour, more stable waves gradually form and progress with the wind.

All waves have the characteristics illustrated in Figure 13.2. The tops of the waves are *crests*, which are separated by *troughs*. The vertical distance between trough and crest is the **wave height.** The horizontal distance separating successive crests is the **wavelength.** The **wave period** is the time interval between the passage of two successive crests at a stationary point.

The height, length, and period that are eventually achieved by a wave depend upon three factors: (1) wind speed; (2) length of time the wind has blown; and (3) *fetch*, the distance that the wind has traveled across the open water. As the quantity of energy transferred from the wind to the water increases, the height and steepness of the waves increase as well. Eventually a critical point is reached where waves grow so tall that they topple over, forming ocean breakers called *whitecaps*.

For a particular wind speed, there is a maximum fetch and duration of wind beyond which waves will no longer increase in size. When the maximum fetch and duration are reached for a given wind velocity, the waves are said to be fully developed. The reason that waves can grow no further is that they are losing as much energy through the breaking of whitecaps as they are receiving from the wind.

A.

B.

Figure 13.1 **A.** The satellite image includes the familiar outline of Cape Cod. Boston is to the upper left. The two large islands off the south shore of Cape Cod are Martha's Vineyard (left) and Nantucket (right). Although the work of waves constantly modifies this coastal landscape, shoreline processes are not responsible for creating it. Rather, the present size and shape of Cape Cod result from the positioning of moraines and other glacial materials deposited during the Pleistocene epoch. (Satellite Image courtesy of Earth Satellite Corporation/Science Photo Library/Photo Researchers, Inc.) **B.** High-altitude image of the Point Reyes area north of San Francisco, California. The 5.5-kilometer-long south-facing cliffs at Point Reyes (bottom of photo) are exposed to the full force of the waves from the Pacific Ocean. Nevertheless, this promontory retreats slowly because the bedrock from which it formed is very resistant. (High-altitude image courtesy of USDA-ASCS.)

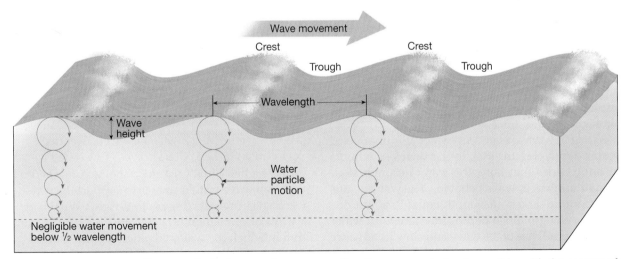

Figure 13.2 This diagram illustrates the basic parts of a wave as well as the movement of water particles with the passage of the wave. Negligible water movement occurs below a depth equal to one half the wavelength (the level of the dashed line).

263

When the wind stops or changes direction, or the waves leave the stormy area where they were created, they continue on without relation to local winds. The waves also undergo a gradual change to *swells* that are lower in height and longer in length and may carry the storm's energy to distant shores. Because many independent wave systems exist at the same time, the sea surface acquires a complex and irregular pattern. Hence, the sea waves we watch from the shore are usually a mixture of swells from faraway storms and waves created by local winds.

Types of Waves

When observing waves, always remember that you are watching *energy* travel through a medium (water). If you make waves by tossing a pebble into a pond, splashing in a pool, or by blowing across the surface of an aquarium, you are imparting *energy* to the water, and the waves you see are just the visible evidence of the energy passing through.

In the open sea, it is the wave energy that moves forward, not the water itself. Each water particle moves in a nearly circular path during the passage of a wave (see Figure 13.2). As a wave passes, a water particle returns almost to its original position. The circular orbits followed by the water particles at the surface have a diameter equal to the wave height. When water is part of the wave crest, it moves in the same direction as the advancing wave form. In the trough, the water moves in the opposite direction. This is demonstrated by observing the behavior of a floating object as a wave passes. The toy boat in Figure 13.3 merely seems to bob up and down and sway slightly to and fro without advancing appreciably from its original position. (The wind does drag the water slightly forward, causing the surface circulation of the ocean.) For this reason, waves in the open sea are called **waves of oscillation.**

The energy contributed by the wind to the water is transmitted not only along the surface of the sea but also downward. However, beneath the surface, the circular motion rapidly diminishes until at a depth equal to about one half the wavelength the movement of water particles becomes negligible. This is shown by the rapidly diminishing diameters of water-particle orbits in Figure 13.2.

As long as a wave is in deep water, it is unaffected by water depth (Figure 13.4, left). However, when a wave approaches the shore, the water becomes shallower and influences wave behavior. The wave begins to "feel bottom" at a water depth equal to about one half its wavelength. Such depths interfere with water movement at the base of the wave and slow its advance (Figure 13.4, center). As a wave continues to advance toward the shore, the slightly faster waves farther out to sea catch up, decreasing the wavelength. As the speed and length of the wave diminish, the wave steadily

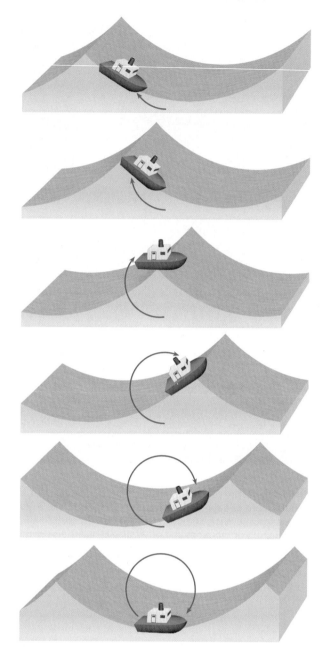

Figure 13.3 The movements of the toy boat illustrate that the wave energy advances but the water itself advances only slightly from its original position. In this sequence, the wave moves from left to right as the boat (and the water in which it is floating) rotates in a circular motion. The boat moves slightly to the left up the front of the approaching wave, then, after reaching the crest, slides to the right down the back of the wave.

grows higher. Finally, a critical point is reached when the steep wave front is unable to support the wave, and it collapses or *breaks* (Figure 13.4, right). What had been a wave of oscillation now becomes a **wave of translation** in which the water advances up the shore.

The turbulent water created by breaking waves is called **surf.** On the landward margin of the surf zone the turbulent sheet of water from collapsing breakers, called *swash*, moves up the slope of the beach. When the

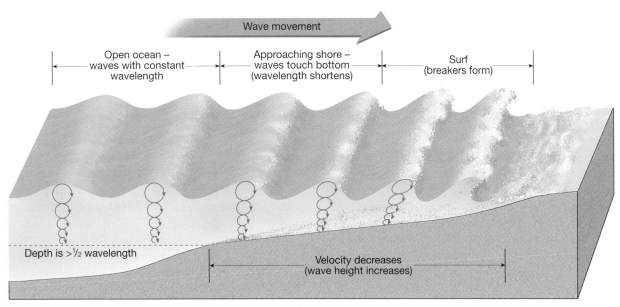

Wave movement

Open ocean –
waves with constant
wavelength

Approaching shore –
waves touch bottom
(wavelength shortens)

Surf
(breakers form)

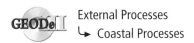
Depth is >½ wavelength

Velocity decreases
(wave height increases)

Figure 13.4   Changes that occur when a wave moves onto shore.

energy of the swash has been expended, the water flows back down the beach toward the surf zone as *backwash*.

Wave Erosion

GEODe   External Processes
↳ Coastal Processes

During calm weather wave action is minimal. However, just as streams do most of their work during floods, so too do waves perform most of their work during storms. The impact of high, storm-induced waves against the shore can be awesome (Figure 13.5). Each breaking wave may hurl thousands of tons of water against the land, sometimes making the ground tremble. The pressures exerted by Atlantic waves in wintertime, for example, average nearly 10,000 kilograms per square meter (more than 2000 pounds per square foot). The force during storms is even greater. During one such storm in Scotland, for instance, a 1350-ton portion of a steel-and-concrete breakwater was ripped from the rest of the structure and moved toward shore. Five years later the 2600-ton unit that replaced the first met a similar fate. There are many such stories that demonstrate the great force of breaking waves.

It is no wonder, then, that cracks and crevices are quickly opened in cliffs, seawalls, breakwaters, and anything else that is subjected to these enormous shocks. Water is forced into every opening, greatly compressing air trapped in the cracks. When the wave subsides, the air expands rapidly, dislodging rock fragments and enlarging and extending fractures.

In addition to the erosion caused by wave impact and pressure, **abrasion,** the sawing and grinding action

Figure 13.5   When waves break against the shore, the force of the water can be powerful and the erosional work that is accomplished can be great. Marin Headlands, Golden Gate National Recreation Area, California. (Photo by Galen Rowell/Mountain Light Photography, Inc.)

of water armed with rock fragments, is also important. In fact, abrasion is probably more intense in the surf zone than in any other environment. Smooth, rounded stones and pebbles along the shore are obvious reminders of the grinding action of rock against rock in the surf zone. Further, such fragments are used as tools by the waves as they cut horizontally into the land (Figure 13.6).

Along shorelines composed of unconsolidated material rather than hard rock, the rate of erosion by breaking waves can be extraordinary. In parts of Britain, where waves have the easy task of eroding glacial deposits of sand, gravel, and clay, the coast has been worn back 3 to 5 kilometers (2 to 3 miles) since Roman times (2000 years ago), sweeping away many villages and ancient landmarks.

Longshore Transport

External Processes
↳ Coastal Processes

Beaches are sometimes called "rivers of sand." The reason is that the energy from breaking waves often causes large quantities of sand to move along the beach and in the surf zone.

Wave Refraction

The bending of waves, called **wave refraction,** plays an important part in shoreline processes. It affects the dis-

Figure 13.6 Cliff undercut by wave erosion along the Oregon coast. (Photo by E. J. Tarbuck)

tribution of energy along the shore and thus strongly influences where and to what degree erosion, sediment transport, and deposition will take place.

Waves seldom approach the shore straight on. Rather, most waves move toward the shore at an angle. However, when they reach the shallow water of a smoothly sloping bottom, they are bent and tend to become parallel to the shore. Such bending occurs because the part of the wave nearest the shore reaches shallow water and slows first, whereas the end that is still in deep water continues forward at its full speed. The net result is a wave front that may approach nearly parallel to the shore regardless of the original direction of the wave.

Because of refraction, wave impact is concentrated against the sides and ends of headlands that project into the water, whereas wave attack is weakened in bays. This differential wave attack along irregular coastlines is illustrated in Figure 13.7. Because the waves reach the shallow water in front of the headland sooner than they do in adjacent bays, they are bent more nearly parallel to the protruding land and strike it from all three sides. By contrast, refraction in the bays causes waves to diverge and expend less energy. In these zones of weakened wave activity, sediments can accumulate and form sheltered sandy beaches. Over a long period, erosion of the headlands and deposition in the bays will straighten an irregular shoreline.

Beach Drift and Longshore Currents

Although waves are refracted, most still reach the shore at an angle, however slight. Consequently, the uprush of water from each breaking wave (the swash) is not head-on, but oblique. However, the backwash is straight down the slope of the beach. The effect of this pattern of water movement is to transport particles of sediment in a zigzag pattern along the beach (Figure 13.8). This movement is called **beach drift,** and it can transport sand and pebbles hundreds or even thousands of meters each day.

Oblique waves also produce currents within the surf zone that flow parallel to the shore and move substantially more sediment than beach drift. Because the water here is turbulent, these **longshore currents** easily move the fine suspended sand and roll larger sand and gravel along the bottom. When the sediment transported by longshore currents is added to the quantity moved by beach drift, the total can be very large. At Sandy Hook, New Jersey, for example, the quantity of sand transported along the shore over a 48-year period averaged almost 750,000 tons annually. For a 10-year period at Oxnard, California, more than 1.5 million tons of sediment moved along the shore each year.

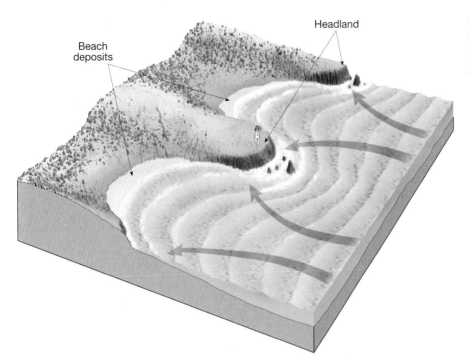

Figure 13.7 Because of wave refraction, the greatest erosional power is concentrated on the headlands. In the bays, the force of the waves is much weaker.

There should be little wonder why beaches are characterized as "rivers of sand." At any point along a beach, there is likely to be more sediment that was derived elsewhere than material eroded from the land immediately behind it. It is also worth noting that much of the sediment comprising beaches is not wave-eroded debris. In many areas the major source of material is sediment-laden rivers that discharge into the ocean (see Box 13.1). For that reason, if it were not for beach drift and longshore currents distributing the sand, many beaches would be nearly sandless.

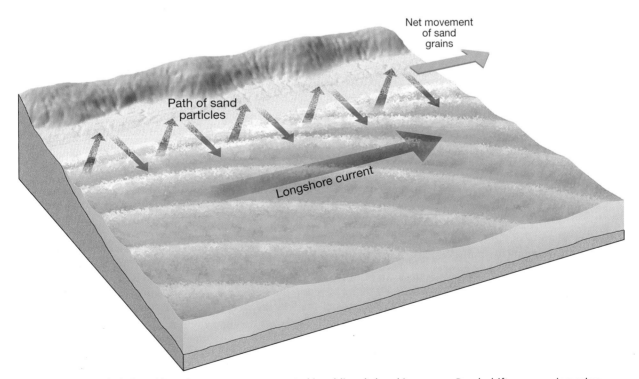

Figure 13.8 Beach drift and longshore currents are created by obliquely breaking waves. Beach drift occurs as incoming waves carry sand obliquely up the beach, while the water from spent waves carries it directly down the slope of the beach. Similar movements occur offshore in the surf zone to create the longshore current. These processes transport large quantities of material along the beach and in the surf zone.

BOX 13.1

Louisiana's Vanishing Coastal Wetlands

Coastal wetlands form in sheltered environments that include swamps, tidal flats, coastal marshes, and bayous. They are rich in wildlife and provide nesting grounds and important stop-overs for waterfowl and migratory birds, as well as spawning areas and valuable habitats for fish.

The delta of the Mississippi River in Louisiana contains about 40 percent of all coastal wetlands in the lower 48 states. Louisiana's wetlands are sheltered by low-lying barrier islands from the damaging wave action of hurricanes and winter storms. Both the wetlands and the protecting islands have formed as a result of the shifting of the Mississippi River during the past 7000 years.

Because the wetlands depend on the Mississippi River for sediment and fresh water, they are vulnerable when changes to the river system occur. Moreover, the reliance on barrier islands for protection from storm waves leaves coastal wetlands vulnerable when these narrow offshore islands are eroded.

Today, the coastal wetlands of Louisiana are disappearing at an alarming rate. Although Louisiana contains 40 percent of the wetlands in the lower 48 states, it accounts for 80 percent of the wetland loss. Between 1956 and 1990, 3460 square kilometers (1350 square miles) of coastal wetlands reverted to open waters. The state continues to lose between 65 and 91 square kilometers (25 to 35 square miles) each year. At this rate another 1800 to 4500 square kilometers (700 to 1750 square miles) will vanish under the Gulf of Mexico by the year 2050.*

Why are Louisiana's wetlands shrinking? The answer is twofold: *natural change* and *human activity*. First, the Mississippi delta and its wetlands are continually changing naturally. As sediment accumulates and builds the delta in one area, erosion and subsidence cause losses elsewhere (Figure 13.A). When the river shifts, the

*See "Louisiana's Vanishing Wetlands: Going, Going . . . ," in *Science*, vol. 289, 15 September 2000, pp. 1860–63.

Figure 13.A Satellite image of the Mississippi delta. For the past 600 years or so, the main flow of the river has been along its present course, extending southeast from New Orleans. During that span, the delta advanced into the Gulf of Mexico at a rate of about 10 kilometers (6 miles) per century. (Image courtesy of NASA)

zones of delta growth and destruction also shift. Second, ever since people arrived, the rate at which the delta and its wetlands are destroyed has accelerated.

Before Europeans settled the delta, the Mississippi River regularly overflowed its banks in seasonal floods. The huge quantities of sediment that were deposited renewed the soil and kept the delta from sinking below sea level. However, with settlement came flood-control efforts and the desire to maintain and improve navigation on the river. Artificial levees were constructed to contain the rising river during flood stage. Over time, the levees were extended all the way to the mouth of the Mississippi to keep the channel open for navigation.

The effects have been straightforward. The levees prevent sediment and fresh water from being dispersed into the wetlands. Instead, the river is forced to carry its load to the deep waters at the mouth. Meanwhile, the processes of compaction, subsidence, and wave erosion continue. Because not enough sediment is added to off-

set these forces, the size of the delta and the extent of its wetlands gradually shrink.

The problem has been aggravated by a decline in the sediment transported by the Mississippi, decreasing by approximately 50 percent over the past century. A substantial portion of the reduction results from the trapping of sediment in large reservoirs created by dams built on tributaries to the Mississippi.

Another factor contributing to wetland decline is the fact that the delta is laced with 13,000 kilometers (8000 miles) of navigation channels and canals. These artificial openings to the sea allow salty Gulf waters to flow far inland. The invasion of saltwater and tidal action cause massive "brownouts" or marsh die-offs.

Understanding and modifying the impact of people is a necessary basis for any plan to reduce the loss of wetlands in the Mississippi delta. Although projects are under way to restore Louisiana's wetlands, plans to cope effectively with the problem have yet to be implemented.

Shoreline Features

External Processes
↳ Coastal Processes

Shoreline features vary depending on the rocks of the shore, currents, wave intensity, and whether the coast is stable, sinking, or rising. This section summarizes several common shoreline features, many of which are shown in Figure 13.13, a large two-page diagram at the end of this section.

Wave-Cut Cliffs and Platforms

Whether along the rugged and irregular New England coast or along the steep shorelines of the U.S. Pacific Coast, the effects of wave erosion are often easy to see. **Wave-cut cliffs,** as their name implies, originate by the cutting action of the surf against the base of coastal land. As erosion progresses, rocks overhanging the notch at the base of the cliff crumble into the surf and the cliff retreats. A relatively flat, benchlike surface, the **wave-cut platform,** is left behind by the receding cliff (Figure 13.9). The platform broadens as wave attack continues. Some debris produced by the breaking waves remains along the water's edge as part of the beach, while the remainder is transported farther seaward.

Arches, Stacks, Spits, and Bars

Headlands that extend into the sea are vigorously attacked by waves because of refraction. The surf erodes the rock selectively, wearing away the softer or more highly fractured rock at the fastest rate. At first, sea caves may form. When two caves on opposite sides of a headland unite, a **sea arch** results (Figure 13.10). Eventually, the arch falls in, leaving an isolated remnant, or **sea stack,** on the wave-cut platform (Figure 13.10). In time, it too will be consumed by the action of the waves.

Where beach drift and longshore currents are active, several features related to the movement of sediment along the shore may develop. **Spits** are elongated ridges of sand that project from the land into the mouth of an adjacent bay. Often the end in the water hooks landward in response to wave-generated currents (Figure 13.11). The term **baymouth bar** is applied to a sandbar that completely crosses a bay, sealing it off from the open ocean (Figure 13.11). Such a feature tends to form across bays where currents are weak, allowing a spit to extend to the other side. A **tombolo,** a ridge of sand that connects an island to the mainland or to another island, forms in much the same manner as does a spit.

Barrier Islands

The Atlantic and Gulf Coastal Plains are relatively flat and slope gently seaward. The shore zone is character-

Figure 13.9 Elevated wave-cut platform along the California coast near San Francisco (right). A new platform is being created at the base of the cliff (left foreground). (Photo by John S. Shelton)

Figure 13.10 Sea arch and sea stack at the tip of Mexico's Baja Peninsula. (Photo by Mark A. Johnson/The Stock Market)

ized by **barrier islands.** These low ridges of sand parallel the coast at distances from 3 to 30 kilometers offshore. From Cape Cod, Massachusetts, to Padre Island, Texas, nearly 300 barrier islands rim the coast (Figure 13.12).

Most barrier islands are from 1 to 5 kilometers wide and between 15 and 30 kilometers long. The tallest features are sand dunes, which usually reach heights of 5 to 10 meters. The lagoons that separate these narrow islands from the shore are zones of relatively quiet water that allow small craft traveling between New York and northern Florida to avoid the rough waters of the North Atlantic.

Barrier islands probably form in several ways. Some originate as spits that were subsequently severed from the mainland by wave erosion or by the general rise in sea level following the last episode of glaciation. Others

are created when turbulent waters in the line of breakers heap up sand that has been scoured from the bottom. Finally, some barrier islands may be former sand-dune ridges that originated along the shore during the last glacial period, when sea level was lower. As the ice sheets melted, sea level rose and flooded the area behind the beach-dune complex.

The Evolving Shore

A shoreline continually undergoes modification regardless of its initial configuration. At first most coastlines are irregular, although the degree of, and reason for, the irregularity may differ considerably from place to place. Along a coastline of varied geology, the pounding surf may initially increase its irregularity because the waves will erode the weaker rocks more easily than

Figure 13.11 High-altitude image of a well-developed spit and baymouth bar along the coast of Martha's Vineyard, Massachusetts. Also notice the tidal delta in the lagoon adjacent to the inlet through the baymouth bar. (Image courtesy of USDA-ASCS)

Figure 13.12 This satellite image of New Jersey shows a large spit, known as Sandy Hook, extending north from the coast into the bay. Across the bay is New York City. Farther south, barrier islands parallel the coast. Atlantic City is located on one of these islands. (LANDSAT image courtesy of Phillips Petroleum Company, Exploration Projects Section)

the stronger ones. However, if a shoreline remains stable, marine erosion and deposition will eventually produce a straighter, more regular coast.

Figure 13.13 illustrates the evolution of an initially irregular coast. As waves erode the headlands, creating cliffs and a wave-cut platform, sediment is carried along the shore. Some material is deposited in the bays, whereas other debris is formed into spits and baymouth bars. At the same time, rivers fill the bays with sediment. Ultimately, a generally straight, smooth coast results.

Shoreline Erosion Problems

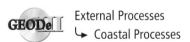 External Processes
↳ Coastal Processes

Compared with natural hazards such as earthquakes, volcanic eruptions, and landslides, shoreline erosion appears to be a more continuous and predictable process that causes relatively modest damage to limited areas. In reality, the shoreline is a dynamic place that can change rapidly in response to natural forces. Exceptional storms are capable of eroding beaches and cliffs at rates that far exceed the long-term average. Such bursts of accelerated erosion not only have a significant impact on the natural evolution of a coast but can also

have a profound impact on people who reside in the coastal zone. Erosion along our coasts causes significant property damage. Huge sums are spent annually not only to repair damage but also to prevent or control erosion. Already a problem at many sites, shoreline erosion is certain to become increasingly serious as extensive coastal development continues.

Although the same processes cause change along every coast, not all coasts respond in the same way. Interactions among different processes and the relative importance of each process depend on local factors, which include (1) the proximity of a coast to sediment-laden rivers, (2) the degree of tectonic activity, (3) the topography and composition of the land, (4) prevailing winds and weather patterns, and (5) the configuration of the coastline and nearshore areas.

During the past 100 years, growing affluence and increasing demands for recreation have brought unprecedented development to many coastal areas. As both the number and the value of structures have increased, so too have efforts to protect property from storm waves. Also, controlling the natural migration of sand is an ongoing struggle in many coastal areas. Such interference can result in unwanted changes that are difficult and expensive to correct.

Groins

To maintain or widen beaches that are losing sand, groins are sometimes constructed. A **groin** is a barrier built at a right angle to the beach to trap sand that is moving parallel to the shore (Figure 13.14). These structures often do their job so effectively that the longshore current beyond the groin becomes sand-starved. As a result, the current erodes sand from the beach on the leeward side of the groin.

To offset this effect, property owners downcurrent from the structure may erect a groin on their property. In this manner, the number of groins multiplies. An example of such proliferation is the shoreline of New Jersey, where hundreds of these structures have been built. Because it has been shown that groins often do not provide a satisfactory solution, they are no longer the preferred method of keeping beach erosion in check.

Breakwaters and Seawalls

In some coastal areas a **breakwater** may be constructed parallel to the shoreline. The purpose of such a structure is to protect boats from the force of large breaking waves by creating a quiet water zone near the shore. However, when this is done, the reduced wave activity along the shore behind the structure may allow sand to accumulate. If this happens, the marina will eventually fill with sand while the downstream beach erodes and retreats. At Santa Monica, California, where the

Figure 13.13   These diagrams illustrate the changes that can take place through time along an initially irregular coastline that remains relatively stable. The coastline shown in part **A** gradually evolves to **B**, then **C**, and then **D**. The diagrams also serve to illustrate many of the features described in the section on shoreline features. (Photos by E. J. Tarbuck)

Spit

Wave-cut cliff

Spit

Sea stack

C.

Spit

Wave-cut cliff

Beach deposits

Longshore current

Baymouth bar

Wave-cut platform

D.

Baymouth bar

Figure 13.14 Groins along the New Jersey shore at Cape May. (Photo by John S. Shelton)

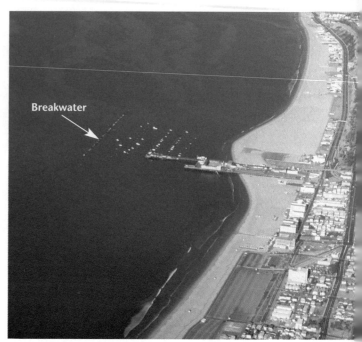

Figure 13.15 Aerial view of a breakwater at Santa Monica, California. The breakwater appears as a faint line in the water behind which many boats are anchored. The construction of the breakwater disrupted longshore transport and caused the seaward growth of the beach. (Photo by John S. Shelton)

building of a breakwater created such a problem, the city had to install a dredge to remove sand from the protected quiet water zone and deposit it down the beach where longshore currents and beach drift could recirculate the sand (Figure 13.15).

As development has moved ever closer to the beach, seawalls are sometimes built to defend property from the force of breaking waves. **Seawalls** are simply massive barriers intended to prevent waves from reaching the areas behind the wall. Waves expend much of their energy as they move across an open beach. Seawalls cut this process short by reflecting the force of unspent waves seaward. As a consequence, the beach to the seaward side of the seawall experiences significant erosion and may, in some instances, be eliminated entirely (Figure 13.16). Once the width of the beach is reduced, the seawall is subjected to even greater pounding by the waves. Eventually this battering will cause the wall to fail, and a larger, more expensive wall must be built to take its place.

The wisdom of building temporary protective structures along shorelines is increasingly questioned. The feelings of many coastal scientists and engineers are expressed in the following excerpt from a position paper that grew out of a conference on America's eroding shoreline:

It is now clear that halting the receding shoreline with protective structures benefits only a few and seriously degrades or destroys the natural beach and the value it holds for the majority. Protective structures divert the ocean's energy temporarily from private properties, but usually refocus the energy on the adjacent natural beaches. Many interrupt the natural sand flow in coastal currents, robbing many beaches of vital sand replacement.*

*Strategy for Beach Preservation Proposed, *Geotimes 30* (No. 12, December 1985): 15.

Beach Nourishment

Beach nourishment represents another approach to stabilizing shoreline sands. As the term implies, this practice simply involves the addition of large quantities of sand to the beach system. By building the beaches seaward, beach quality and storm protection are both improved. Beach nourishment, however, is not a permanent solution to the problem of shrinking beaches. The same processes that removed the sand in the first place will eventually remove the replacement sand as well. In addition, beach nourishment is very expensive. When beach nourishment was used to renew 24 kilometers (15 miles) of Miami Beach, the cost was $64 million. Here the restoration must be redone every 10 to 12 years (Figure 13.17).

By the late 1990s more than 600 kilometers (370 miles) of U.S. coastline had received some sort of beach nourishment. Not every project has been successful. For example, when Ocean City, New Jersey, restored a stretch of beach, the sand lasted less than three months before a series of storms eroded away the $5 million investment.

In some instances, beach nourishment can lead to unwanted environmental effects. For example, beach replenishment at Waikiki Beach, Hawaii, involved replacing coarse calcareous sand with softer,

Figure 13.16 Seabright in northern New Jersey once had a broad, sandy beach. A seawall 5 to 6 meters (16 to 20 feet) high and 8 kilometers (5 miles) long was built to protect the town and the railroad that brought tourists to the beach. As you can see, after the wall was built, the beach narrowed dramatically. (Photo by Rafael Macia/Photo Researchers, Inc.)

muddier calcareous sand. Destruction of the soft beach sand by breaking waves increased the water's turbidity and killed offshore coral reefs. At Miami Beach, increased turbidity also damaged local coral communities.

Beach nourishment appears to be an economically viable long-range solution to the beach preservation problem only in areas where there exists dense development, large supplies of sand, relatively low wave energy, and reconcilable environmental issues. Unfortunately, few areas possess all these attributes.

Abandonment and Relocation

So far, two basic responses to shoreline erosion problems have been considered: (1) the building of structures, such as groins and seawalls, to hold the shoreline in place; and (2) the addition of sand to replenish eroding beaches. However, a third option is also available. Many coastal scientists and planners are calling for a policy shift from defending and rebuilding beaches and coastal property in high hazard areas to removing storm-damaged buildings in those

A.

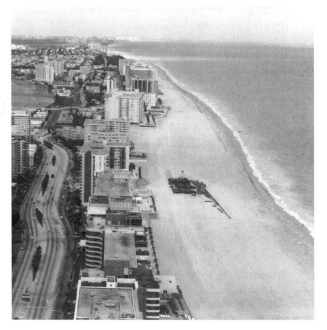

B.

Figure 13.17 Miami Beach. **A.** Before beach nourishment and **B.** After beach nourishment. (Courtesy of the U.S. Army Corps of Engineers, Vicksburg District)

places and letting nature reclaim the beach. This approach is similar to that adopted by the federal government for river floodplains following the devastating 1993 Mississippi River floods in which vulnerable structures are abandoned and relocated on higher, safer ground.

Such proposals, of course, are controversial. People with significant nearshore investments shudder at the thought of not rebuilding and not defending coastal developments from the erosional wrath of the sea. Others, however, argue that with sea level rising, the impact of coastal storms will only get worse in the decades to come (see Box 13.2). This group advocates that oft-damaged structures be abandoned or relocated to improve personal safety and to reduce costs.

Contrasting the Atlantic and Pacific Coasts

The shoreline along the Pacific Coast of the United States is strikingly different from that characterizing the Atlantic and Gulf Coast regions. Some of the differences are related to plate tectonics. The West Coast represents the leading edge of the North American plate, and because of this, it experiences active uplift and deformation. By contrast, the East Coast is a tectonically quiet region that is far from any active plate margin. Because of this basic geological difference, there are differences in the nature of shoreline erosion problems along America's opposite coasts.

Atlantic and Gulf Coasts. Much of the development along the Atlantic and Gulf coasts has occurred on bar-

BOX 13.2 Coastal Vulnerability to Sea-Level Rise

Human activities, especially the combustion of fossil fuels, have been adding vast amounts of carbon dioxide and other trace gases to the atmosphere for 200 years or more. The prospect is that emissions of these gases will continue to increase during the twenty-first century. One consequence of this change in the composition of the atmosphere is an enhancement of Earth's greenhouse effect with a resulting increase in global temperatures. During the twentieth century, average global temperatures increased by about 0.6°C. During the twenty-first century, the increase is projected to be considerably greater.

One probable impact of a human-induced global warming is a rise in sea level. How is a warmer atmosphere related to a global rise in sea level? The most obvious connection—the melting of glaciers—is important but *not* the most significant factor. More significant is that a warmer atmosphere causes an increase in ocean volume due to thermal expansion. Higher air temperatures warm the adjacent upper layers of the ocean, which in turn causes the water to expand and sea level to rise.

Research indicates that sea level has risen between 20 and 25 centimeters (4 and 8 inches) over the past century and that the trend will continue. Some models indicate that the rise in the next

A.

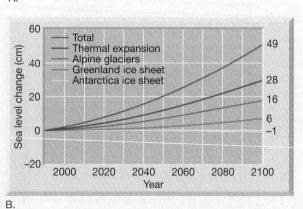

B.

Figure 13.B *A.* High, middle, and low projections of global sea-level rise, 1990–2100. *B.* Projected individual contributions to global sea-level change, 1990–2100, for the middle projection in part A. Thermal expansion is responsible for 28 centimeters of the sea-level rise. Melting alpine glaciers and ice caps contribute another 16 centimeters. The impact shown for Antarctica indicates that this massive ice sheet is projected to grow larger during this period.

rier islands. Typically, barrier islands consist of a wide beach that is backed by dunes and separated from the mainland by marshy lagoons. The broad expanses of sand and exposure to the ocean have made barrier islands exceedingly attractive sites for development. Unfortunately, development has taken place more rapidly than our understanding of barrier island dynamics.

Because barrier islands face the open ocean, they receive the full force of major storms that strike the coast. When a storm occurs, the barriers absorb the energy of the waves primarily through the movement of sand. This process and the dilemma that results have been described as follows:

Waves may move sand from the beach to offshore areas or, conversely, into the dunes; they may erode the dunes, de-

positing sand onto the beach or carrying it out to sea; or they may carry sand from the beach and the dunes into the marshes behind the barrier, a process known as overwash. The common factor is movement. Just as a flexible reed may survive a wind that destroys an oak tree, so the barriers survive hurricanes and nor'easters not through unyielding strength but by giving before the storm.

This picture changes when a barrier is developed for homes or a resort. Storm waves that previously rushed harmlessly through gaps between the dunes now encounter buildings and roadways. Moreover, because the dynamic nature of the barrier is readily perceived only during storms, homeowners tend to attribute damage to a particular storm, rather than to the basic mobility of coastal barriers. With their homes or investments at stake, local residents are more likely to seek to hold the sand in place and the waves at bay

100 years may approach or even exceed 50 centimeters (20 inches) (Figure 13.B). Such a change may seem modest, but scientists realize that any rise in sea level along a *gently* sloping shoreline, such as the Atlantic and Gulf coasts of the United States, will lead to significant erosion and severe permanent inland flooding (Figure 13.C). If this happens, many beaches and wetlands will be eliminated, and coastal civilization would be severely disrupted.

Because rising sea level is a gradual phenomenon, it may be overlooked by coastal residents as an important contributor to shoreline erosion problems. Rather, the blame may be assigned to other forces, especially storm activity. Although a given storm may be the immediate cause, the magnitude of its destruction may result from the relatively small sea-level rise that allowed the storm's power to cross a much greater land area.

One of the most challenging problems for coastal scientists today is determining the physical response of the coastline to sea-level rise. Predicting shoreline retreat and land-loss rates is critical to formulating coastal management strategies. To date, long-term planning for our shorelines has been piecemeal, if at all. Consequently, development continues without adequate consideration of the potential costs of erosion, flooding, and storm damage.

Figure 13.C The slope of a shoreline is critical to determining the degree to which sea-level changes will affect it. *A.* When the slope is gentle, small changes in sea level cause a substantial shift. *B.* The same sea-level rise along a steep coast results in only a small shoreline shift.

than to admit that development was improperly placed to begin with.*

Pacific Coast.   In contrast to the broad, gently sloping coastal plains of the East, much of the Pacific Coast is characterized by relatively narrow beaches that are backed by steep cliffs and mountain ranges. Recall that America's western margin is a more rugged and tectonically active region than the East. Because uplift continues, a rise in sea level in the West is not so readily apparent. Nevertheless, like the shoreline erosion problems facing the East's barrier islands, West Coast difficulties also stem largely from the alteration of a natural system by people.

A major problem facing the Pacific shoreline, and especially portions of southern California, is a significant narrowing of many beaches. The bulk of the sand on many of these beaches is supplied by rivers that transport it from the mountainous regions to the coast. Over the years this natural flow of material to the coast has been interrupted by dams built for irrigation and flood control. The reservoirs effectively trap the sand that would otherwise nourish the beach environment. When the beaches were wider, they served to protect the cliffs behind them from the force of storm waves. Now, however, the waves move across the narrowed beaches without losing much energy and cause more rapid erosion of the sea cliffs.

Although the retreat of the cliffs provides material to replace some of the sand impounded behind dams, it also endangers homes and roads built on the bluffs. In addition, development atop the cliffs aggravates the problem. Urbanization increases runoff, which, if not carefully controlled, can result in serious bluff erosion. Watering lawns and gardens adds significant quantities of water to the slope. This water percolates downward toward the base of the cliff, where it may emerge in small seeps. This action reduces the slope's stability and facilitates mass wasting.

Shoreline erosion along the Pacific Coast varies considerably from one year to the next, largely because of the sporadic occurrence of storms. As a consequence, when the infrequent but serious episodes of erosion occur, the damage is often blamed on the unusual storms and not on coastal development or the sediment-trapping dams that may be great distances away. If, as predicted, sea level rises at an increasing rate in the years to come, increased shoreline erosion and sea-cliff retreat should be expected along many parts of the Pacific Coast.

*Frank Lowenstein, Beaches or Bedrooms—The Choice as Sea Level Rises, *Oceanus* 28 (No. 3, Fall 1985): 22.

Emergent and Submergent Coasts

The great variety of shorelines demonstrates their complexity. Indeed, to understand any particular coastal area, you must consider rock types, size and direction of waves, frequency of storms, range between high and low tides, and submarine profile. Moreover, recent tectonic events and changes in sea level must also be taken into account. These many variables make shoreline classification difficult.

Many geologists classify coasts based on changes that have occurred with respect to sea level. This commonly used classification divides coasts into two very general categories: emergent and submergent. **Emergent coasts** develop either because an area experiences uplift or as a result of a drop in sea level. Conversely, **submergent coasts** are created when sea level rises or the land adjacent to the sea subsides.

In some areas the coast is clearly emergent because rising land or a falling water level exposes wave-cut cliffs and platforms above sea level. Excellent examples include portions of coastal California where uplift has occurred in the recent geological past. The elevated wave-cut platform shown in Figure 13.9 illustrates this. In the case of the Palos Verdes Hills, south of Los Angeles, seven different terrace levels exist, indicating seven episodes of uplift. The ever persistent sea is now cutting a new platform at the base of the cliff. If uplift follows, it too will become an elevated marine terrace.

Other examples of emergent coasts include regions that were once buried beneath great ice sheets. When glaciers were present, their weight depressed the crust, and when the ice melted, the crust began gradually to spring back. Consequently, prehistoric shoreline features today are found high above sea level. The Hudson Bay region of Canada is one such area, portions of which are still rising at a rate of more than 1 centimeter per year.

In contrast to the preceding examples, other coastal areas show definite signs of submergence. The shoreline of a coast that has been submerged in the relatively recent past is often highly irregular because the sea typically floods the lower reaches of river valleys flowing into the ocean. The ridges separating the valleys, however, remain above sea level and project into the sea as headlands. These drowned river mouths, which are called **estuaries,** characterize many coasts today. Along the Atlantic Coast, the Chesapeake and Delaware bays are examples of estuaries created by submergence (Figure 13.18). The picturesque coast of Maine, particularly in the vicinity of Acadia National Park, is another excellent example of an area that was flooded by the postglacial rise in sea level and transformed into a highly irregular submerged coastline.

Keep in mind that most coasts have a complicated geologic history. With respect to sea level, many have at

Figure 13.18 Estuaries along the East Coast of the United States. The lower portions of many river valleys were submerged by the rise in sea level that followed the end of the Ice Age. Chesapeake Bay and Delaware Bay are especially prominent examples.

various times emerged and then submerged again. Each time, they retain some of the features created during the previous situation.

Tides

Tides are daily changes in the elevation of the ocean surface. Their rhythmic rise and fall along coastlines have been known since antiquity. Other than waves, they are the easiest ocean movements to observe. An exceptional example of extreme daily tides is shown in Figure 13.19.

Although known for centuries, tides were not explained satisfactorily until Sir Isaac Newton applied the law of gravitation to them. Newton showed that there is a mutual attractive force between two bodies, as be-tween Earth and the Moon. Because the atmosphere and the ocean both are fluids and thus free to move, both are deformed by this force. Hence, ocean tides result from the gravitational attraction exerted upon Earth by the Moon and, to a lesser extent, by the Sun.

Causes of Tides

To illustrate how tides are produced, we will assume Earth is a rotating sphere covered to a uniform depth with water (Figure 13.20). It is easy to see how the Moon's gravitational force can cause the water to bulge on the side of Earth nearest the Moon. In addition, however, an equally large tidal bulge is produced on the side of Earth directly opposite the Moon.

Both tidal bulges are caused, as Newton discovered, by the pull of gravity. Gravity is inversely proportional to the square of the distance between two objects, meaning simply that it quickly weakens with distance. In this case, the two objects are the Moon and Earth. Because the force of gravity decreases with distance, the Moon's gravitational pull on Earth is slightly greater on the near side of Earth than on the far side. The result of this differential pulling is to stretch (elongate) the solid Earth very slightly. In contrast, the world ocean, which is mobile, is deformed quite dramatically by this effect to produce the two opposing tidal bulges.

Because the position of the Moon changes only moderately in a single day, the tidal bulges remain in place while Earth rotates through them. For this reason, if you stand on the seashore for 24 hours, Earth will rotate you through alternating areas of deeper and shallower water. As you are carried into regions of deeper water, the tide rises, and as you are carried away, the tide falls. Therefore, during one day you experience two high tides and two low tides.

Further, the tidal bulges migrate as the Moon revolves around Earth every 29 days. As a result, the tides, like the time of moonrise, shift about 50 minutes later each day. After 29 days the cycle is complete and a new one begins.

There may be an inequality between the high tides during a given day. Depending on the Moon's position, the tidal bulges may be inclined to the equator as in Figure 13.20. This figure illustrates that the first high tide experienced by an observer in the Northern Hemisphere is considerably lower than the high tide half a day later. Conversely, a Southern Hemisphere observer would experience the opposite effect.

Spring and Neap Tides

The Sun also influences the tides. It is far larger than the Moon, but because the Sun is so far away, the effect is considerably less than that of the Moon. In fact, the tide-generating potential of the Sun is slightly less than half that of the Moon.

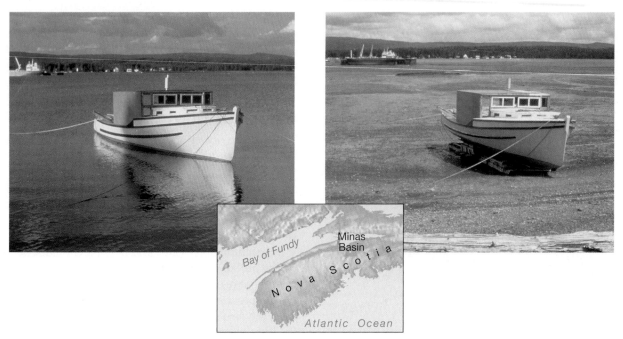

Figure 13.19 High tide and low tide on Nova Scotia's Minas Basin in the Bay of Fundy. Tital flats are exposed during low tide. (Photos courtesy of Nova Scotia Department of Tourism)

Near the times of new and full moons, the Sun and Moon are aligned and their forces are added together (Figure 13.21A). Accordingly, the combined gravitational forces of these two tide-producing bodies cause higher tidal bulges (high tides) and lower tidal troughs (low tides). These are called the **spring tides.** Spring tides create the largest daily tidal range—that is, the largest variation between high and low tides. Con-

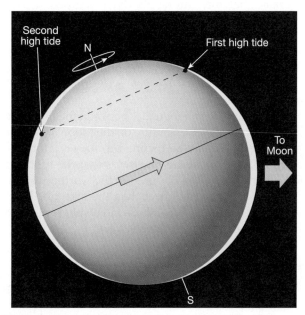

Figure 13.20 Tides on an Earth that is covered to a uniform depth with water. Depending on the Moon's position, tidal bulges may be inclined to the equator. In this situation an observer will experience two unequal high tides.

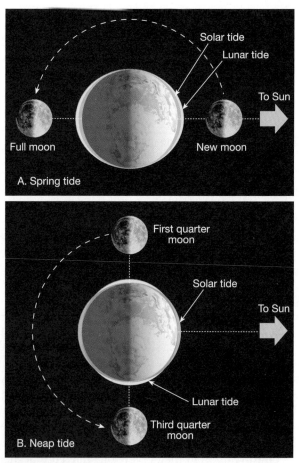

Figure 13.21 Relationship of the Moon to Earth during **A.** spring tides and **B.** neap tides.

versely, at about the time of the first and third quarters of the Moon, the gravitational forces of the Moon and Sun act on Earth at right angles, and each partially offsets the influence of the other (Figure 13.21B). As a result, the daily tidal range is less. These are called **neap tides.**

So far we have explained the basic causes and patterns of tides. But keep in mind that these theoretical considerations cannot be used to predict either the height or the time of actual tides at a particular place. The shape of coastlines and the configuration of ocean basins greatly influence the tides. Consequently, tides at various locations respond differently to the tide-producing forces. This being the case, the nature of the tide at any place can be determined most accurately by actual observation. The predictions in tidal tables and the tidal data on nautical charts are based on such observations.

Tidal Currents

Tidal current is the term used to describe the *horizontal* flow of water accompanying the rise and fall of the tide. These water movements induced by tidal forces can be important in some coastal areas. Tidal currents that advance into the coastal zone as the tide rises are called *flood currents*. As the tide falls, seaward-moving water generates *ebb currents*. Periods of little or no current, called *slack water*, separate flood and ebb. The areas affected by these alternating tidal currents are called **tidal flats** (see Figure 13.19, right). Depending on the nature of the coastal zone, tidal flats vary from narrow strips seaward of the beach to zones that may extend for several kilometers.

Although tidal currents are not important in the open sea, they can be rapid in bays, river estuaries, straits, and other narrow places. Off the coast of Brittany in France, for example, tidal currents that accom-

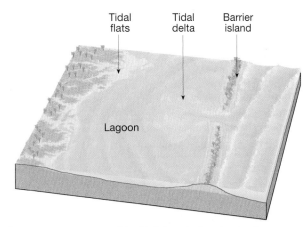

Figure 13.22   Because this tidal delta is forming in the relatively quiet waters on the landward side of a barrier island, it is termed a flood delta. As a rapidly moving tidal current emerges from the inlet, it slows and deposits sediment. The shapes of tidal deltas are variable.

pany a high tide of 12 meters (40 feet) may attain a speed of 20 kilometers (12 miles) per hour. Although tidal currents are not generally considered to be major agents of erosion and sediment transport, notable exceptions occur where tides move through narrow inlets. Here they scour the small entrances to many good harbors that would otherwise be blocked.

Sometimes deposits called **tidal deltas** are created by tidal currents (Figure 13.22). They may develop either as *flood deltas* landward of an inlet or as *ebb deltas* on the seaward side of an inlet. Because wave activity and longshore currents are reduced on the sheltered landward side, flood deltas are more common and actually more prominent (see Figure 13.11). They form after the tidal current moves rapidly through an inlet. As the current emerges into more open waters from the narrow passage, it slows and deposits its load of sediment.

The Chapter in Review

- The three factors that influence the *height, wavelength,* and *period* of a wave are (1) *wind speed,* (2) *length of time the wind has blown,* and (3) *fetch,* the distance that the wind has traveled across the open water.

- The two types of wind-generated waves are (1) *waves of oscillation,* which are waves in the open sea in which the wave form advances as the water particles move in circular orbits, and (2) *waves of translation,* the turbulent advance of water formed near the shore as waves of oscillation collapse, or *break,* and form *surf.*

- Wave erosion is caused by *wave impact pressure* and *abrasion* (the sawing and grinding action of water armed with rock fragments). The bending of waves is called *wave refraction.* Owing to refraction, wave impact is concentrated against the sides and ends of headlands.

- Most waves reach the shore at an angle. The uprush (swash) and backwash of water from each breaking wave moves the sediment in a zigzag pattern along the beach. This movement, called *beach drift,* can transport sand hundreds or even thousands of meters each day. Oblique waves also produce *longshore currents* within the surf zone that flow parallel to the shore.

- Features produced by *shoreline erosion* include *wave-cut cliffs* (which originate from the cutting action of the surf against the base of coastal land), *wave-cut platforms* (relatively flat, benchlike surfaces left behind by receding cliffs), *sea arches* (formed when a headland is eroded and two caves from opposite sides unite), and *sea stacks* (formed when the roof of a sea arch collapses).

- Some of the depositional features formed when sediment is moved by beach drift and longshore currents are *spits* (elongated ridges of sand that project from the land into the mouth of an adjacent bay), *baymouth bars* (sandbars that completely cross a bay), and *tombolos* (ridges of sand that connect an island to the mainland or to another island). Along the Atlantic and Gulf coastal plains, the shore zone is characterized by *barrier islands*, low ridges of sand that parallel the coast at distances from 3 to 30 kilometers offshore.

- Local factors that influence shoreline erosion are (1) the proximity of a coast to sediment-laden rivers, (2) the degree of tectonic activity, (3) the topography and composition of the land, (4) prevailing winds and weather patterns, and (5) the configuration of the coastline and nearshore areas.

- Three basic responses to shoreline erosion problems are (1) building *structures* such as *groins* (short walls built at a right angle to the shore to trap moving sand), *breakwaters* (structures built parallel to the shoreline to protect it from the force of large breaking waves), and *seawalls* (barriers constructed to prevent waves from reaching the area behind the wall) to hold the shoreline in place; (2) *beach nourishment*, which involves the addition of sand to replenish eroding beaches; and (3) *relocating* buildings away from the beach.

- One frequently used classification of coasts is based upon changes that have occurred with respect to sea level. *Emergent coasts*, often with wave-cut cliffs and wave-cut platforms above sea level, develop either because an area experiences uplift or as a result of a drop in sea level. Conversely, *submergent coasts*, with their drowned river mouths, called *estuaries*, are created when sea level rises or the land adjacent to the sea subsides.

- *Tides*, the daily rise and fall in the elevation of the ocean surface, are caused by the *gravitational attraction* of the Moon and, to a lesser extent, by the Sun. Near the times of new and full moons, the Sun and Moon are aligned, and their gravitational forces are added together to produce especially high and low tides. These are called the *spring tides*. Conversely, at about the times of the first and third quarters of the Moon, when the gravitational forces of the Moon and Sun are at right angles, the daily tidal range is less. These are called *neap tides*.

- *Tidal currents* are horizontal movements of water that accompany the rise and fall of tides. *Tidal flats* are the areas that are affected by the advancing and retreating tidal currents. When tidal currents slow after emerging from narrow inlets, they deposit sediment that may eventually create *tidal deltas*.

Key Terms

abrasion (p. 265)
barrier island (p. 270)
baymouth bar (p. 269)
beach drift (p. 266)
beach nourishment (p. 274)
breakwater (p. 271)
emergent coast (p. 278)
estuary (p. 278)

groin (p. 271)
longshore current (p. 266)
neap tide (p. 281)
sea arch (p. 269)
sea stack (p. 269)
seawall (p. 274)
spit (p. 269)
spring tide (p. 280)

submergent coast (p. 278)
surf (p. 264)
tidal current (p. 281)
tidal deltas (p. 281)
tidal flats (p. 281)
tide (p. 279)
tombolo (p. 269)
wave-cut cliff (p. 269)

wave-cut platform (p. 269)
wave height (p. 262)
wavelength (p. 262)
wave of oscillation (p. 264)
wave of translation (p. 264)
wave period (p. 262)
wave refraction (p. 266)

Questions for Review

1. List three factors that determine the height, length, and period of a wave.

2. Describe the motion of a water particle as a wave passes.

3. Explain what happens when a wave breaks.

4. Describe two ways in which waves cause erosion.

5. What is wave refraction? What is the effect of this process along irregular coastlines?

6. Why are beaches often called "rivers of sand"?

7. How has the construction of artificial levees and dams on the Mississippi River and its tributaries contributed to the shrinking of the Mississippi delta and its extensive wetlands? (See Box 13.1.)

8. Describe the formation of the following features: wave-cut cliff, wave-cut platform, sea stack, spit, baymouth bar, tombolo.

9. List three possible ways that barrier islands form.

10. In what direction are beach drift and longshore currents moving sand in Figure 13.14, p. 274? Is it moving toward the top or toward the bottom of the photo?

11. How might a seawall increase beach erosion?

12. How is a warmer atmosphere related to a global rise in sea level? (See Box 13.2.)

13. Relate the damming of rivers to the shrinking of beaches at some locations along the West Coast of the United States. Why do narrower beaches lead to accelerated sea-cliff retreat?

14. What observable features would lead you to classify a coastal area as emergent?

15. Are estuaries associated with submergent or emergent coasts? Why?

16. Discuss the origin of ocean tides.

17. Explain why an observer can experience two unequal high tides during one day.

18. How does the Sun influence tides?

19. Distinguish between flood current and ebb current.

Web Resources

The *Essentials of Geology* Website uses the resources and flexibility of the Internet to aid in your study of the topics in this chapter. Written and developed by geology instructors, this site will help improve your understanding of geology. Visit **www.prenhall.com/lutgens** and click on the cover of *Essentials of Geology, 8e* to find:

- On-line review quizzes.
- Web-based critical thinking and writing exercises.
- Links to chapter-specific Web resources.
- Internet-wide key term searches.

http://www.prenhall.com/lutgens

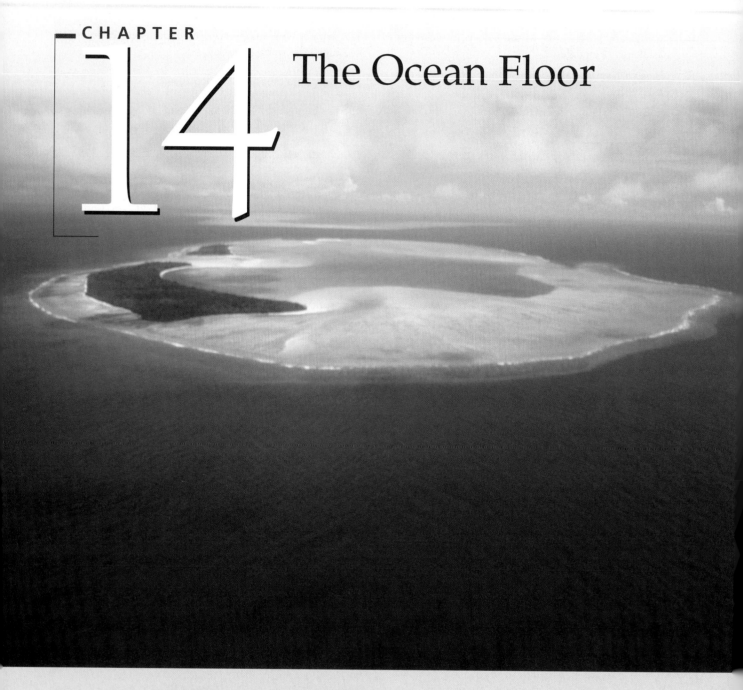

Focus on Learning

To assist you in learning the important concepts in this chapter, you will find it helpful to focus on the following questions:

• What is the extent of the world ocean?

• What are the three major topographic units of the ocean basins and the major features associated with each?

• What is the relation among submarine canyons, turbidity currents, and turbidites?

• How do coral reefs and atolls form?

• How do the three broad categories of seafloor sediments originate?

Aerial view of Kayangel Atoll in the Pacific Ocean. (Photo by Jeffrey L. Rotman/CORBIS)

How deep is the ocean? How much of Earth is covered by the global sea? These are basic questions, but they went unanswered for thousands of years. Interest in the oceans undoubtedly dates back to ancient times, but it was not until rather recently that these seemingly simple questions began to be answered (Figure 14.1). The beginning of this chapter deals with these answers; the remainder provides a look at Earth beneath the sea. Suppose that all of the water was drained from the ocean. What would we see? Plains? Mountains? Canyons? Plateaus? Indeed, the ocean conceals all of these features, and more. In fact, the topography of the ocean floor is nearly as varied as that of the continents. And what about the carpet of sediment that covers much of the seafloor? Where did it come from, and what can we learn by examining it?

The Vast World Ocean

A glance at a globe or a view of Earth from space shows a planet dominated by the world ocean (Figure 14.2). Indeed, it is for this reason that Earth is often referred to as the *blue planet*. The area of Earth is about 510 million square kilometers (197 million square miles). Of this total, approximately 360 million square kilometers (140 million square miles), or 71 percent, is represented by oceans and marginal seas (meaning seas around the ocean's margin, like the Mediterranean Sea and Caribbean Sea). Continents and islands comprise the remaining 29 percent, or 150 million square kilometers (58 million square miles).

By studying a globe or world map, it is readily apparent that the continents and oceans are not evenly divided between the Northern and Southern hemispheres (Figure 14.2). When we compute the percentages of land and water in the Northern Hemisphere, we find that nearly 61 percent of the surface is water, whereas about 39 percent is land. In the Southern Hemisphere, in contrast, almost 81 percent of the surface is water, and only 19 percent is land. It is no wonder, then, that the Northern Hemisphere is called the *land hemisphere* and the Southern Hemisphere the *water hemisphere*.

Figure 14.3A shows the distribution of land and water in the Northern and Southern hemispheres by way of a graph. Between latitudes 45 degrees north and 70 degrees north, there is actually more land than water, whereas between 40 degrees south and 65 degrees south there is almost no land to interrupt the oceanic and atmospheric circulation.

The volume of the ocean basins is many times greater than the volume of the continents above sea level. In fact, the volume of all land above sea level is only 1/18 that of the ocean.

An obvious difference between continents and the ocean basins is their relative levels. The average elevation of the continents above sea level is about 840 meters (2750 feet), whereas the average depth of the oceans is more than 4.5 times this figure—3800 meters (12,500 feet). If Earth's solid mass were perfectly smooth (level) and spherical, the oceans would cover it to a uniform depth of more than 2000 meters (1.2 miles).

A comparison of the three major oceans shows that the Pacific is by far the largest; it is nearly as large as the Atlantic and Indian oceans combined (Figure 14.3). The Pacific contains slightly more than half of the water in the world ocean, and because it includes few shallow seas along its margins, it has the greatest average depth—3940 meters (12,900 feet) (see Box 14.1).

Figure 14.1   The H.M.S. *Challenger* made the first comprehensive study of the ocean in an historic three-and-a-half-year voyage that began in December 1872. (From C.W. Thomson and Sir John Murray, *Report on the Scientific Results of the Voyage of the H.M.S. Challenger*, Vol. 1. Great Britain: Challenger Office, 1895, Plate 1)

Figure 14.2 These views of Earth show the uneven distribution of land and water between the Northern and Southern hemispheres. Almost 81 percent of the Southern Hemisphere is covered by the oceans—20 percent more than the Northern Hemisphere.

The Atlantic, bounded by almost parallel continental margins, is a relatively narrow ocean when compared to the Pacific. When the Arctic Ocean is included, the Atlantic has the greatest north–south extent and connects the two polar regions. Because the Atlantic has many shallow adjacent seas, including the Caribbean, Gulf of Mexico, Baltic, and Mediterranean, as well as wide continental shelves along its borders, it is the shallowest of the three oceans, with an average depth of 3310 meters (10,860 feet).

As Figure 14.3 shows, the Indian Ocean covers less area than the Pacific or the Atlantic: Unlike the others, it is largely a Southern Hemisphere water body.

Mapping the Ocean Floor

 Introduction
↳ Floor of the Ocean

If all water were removed from the ocean basins, what kind of surface would be revealed? It would not be quiet, subdued topography as once thought, but a surface characterized by mountains, deep canyons, and flat plains. In fact, the scenery would be just as varied as that on the continents.

The ocean's vast expanse first became apparent through voyages of discovery in the fifteenth and sixteenth centuries. An understanding of the ocean floor's varied topography did not unfold until much later with the historic 3-year voyage of the H.M.S. *Challenger* (see Figure 14.1). From December 1872 to May 1876, the *Challenger* expedition made the first, and perhaps still most comprehensive, study of the global ocean ever attempted by one agency. The 110,000-kilometer (68,000-mile) trip took the ship and its crew of scientists to every ocean except the Arctic. Throughout the voyage, they

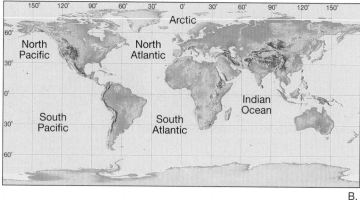

Figure 14.3 Distribution of land and water. **A.** The graph shows the amount of land and water in each 5-degree latitude belt. **B.** The world map provides a more familiar view.

Drilling Deep into the Ocean Floor

The *JOIDES Resolution* is the drilling ship of the Ocean Drilling Program. The *JOIDES* in the ship's name stands for Joint Oceanographic Institutions for Deep Earth Sampling and reflects the international commitment from the program's 22 member countries. The *Resolution* honors the H.M.S. *Resolution*, which was commanded more than 200 years ago by the well-known explorer Captain James Cook.

During cruises, holes are drilled deep into the seafloor. The cores of sediment and rock that are recovered represent millions of years of Earth history and are used by scientists to study many aspects of Earth science, including changes in global climate.

The ship is 143 meters long and 21 meters wide, with a derrick that towers 62 meters above the waterline. A computer-controlled positioning system maintains the ship over a specific location. The ship can drill in water depths up to 8200 meters (about 27,000 feet) and can deploy as much as 9100 meters (about 30,000 feet) of drill pipe (Figure 14.A).

The *JOIDES Resolution* has living quarters for 50 scientists and technicians, plus a crew of 65. The 12 onboard laboratories are equipped with the largest and most varied array of seagoing research equipment in the world.

Figure 14.A *JOIDES Resolution* can reenter holes in the seafloor years after initial drilling. The ship's dynamic positioning system consists of powerful thrusters (small propellers) that allow it to remain in place. Previous drill sites are located by bouncing sound waves between the ship's hydrophones and sonar beacons. A remote television camera aids in positioning the drill pipe into the reentry cone.

Since 1985, the ship has drilled in the Atlantic, Pacific, Indian, and Arctic oceans as well as several seas. The result has been the recovery of more than 168,000 meters (more than 104 miles!) of core samples.

The ODP is scheduled to be replaced in 2003 by the Integrated Ocean Drilling Program (IODP), which will be led by the United States and Japan. Two new drill ships will be constructed: one for drilling shallow high-resolution cores, and one with advanced technology that can drill up to 4000 meters (13,000 feet) below the seafloor. The primary objective of the new program is to collect cores that will allow scientists to better understand Earth history and Earth system processes, including the properties of the deep crust, climate change patterns, earthquake mechanisms, and the microbiology of the deep ocean floor.

sampled the water depth by laboriously lowering a weighted line overboard. Not many years later, the knowledge gained by the *Challenger* of the ocean's great depths and varied topography was further expanded with the laying of transatlantic cables. However, as long as ocean depths had to be measured with weighted lines, our knowledge of the seafloor remained limited.

In the 1920s a technological breakthrough occurred with the invention of electronic depth-sounding equipment (**echo sounder**). (The echo sounder is also referred to as *sonar*, an acronym for *so*und *na*vigation and *r*anging.) The echo sounder works by transmitting sound waves toward the ocean bottom (Figure 14.4A). A delicate receiver intercepts the echo reflected from the bottom, and a clock precisely measures the time interval to fractions of a second. By knowing the velocity of

sound waves in water (about 1500 meters per second) and the time required for the energy pulse to reach the ocean floor and return, depth can be established. The depths determined from continuous monitoring of these echoes are normally plotted so that a profile of the ocean floor is obtained. By laboriously combining profiles from several adjacent traverses, a chart of the seafloor is produced.

In the last few decades, researchers have designed even more sophisticated echo sounders to map the ocean floor. In contrast to simple echo sounders, *multibeam sonar* employs an array of sound sources and listening devices. Thus, rather than obtaining the depth of a single point every few seconds, this technique obtains a profile of a narrow strip of seafloor (Figure 14.4B). These profiles are recorded every few seconds

Outgoing signal

Reflected signal

Sea floor

A.

Sea floor

B.

Figure 14.4   Echo sounders. **A.** An echo sounder determines the water depth by measuring the time interval required for an acoustic wave to travel from a ship to the seafloor and back. The speed of sound in water is 1500 m/sec. Therefore, depth = 1/2 (1500 m/sec × echo travel time). **B.** Modern multibeam sonar obtains a profile of a narrow swath of seafloor every few seconds.

as the research vessel advances, building a continuous swath of relatively detailed coverage. (A ship will map a section of seafloor traveling in a back-and-forth pattern like that used when mowing a lawn.)

Despite their greater efficiency and enhanced detail, research vessels equipped with multibeam sonar travel at a mere 10 to 20 kilometers per hour. It would take at least 100 vessels outfitted with this equipment hundreds of years to map the entire seafloor.

Another technological breakthrough that has enhanced our understanding of the seafloor involves measuring the shape of the ocean surface from space. The ocean surface is not perfectly flat. Gravity attracts water toward regions where massive features occur. Therefore, mountains and ridges produce elevated areas on the ocean surface. Conversely, canyons and trenches cause slight depressions. Satellites equipped with radar altimeters are able to measure these subtle differences by bouncing microwaves off the sea surface. These devices can measure variations as small as 3 to 6 centimeters. Such data have added greatly to our knowledge of ocean-floor topography. Coupled with traditional sonic depth measurements, more detailed ocean-floor maps are being produced.

Oceanographers studying the topography of the ocean basins have delineated three major units: *continental margins*, the *deep ocean basin*, and *oceanic ridges*. The map in Figure 14.5 shows these provinces for the North Atlantic, and the profile at the bottom of the illustration shows the varied topography. Such profiles usually have their vertical dimension exaggerated many times to make topographic features more conspicuous. Because of this, the slopes in the profile of the seafloor in Figure 14.5 appear to be much steeper than they actually are.

Continental Margins

GEODe ∫

Introduction
↳ Floor of the Ocean

Two main types of *continental margins* have been identified—passive and active. Passive margins are found along most of the coastal areas that surround the Atlantic Ocean, including the east coasts of North and South America, as well as the coastal areas of western Europe and Africa. Passive margins are *not* associated with plate boundaries and therefore experience very little volcanism and few earthquakes. Here, weathered materials eroded from the adjacent landmass accumulate to form a thick, broad wedge of relatively undisturbed sediments.

By contrast, active continental margins occur where oceanic lithosphere is being subducted beneath the edge of a continent. The result is a relatively narrow margin, consisting of highly deformed sediments that were scraped from the descending lithospheric slab. Active continental margins are common around the Pacific Rim, where they parallel deep oceanic trenches.

Passive Continental Margins

The features comprising a **passive continental margin** include the continental shelf, the continental slope, and the continental rise (Figure 14.6).

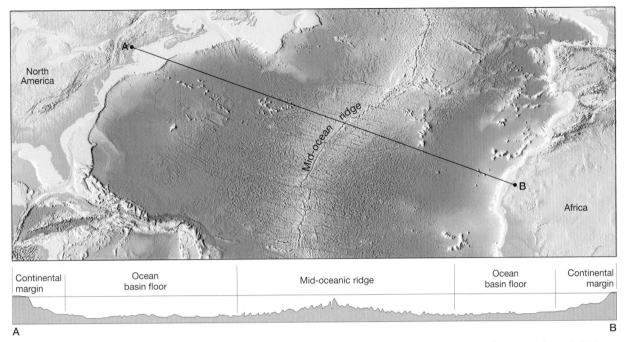

Figure 14.5   Major topographic divisions of the North Atlantic and a profile from New England to the coast of North Africa.

Continental Shelf.   The **continental shelf** is a gently sloping submerged surface extending from the shoreline toward the deep-ocean basin. Because it is underlain by continental crust, it is clearly a flooded extension of the continents. The continental shelf varies greatly in width. Almost nonexistent along some continents, the shelf may extend seaward as far as 1500 kilometers (930 miles) along others. On the average, the continental shelf is about 80 kilometers (50 miles) wide and 130 meters (430 feet) deep at the seaward edge. The average inclination of the continental shelf is only about one-tenth of 1 degree, a drop of about 2 meters per kilometer (10 feet per mile). The slope is so slight that it would appear to an observer to be a horizontal surface.

Although continental shelves represent only 7.5 percent of the total ocean area, they have economic and political significance because they contain important mineral deposits, including large reservoirs of petroleum and natural gas, as well as huge sand and gravel deposits. The waters of the continental shelf also contain many important fishing grounds that are significant sources of food.

Even though the continental shelf is relatively featureless, some areas are mantled by extensive glacial deposits and are thus quite rugged. The most profound

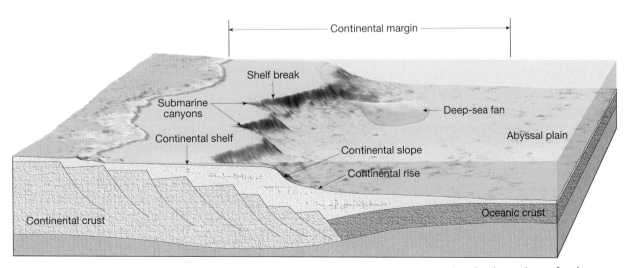

Figure 14.6   Schematic view showing the provinces of a passive continental margin. Note that the slopes shown for the continental shelf and continental slope are greatly exaggerated. The continental shelf has an average slope of one-tenth of 1 degree, while the continental slope has an average slope of about 5 degrees.

features are long valleys running from the coastline into deeper waters. Many of these valleys are the seaward extensions of river valleys on the adjacent landmass. Such valleys appear to have been excavated during the Pleistocene epoch (Ice Age).

During this time, great quantities of water were tied up in vast ice sheets on the continents. This caused sea level to drop by 100 meters or more, exposing large areas of the continental shelves (see Figure 11.A, p. 223). Because of this drop in sea level, rivers extended their courses, and land-dwelling plants and animals inhabited the newly exposed portions of the continents. Dredging off the eastern coast of North America has produced the remains of numerous land dwellers, including mammoths, mastodons, and horses, adding to the evidence that portions of the continental shelves were once above sea level.

Most passive continental shelves, such as those along the eastern coast of the United States, consist of thick accumulations of shallow-water sediments. These sediments are frequently several kilometers thick and are interbedded with limestones that formed during earlier periods of coral reef building, a process that occurs only in shallow water. Such evidence led researchers to conclude that these thick accumulations of sediment are produced along a gradually subsiding continental margin.

Continental Slope. Marking the seaward edge of the continental shelf is the **continental slope**, a relatively steep structure (as compared with the shelf) that marks the boundary between continental crust and oceanic crust (Figure 14.6). Although the inclination of the continental slope varies greatly from place to place, it averages about 5 degrees and in some areas may exceed 25 degrees. Further, the continental slope is a relatively narrow feature, averaging only about 20 kilometers in width.

Continental Rise. In regions where trenches do not exist, the continental slope merges into a more gradual incline known as the **continental rise**. Here the slope drops to about one-third degree, or about 6 meters per kilometer. Whereas the width of the continental slope averages about 20 kilometers, the continental rise may extend for hundreds of kilometers into the deep-ocean basin.

The continental rise consists of a thick accumulation of sediment that moved downslope from the continental shelf to the deep-ocean floor. The sediments are delivered to the base of the continental slope by *turbidity currents* that follow submarine canyons (see Figure 14.8). When these muddy currents emerge from the mouth of a canyon onto the relatively flat ocean floor, they deposit sediment that forms a **deep-sea fan**. Deep-sea fans have the same basic shape as alluvial fans, which form at the foot of steep mountain slopes. As fans from adjacent submarine canyons grow, they coalesce to produce

the continuous apron of sediment at the base of the continental slope that we call the continental rise.

Active Continental Margins

Along some coasts the continental slope descends abruptly into a deep-ocean trench located between the continent and ocean basin. Thus, the landward wall of the trench and the continental slope are essentially the same feature. In such locations, the continental shelf is very narrow, if it exists at all.

Active continental margins are located primarily around the Pacific Ocean in areas where the leading edge of a continent is overrunning oceanic lithosphere (Figure 14.7). Here sediments from the ocean floor and pieces of oceanic crust are scraped from the descending oceanic plate and plastered against the edge of the overriding continent. This chaotic accumulation of deformed sediment and scraps of oceanic crust is called an **accretionary wedge**. Prolonged subduction, along with the accretion of sediments on the landward side of the trench, can produce a large accumulation of sediments along a continental margin. For example, a large accretionary wedge is found along the northern coast of Japan's Honshu island.*

Some subduction zones have little or no accumulation of sediments, indicating that ocean sediments are being carried into the mantle with the subducting plate. Here the continental margin is very narrow, as the trench may lie a mere 50 kilometers offshore.

Submarine Canyons and Turbidity Currents

Deep, steep-sided valleys known as **submarine canyons** are cut into the continental slope and may extend across the entire continental rise to the deep-ocean basin (Figure 14.8). Although some of the canyons appear to be the seaward extensions of river valleys such as the Hudson Valley and Amazon River, others are not directly associated with existing river systems. Furthermore, because these canyons extend to depths far below the lowest level of the sea during the Ice Age, we cannot attribute their formation to stream erosion. These features must be created by some process that operates far below the ocean surface.

Most available information favors the view that submarine canyons have been eroded at least in part, by turbidity currents (Figure 14.8). **Turbidity currents** are downslope movements of dense, sediment-laden water. They are created when sand and mud on the continental shelf and slope are dislodged, perhaps by an earthquake, and are thrown into suspension. Because

*For more on accretionary wedges, see *Mountain Building at Convergent Boundaries* in Chapter 17.

Figure 14.7 Active continental margin. Here sediments from the ocean floor are scraped from the descending plate and added to the continental crust as an accretionary wedge.

such mud-choked water is denser than normal seawater, it flows downslope, eroding and accumulating more sediment. The erosional work repeatedly carried on by these muddy torrents is thought to be the major force in the growth of most submarine canyons.

Narrow continental margins, such as the one located along the California coast, are dissected by numerous submarine canyons. Here, erosion has extended many of these canyons landward into shallow water where longshore currents are active. As a result, sediments carried to the coasts by rivers are transported along the shore until they reach a submarine canyon. This steady supply of sediment collects until it becomes unstable and moves as a massive landslide (turbidity current) to the deep-ocean floor.

Turbidity currents eventually lose momentum and come to rest along the floor of the ocean basin (Figure 14.8). As these currents slow, the suspended sediments begin to settle out. First, the coarser, heavier sand is dropped, followed by successively finer deposits of silt and then clay. Consequently, these deposits, called **turbidites**, are characterized by a decrease in sediment grain size from bottom to top, a phenomenon known as **graded bedding** (Figure 14.8).

Although there is still more to be learned about the complex workings of turbidity currents, it has been well established that they are an important mechanism of sediment transport in the ocean. By the action of turbidity currents, submarine canyons are created and sediments are carried to the deep-ocean floor.

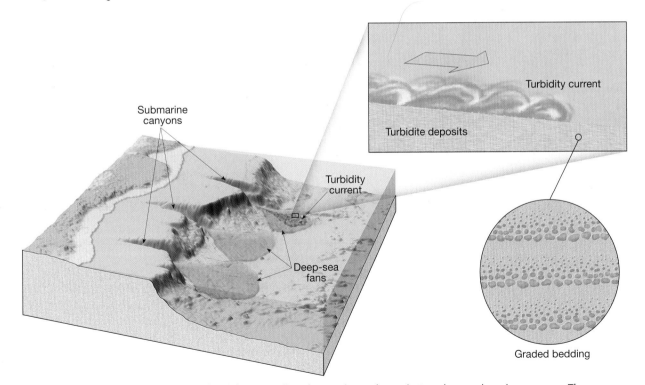

Figure 14.8 Turbidity currents move downslope, eroding the continental margin to enlarge submarine canyons. These sediment-laden density currents eventually lose momentum and deposit their loads of sediment as deep-sea fans. Beds deposited by these currents are called turbidites. Each event produces a single bed characterized by a decrease in sediment size from bottom to top, a feature known as a graded bed.

Figure 14.9 The topography of Earth's solid surface is shown on these two pages.

Features of the Deep-Ocean Basin

 Introduction
↳ Floor of the Ocean

Between the continental margin and the oceanic ridge system lies the **deep-ocean basin**. The size of this region—almost 30 percent of Earth's surface—is roughly comparable to the percentage of the surface that projects above sea level as land. Here we find remarkably flat regions known as abyssal plains, broad volcanic peaks called seamounts, and deep-ocean trenches, which are extremely deep linear depressions in the ocean floor (Figure 14.9).

Deep-Ocean Trenches

Deep-ocean **trenches** are long, relatively narrow features that form the deepest parts of the ocean (Figure

Puerto-Rico
trench

Mid-Atlantic Ridge

Peru-Chile
trench

Mid-Indian Ridge

South Sandwich
trench

14.9). Most trenches are located in the Pacific Ocean where some approach or exceed 10,000 meters in depth, and at least a portion of one, the Challenger Deep in the Mariana trench, is more than 11,000 meters below sea level. Table 14.1 presents dimensional characteristics of some of the larger trenches.

Although deep-ocean trenches represent only a very small portion of the area of the ocean floor, they are significant geologic features. Trenches are the sites where moving lithospheric plates plunge back into the mantle.

In addition to the earthquakes created as one plate descends beneath another, volcanic activity is also associated with these regions. Thus, trenches are often paralleled by volcanic island arcs. Furthermore, volcanic mountains, such as those making up portions of the Andes, are located parallel to trenches that lie adjacent to continental margins. The release of volatiles (water) from a descending plate triggers melting in the wedge of asthenosphere above it. This buoyant material slowly migrates upward and gives rise to volcanic activity at the surface.

Table 14.1 Dimensions of Some Deep-Ocean Trenches

Trench	Depth (kilometers)	Average Width (kilometers)	Length (kilometers)
Aleutian	7.7	50	3700
Japan	8.4	100	800
Java	7.5	80	4500
Kurile–Kamchatka	10.5	120	2200
Mariana	11.0	70	2550
Central America	6.7	40	2800
Peru–Chile	8.1	100	5900
Philippine	10.5	60	1400
Puerto Rico	8.4	120	1550
South Sandwich	8.4	90	1450
Tonga	10.8	55	1400

Abyssal Plains

Abyssal plains are incredibly flat features. In fact, these regions are likely the most level places on Earth. The abyssal plain found off the coast of Argentina, for example, has less than 3 meters of relief over a distance exceeding 1300 kilometers (Figure 14.9, dark area off southeast coast of South America). The monotonous topography of abyssal plains will occasionally be interrupted by the protruding summit of a partially buried volcanic structure.

Using seismic profilers, instruments whose signals penetrate far below the ocean floor, researchers have determined that abyssal plains owe their relatively featureless topography to thick accumulations of sediment that have buried an otherwise rugged ocean floor. The nature of the sediment indicates that these plains consist primarily of sediments transported far out to sea by turbidity currents.

Abyssal plains are found in all the oceans. However, because the Atlantic Ocean has fewer trenches to act as traps for the sediments carried down the continental slope, it has more extensive abyssal plains than does the Pacific.

Seamounts

Dotting the ocean floors are isolated volcanic peaks called **seamounts.** These features can rise hundreds of meters above the surrounding topography. Although these broad conical peaks have been discovered in all the oceans, the greatest number have been identified in the Pacific.

Many of these undersea volcanoes form near oceanic ridges, regions of seafloor spreading. If a volcano grows rapidly, it may emerge as an island. Examples of volcanic islands in the Atlantic include the Azores, Ascension, Tristan da Cunha, and St. Helena.

While they exist as islands, some of these volcanoes are eroded to near sea level by running water and wave action. Over a span of millions of years, the islands gradually sink as the moving plate slowly carries them from the oceanic ridge area. These submerged, flat-topped seamounts are called **tablemounts,** or **guyots** (pronounced "GEE-oh"). In other instances, guyots may be remnants of eroded volcanic islands that were formed away from the ridge crest, possibly by hot-spot activity. Here subsidence occurs after the volcanic activity ceases and the seafloor cools and contracts.

Oceanic Ridges

GEODe   Introduction
   ↳ Floor of the Ocean

Our knowledge of **oceanic ridges** comes from soundings taken of the ocean floor, core samples obtained from deep-sea drilling, and visual inspection using deep-diving submersibles (Figure 14.10). We also have firsthand inspection of slices of ocean floor that have been shoved up onto dry land. The ocean ridge system is characterized by an elevated position, extensive faulting, and numerous volcanic structures that have developed on the newly formed crust.

The interconnected oceanic ridge system is the longest topographic feature on Earth, exceeding 70,000 kilometers (43,000 miles) in length (see Figure 14.9). Representing 20 percent of Earth's surface, the oceanic ridge system winds through all major oceans in a manner similar to the seam on a baseball (see Figure 14.9). The term *ridge* may be misleading, as these features are not narrow but have widths between 1000 and 4000 kilometers and in places may occupy as much as one half the total area of the ocean floor. An examination of Figure 14.9 shows that the ridge system is broken into segments that are offset by large transform faults. Further, along the axis of some segments are deep down-faulted structures called **rift valleys.** One of the unique features of these rift zones are hydrothermal vents that

Figure 14.10 The deep-diving submersible *Alvin* is 7.6 meters long, weighs 16 tons, has a cruising speed of 1 knot, and can reach depths as great as 4000 meters. A pilot and two scientific observers are along during a normal 6- to 10-hour dive. (Courtesy of Rod Catanach/Woods Hole Oceanographic Institution)

sometimes support exotic communities of living organisms found nowhere else (see Box 14.2).

Although ocean ridges stand 2 to 3 kilometers above the adjacent deep-ocean basins, they are much different from mountains found on the continents. Rather than consisting of thick sequences of folded and faulted sedimentary rocks, oceanic ridges consist of layer upon layer of basaltic rocks that have been faulted and uplifted.

The flanks of most ridges are relatively featureless and rise very gradually (slope less than 1 degree) toward the ridge crest. Approaching the ridge crest, the topography begins to exhibit greater local relief as volcanic structures and faulted valleys become more prominent. The most rugged topography is found on those ridges that exhibit large rift valleys.

Partly because of its accessibility to both American and European scientists, the Mid-Atlantic Ridge has been studied more thoroughly than other ridge systems (see Figure 14.9). The Mid-Atlantic Ridge is a broad, submerged structure standing 2500 to 3000 meters above the adjacent ocean basin floor. In a few places, such as Iceland, the ridge has actually grown above sea level. Throughout most of its length, however, this divergent plate boundary lies 2500 meters below sea level.

Another prominent feature of the Mid-Atlantic Ridge is its deep linear rift valley extending along the ridge axis. In places this rift valley is deeper than the Grand Canyon of the Colorado River and two or three times as wide. The name *rift valley* has been applied to this feature because it is so strikingly similar to continental rift valleys such as the East African Rift Valley.

Coral Reefs and Atolls

Coral reefs are among the most picturesque features in the ocean (Figure 14.11D and E). They are constructed primarily from the skeletal remains and secretions of corals and certain algae, built up over thousands of years. The term *coral reef* is somewhat misleading in that it makes no mention of the skeletons of many other small animals and plants found inside the branching framework built by the corals, or the fact that limey secretions of algae help bind the entire structure together.

Coral reefs are confined largely to the warm, clear waters of the Pacific and Indian oceans, although a few occur elsewhere. Reef-building corals grow best in waters with an average annual temperature of about 24°C (75°F). They can survive neither sudden temperature changes nor prolonged exposure to temperatures below 18°C (65°F). In addition, these reef-builders require clear sunlit water. For this reason, the limiting depth of active reef growth is about 45 meters (150 feet).

In 1831, the naturalist Charles Darwin set out aboard the British ship H.M.S. *Beagle* on its famous five-year expedition that circumnavigated the globe. One outcome of Darwin's studies during the voyage was his hypothesis on the formation of coral islands, called **atolls.** As Figure 14.11 illustrates, atolls consist of a continuous or broken ring of coral reef surrounding a central lagoon.

Darwin's hypothesis explained what seemed to be a paradox; that is, how can corals, which require warm, shallow, sunlit water no deeper than 45 meters, create structures that reach thousands of meters to the floor of the ocean? The essence of Darwin's hypothesis was that coral reefs form on the flanks of sinking volcanic islands. As an island slowly sinks, the corals continue to build the reef complex upward (Figure 14.11).

Thus atolls, like guyots, are thought to owe their existence to the gradual sinking of oceanic crust. Drilling on atolls has revealed that volcanic rock does indeed underlie the thick coral reef structure. This finding confirmed Darwin's explanation.

Seafloor Sediments

Except for a few areas, such as near the crests of mid-ocean ridges, the ocean floor is manded with sediment. Part of this material has been deposited by turbidity currents, and the rest has slowly settled to the bottom from above. The thickness of this carpet of debris varies greatly. In some trenches, which act as traps for sediments originating on the continental margin, accumu-

BOX 14.2

Hydrothermal Vent Communities

David Cox*

Hot springs called *hydrothermal vents* form along active *rift zones* of the oceanic ridge. Here seawater percolates a kilometer or two down into the hot and newly formed oceanic crust. During its journey the water becomes saturated with minerals before boiling out at a hydrothermal vent. At some vents, metal sulfide minerals precipitate as the hot vent water enters the cold seawater, forming cone-shaped chimneys called *black smokers* (Figure 14.B).

Water temperatures at some vents reach as high as 660°C, far too hot for anything to live. But water temperatures of 100°C or lower at other vents nourish dense, exotic communities of living organisms found nowhere else. Over 400 species *new to science* have been discovered surrounding these deep-sea habitats.

Hydrothermal vent communities were discovered in 1977 by deep-sea submersibles. The best-studied hydrothermal vent regions are along the Galapagos Rift, East Pacific Rise, and Mid-Atlantic Ridge. One large vent area off the Azores covers over 15 acres.

Photosynthesis was previously thought to be the basis of all food chains on Earth. But study of hydrothermal vents reveals that the Sun does not provide energy and nutrients to this dark, hot, sulfur-rich but oxygen-deficient environment. Instead, *archaebacteria* ("ancient bacteria") living in and near the vents do *chemosynthesis*. The hydrothermal vent provides heat energy for these bacteria to oxidize the hydrogen sulfide (H_2S) formed by the reaction of hot water with dissolved sulfate (SO_4^{-2}). The chemosynthesis joins CO_2 molecules to produce sugars and other foods that

Figure 14.C   Tube worms are among the organisms found in the extreme environment of oceanic ridges where sunlight is nonexistent. These organisms derive nourishment and energy from processes associated with hydrothermal vents. (Photo by Al Gaddings Images, Inc.)

Figure 14.B   A black smoker spewing hot, mineral-rich water along the East Pacific Rise. As heated solutions meet cold seawater, sulfides of copper, iron, and zinc precipitate immediately, forming mounds of minerals around these vents. (Photo by Dudley Foster, Woods Hole Oceanographic Institution)

*David Cox is a professor emeritus of biology at Illinois Central College.

enable the bacteria as well as other organisms to live in this very unusual and extreme environment.

Some archaebacteria species live inside giant gutless tube-dwelling worms (Figure 14.C). The bacteria provide food for the tube worms to grow as rapidly as 1 meter each year and over 3 meters long. Other chemosynthetic bacteria are consumed by specialized yellow mussels, giant white clams, and pink sea urchins. These, in turn, are eaten by a variety of unique crabs and fishes. The bacteria are the foundation of a living ecosystem that *does not* require sunlight. Some scientists suggest the chemistry of vent communities may be similar to how life on Earth first began.

lations can approach 10 kilometers (6 miles). In general, however, sediment accumulations are considerably less. In the Pacific Ocean, uncompacted sediment measures about 600 meters (2000 feet) or less, whereas on the floor of the Atlantic, the thickness varies from 500 to 1000 meters (1500 to 3000 feet).

Although deposits of sand-size particles are found on the deep-ocean floor, mud is the most common sediment covering this region. Muds also predominate on the continental shelf and slope, but the sediments in these areas are coarser overall because of greater quantities of sand.

D. Barrier reef

E. Atoll

Figure 14.11   Formation of a coral atoll due to the gradual sinking of oceanic crust and upward growth of the coral reef. **A.** Fringing reef forms around volcanic island. **B.** As the volcanic island sinks, the fringing reef gradually becomes a barrier reef. **C.** Eventually, the volcano is completely submerged and an atoll remains. **D.** Aerial view of Bora Bora with surrounding barrier reef. (Photo by Tom Till/DRK Photo) **E.** An aerial view of Tetiaroa Atoll in the Pacific. The light blue waters of the relatively shallow lagoon contrast with the dark blue color of the deep ocean surrounding the atoll. (Photo by Douglas Peebles Photography)

Seafloor sediments can be classified according to their origin into three broad categories: (1) terrigenous ("derived from land") sediment, (2) biogenous ("derived from organisms") sediment, and (3) hydrogenous ("derived from water") sediment. Although we discuss each category separately, remember that all seafloor sediments are mixtures. No body of sediment comes entirely from a single source.

Terrigenous Sediment

Terrigenous sediment consists primarily of mineral grains that were weathered from continental rocks and transported to the ocean. The sand-size particles settle near shore. Because the very smallest particles take years to settle to the ocean floor, they may be carried for thousands of kilometers by ocean currents. As a consequence, virtually every area of the ocean receives some terrigenous sediment. The rate at which this sediment accumulates on the deep-ocean floor is indeed very slow. From 5000 to 50,000 years are necessary for a 1-centimeter layer to form. Conversely, on the continental margins near the mouths of large rivers, terrige-

nous sediment accumulates rapidly. In the Gulf of Mexico, for example, the sediment carried by the Mississippi River has reached a depth of many kilometers.

Because fine particles remain suspended in the water for a very long time, there is ample opportunity for chemical reactions to occur. Because of this, the colors of deep-sea sediments are often red or brown. This results when iron in the particle or in the water reacts with dissolved oxygen in the water and produces a coating of iron oxide (rust).

Biogenous Sediment

Biogenous sediment consists of shells and skeletons of marine animals and plants (Figure 14.12). This debris is produced mostly by microscopic organisms living in the sunlit waters near the ocean surface. Once these organisms die, their remains continually "rain" down upon the seafloor.

The most common biogenous sediments are known as *calcareous* ($CaCO_3$) *oozes*, and as the name implies, they have the consistency of thick mud. These sediments are produced by organisms that inhabit warm

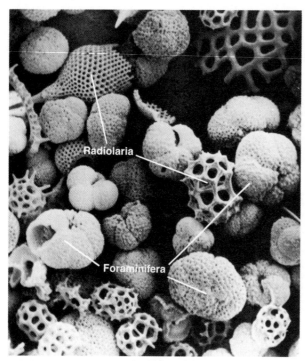

Figure 14.12 Microscopic hard parts of radiolarians and foraminifera are examples of biogenous sediments. This photomicrograph has been enlarged hundreds of times. (Photo courtesy of Deep Sea Drilling Project, Scripps Institution of Oceanography, University of California, San Diego)

Figure 14.13 Manganese nodules photographed at a depth of 2909 fathoms (5323 meters) beneath the *Robert Conrad* south of Tahiti. (Photo courtesy of Lawrence Sullivan, Lamont-Doherty Earth Observatory/Columbia University)

surface waters. When calcareous hard parts slowly sink through a cool layer of water, they begin to dissolve. This results because cold seawater contains more carbon dioxide and is thus more acidic than warm water. In seawater deeper than about 4500 meters (15,000 feet), calcareous shells will completely dissolve before they reach bottom. Consequently, calcareous ooze does not accumulate where depths are great.

Other biogenous sediments include *siliceous* (SiO_2) *oozes*. They are composed primarily of skeletons of diatoms (single-celled algaelike plants) and radiolarians (single-celled animals). Other sediments are derived from the bones, teeth, and scales of fish and other marine organisms.

Hydrogenous Sediment

Hydrogenous sediment consists of minerals that crystallize directly from seawater through various chemical reactions. For example, some limestones are formed

when calcium carbonate precipitates directly from the water; however, most limestone is composed of biogenous sediment.

Evaporation is often the mechanism triggering deposition of hydrogenous sediments. Minerals commonly precipitated in this fashion include halite (sodium chloride), the chief component of rock salt; gypsum (hydrous calcium sulfate); and the mineral sylvite, which is used as a source of potassium for fertilizer.

Among the more interesting sediments on the ocean floor in terms of economic potential are **manganese modules** (Figure 14.13). These rounded blackish lumps are composed of a complex mixture of minerals that form very slowly on the floors of the ocean basins. *Sulfide deposits* are another potential source of deep-ocean mineral resources. The richest submarine metallic sulfide deposits yet known have been discovered in a series of deep basins along the axis of the Red Sea. The largest deposit ranks with the largest deposits on land. It may contain 100 million metric tons of potential ore with 29 percent iron, over 3 percent zinc, 1 percent copper, and substantial gold and silver.

The Chapter in Review

• *Earth is a planet dominated by oceans.* Seventy-one percent of Earth's surface area consists of oceans and marginal seas. In the Southern Hemisphere, often called the *water hemisphere*, about 81 percent of the surface is water. Of the three

major oceans, Pacific, Atlantic, and Indian, the *Pacific Ocean* is the *largest*, contains slightly *more than half of the water* in the world ocean, and has the *greatest average depth*—3940 meters (12,900 feet).

- Ocean depths are determined using *echo sounders* and *multibeam sonars*. These devices bounce sonic signals off the ocean floor. Delicate receivers intercept the reflected echos. The time interval of the signal must also be carefully measured. Ocean depths are plotted to produce ocean-floor profiles. Recently, satellite measurements of the shape of the ocean surface have added data for mapping ocean-floor features.

- Oceanographers studying the topography of the ocean basins have delineated three major units: *continental margins*, the *deep ocean basin*, and *oceanic ridges*.

- The zones that collectively make up a passive continental margin include the *continental shelf* (a gently sloping, submerged surface extending from the shoreline toward the deep-ocean basin), *continental slope* (the true edge of the continent, which has a steep slope that leads from the continental shelf into deep water), and in regions where trenches do not exist, the steep continental slope merges into a more gradual incline known as the *continental rise*. The continental rise consists of sediments that have moved downslope from the continental shelf to the deep-ocean floor.

- *Active continental margins* are located primarily around the Pacific Ocean in areas where the leading edge of a continent is overrunning oceanic lithosphere. Here sediment scraped from the descending oceanic plate is plastered against the continent to form a collection of sediments called an *accretionary wedge*. An active continental margin generally has a narrow continental shelf, which grades into a deep-ocean trench.

- *Submarine canyons* are deep, steep-sided valleys that originate on the continental slope and may extend to depths of 3 kilometers. Some of these canyons appear to be the seaward extensions of river valleys. However, most information seems to favor the view that many submarine canyons have been excavated by *turbidity currents* (downslope movements of dense, sediment-laden water). *Turbidites*, sediments deposited by turbidity currents, are characterized by a decrease in sediment grain size from bottom to top, a phenomenon known as *graded bedding*.

- The deep ocean basin lies between the continental margin and the oceanic ridge system. The features of the deep ocean basin include *deep-ocean trenches* (long, narrow troughs that are the deepest parts of the ocean, and where moving crustal plates descend back into the mantle), *abyssal plains* (among the most level places on Earth, consisting of thick accumulations of sediments that were deposited atop the low, rough portions of the ocean floor by turbidity currents), and *seamounts* (isolated, steep-sided volcanic peaks on the ocean floor that originate near oceanic ridges or in association with volcanic hot spots).

- *Oceanic ridges*, the sites of seafloor spreading, are found in all major oceans and represent more than 20 percent of Earth's surface. These broad features are certainly the most prominent features in the oceans, for they form an almost continuous mountain range. Ridges are characterized by an *elevated position, extensive faulting*, and *volcanic structures* that have developed on newly formed oceanic crust. Most of the geologic activity associated with ridges occurs along a narrow region on the ridge crest, called the *rift valley*, where magma from the asthenosphere moves upward to create new slivers of oceanic crust.

- *Coral reefs*, which are confined largely to the warm, sunlit waters of the Pacific and Indian oceans, are constructed over thousands of years primarily from the accumulation of skeletal remains and secretions of corals and certain algae. A coral island, called an *atoll*, consists of a continuous or broken ring of coral reef surrounding a central lagoon. Atolls form from corals that grow on the flanks of sinking volcanic islands, where the corals continue to build the reef complex upward as the island slowly sinks.

- There are three broad categories of seafloor sediments. *Terrigenous sediment* consists primarily of mineral grains that were weathered from continental rocks and transported to the ocean. *Biogenous sediment* consists of shells and skeletons of marine animals and plants. *Hydrogenous sediment* includes minerals that crystallize directly from seawater through various chemical reactions.

Key Terms

abyssal plain (p. 294)
accretionary wedge (p. 290)
active continental margin (p. 290)
atoll (p. 295)
biogenous sediment (p. 297)
continental rise (p. 290)
continental shelf (p. 289)

continental slope (p. 290)
coral reef (p. 295)
deep-ocean basin (p. 292)
deep-sea fan (p. 290)
echo sounder (p. 287)
graded bedding (p. 291)
guyot (p. 294)
hydrogenous sediment (p. 298)

manganese nodule (p. 298)
oceanic ridge (p. 294)
passive continental margin (p. 288)
rift valley (p. 294)
seamount (p. 294)
submarine canyon (p. 290)

tablemount (p. 294)
terrigenous sediment (p. 297)
trench (p. 292)
turbidite (p. 291)
turbidity current (p. 290)

Questions for Review

1. How does the area covered by the oceans compare with that of the continents? Describe the distribution of land and water on Earth.

2. Answer the following questions about the Atlantic, Pacific, and Indian oceans.
   a. Which is the largest in area? Which is smallest?
   b. Which has the greatest north–south extent?
   c. Which is shallowest? Explain why its average depth is least.

3. How does the average depth of the ocean compare to the average elevation of the continents?

4. Assuming that the average speed of sound waves in water is 1500 meters per second, determine the water depth if the signal sent out by an echo sounder requires 6 seconds to strike bottom and return to the recorder (see Figure 14.4).

5. List the three major subdivisions of a passive continental margin. Which subdivision is considered a flooded extension of the continent? Which has the steepest slope?

6. How does an active continental margin differ from a passive continental margin? Give examples of each type.

7. Defend or rebut the statement "Most submarine canyons found on the continental slope and rise were formed during the Ice Age when rivers extended their valleys seaward."

8. What are turbidites? What is meant by the term *graded bedding*?

9. Why are abyssal plains more extensive on the floor of the Atlantic than on the floor of the Pacific?

10. How are oceanic ridges and deep-ocean trenches related to plate tectonics?

11. What is an atoll? Describe Darwin's proposal on the origin of atolls. Was it ever confirmed?

12. Distinguish among the three basic types of seafloor sediment.

Web Resources

The *Essentials of Geology* Website uses the resources and flexibility of the Internet to aid in your study of the topics in this chapter. Written and developed by geology instructors, this site will help improve your understanding of geology. Visit **www.prenhall.com/lutgens** and click on the cover of *Essentials of Geology, 8e* to find:

- On-line review quizzes.
- Web-based critical thinking and writing exercises.
- Links to chapter-specific Web resources.
- Internet-wide key term searches.

http://www.prenhall.com/lutgens

CHAPTER

Earthquakes and Earth's Interior

15

Focus on Learning

To assist you in learning the important concepts in this chapter, you will find it helpful to focus on the following questions:

• What is an earthquake?
• What are the types of earthquake waves?
• How is the epicenter of an earthquake determined?
• Where are the principal earthquake zones on Earth?
• How is earthquake strength expressed?
• What are the major zones of Earth's interior?
• How do continental crust and oceanic crust differ?

Collapsed building in Ahmedabad, India, after a destructive earthquake on January 26, 2001. (Photo by Paras Shah/Reuters)

301

What is an earthquake? How does a seismograph record the location of a quake? Can earthquakes ever be predicted? If we could view Earth's interior, what would it look like? In this chapter, we shall answer these and other questions. The study of earthquakes is important, not only because of the devastating effect that some earthquakes have on us but also because they furnish clues about the structure of Earth's interior.

On October 17, 1989, at 5:04 p.m. Pacific Daylight Time, millions of television viewers around the world were settling in to watch the third game of the World Series. Instead, they saw their television sets go black as tremors hit San Francisco's Candlestick Park. Although the earthquake was centered in a remote section of the Santa Cruz Mountains, 100 kilometers to the south, major damage occurred in the Marina District of San Francisco.

The most tragic result of the violent shaking was the collapse of some double-decked sections of Interstate 880, also known as the Nimitz Freeway. The ground motions caused the upper deck to sway, shattering the concrete support columns along a mile-long section of the freeway. The upper deck then collapsed onto the lower roadway, flattening cars as if they were aluminum beverage cans. This earthquake, named the Loma Prieta quake for its point of origin, claimed 67 lives.

In mid-January 1994, less than five years after the Loma Prieta earthquake devastated portions of the San Francisco Bay Area, a major earthquake struck the Northridge area of Los Angeles. Although not the fabled "Big One," this moderate 6.7-magnitude earthquake left 57 dead, over 5000 injured, and tens of thousands of households without water and electricity. The damage exceeded $40 billion and was attributed to an apparently unknown fault that ruptured 18 kilometers (11 miles) beneath Northridge.

The Northridge earthquake began at 4:31 a.m. and lasted roughly 40 seconds. During this brief period, the quake terrorized the entire Los Angeles area. In the three-story Northridge Meadows apartment complex, 16 people died when sections of the upper floors collapsed onto the first-floor units. Nearly 300 schools were seriously damaged, and a dozen major roadways buckled. Among these were two of California's major arteries—the Golden State Freeway (Interstate 5), where an overpass collapsed completely and blocked the roadway, and the Santa Monica Freeway. Fortunately, these roadways had practically no traffic at this early morning hour.

In nearby Granada Hills, broken gas lines were set ablaze while the streets flooded from broken water mains. Seventy homes burned in the Sylmar area. A 64-car freight train derailed, including some cars carrying hazardous cargo. But it is remarkable that the de-

Figure 15.1   Damaged apartment complex near Ismit, Turkey, following the August 17, 1999 earthquake. (Photo by CORBIS/SYGMA)

struction was not greater. Unquestionably, the upgrading of structures to meet the requirements of building codes developed for this earthquake-prone area helped minimize what could have been a much greater human tragedy.

Over 30,000 earthquakes that are strong enough to be felt occur worldwide annually. Fortunately, most are minor tremors and do very little damage. Generally, only about 75 significant earthquakes take place each year, and many of these occur in remote regions. However, occasionally a large earthquake occurs near a large population center. Under these conditions, an earthquake is among the most destructive natural forces on Earth (Figure 15.1).

The shaking of the ground, coupled with the liquefaction of some soils, wreaks havoc on buildings and other structures. In addition, when a quake occurs in a populated area, power and gas lines are often ruptured, causing numerous fires. In the famous 1906 San Francisco earthquake, much of the damage was caused by fires (Figure 15.2). They quickly became uncontrollable when broken water mains left firefighters with only trickles of water.

What Is an Earthquake?

Internal Processes
  ↳ Earthquakes

An **earthquake** is the vibration of Earth produced by the rapid release of energy. Most often, earthquakes are caused by slippage along a fault in Earth's crust. The energy released radiates in all directions from its source, the **focus**, in the form of waves. These waves are analogous to those produced when a stone is dropped into a calm pond (Figure 15.3). Just as the impact of the stone sets water waves in motion, an earthquake generates seismic waves that radiate throughout Earth. Even though the energy dissipates rapidly with increasing distance from the focus, sensitive instruments located throughout the world record the event.

Earthquakes and Faults

The tremendous energy released by atomic explosions or by volcanic eruptions can produce an earthquake, but these events are relatively weak and infrequent. What mechanism produces a destructive earthquake? Ample evidence exists that Earth is not a static planet. We know that Earth's crust has been uplifted at times, because we have found numerous ancient wave-cut benches many meters above the level of the highest tides. Other regions exhibit evidence of extensive subsidence. In addition to these vertical displacements, offsets in fence lines, roads, and other structures indicate that horizontal displacement between adjacent segments of Earth's crust is also common (Figure 15.4). These movements are usually associated with large fractures in Earth's crust called **faults.**

Most of the displacement along faults can be satisfactorily explained by the plate tectonics theory. This theory states that large slabs of Earth's crust are in continual slow motion. These mobile plates interact with neighboring plates, straining and deforming the rocks at their edges. In fact, it is along faults associated with plate boundaries that most earthquakes occur. Furthermore, earthquakes are repetitive: As soon as one is over,

Figure 15.2   San Francisco in flames after the 1906 earthquake. (Reproduced from the collection of the Library of Congress)

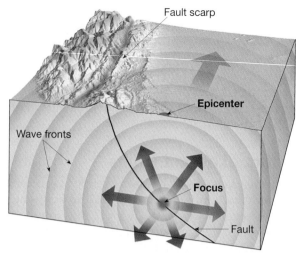

Figure 15.3 The focus of all earthquakes is located at depth. The surface location directly above it is called the epicenter.

the continuous motion of the plates resumes, adding strain to the rocks until they fail again.

Elastic Rebound

The actual mechanism of earthquake generation eluded geologists until H. F. Reid of Johns Hopkins University conducted a study following the great 1906 San Francisco earthquake. The earthquake was accompanied by horizontal surface displacements of several meters along the northern portion of the San Andreas Fault. This 1300-kilometer (780-mile) fracture runs north-south through southern California. It is a large fault zone that separates two great sections of Earth's crust—the North American plate and the Pacific plate. Field investigations determined that during this single earthquake, the Pacific plate lurched as much as 4.7 meters (15 feet) northward past the adjacent North American plate.

Figure 15.4 Slippage along a fault produced an offset in this orange grove east of Calexico, California. (Photo by John S. Shelton)

The mechanism for earthquake formation that Reid deduced from this information is illustrated in Figure 15.5. In part A of the figure, you see an existing fault, or break in the rock. In part B, tectonic forces ever so slowly deform the crustal rocks on both sides of the fault, as demonstrated by the bent features. Under these conditions, rocks are bending and storing elastic energy, much like a wooden stick does if bent. Eventually, the frictional resistance holding the rocks together is overcome. As slippage occurs at the weakest point (the focus), displacement will exert stress farther along the fault, where additional slippage will occur until most of the built-up strain is released (Figure 15.5C). This slippage allows the deformed rock to snap back. The vibrations we know as an earthquake occur as the rock elastically returns to its original shape. The springing back of the rock was termed **elastic rebound** by Reid, because the rock behaves elastically, much like a stretched rubber band does when it is released.

In summary, most earthquakes are produced by the rapid release of elastic energy stored in rock that has been subjected to great stress. Once the strength of the rock is exceeded, it suddenly ruptures, causing the vibrations of an earthquake. Earthquakes most often occur along existing faults whenever the frictional forces on the fault surfaces are overcome.

Foreshocks and Aftershocks

The intense vibrations of the 1906 San Francisco earthquake lasted about 40 seconds. Although most of the displacement along the fault occurred in this rather short period, additional movements along this and other nearby faults lasted for several days following the main quake. The adjustments that follow a major earthquake often generate smaller earthquakes called **aftershocks.** Although these aftershocks are usually much weaker than the main earthquake, they can sometimes destroy already badly weakened structures. This occurred, for example, during a 1988 earthquake in Armenia. A large aftershock of magnitude 5.8 collapsed many structures that had been weakened by the main tremor.

In addition, small earthquakes called **foreshocks** often precede a major earthquake by days or, in some cases, by as much as several years. Monitoring of these foreshocks has been used as a means of predicting forthcoming major earthquakes, with mixed success. We will consider the topic of earthquake prediction in a later section of this chapter.

San Andreas Fault: An Active Earthquake Zone

The San Andreas is undoubtedly the most studied fault system in the world. Over the years, investigations have shown that displacement occurs along discrete seg-

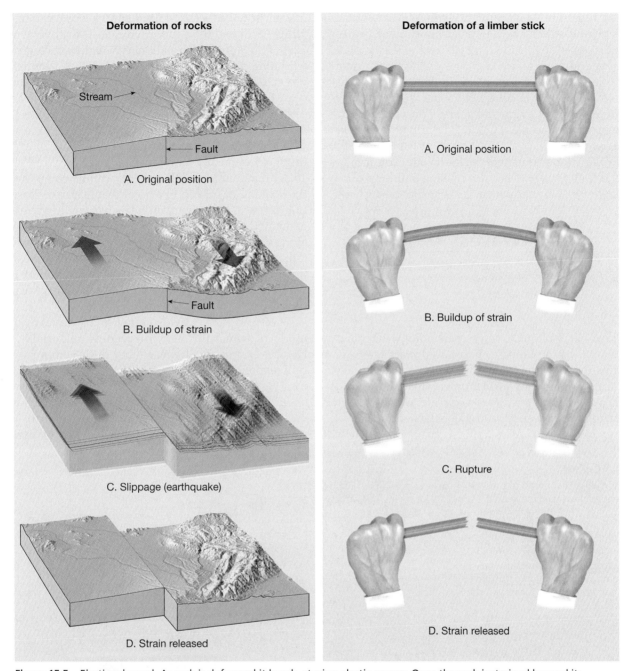

Deformation of rocks

Stream →

← Fault

A. Original position

← Fault

B. Buildup of strain

C. Slippage (earthquake)

D. Strain released

Deformation of a limber stick

A. Original position

B. Buildup of strain

C. Rupture

D. Strain released

Figure 15.5 Elastic rebound. As rock is deformed it bends, storing elastic energy. Once the rock is strained beyond its breaking point it ruptures, releasing the stored-up energy in the form of earthquake waves.

ments that are 100 to 200 kilometers long. Further, each fault segment behaves somewhat differently from the others. Some portions of the San Andreas exhibit a slow, gradual displacement known as **fault creep,** which occurs relatively smoothly and therefore with little noticeable seismic activity. Other segments regularly slip, producing small earthquakes.

Still other segments remain locked and store elastic energy for hundreds of years before rupturing in great earthquakes. The latter process is described as *stick-slip motion*, because the fault exhibits alternating periods of locked behavior followed by sudden slippage. It is estimated that great earthquakes should

occur about every 50 to 200 years along those sections of the San Andreas Fault that exhibit stick-slip motion. This knowledge is useful when assigning a potential earthquake risk to a given segment of the fault zone.

The tectonic forces along the San Andreas fault zone that were responsible for the 1906 San Francisco earthquake are still active. Currently, laser beams are used to measure the relative motion between the opposite sides of this fault. These measurements reveal a displacement of 2 to 5 centimeters (1 to 2 inches) per year. Although this seems slow, it produces substantial movement over millions of years.

To illustrate, in 30 million years, this rate of displacement would slide the western portion of California northward so that Los Angeles, on the Pacific plate, would be adjacent to San Francisco on the North American plate! More important in the short term, a displacement of just 2 centimeters per year produces 2 meters of offset every 100 years. Consequently, the 4 meters of displacement produced during the 1906 San Francisco earthquake should occur at least every 200 years along this segment of the fault zone. This fact lies behind California's concern for making buildings earthquake-resistant in anticipation of the inevitable "Big One."

Earthquake Waves

Internal Processes
↳ Earthquakes

The study of earthquake waves, **seismology,** dates back to attempts by the Chinese almost 2000 years ago to determine the direction to the source of each earthquake. Modern **seismographs** are instruments that record earthquake waves. Their principle is simple. A weight is freely suspended from a support that is attached to bedrock (Figure 15.6). When waves from an earthquake reach the instrument, the inertia of the weight keeps it stationary, while Earth and the support vibrate. The movement of Earth in relation to the stationary weight is recorded on a rotating drum. (Inertia is the tendency of a stationary object to hold still, or a moving object to stay in motion.)

Modern seismographs amplify and record ground motion, producing a trace as shown in Figure 15.7. These records, called **seismograms,** reveal that seismic waves are elastic energy. This energy radiates outward in all directions from the focus, as you saw in Figure 15.3. Transmission of this energy can be compared to the shaking of gelatin in a bowl that is jarred. Seismograms reveal that two main types of seismic waves are generated by the slippage of a rock mass. Some travel along Earth's outer layer and are called **surface waves.** Others travel through Earth's interior and are called **body waves.** Body waves are further divided into **primary waves (P waves)** and **secondary waves (S waves).**

Body waves are divided into P and S waves by their mode of travel through intervening materials. The P waves are push-pull waves—they push (compress) and pull (expand) rocks in the direction the wave is traveling (Figure 15.8A). Imagine holding someone by the shoulders and shaking that person. This push-pull movement is how P waves move through Earth. This wave motion is analogous to that generated by human

Figure 15.6 Principle of the seismograph. The inertia of the suspended mass tends to keep it motionless, while the recording drum, which is anchored to bedrock, vibrates in response to seismic waves. Thus, the stationary mass provides a reference point from which to measure the amount of displacement occurring as the seismic wave passes through the ground.

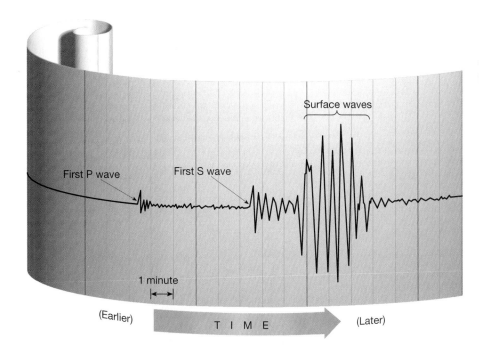

vocal cords as they move air to create sound. Solids, liquids, and gases resist a change in volume when compressed and will elastically spring back once the force is removed. Therefore, P waves, which are compressional waves, can travel through all these materials.

Conversely, S waves shake the particles at right angles to their direction of travel. This can be illustrated by fastening one end of a rope and shaking the other end, as shown in Figure 15.8B. Unlike P waves, which temporarily change the volume of the intervening material by alternately compressing and expanding it, S waves temporarily change the shape of the material that transmits them. Because fluids (gases and liquids) do not respond elastically to changes in shape, they will not transmit S waves.

The motion of surface waves is somewhat more complex. As surface waves travel along the ground, they cause the ground and anything resting upon it to move, much like ocean swells toss a ship (Figure 15.8D). In addition to their up-and-down motion, surface waves have a side-to-side motion similar to an S wave oriented in a horizontal plane (Figure 15.8C). This latter motion is particularly damaging to the foundations of structures.

By observing a typical seismic record, as shown in Figure 15.7, you can see a major difference among these seismic waves: P waves arrive at the recording station first, then S waves, and then surface waves. This is a consequence of their speeds. To illustrate, the velocity of P waves through granite within the crust is about 6 kilometers per second. Under the same conditions, S waves travel at 3.5 kilometers per second. Differences in density and elastic properties of the rock greatly influence the velocities of these waves. Generally, in any solid material, P waves travel about 1.7 times faster than

S waves, and surface waves can be expected to travel at 90 percent of the velocity of the S waves.

As you shall see, seismic waves allow us to determine the location and magnitude of earthquakes. In addition, seismic waves provide us with a tool for probing Earth's interior.

Finding Earthquake Epicenters

Internal Processes
↳ Earthquakes

Recall that the *focus* is the place within Earth where earthquake waves originate. The **epicenter** is the location on the surface directly above the focus (see Figure 15.3).

The difference in velocities of P and S waves provides a method for locating the epicenter. The principle used is analogous to a race between two autos, one faster than the other. The P wave always wins the race, arriving ahead of the S wave. But the greater the length of the race, the greater the difference in the arrival times at the finish line (the seismic station). Therefore, the greater the interval measured on a seismogram between the arrival of the first P wave and the first S wave, the greater the distance to the earthquake source.

A system for locating earthquake epicenters was developed by using seismograms from earthquakes whose epicenters could be easily pinpointed from physical evidence. From these seismograms, travel-time graphs were constructed (Figure 15.9). The first travel-time graphs were greatly improved when seismograms

A. P wave

P wave

B. S wave

S wave

C. Surface wave

D. Surface wave

Figure 15.8 Types of seismic waves and their characteristic motion. (Note that during a strong earthquake, ground shaking consists of a combination of various kinds of seismic waves.) **A.** As illustrated by a slinky, P waves are compressional waves that alternately compress and expand the material through which they pass. The back-and-forth motion produced as compressional waves travel along the surface can cause the ground to buckle and fracture, and may cause power lines to break. **B.** S waves cause material to oscillate at right angles to the direction of wave motion. Because S waves can travel in any plane, they produce up-and-down and sideways shaking of the ground. **C.** One type of surface wave is essentially the same as that of an S wave that exhibits only horizontal motion. This kind of surface wave moves the ground from side to side and can be particularly damaging to the foundations of buildings. **D.** Another type of surface wave travels along Earth's surface much like rolling ocean waves. The arrows show the elliptical movement of rock as the wave passes.

Figure 15.9 A travel-time graph is used to determine the distance to the epicenter. The difference in arrival time of the first P wave and the first S wave in the example is 5 minutes. Thus, the epicenter is roughly 3800 kilometers (2350 miles) away.

became available from nuclear explosions, because the precise location and time of detonation were known.

Using the sample seismogram in Figure 15.7 and the travel-time curves in Figure 15.9, we can determine the distance separating the recording station from the earthquake in two steps: (1) Determine the time interval between the arrival of the first P wave and the first S wave, and (2) find on the travel-time graph the equivalent time spread between the P and S wave curves. From this information, we can determine that this earthquake occurred 3800 kilometers (2350 miles) from the recording instrument.

Now we know the *distance*, but what *direction*? The epicenter could be in any direction from the seismic station. As shown in Figure 15.10, the precise location can be found when the distance is known from three or more different seismic stations. On a globe, we draw a circle around each seismic station. Each circle represents the epicenter distance for each station. The point where the three circles intersect is the epicenter of the quake. This method is called *triangulation*.

About 95 percent of the energy released by earthquakes originates in a few relatively narrow zones (Figure 15.11). The greatest energy is released along a path around the outer edge of the Pacific Ocean known as the *circum-Pacific belt*. Included in this zone are regions of great seismic activity, such as Japan, the Philip-

pines, Chile, and numerous volcanic island chains, as exemplified by Alaska's Aleutian Islands.

Figure 15.11 reveals another continuous belt that extends for thousands of kilometers through the world's oceans. This zone coincides with the oceanic ridge system, an area of frequent but low-intensity seismic activity. By comparing this figure with Figure 16.8, you can see a close correlation between the location of earthquake epicenters and plate boundaries (see Box 15.1).

Measuring the Size of Earthquakes

Historically, seismologists have employed a variety of methods to obtain two fundamentally different measures that describe the size of an earthquake—intensity and magnitude. The first of these to be used was **intensity**—a measure of the degree of earthquake shaking at a given locale based on the amount of damage. With the development of seismographs, it became clear that a quantitative measure of an earthquake based on seismic records rather than uncertain personal estimates of damage was desirable. The measurement that was developed, called **magnitude,** relies on calculations that use data provided by seismic records (and other techniques) to estimate the amount of energy released at the source of the earthquake.

At it turns out, both intensity and magnitude provide useful, although quite different, information about earthquake strength. Consequently, both measures are still used to describe the relative sizes of earthquakes.

Intensity Scales

Until a little over a century ago, historical records provided the only accounts of the severity of earthquake shaking and destruction. Using these descriptions—which were compiled without any established standards for reporting—made accurate comparisons of earthquake sizes difficult, at best.

In order to standardize the study of earthquake severity, workers developed various intensity scales that considered damage done to buildings, as well as individual descriptions of the event, and secondary effects—landslides and the extent of ground rupture. By 1902, Guiseppe Mercalli had developed a relatively reliable intensity scale, which in a modified form is still used today. The **Modified Mercalli Intensity Scale** shown in Table 15.1 was developed using California buildings as its standard, but it is appropriate for use throughout most of the United States and Canada to estimate the strength of an earthquake. For example, if some well-built wood structures and most masonry buildings are destroyed by an earthquake, a region would be assigned an intensity of X on the Mercalli scale (Table 15.1).

Figure 15.10 Earthquake epicenter is located using the distance obtained from three seismic stations.

Figure 15.11 Distribution of the 14.229 earthquakes with magnitudes equal to or greater than 5 for the period 1980–1990. (Data from National Geophysical Data Center/NOAA)

BOX 15.1

Damaging Earthquakes East of the Rockies

When you think "earthquake," you probably think of California and Japan. However, six major earthquakes have occurred in the central and eastern United States since Colonial times. Three of these had estimated Richter magnitudes of 7.5, 7.3, and 7.8, and they were centered near the Mississippi River Valley in southeastern Missouri. Occurring on December 16, 1811; January 23, 1812; and February 7, 1812, these earthquakes, plus numerous smaller tremors, destroyed the town of New Madrid, Missouri, triggered massive landslides, and caused damage over a six-state area. The course of the Mississippi River was altered, and Tennessee's Reelfoot Lake was enlarged. The distances over which these earthquakes were felt are truly remarkable, Chimneys were reported downed in Cincinnati, Ohio, and Richmond, Virginia, while Boston residents, located 1770 kilometers (1100 miles) to the northeast, felt the tremor.

Despite the history of the New Madrid earthquake, Memphis, Tennessee, the largest population center in the area, does not have adequate earthquake provisions in its building code. Further, because Memphis is located on unconsolidated floodplain deposits, buildings are more susceptible to damage than are similar structures built on bedrock. It has been estimated that if an earthquake the size of the 1811–1812 New Madrid event were to strike in the next decade, it would result in casualties in the thousands and damages in tens of billions of dollars.

Damaging earthquakes that occurred in Aurora, Illinois (1909), and Valentine, Texas (1931), remind us that other areas in the central United States are vulnerable.

The greatest historical earthquake in the eastern states occurred August 31, 1886, in Charleston, South Carolina. The event, which spanned one minute, caused 60 deaths, numerous injuries, and great economic loss within a radius of 200 kilometers (120 miles) of Charleston. Within eight minutes, effects were felt as far away as Chicago and St. Louis, where strong vibrations shook the upper floors of buildings, causing people to rush outdoors. In Charleston alone, over 100 buildings were destroyed and 90 percent of the remaining structures damaged. It was difficult to find a chimney still standing (Figure 15.A).

Numerous other strong earthquakes have been recorded in the eastern United States. New England and adjacent areas have experienced sizable shocks ever since Colonial times. The first reported earthquake in the Northeast took place in Plymouth, Massachusetts, in 1683 and was followed in 1755 by the destructive Cambridge, Massachusetts, quake. Moreover, from the time that records have been kept, New York State alone has experienced over 300 earthquakes large enough to be felt.

Earthquakes in the central and eastern United States occur far less frequently than in California, yet history indicates that the East is vulnerable. Further, these shocks east of the Rockies have generally produced structural damage over a larger area than counterparts of similar magnitude in California. The reason is that the underlying bedrock in the central and eastern United States is older and more rigid. As a result, seismic waves are able to travel greater distances with less attenuation than in the western United States. It is estimated that for earthquakes of similar magnitude, the region of maximum ground motion in the East may be up to 10 times larger than in the West. Consequently, the higher rate of earthquake occurrence in the western United States is balanced somewhat by the fact that central and eastern U.S. quakes can damage larger areas.

Figure 15.A   Damage to Charleston, South Carolina, caused by the August 31, 1886, earthquake. Damage ranged from toppled chimneys and broken plaster to total collapse. (Photo courtesy of U.S. Geological Survey)

Table 15.1 Modified Mercalli Intensity Scale

I   Not felt except by a very few under especially favorable circumstances.

II Felt only by a few persons at rest, especially on upper floors of buildings.

III Felt quite noticeably indoors, especially on upper floors of buildings, but many people do not recognize it as an earthquake.

IV During the day felt indoors by many, outdoors by few. Sensation like heavy truck striking building.

V   Felt by nearly everyone, many awakened. Disturbances of trees, poles, and other tall objects sometimes noticed.

VI Felt by all; many frightened and run outdoors. Some heavy furniture moved; few instances of fallen plaster or damaged chimneys. Damage slight.

VII Everybody runs outdoors. Damage negligible in buildings of good design and construction; slight-to-moderate in well-built ordinary structures; considerable in poorly built or badly designed structures.

VIII Damage slight in specially designed structures; considerable in ordinary substantial buildings with partial collapse; great in poorly built structures. (Fall of chimneys, factory stacks, columns, monuments, walls.)

IX Damage considerable in specially designed structures. Buildings shifted off foundations. Ground cracked conspicuously.

X   Some well-built wooden structures destroyed. Most masonry and frame structures destroyed. Ground badly cracked.

XI Few, if any, (masonry) structures remain standing. Bridges destroyed. Broad fissures in ground.

XII Damage total. Waves seen on ground surfaces. Objects thrown upward into air.

Despite their usefulness in providing seismologists with a tool to compare earthquake severity, particularly in regions where there are no seismographs, intensity scales have severe drawbacks. In particular, intensity scales are based on effects (largely destruction) of earthquakes that depend not only on the severity of ground shaking but also on factors, such as population density, building design, and the nature of surface materials. The modest 6.9-magnitude earthquake in Armenia in 1988 was extremely destructive, mainly because of inferior building construction, whereas the 1985 Mexico City quake was deadly because of the soft sediment upon which the city rests. Thus, the destruction wrought by earthquakes may not be a true measure of the earthquake's actual size.

Magnitude Scales

In order to compare earthquakes across the globe, a measure was needed that does not rely on parameters that vary considerably from one part of the world to another, such as construction practices. As a consequence, a number of magnitude scales were developed.

Richter Magnitude. In 1935 Charles Richter of the California Institute of Technology developed the first magnitude scale using seismic records to estimate the relative sizes of earthquakes. As shown in Figure 15.12 (top), the **Richter scale** is based on the amplitude of the largest seismic wave (P, S, or surface wave) recorded on a seismogram. Because seismic waves weaken as the distance between the earthquake focus and the seismograph increases (in a manner similar to light), Richter developed a method that accounted for the decrease in wave amplitude with increased distance. Theoretically, as long as equivalent instruments were used, monitoring stations at various locations would obtain the same Richter magnitude for each recorded earthquake. (Richter selected the Wood-Anderson seismograph as the standard recording device.)

Although the Richter scale has no upper limit, the largest magnitude recorded on a Wood-Anderson seismograph was 8.9. These great shocks released approximately 10^{26} ergs of energy—roughly equivalent to the detonation of 1 billion tons of TNT. Conversely, earthquakes with a Richter magnitude of less than 2.0 are not felt by humans. With the development of more sensitive instruments, tremors of a magnitude of minus 2 were recorded. Table 15.2 shows how Richter magnitudes and their effects are related.

Earthquakes vary enormously in strength, and great earthquakes produce wave amplitudes that are thousands of times larger than those generated by weak tremors. To accommodate this wide variation, Richter used a *logarithmic scale* to express magnitude, where a *tenfold* increase in wave amplitude corresponds to an increase of 1 on the magnitude scale. Thus, the amount of ground shaking for a 5-magnitude earthquake is 10 times greater than that produced by an earthquake having a Richter magnitude of 4.

In addition, each unit of Richter magnitude equates to roughly a *32-fold energy increase*. Thus, an earthquake with a magnitude of 6.5 releases 32 times more energy than one with a magnitude of 5.5, and roughly 1000 times more energy than a 4.5-magnitude quake. A major earthquake with a magnitude of 8.5 releases millions of times more energy than the smallest earthquakes felt by humans.

Other Magnitude Scales. Richter's original goal was modest in that he only attempted to rank the earthquakes of southern California (shallow-focus

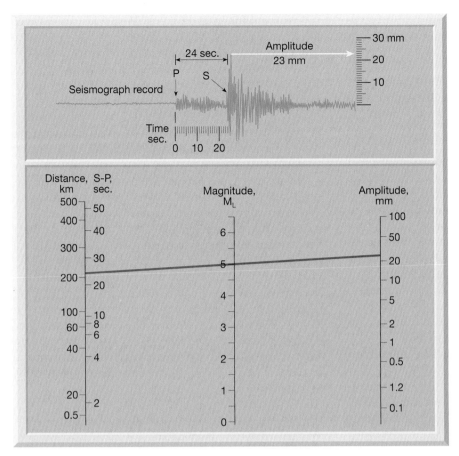

Figure 15.12 Illustration showing how the Richter magnitude of an earthquake can be determined graphically using a seismograph record from a Wood-Anderson instrument. First, measure the height (amplitude) of the largest wave on the seismogram (23 mm) and then the distance to the focus using the time interval between S and P waves (24 seconds). Next, draw a line between the distance scale (left) and the wave amplitude scale (right). By doing this, you should obtain the Richter magnitude (M_L) of 5. (Data from California Institute of Technology)

earthquakes) into groups of large, medium, and small magnitude. Hence, Richter magnitude was designed to study nearby (or local) earthquakes and is denoted by the symbol (M_L)—where M is for *magnitude* and L is for *local*.

The convenience of describing the size of an earthquake by a single number that could be calculated quickly from seismographs makes the Richter scale a powerful tool. Further, unlike intensity scales that could only be applied to populated areas of the globe, Richter magnitudes could be assigned to earthquakes in more remote regions and even to events that occurred in the ocean basins. As a result, the method devised by Richter

was adapted to a number of different seismographs located throughout the world. In time, seismologists modified Richter's work and developed new magnitude scales.

However, despite their usefulness, none of these "Richter-like" magnitude scales are adequate for describing very large earthquakes. For example, the 1906 San Francisco earthquake and the 1964 Alaskan earthquake have roughly the same Richter magnitudes of about 8.3. However, the Alaskan earthquake released considerably more energy than the San Francisco quake, based on the size of the fault zone and the amount of displacement observed. Thus, the Richter scale (as well

Table 15.2 Earthquake magnitudes and expected world incidence.

Richter Magnitudes	Effects Near Epicenter	Estimated Number per Year
< 2.0	Generally not felt, but recorded.	600,000
2.0–2.9	Potentially perceptible.	300,000
3.0–3.9	Felt by some.	49,000
4.0–4.9	Felt by most.	6200
5.0–5.9	Damaging shocks.	800
6.0–6.9	Destructive in populous regions.	266
7.0–7.9	Major earthquakes. Inflict serious damage.	18
≥ 8.0	Great earthquakes. Destroy communities near epicenter.	1.4

SOURCE: *Earthquake Information Bulletin* and others.

as the other related magnitude scales) are said to be *saturated* for large earthquakes because they cannot distinguish between the size of these events.

Moment Magnitude. In recent years seismologists have been employing a more precise measure called **moment magnitude** (M_W), which is derived from the amount of displacement that occurs along a fault zone rather than by measuring the ground motion at some distant point. The moment magnitude is calculated using a combination of factors that include the average amount of displacement along the fault, the area of the rupture surface, and the shear strength of the faulted rock—a measure of how much strain energy a rock can store before it suddenly slips and releases this energy in the form of an earthquake (and heat). For example, the energy involved in a 3-meter displacement of a rock body along a rupture a few hundred kilometers long would be much larger than that produced by a 1-meter displacement along a 10-kilometer-long rupture (assuming comparable rupture depths).

The moment magnitude can also be calculated from seismograms by examining very long period seismic waves. The values obtained have been calibrated so that small- and moderate-sized earthquakes have moment magnitudes that are roughly equivalent to Richter magnitudes. However, moment magnitudes are much better for describing very large earthquakes. For example, on the moment magnitude scale, the 1906 San Francisco earthquake, which had a surface-wave magnitude of 8.3, would be demoted to 7.9 on the moment magnitude scale, whereas the 1964 Alaskan earthquake with an 8.3 Richter magnitude would be increased to 9.2. The strongest earthquake on record is the 1960 Chilean earthquake with a moment magnitude of 9.6.

Moment magnitude has gained wide acceptance among seismologists and engineers because: (1) it is the only magnitude scale that estimates adequately the size of very large earthquakes; (2) it is a measure established from the size of the rupture surface and the amount of displacement, thus it better reflects the total energy released during an earthquake; and (3) it can be verified by two independent methods—field studies that are based on measurements of fault displacement and by seismographic methods using long-period waves.

In summary, the Richter scale was designed to differentiate the sizes of shallow-focus earthquakes in southern California occurring near a particular type of seismograph. Richter's original magnitude scale (M_L) was extended to measure earthquakes at any distance and focal depth, as well as taking into account geological variations from place to place. All of these "Richter-like" magnitude scales were designed so that the final computed magnitude is reasonably consistent with the Richter scale. Because none of these magnitude scales estimate adequately the size of very large earthquakes, the moment magnitude scale (M_W), derived from the amount of displacement that occurs along a fault zone, was developed.

Destruction from Earthquakes

The most violent earthquake to jar North America this century—the Good Friday Alaskan Earthquake—occurred in 1964. Felt throughout that state, the earthquake had a Richter magnitude of 8.3 and reportedly lasted 3 to 4 minutes. This event left 131 people dead, thousands homeless, and the economy of the state badly disrupted because it occurred near major towns and seaports (Figure 15.13). Had the schools and business districts been open on this holiday, the toll surely would

Figure 15.13 Region most affected by the Good Friday earthquake of 1964. Note the epicenter (red dot). (After U.S. Geological Survey)

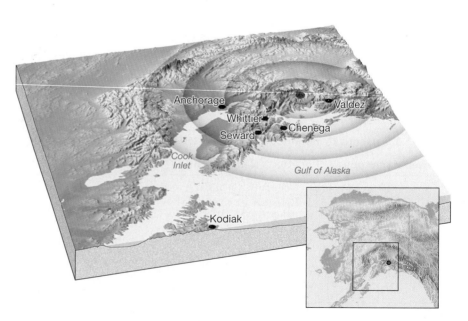

have been higher. Within 24 hours of the initial shock, 28 aftershocks were recorded, 10 of which exceeded a Richter magnitude of 6.

Destruction from Seismic Vibrations

The 1964 Alaskan earthquake provided geologists with new insights into the role of ground shaking as a destructive force. The structural damage from earthquake waves depends on several factors, including (1) the intensity of the vibrations, (2) the duration of the vibrations, (3) the nature of the material upon which the structure rests, and (4) the design of the structure.

All multistory structures in Anchorage were damaged by the vibrations, but the more flexible wood-frame residential buildings fared best. Figure 15.14 offers a striking example of how construction variations affect earthquake damage. You can see that the steel-frame structure on the left withstood the vibrations, whereas the poorly designed J.C. Penney building was badly damaged. Engineers have learned that unreinforced masonry buildings are the most serious safety threats in earthquakes.

Most large structures in Anchorage were damaged even though they were built to conform to the earthquake provisions of the Uniform Building Code. Perhaps some of that destruction can be attributed to the unusually long duration of this earthquake (estimated at 3 to 4 minutes). (The San Francisco earthquake of 1906 was felt for about 40 seconds; the strong vibrations of the 1989 Loma Prieta earthquake lasted less than 15 seconds.)

The buildings in Anchorage are situated on soft, unconsolidated sediments, which amplify vibrations more than does solid bedrock. Thus, damage in Anchorage was more severe than in Whittier, which rests on a firm foundation of granite. However, Whittier was damaged by a seismic sea wave (described in the next section).

The 1985 Mexican earthquake gave seismologists and engineers a vivid reminder of what had been learned from the 1964 Alaskan earthquake. In central Mexico City, nearly 400 kilometers (250 miles) from the epicenter, the vibrations intensified to five times that experienced in outlying districts, owing to the city's construction on soft sediments, remnants of an ancient lake bed (Figure 15.15).

Where unconsolidated materials are saturated with water, earthquakes can generate a phenomenon known as **liquefaction.** Under these conditions, what had been a stable soil turns into a fluid that is no longer capable of supporting buildings or other structures (Figure 15.16). As a result, underground objects such as storage tanks and sewer lines may float toward the surface of their newly liquefied environment. Buildings and other structures may settle and collapse. During the Loma Prieta earthquake, in San Francisco's Marina District, foundations failed and geysers of sand and water shot from the ground, indicating that liquefaction had occurred (Figure 15.17).

Tsunami

Most deaths associated with the 1964 Alaskan quake were caused by **seismic sea waves,** or **tsunami.** These destructive waves often are called tidal waves by the media. However, this name is inappropriate, for these waves are generated by earthquakes, not by the tidal effect of the Moon or Sun. The name *tsunami* is Japanese for harbor wave, for Japanese harbors have suffered from many of them.

Most tsunami result from displacement along a fault located on the ocean floor, or from a large underwater landslide triggered by an earthquake (Figure 15.18.) Once formed, tsunami resemble ripples created when a pebble is dropped into a pond. In contrast to ripples, tsunami advance across the ocean at speeds between 500 and 950 kilometers (300 and 600 miles) per hour. Despite this, a tsunami in the open ocean can pass undetected because its height is usually less than a

Figure 15.14 Damage to the five-story JC Penney Co. building, Anchorage, Alaska. Very little structural damage was incurred by the adjacent building. (Courtesy of NOAA)

Figure 15.15 During the 1985 Mexican earthquake, multistory buildings swayed back and forth as much as 1 meter. Many, including the hotel shown here, collapsed or were seriously damaged. (Photo by James L. Beck)

meter and the distance between wave crests is great, ranging from 100 to 700 kilometers. However, upon entering shallower coastal water, these destructive waves are slowed and the water begins to pile up to heights that occasionally exceed 30 meters (100 feet), as shown in Figure 15.18. As the crest of a tsunami approaches shore, it appears as a rapid rise in sea level with a turbulent and chaotic surface.

Usually the first warning of an approaching tsunami is a rather rapid withdrawal of water from beaches. Coastal residents have learned to heed this warning and move to higher ground, for about 5 to 30 minutes later the retreat of water is followed by a surge capable of extending hundreds of meters inland (Figure 15.19). In a successive fashion, each surge is followed by a rapid oceanward retreat of the water. These waves are separated by intervals of between 10 and 60 minutes. They are able to traverse thousands of kilometers of the ocean before their energy is dissipated.

The tsunami generated in the 1964 Alaskan earthquake heavily damaged communities along the Gulf of Alaska and killed 107. By contrast, only nine people died in Anchorage as a direct result of the vibrations. Tsunami damage following the Alaskan earthquake extended along much of the western coast of North America. Despite a one-hour warning, 12 persons per-

ished in Crescent City, California, from the fifth wave. The first wave crested about 4 meters (13 feet) above low tide and was followed by three progressively smaller waves. Believing that the tsunami had ceased, people returned to the shore, only to be met by the fifth and most devastating wave. Superimposed upon high tide, it crested about 6 meters higher than the level of low tide.

On July 17, 1998, four villages on New Guinea's north coast were almost entirely swept away. Here an otherwise ordinary earthquake that measured 7.1 on the magnitude scale is thought to have triggered a massive underwater landslide. (Tremors of at least this size strike the globe every three weeks.) Within 5 to 10 minutes a plateau of water averaging 10 meters (30 feet) high and perhaps 1 or 2 kilometers across swept over that shore for more than a minute. Two similar waves followed at intervals of several minutes. Officially, the tsunami, the worst in more than two decades, killed 2134 people, with many still unaccounted for.

Landslides and Ground Subsidence

In the 1964 Alaskan earthquake, the greatest damage to structures was from landslides and ground subsidence triggered by the vibrations. At Valdez and Seward, the violent shaking caused river-delta materials to experi-

Figure 15.16 Effects of liquefaction. This tilted building rests on unconsolidated sediment that imitated quicksand during the 1985 Mexican earthquake. (Photo by James L. Beck)

Figure 15.17 These "mud volcanoes" were produced by the Loma Prieta earthquake of 1989. They formed when geysers of sand and water shot from the ground, an indication that liquefaction occurred. (Photo by Richard Hilton, courtesy of Dennis Fox)

ence liquefaction; the subsequent slumping carried both waterfronts away. Because the disaster could happen again, the entire town of Valdez was relocated about 7 kilometers away on more stable ground. In Valdez, 31 persons on a dock perished when it slid into the sea.

Most of the damage in Anchorage was attributed to landslides. Many homes were destroyed in Turnagain Heights when a layer of clay lost its strength and over 200 acres of land slid toward the ocean (Figure 15.20). A portion of this landslide has been left in its natural condition as a reminder of this destructive event. The site was named Earthquake Park. Downtown Anchorage was also disrupted as sections of the main business district dropped by as much as 3 meters (10 feet).

Fire

The 1906 San Francisco earthquake reminds us of the formidable threat of fire. The central city contained mostly large, older wooden structures and brick buildings. The greatest destruction was caused by fires that started when gas and electrical lines were severed. The fires raged uncontrolled for three days and devastated over 500 city blocks (see Figure 15.2). The problem was compounded by the initial ground shaking, which broke the city's water lines into hundreds of unconnected pieces.

The fire was finally contained when buildings were dynamited along a wide boulevard to create a fire break, the same strategy used in fighting a forest fire. Although only a few deaths were attributed to the fires, such is not always the case. A 1923 earthquake in Japan (their worst quake prior to the 1995 Kobe tremor) triggered an estimated 250 fires, which devastated the city of Yokohama and destroyed more than half the homes in Tokyo. Over 100,000 deaths were attributed to the fires, which were driven by unusually high winds.

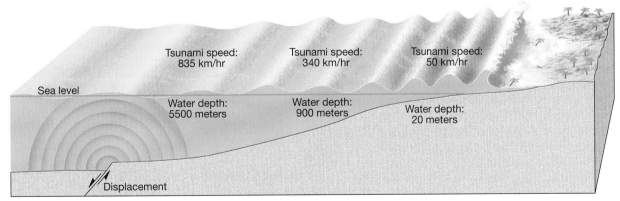

Figure 15.18 Schematic drawing of a tsunami generated by displacement of the ocean floor. The speed of a wave moving across the ocean correlates with ocean depth. As shown, waves moving in deep water advance at speeds in excess of 800 kilometers per hour. Speed gradually slows to 50 kilometers per hour at depths of 20 meters. Decreasing depth slows the movement of the wave column. As waves slow in shallow water, they grow in height until they topple and rush onto shore with tremendous force. The size and spacing of these swells are not to scale.

Figure 15.19 A man stands before a wall of water about to engulf him at Hilo, Hawaii, on April 1, 1946. This tsunami, which originated in the Aleutian Islands near Alaska, was still powerful enough when it hit Hawaii to rise 9 to 16 meters (30 to 55 feet). The S.S. *Brigham Victory*, from which this photograph was taken, managed to survive the onslaught, but 159 people in Hawaii, including the man seen here, were killed. (Photo courtesy of Water Resources Center Archives, University of California, Berkeley)

Can Earthquakes Be Predicted?

The vibrations that shook Northridge, California, in 1994 inflicted 57 deaths and about $40 billion in damage (Figure 15.21). This was from a brief earthquake (about 40 seconds) of moderate rating (M_W 6.7). Seismologists warn that earthquakes of comparable or greater strength will occur along the San Andreas Fault, which cuts a 1300-kilometer (800-mile) path through the state. The obvious question is: Can earthquakes be predicted?

Short-Range Predictions

The goal of short-range earthquake prediction is to provide a warning of the location and magnitude of a large earthquake within a narrow time frame. Substantial efforts to achieve this objective are being advanced in Japan, the United States, China, and Russia—countries where earthquake risks are high (Table 15.3). This research has concentrated on monitoring possible *precursors*—phenomena that precede, and thus provide a warning of, a forthcoming earthquake. In California, for example, seismologists are measuring uplift, subsidence, and strain in the rocks near active faults (Figure 15.22). Some Japanese scientists are studying peculiar anomalous behavior that may precede a quake.

One claim of a successful short-range prediction was made by Chinese seismologists after the February 4, 1975, earthquake in Liaoning Province. According to reports, very few people were killed, although more than 1 million lived near the epicenter, because the earthquake was predicted and the population was evacuated. Recently, some Western seismologists have questioned this claim and suggest instead that an intense swarm of foreshocks that began 24 hours before the main earthquake may have caused many people to evacuate spontaneously. Further, an official Chinese government report issued 10 years later stated that 1328 people died and 16,980 injuries resulted from this quake.

One year after the Liaoning earthquake, at least 240,000 people died in the Tangshan, China, earthquake, which was not predicted. The Chinese have also issued false alarms. In a province near Hong Kong, people reportedly left their dwellings for over a month, but no earthquake followed. Clearly, whatever method the Chinese employ for short-range predictions, it is *not* reliable.

For a short-range prediction scheme to warrant general acceptance, it must be both accurate and reliable. Thus, *it must have a small range of uncertainty as to location and timing, and it must produce few failures, or false alarms.* Can you imagine the debate that would precede an order to evacuate a large city in the United States, such as Los Angeles or San Francisco? The cost of evacuating millions of people, arranging for living accommodations, and providing for their lost work time and wages would be staggering.

Long-Range Forecasts

In contrast to short-range predictions, which aim to forecast earthquakes within a time frame of hours or, at most, days, long-range forecasts give the probability of a certain magnitude earthquake occurring on a time

Figure 15.20 Turnagain Heights slide caused by the 1964 Alaskan earthquake. **A.** Vibrations from the earthquake caused cracks to appear near the edge of the bluff. **B.** Within seconds blocks of land began to slide toward the sea on a weak layer of clay. In less than 5 minutes, as much as 200 meters of the Turnagain Heights bluff area had been destroyed. **C.** Photo of a small portion of the Turnagain Heights slide. (Photo courtesy of U.S. Geological Survey)

Figure 15.21   Damage to Interstate 5 during the January 17, 1994, Northridge earthquake. (Photo by John T. Barr/Liaison Agency, Inc.)

BOX 15.2

A Major Earthquake in Turkey

On August 17, 1999, at 3:02 a.m. local time, northwestern Turkey was shaken by a moment magnitude (M_W) 7.4 earthquake, catching most people in their sleep. The epicenter was 10 kilometers southeast of Ismit in a region that is the industrial center and most densely populated part of the country. Istanbul and its 13 million people is just 70 kilometers (43 miles) to the west.

According to official government estimates, the earthquake killed more than 17,000 people and injured nearly 44,000. More than 250,000 people were forced out of their damaged homes and were sheltered in 120 makeshift "tent cities." Estimates of property losses by the World Bank approached $7 billion. Liquefaction and ground shaking were the dominant causes of damage, but surface faulting and landslides were also responsible for substantial death and destruction. It was the most devastating earthquake to strike Turkey in 60 years.

Turkey is a geologically active region that frequently experiences large earthquakes. Most of the country is part of a small block of continental lithosphere known as the Turkish microplate. This small plate is caught between the northward-moving Arabian and African plates and the relatively stable Eurasian plate (Figure 15.B). The August 1999 earthquake occurred along the western end of the 1500-

Figure 15.B Earthquakes in Turkey are caused by the northward movement of the Arabian and African plates against the Eurasian plate, squeezing the small Turkish microplate westward. Movement is accommodated along two major strike-slip faults—the North Anatolian Fault and the East Anatolian Fault.

scale of 30 to 100 years or more. Stated another way, these forecasts give statistical estimates of the expected intensity of ground motion for a given area over a specified time frame. Although long-range forecasts may not be as informative as we would prefer, the data are important for updating the Uniform Building Code, which contains nationwide standards for designing earthquake-resistant structures.

Long-range predictions are based on the premise that earthquakes are repetitive or cyclical, like the weather. In other words, as soon as one earthquake is over, the continuing motions of Earth's plates begin to build strain in the rocks again until they fail once more. This has led seismologists to study historical records of earthquakes to see if there are any discernable patterns so that the probability of recurrence might be established (see Box 15.2).

One study conducted by the U.S. Geological Survey gives the probability of a rupture occurring along various segments of the San Andreas Fault for the 30 years between 1988 and 2018 (Figure 15.23). From

this investigation, the area around the Santa Cruz Mountains was given a 30 percent probability of producing a 6.5-magnitude earthquake during this time period. In fact, it produced the Loma Prieta quake in 1989, of 7.1 magnitude.

The region along the San Andreas Fault given the highest probability (90 percent) of generating a quake is the Parkfield section. This area has been called the "Old Faithful" of earthquake zones because activity here has been very regular since record-keeping began in 1857. (Although this section has had an earthquake on the average of once every 22 years, the last one occurred in 1966—it is overdue by more than 16 years and counting!) Another section between Parkfield and the Santa Cruz Mountains is given a very low probability of generating an earthquake. This area has experienced very little seismic activity in historical times; rather, it exhibits a slow, continual movement known as *fault creep*. Such movement is beneficial because it prevents strain from building to high levels in the rocks.

kilometer- (930-mile-) long North Anatolian fault system. This fault has much in common with California's San Andreas Fault. Both are right-lateral strike-slip faults having similar lengths and similar long-term rates of movement.* Also like its North American counterpart, the North Anatolian Fault is a transform plate boundary.

The fact that a large earthquake occurred along this portion of the North Anatolian Fault did not come as a complete surprise. Based on historical records, the region of the epicenter had been identified as a *seismic gap*, a "quiet zone" along the fault where strain had been building for perhaps 300 years. Moreover, during the preceding 60 years an interesting pattern of seismic activity had developed. Beginning in 1939 with a (M_w) 7.9 quake that produced about 350 kilometers of ground rupture, seven earthquakes had broken the fault progressively from east to west as shown in Figure 15.C

*If a person is looking across a right-lateral strike-slip fault during an earthquake, that person would see the opposite side move to the right.

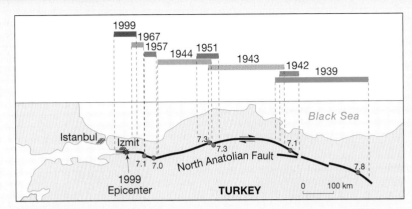

Figure 15.C This map depicts the sequential westward progression of large earthquakes along the North Anatolian Fault between 1939 and 1999. The epicenter and magnitude of each is noted. The length of each colored segment shows the extent of surface rupture along the fault for each event.

Researchers now understand that as each earthquake occurred, it loaded the zone to the west with additional stress. That is, as a quake released stress on the section of the fault it broke, it transferred stress to adjacent segments. The next segment in line to break is west of Ismit, near Istanbul. It could happen relatively soon. In the sequence since 1939, no earthquake waited longer than 22 years, and some came within a year of the one before.

The 1999 earthquake near Ismit, Turkey, demonstrated the awesome power of a large earthquake and the immense human suffering that can occur when an earthquake strikes an urban area. Although no one knows for sure where or when the next major quake will occur in the region, it appears that the 1999 earthquake near Ismit increased the risk for those living near Istanbul.

In summary, it appears that the best prospects for making useful earthquake predictions involve forecasting magnitudes and locations on time scales of years or perhaps even decades. These forecasts are important because they provide information used to develop the Uniform Building Code and assist in land-use planning.

Earthquakes and Earth's Interior

  Introduction
  ↳ Earth's Layered Structure

Earth's interior lies just below us; however, direct access is very limited. The deepest well yet drilled has penetrated Earth's crust only 12 kilometers (7.5 miles), less than 0.2 percent of the distance to the planet's center. Consequently, most knowledge of our planet's interior comes from the study of earthquake waves that

travel through Earth and vibrate the surface at some distant point.

If Earth were a perfectly homogeneous body, seismic waves would spread through it in a straight line at a constant speed. However, this is not the case for Earth. It so happens that the seismic waves reaching seismographs located farther from an earthquake travel at faster average speeds than those recorded at locations closer to the event. This general increase in speed with depth is a consequence of increased pressure, which enhances the elastic properties of deeply buried rock. As a result, the paths of seismic rays through Earth are refracted (bent) as they travel (Figure 15.24).

As more sensitive seismographs were developed, it became apparent that in addition to gradual changes in seismic-wave velocities, rather abrupt velocity changes also occur at particular depths. Because these changes were detected worldwide, seismologists concluded that Earth must be composed of distinct shells having varying compositions and/or mechanical properties (Figure 15.24).

Table 15.3 Some notable earthquakes.

Year	Location	Deaths (est.)	Magnitude[†]	Comments
1556	Shensi, China	830,000		Possibly the greatest natural disaster.
1755	Lisbon, Portugal	70,000		Tsunami damage extensive.
*1811–1812	New Madrid, Missouri	Few	7.9	Three major earthquakes.
*1886	Charleston, South Carolina	60		Greatest historical earthquake in the eastern United States.
*1906	San Francisco, California	1500	7.8	Fires caused extensive damage.
1908	Messina, Italy	120,000		
1923	Tokyo, Japan	143,000	7.9	Fire caused extensive destruction.
1960	Southern Chile	5700	9.6	Possibly the largest-magnitude earthquake ever recorded.
*1964	Alaska	131	9.2	Greatest North American earthquake.
1970	Peru	66,000	7.8	Great rockslide.
*1971	San Fernando, California	65	6.5	Damage exceeded $1 billion.
1975	Liaoning Province, China	1328	7.5	First major earthquake to be predicted.
1976	Tangshan, China	240,000	7.6	Not predicted.
1985	Mexico City	9500	8.1	Major damage occurred 400 km from epicenter.
1988	Armenia	25,000	6.9	Poor construction practices.
*1989	Loma Prieta	62	6.9	Damages exceeded $6 billion.
1990	Iran	50,000	7.3	Landslides and poor construction practices caused great damage.
1993	Latur, India	10,000	6.4	Located in stable continental interior.
*1994	Northridge, California	57	6.7	Damages in excess of $40 billion.
1995	Kobe, Japan	5472	6.9	Damage estimated to exceed $100 billion.
1999	Izmit, Turkey	17,127	7.4	Nearly 44,000 injured and more than 250,000 displaced.
1999	Chi-Chi, Taiwan	2300	7.6	Severe destruction; 8700 injuries.
2001	El Salvador	1000	7.6	Triggered many landslides.
2001	Bhuj, India	20,000[†]	7.9	1 million or more homeless.

*U.S. earthquakes.
[†]Widely differing magnitudes have been estimated for some of these earthquakes. When available, moment magnitudes are used.
SOURCE: U.S. Geological Survey.

Layers Defined by Composition

Compositional layering in Earth's interior likely resulted from density sorting that took place during an early period of partial melting. During this period the heavier elements, principally iron and nickel, sank as the lighter rocky components floated upward. This segregation of material is still occurring, but at a much reduced rate. Because of this chemical differentiation, Earth's interior is not homogeneous. Rather, it consists of three major regions that have markedly different chemical compositions (Figure 15.25).

The principal compositional layers of Earth include:

- the **crust,** Earth's comparatively thin outer skin that ranges in thickness from 3 kilometers (2 miles) at the oceanic ridges to over 70 kilometers (40 miles) in some mountain belts, such as the Andes and Himalayas;

- the **mantle,** a solid rocky (silica-rich) shell that extends to a depth of about 2900 kilometers (1800 miles);

- the **core,** an iron-rich sphere having a radius of 3486 kilometers (2161 miles).

We will consider the composition and structure of these major divisions of Earth's interior later in the chapter.

Figure 15.22   Lasers used to measure movement along the San Andreas Fault. (Photo by John K. Nakata/U.S. Geological Survey)

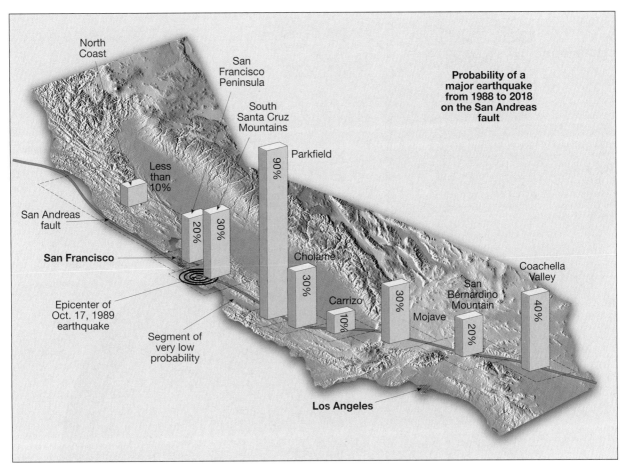

Figure 15.23   Probability of a major earthquake from 1988 to 2018 on the San Andreas Fault.

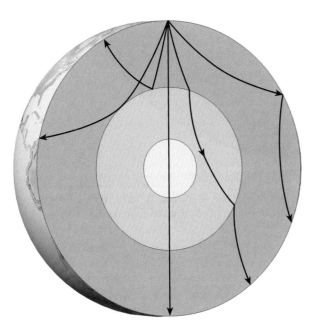

Figure 15.24   A few of the many possible paths that seismic rays take through Earth.

Layers Defined by Physical Properties

Earth's interior is characterized by a gradual increase in temperature, pressure, and density with depth. Estimates put the temperature at a depth of 100 kilometers at between 1200°C and 1400°C, whereas the temperature at Earth's center may exceed 6700°C. Clearly, Earth's interior has retained much of the energy acquired during its formative years, despite the fact that heat is continuously flowing toward the surface, where it is lost to space. The increase in pressure with depth causes a corresponding increase in rock density.

The gradual increase in temperature and pressure with depth affects the physical properties and hence the mechanical behavior of Earth materials. When a substance is heated, its chemical bonds weaken and its mechanical strength (resistance to deformation) is reduced. If the temperature exceeds the melting point of an Earth material, the material's chemical bonds break and melting ensues. If temperature were the only factor that determined whether a substance melted, our planet would be a molten ball covered with a thin, solid outer shell. However, pressure also increases with depth and tends

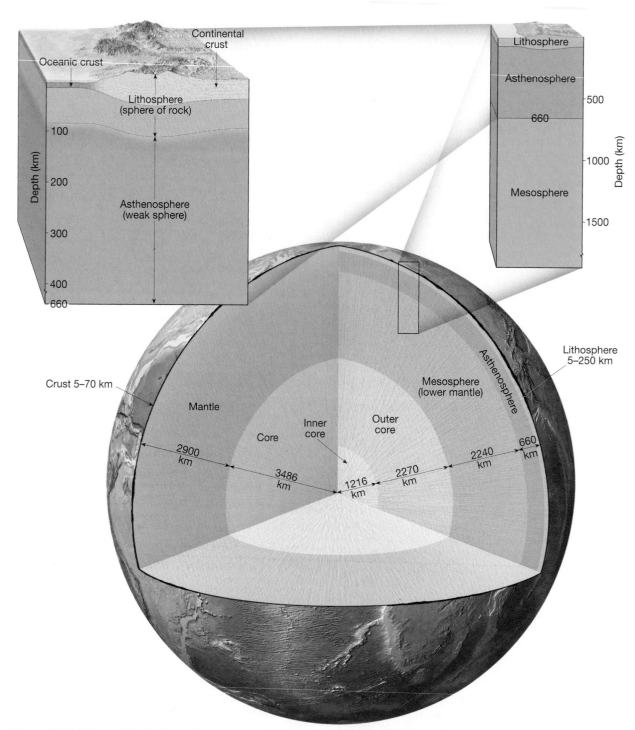

Figure 15.25 Views of Earth's layered structure. The left side of the large cross section shows that Earth's interior is divided into three different layers based on compositional differences—the crust, mantle, and core. The right side of the large cross section depicts the five main layers of Earth's interior based on physical properties and hence mechanical strength—the lithosphere, asthenosphere, mesosphere, outer core, and inner core. The block diagrams above the large cross section show an enlarged view of the upper portion of Earth's interior.

to increase rock strength. Furthermore, because melting is accompanied by an increase in volume, it occurs at higher temperatures at depth because of greater confining pressure. Thus, depending on the physical environment (temperature and pressure), a particular Earth material may behave like a brittle solid, deform in a puttylike manner, or even melt and become liquid.

Earth can be divided into five main layers based on their physical properties and hence mechanical strength—the *lithosphere, asthenosphere, mesosphere (lower mantle), outer core,* and *inner core.*

Lithosphere and Asthenosphere. Based on physical properties, Earth's outermost layer consists of the crust and uppermost mantle and forms a relatively cool, rigid shell. Although this layer is composed of materials with markedly different chemical compositions, it tends to act as a unit that exhibits rigid behavior—mainly because it is cool and thus strong. This layer, called the **lithosphere** (*sphere of rock*), averages about 100 kilometers in thickness but may be 250 kilometers thick or more below the older portions of the continents (Figure 15.25). Within the ocean basins, the lithosphere is only a few kilometers thick along the oceanic ridges but increases to perhaps 100 kilometers in regions of older and cooler oceanic crust.

Beneath the lithosphere in the upper mantle (to a depth of about 660 kilometers) lies a soft, comparatively weak layer known as the **asthenosphere** (*weak sphere*). The top portion of the asthenosphere has a temperature/pressure regime that results in a small amount of melting. Within this very weak zone, the lithosphere is mechanically detached from the layer below. The result is that the lithosphere is able to move independently of the asthenosphere, a topic we will consider in the next chapter.

It is important to emphasize that the strength of various Earth materials is a function of both their composition and of the temperature and pressure of their environment. You should not get the idea that the entire lithosphere behaves like a brittle solid similar to rocks found on the surface. Rather, the rocks of the lithosphere get progressively hotter and weaker (more easily deformed) with increasing depth. At the depth of the uppermost asthenosphere, the rocks are close enough to their melting temperatures (some melting may actually occur) that they are very easily deformed. Thus, the uppermost asthenosphere is weak because it is near its melting point, just as hot wax is weaker than cold wax.

Mesosphere or Lower Mantle.   Below the zone of weakness in the uppermost asthenosphere, increased pressure counteracts the effects of higher temperature and the rocks gradually strengthen with depth. Between the depths of 660 kilometers and 2900 kilometers a more rigid layer called the **mesosphere** (*middle sphere*) or **lower mantle** is found (Figure 15.25). Despite their strength, the rocks of the mesosphere are still very hot and capable of very gradual flow.

Inner and Outer Core.   The core, which is composed mostly of an iron-nickel alloy, is divided into two regions that exhibit very different mechanical strengths (Figure 15.25). The **outer core** is a *liquid layer* 2270 kilometers (1410 miles) thick. It is the convective flow of metallic iron within this zone that generates Earth's magnetic field. The **inner core** is a sphere having a radius of 1216 kilometers (754 miles). Despite its higher temperature, the material in the inner core is stronger (because of immense pressure) than the outer core and behaves like a *solid*.

Discovering Earth's Major Layers

The story of how seismologists discovered Earth's core and layers is interesting. In 1909, a pioneering Yugoslavian seismologist, Andrija Mohorovičić, presented the first convincing evidence for layering within Earth. By studying seismic records, he found that the velocity of seismic waves increases abruptly below about 50 kilometers of depth. This boundary separates the crust from the underlying mantle and is known as the **Mohorovičić discontinuity** in his honor. For reasons that are obvious, the name for this boundary was quickly shortened to **Moho.**

A few years later another boundary was discovered, by the German seismologist Beno Gutenberg. Generally, seismic waves from even small earthquakes are strong enough to travel around the world. This is why a seismograph in Antarctica can record earthquakes in California or Italy. But Gutenberg observed that P waves diminish and eventually die out about 105 degrees around the globe from an earthquake. Then, about 140 degrees away, the P waves reappear, but about 2 minutes later than would be expected, based on the distance traveled (Figure 15.26). This belt, where direct P waves are absent, is about 35 degrees wide and has been named the **shadow zone.**

Gutenberg realized that the shadow zone could be explained if Earth contained a core composed of material unlike the overlying mantle. The core must somehow hinder the transmission of P waves in a manner similar to the light rays being blocked by an opaque object that casts a shadow. However, rather than actually stopping the P waves, the shadow zone bends them, as shown in Figure 15.26. It was further learned that S waves could not travel through the core. Therefore, geologists concluded that at least a portion of this region is liquid.

In 1936, the last major subdivision of Earth's interior was predicted by Inge Lehmann, a Danish seismologist. Lehmann discovered a new region of seismic reflection within the core. Hence, Earth has a core within a core. The size of the inner core was not accurately calculated until the early 1960s, when underground nuclear tests were conducted in Nevada. Because the precise locations and times of the explosions were known, echoes from seismic waves that bounced off the inner core provided an accurate means of determining its size.

Over the past few decades, advances in seismology have allowed for much refinement of the gross view of Earth's interior presented so far. Of major importance was the discovery of the lithosphere and asthenosphere.

Discovering Earth's Composition

We have examined Earth's structure, so let us now look at the composition of each layer. Composition tells us much about how our planet has developed over its estimated age of 4.5 billion years.

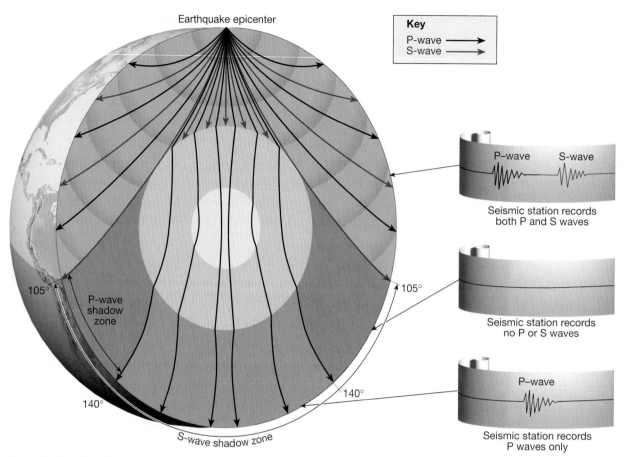

Figure 15.26 The abrupt change in physical properties at the mantle–core boundary causes the wave paths to bend sharply. This abrupt change in wave direction results in a shadow zone for P waves between about 105 and 140 degrees. Further, any location more than 105° from the earthquake epicenter will not receive direct S waves since the outer core will not transmit them.

Earth's crust varies in thickness, exceeding 70 kilometers (40 miles) in some mountainous regions and being thinner than 3 kilometers (2 miles) in some oceanic areas (see Figure 15.25). Early seismic data indicated that the continental crust, which is mostly made of lighter, granitic rocks, is quite different in composition from denser oceanic crust. Until recently, however, scientists had only seismic evidence from which to determine the composition of oceanic crust, because it lies beneath an average of 3 kilometers of water as well as hundreds of meters of sediment. With the development of deep-sea drilling technology, the recovery of ocean floor samples was made possible. The samples were of basaltic composition—very different from the rocks that make up the continents.

Our knowledge of the rocks of the mantle and core is much more speculative. However, we do have some clues. Recall that some of the lava that reaches Earth's surface originates in the partially melted asthenosphere, within the mantle. In the laboratory, experiments have shown that partial melting of a rock called *peridotite* results in a melt that has a basaltic composition similar to lava that emerges during volcanic activity of oceanic islands. Denser rocks like peridotite are thought to

make up the mantle and provide the lava for oceanic eruptions.

Surprisingly, meteorites or "shooting stars" that collide with Earth provide evidence of Earth's inner composition. Because meteorites are part of the solar system, they are assumed to be representative samples. Their composition ranges from metallic meteorites made of iron and nickel to stony meteorites composed of dense rock similar to peridotite.

Because Earth's crust contains a much smaller percentage of iron than do meteorites, geologists believe that the dense iron and other dense metals literally sank toward Earth's center during the planet's early history. By the same token, lighter substances may have floated to the surface, creating the less dense crust. Thus, Earth's core is thought to be mainly dense iron and nickel, similar to metallic meteorites, whereas the surrounding mantle is believed to be composed of rocks similar to stony meteorites.

The concept of a molten iron outer core is further supported by Earth's magnetic field. Our planet acts as a large magnet. The most widely accepted mechanism explaining why Earth has a magnetic field requires that the core be made of a material that conducts electricity, such as iron, and that it is mobile enough to allow circulation. Both of these conditions are met by the model of Earth's core that was established on the basis of seismic data.

Not only does an iron core explain Earth's magnetic field, but it also explains the high density of inner Earth—about 14 times that of water at Earth's center. Even under the extreme pressure at those depths, average crustal rocks with densities 2.8 times that of water would not have the density calculated for the core. But iron, which is three times more dense than crustal rocks, has the required density.

In summary, although earthquakes can be very destructive, much of our knowledge about Earth's interior comes from the study of these phenomena. As our knowledge of earthquakes and their causes improves, we learn more of our planet's internal workings and how to reduce the consequences of tremors. In the next chapter, you will see that most earthquakes originate at the boundaries of Earth's great lithospheric plates.

The Chapter in Review

- *Earthquakes* are vibrations of Earth produced by the rapid release of energy from rocks that rupture because they have been subjected to stresses beyond their limit. This energy, which takes the form of waves, radiates in all directions from the earthquake's source, called the *focus*. The movements that produce most earthquakes occur along large fractures, called *faults*, that are usually associated with plate boundaries.

- Along a fault, rocks store energy as they are bent. As slippage occurs at the weakest point (the focus), displacement will exert stress farther along a fault, where additional slippage will occur until most of the built-up strain is released. An earthquake occurs as the rock elastically returns to its original shape. The springing back of the rock is termed *elastic rebound*. Small earthquakes, called *foreshocks*, often precede a major earthquake. The adjustments that follow a major earthquake often generate smaller earthquakes called *aftershocks*.

- Two main types of *seismic waves* are generated during an earthquake: (1) *surface waves*, which travel along the outer layer of Earth; and (2) *body waves*, which travel through Earth's interior. Body waves are further divided into *primary*, or *P, waves*, which push (compress) and pull (expand) rocks in the direction the wave is traveling, and *secondary*, or *S, waves*, which shake the particles in rock at right angles to their direction of travel. The P waves can travel through solids, liquids, and gases. Fluids (gases and liquids) will not transmit S waves. In any solid material, P waves travel about 1.7 times faster than do S waves.

- The location on Earth's surface directly above the focus of an earthquake is the *epicenter*. An epicenter is determined using the difference in velocities of P and S waves. Using the difference in arrival times between P and S waves, the distance separating a recording station from the earthquake can be determined. When the distances are known from three or more seismic stations, the epicenter can be located using a method called *triangulation*.

- *A close correlation exists between earthquake epicenters and plate boundaries.* The principal earthquake epicenter zones are along the outer margin of the Pacific Ocean, known as the *circum-Pacific belt*, and through the world's oceans along the *oceanic ridge system*.

- Seismologists use two fundamentally different measures to describe the size of an earthquake—intensity and magnitude. *Intensity* is a measure of the degree of ground shaking at a given locale based on the amount of damage. The *Modified Mercalli Intensity Scale* uses damages to buildings in California as the basis for estimating the intensity of ground shaking for a local earthquake. *Magnitude* is calculated from seismic records and estimates the amount of energy released at the source of an earthquake. Using the *Richter scale*, the magnitude of an earthquake is estimated by measuring the *amplitude* (maximum displacement) of the largest seismic wave recorded. A logarithmic scale is used to express magnitude, in which a tenfold increase in ground shaking corresponds to an increase of 1 on the magnitude scale. *Moment magnitude* is currently used to estimate the size of moderate and large earthquakes. It is calculated using the average displacement of the fault, the area of the fault surface, and the sheer strength of the faulted rock.

- The most obvious factors determining the amount of destruction accompanying an earthquake are the magnitude of the earthquake and the proximity of the quake to a populated area. Structural damage attributable to earthquake vibrations depends on several factors, including (1) wave amplitudes, (2) the duration of the vibrations, (3) the nature of the material upon which the structure rests, and (4) the design of the structure. Secondary effects of earthquakes include *tsunamis*, landslides, ground subsidence, and fire.

- Substantial research to predict earthquakes is under way in Japan, the United States, China, and Russia—countries where earthquake risk is high. No reliable method of short-range prediction has yet been devised. Long-range forecasts are based on the premise that earthquakes are repetitive or cyclical. Seismologists study the history of earthquakes for patterns so their occurrences might be predicted.

- Earth's internal structure is divided into layers based on differences in chemical composition and on the basis of changes in physical properties. Compositionally, Eath is divided into a thin outer *crust,* a soild rocky *mantle,* and a dense *core.* Based on physical properties, the layers of Earth are (1) the *lithosphere*—the cool, rigid, outermost layer that averages about 100 kilometers thick, (2) the *asthenosphere,* a relatively weak layer located in the mantle beneath the lithosphere, (3) the more rigid *mesosphere,* where rocks are very hot and capable of very gradual flow,

- (4) the liquid *outer core,* where Earth's magentic field is generated, and (5) the solid *inner core.*

- The *continental crust* is primarily made of *granitic* rock, whereas the *oceanic crust* has a *basaltic* composition. Rocks similar to *peridotite* make up the *mantle.* The *core* is made up mainly of *iron* and *nickel.* An iron core explains the high density of Earth's interior as well as Earth's magnetic field.

Key Terms

aftershock (p. 304)
asthenosphere (p. 325)
body wave (p. 306)
core (p. 322)
crust (p. 322)
earthquake (p. 302)
elastic rebound (p. 304)
epicenter (p. 307)
fault (p. 302)
fault creep (p. 305)

focus (p. 302)
foreshock (p. 304)
inner core (p. 325)
intensity (p. 309)
liquefaction (p. 315)
lithosphere (p. 325)
lower mantle (p. 325)
magnitude (p. 309)
mantle (p. 322)

mesosphere (p. 355)
Modified Mercalli
   Intensity Scale (p. 309)
Mohorovičić discontinuity
   (Moho) (p. 325)
moment magnitude
   (p. 314)
outer core (p. 325)
primary (P) wave (p. 306)

Richter scale (p. 312)
secondary (S) wave (p. 306)
seismic sea wave
   (tsunami) (p. 315)
seismogram (p. 306)
seismograph (p. 306)
seismology (p. 306)
shadow zone (p. 325)
surface wave (p. 306)

Questions for Review

1. What is an earthquake? Under what circumstances do earthquakes occur?

2. How are faults, foci, and epicenters related?

3. Who was first to explain the actual mechanism by which earthquakes are generated?

4. Explain what is meant by elastic rebound.

5. Faults that are experiencing no active creep may be considered safe. Rebut or defend this statement.

6. Describe the principle of a seismograph.

7. Using Figure 15.9, determine the distance between an earthquake and a seismic station if the first S wave arrives three minutes after the first P wave.

8. List the major differences between P and S waves.

9. Which type of seismic wave causes the greatest destruction to buildings?

10. Most strong earthquakes occur in a zone on the globe known as the _____

11. What factor contributed most to the extensive damage that occurred in the central portion of Mexico City during the 1985 earthquake?

12. The 1988 Armenian earthquake had a Richter magnitude of 6.9, less than the 1994 Northridge California earthquake. Nevertheless, the loss of life was far greater in the Armenian event. Why?

13. An earthquake measuring 7 on the Richter scale releases about _____ times more energy than an earthquake with a magnitude of 6.

14. List three reasons why the moment magnitude scale has gained popularity among seismologists.

15. List four factors that affect the amount of destruction caused by seismic vibrations.

16. In addition to the destruction created directly by seismic vibrations, list three other types of destruction associated with earthquakes.

17. Distinguish between the Mercalli scale and the Richter scale.

18. What is a tsunami? How is one generated?

19. Cite some reasons why an earthquake with a moderate magnitude might cause more extensive damage than a quake with a high magnitude.

20. What evidence do we have that Earth's outer core is molten?

21. Contrast the physical makeup of the asthenosphere and the lithosphere.

22. Why are meteorites considered important clues to the composition of Earth's interior?

23. Describe the composition of the following:
a. continental crust
b. oceanic crust
c. mantle
d. core

Web Resources

The *Essentials of Geology* Website uses the resources and flexibility of the Internet to aid in your study of the topics in this chapter. Written and developed by geology instructors, this site will help improve your understanding of geology. Visit **www.prenhall.com/lutgens** and click on the cover of *Essentials of Geology, 8e* to find:

- On-line review quizzes.
- Web-based critical thinking and writing exercises.
- Links to chapter-specific Web resources.
- Internet-wide key term searches.

www.prenhall.com/lutgens

16 Plate Tectonics

Focus on Learning

To assist you in learning the important concepts in this chapter, you will find it helpful to focus on the following questions:

- What evidence was used to support the continental drift hypothesis?
- What was one of the main objections to the continental drift hypothesis?
- What is the theory of plate tectonics?
- In what major way does the plate tectonics theory depart from the continental drift hypothesis?
- What are the three types of plate boundaries?
- What is the evidence used to support the plate tectonics theory?
- What models have been proposed to explain the driving mechanism for plate motion?

Aerial view of the San Andreas Fault where it crosses the Carrizo Plain. (Photo by David Parker/Photo Researchers, Inc.)

Will California eventually slide into the ocean, as some predict? Have continents really drifted apart over the centuries? Answers to these questions and many others that have intrigued geologists for decades are now being provided by an exciting theory on large-scale movements taking place within Earth. This theory, called plate tectonics, represents the real frontier of geology, and its implications are so far-reaching that it can be considered the framework from which most other geological processes should be viewed.

Early in the twentieth century, most geologists believed that the geographic positions of the ocean basins and continents were fixed. During the last few decades, however, vast amounts of new data have dramatically changed our understanding of the nature and workings of our planet. Earth scientists now realize that the continents gradually migrate across the globe. Where landmasses split apart, new ocean basins are created between the diverging blocks. Meanwhile, older portions of the seafloor are carried back into the mantle in regions where trenches occur in the deep-ocean floor. Because of these movements, blocks of continental material eventually collide and form Earth's great moun-

tain ranges (Figure 16.1). In short, a revolutionary new model of Earth's tectonic* processes has emerged.

This profound reversal of scientific understanding has been appropriately described as a scientific revolution. Like other scientific revolutions, considerable time elapsed between the idea's inception and its general acceptance. The revolution began early in the twentieth century as a relatively straightforward proposal that the continents drift about the face of Earth. After many years of heated debate, the idea of drifting continents was rejected by the vast majority of geologists. However, during the 1950s and 1960s, new evidence rekindled interest in this proposal. By 1968, these new developments led to the unfolding of a far more encompassing theory than continental drift—a theory known as *plate tectonics*.

---

*Tectonics refers to the deformation of Earth's crust and results in the formation of structural features such as mountains.

Figure 16.1   Converging lithospheric plates were responsible for creating the Sierra Nevada. Shown is Mount Tom, California. (Photo by Dennis Flaherty/Photo Researchers, Inc.)

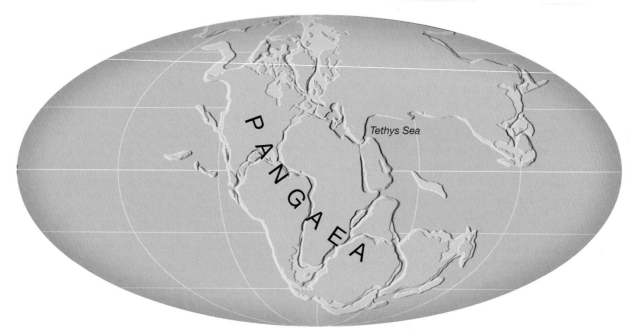

Figure 16.2   Reconstruction of Pangaea as it is thought to have appeared 200 million years ago. (After R. S. Deitz and J. C. Holden. *Journal of Geophysical Research 75*:4943. Copyright by American Geophysical Union)

Continental Drift: An Idea Before Its Time

The idea that continents, particularly South America and Africa, fit together like pieces of a jigsaw puzzle originated with improved world maps. However, little significance was given to this idea until 1915, when Alfred Wegener, a German meteorologist and geophysicist, published *The Origin of Continents and Oceans*. In this book, Wegener set forth his radical hypothesis of **continental drift.***

Wegener suggested that a supercontinent he called **Pangaea** (meaning "all land") once existed (Figure 16.2). He further hypothesized that about 200 million years ago this supercontinent began breaking into smaller continents, which then drifted to their present positions (see Figure 16.A).

Wegener and others collected substantial evidence to support these claims. The fit of South America and Africa and the geographic distribution of fossils, rock structures, and ancient climates all seemed to support the idea that these now separate landmasses were once joined. Let us examine their evidence.

*Wegener's ideas were actually preceded by those of an American geologist, E. B. Taylor, who in 1910 published a paper on continental drift. Taylor's paper provided little supporting evidence for continental drift, which may have been the reason that it had a relatively small impact on the scientific community.

Evidence: The Continental Jigsaw Puzzle

Like a few others before him, Wegener first suspected that the continents might have been joined when he noticed the remarkable similarity between the coastlines on opposite sides of the South Atlantic. However, his use of present-day shorelines to make a fit of the continents was challenged immediately by other Earth scientists. These opponents correctly argued that shorelines are continually modified by erosional processes, and even if continental displacement had taken place, a good fit today would be unlikely. Wegener appeared to be aware of this problem, and, in fact, his original jigsaw of the continents was only very crude.

A much better approximation of the true outer boundary of the continents is the continental shelf. Today, the seaward edge of the continental shelf lies submerged, several hundred meters below sea level. In the early 1960s, scientists produced a map that attempted to fit the edges of the continental shelves at a depth of 900 meters. The remarkable fit that was obtained is shown in Figure 16.3. Although the continents overlap in a few places, these are regions where streams have deposited large quantities of sediment, thus enlarging the continental shelves. The overall fit was even better than the supporters of continental drift suspected it would be.

Evidence: Fossils Match Across the Seas

Although Wegener was intrigued by the jigsaw fit of the continental margins that lie on opposite sides of the

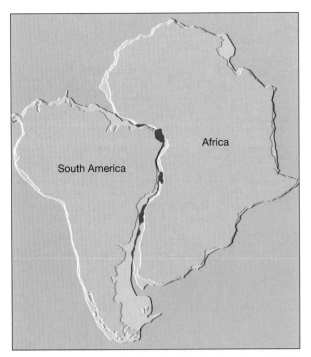

Figure 16.3 This shows the best fit of South America and Africa along the continental slope at a depth of 500 fathoms (about 900 meters). The areas where continental blocks overlap appear in brown.

Atlantic, he at first thought the idea of a mobile Earth improbable. Not until he came across an article citing fossil evidence for the existence of a land bridge connecting South America and Africa did he begin to take his own idea seriously. Through a search of the literature, Wegener learned that most paleontologists were in agreement that some type of land connection was needed to explain the existence of identical fossils on the widely separated landmasses.

Mesosaurus. To add credibility to his argument for the existence of the supercontinent of Pangaea, Wegener cited documented cases of several fossil organisms that had been found on different landmasses but could not have crossed the vast oceans presently separating the continents. The classic example is *Mesosaurus*, a presumably aquatic, snaggle-toothed reptile whose fossil remains are limited to eastern South America and southern Africa (Figure 16.4). If *Mesosaurus* had been able to swim well enough to cross the vast South Atlantic Ocean, its remains would be more widely distributed. As this is not the case, Wegener argued that South America and Africa must have been joined somehow.

How did scientists explain the discovery of identical fossil organisms separated by thousands of kilometers of open ocean? The idea of land bridges was the most widely accepted solution to the problem of migration (Figure 16.5). We know, for example, that during the most recent glacial period, the lowering of sea level allowed animals to cross the narrow Bering Strait between Asia and North America. Was it possible, then, that one or more land bridges once connected Africa and South America? We are now quite certain that land bridges of this magnitude did not exist, for their remnants should still lie below sea level. But they are nowhere to be found.

Present-Day Organisms. In his book, Wegener also cited the distribution of present-day organisms as evidence to support the concept of drifting continents. For example, modern organisms with similar ancestries clearly had to evolve in isolation during the last few tens of millions of years. Most obvious of these are the Australian marsupials, which have a direct fossil link to the marsupial opossums found in the Americas.

Evidence: Rock Types and Structures Match

Anyone who has worked a picture puzzle knows that in addition to the pieces fitting together, the picture must

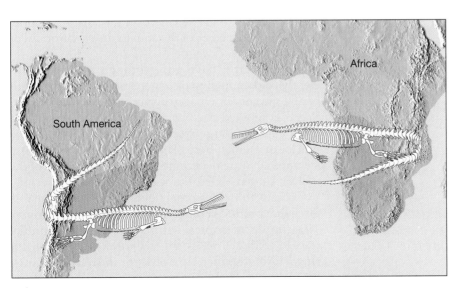

Figure 16.4 Fossils of *Mesosaurus* have been found on both sides of the South Atlantic and nowhere else in the world. Fossil remains of this and other organisms on the continents of Africa and South America appear to link these landmasses during the late Paleozoic and early Mesozoic eras.

Figure 16.5 These sketches by John Holden illustrate various explanations for the occurrence of similar species on landmasses that are presently separated by vast oceans. (Reprinted with permission of John Holden)

RAFTING

ISTHMIAN LINKS

ISLAND STEPPING STONES

CONTINENTAL DRIFT

be continuous as well. The "picture" that must match in the "Continental Drift Puzzle" is one of rock types and mountain belts. If the continents were once together, as the supercontinent of Pangaea, the rocks found in a particular region on one continent should closely match in age and type those in adjacent positions on the adjoining continent.

Such evidence exists in the form of several mountain belts that terminate at one coastline, only to reappear on a landmass across the ocean. For instance, the mountain belt that includes the Appalachians trends northeastward through the eastern United States and disappears off the coast of Newfoundland (Figure 16.6A). Mountains of comparable age and structure are found in the British Isles and Scandinavia. When these landmasses are reassembled, as in Figure 16.6B, the mountain chains form a nearly continuous belt. Numerous other rock structures exist that appear to have formed at the same time and were subsequently split apart.

Wegener was very satisfied that the similarities in rock structure on both sides of the Atlantic linked these landmasses. In his own words, "It is just as if we were to refit the torn pieces of a newspaper by matching their edges and then check whether the lines of print run smoothly across. If they do, there is nothing left but to conclude that the pieces were in fact joined in this way."*

*Alfred Wegener, *The Origin of Continents and Oceans.* Translated from the 4th revised German edition of 1929 by J. Birman (London: Methuen, 1966).

Evidence: Ancient Climates

Because Alfred Wegener was a meteorologist by training, he was keenly interested in obtaining paleoclimatic (ancient climatic) data in support of continental drift. His efforts were rewarded when he found evidence for dramatic global climatic changes. For instance, glacial deposits indicate that near the end of the Paleozoic era (between 220 million and 300 million years ago), ice sheets covered extensive areas of the Southern Hemisphere. Layers of glacial till were found in southern Africa and South America, as well as in India and Australia. Below these beds of glacial debris lay striated and grooved bedrock. In some locations, the striations and grooves indicated that the ice had moved from what is now the sea onto land. Much of the land area containing evidence of this late Paleozoic glaciation presently lies within 30 degrees of the equator in a subtropical or tropical climate.

Could Earth have gone through a period sufficiently cold enough to have generated extensive continental glaciers in what is presently a tropical region? Wegener rejected this explanation because, during the late Paleozoic, large tropical swamps existed in the Northern Hemisphere. The lush vegetation of these swamps eventually became the major coal fields of the eastern United States, Europe, and Siberia.

Fossils from these coal fields indicate that the tree ferns that produced the coal deposits had large fronds. This implies a tropical setting. Furthermore, unlike trees in colder climates, the tree trunks lacked growth rings. Growth rings do not form in tropical plants,

Figure 16.6 Matching mountain ranges across the North Atlantic. **A.** The Appalachian Mountains trend along the eastern flank of North America and disappear off the coast of Newfoundland. Mountains of comparable age and structure are found in the British Isles and Scandinavia. **B.** When these landmasses are placed in their predrift locations, these ancient mountain chains form a nearly continuous belt. These folded mountain belts formed roughly 300 million years ago as the landmasses collided during the formation of the supercontinent of Pangaea.

because there are minimal seasonal fluctuations in temperature.

Wegener believed that a better explanation for the paleoclimatic regimes he observed is provided by fitting together the landmasses as a supercontinent, with South Africa centered over the South Pole (Figure 16.7). This would account for the conditions necessary to generate extensive expanses of glacial ice over much of the Southern Hemisphere. At the same time, this geography would place the northern landmasses nearer the tropics and account for their vast coal deposits.

Wegener was so convinced that his explanation was correct that he wrote, "This evidence is so compelling that by comparison all other criteria must take a back seat."

How does a glacier develop in hot, arid Australia? How do land animals migrate across wide expanses of

open water? As compelling as this evidence may have been, 50 years passed before most of the scientific community would accept it as well as the logical conclusions to which it led.

The Great Debate

Wegener's proposal did not attract much open criticism until 1924 when his book was translated into English. From this time on, until his death in 1930, his drift hypothesis encountered a great deal of hostile criticism. To quote the respected American geologist T. C. Chamberlin, Wegener's hypothesis "... takes considerable liberty with our globe, and is less bound by restrictions or tied down by awkward, ugly facts than most of its rival theories. Its appeal seems to lie in the fact that it plays a game in which there are few restrictive rules and no sharply drawn code of conduct."

One of the main objections to Wegener's hypothesis stemmed from his inability to provide a mechanism that was capable of moving the continents across the globe. Wegener proposed that the tidal influence of the Moon was strong enough to give the continents a westward motion. However, the prominent physicist Harold Jeffreys quickly countered with the argument that tidal friction of the magnitude needed to displace the continents would bring Earth's rotation to a halt in a matter of a few years.

Wegener also proposed that the larger and sturdier continents broke through the oceanic crust, much like ice breakers cut through ice. However, no evidence existed to suggest that the ocean floor was weak enough to permit passage of the continents without themselves being appreciably deformed in the process.

Although most of Wegener's contemporaries opposed his views, even to the point of open ridicule, a few considered his ideas plausible. For these few geologists who continued the search for additional evidence, the exciting concept of continents in motion held their interest. Others viewed continental drift as a solution to previously unexplainable observations.

Plate Tectonics: The New Paradigm

Internal Processes
↳ Plate Tectonics

During the years that followed Wegener's proposal, major strides in technology permitted mapping of the ocean floor. Moreover, extensive data on seismic activity and Earth's magnetic field became available. By 1968, these developments led to the unfolding of a far more

Figure 16.7 **A.** The supercontinent Pangaea showing the area covered by glacial ice 300 million years ago. **B.** The continents as they are today. The white areas outline where evidence of the old ice sheets exists.

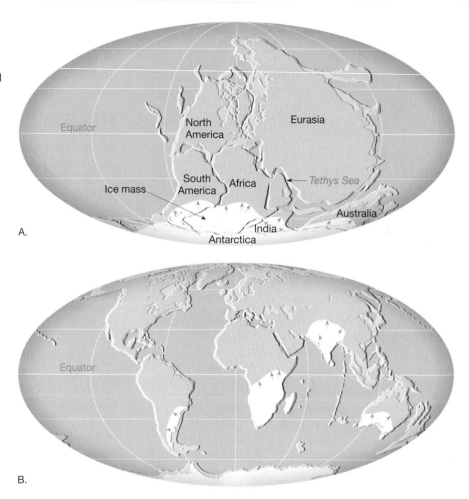

encompassing theory than continental drift, known as **plate tectonics.** The implications of plate tectonics are so far-reaching that this theory is today the framework within which to view most geologic processes.

Earth's Major Plates

According to the plate tectonics model, the uppermost mantle, along with the overlying crust, behaves as a strong, rigid layer, known as the **lithosphere.** This outermost shell overlies a weaker region in the mantle known as the **asthenosphere.** Further, the lithosphere is broken into numerous segments, called **plates,** that are in motion and are continually changing in shape and size. As shown in Figure 16.8, seven major lithospheric plates are recognized. They are the North American, South American, Pacific, African, Eurasian, Australian, and Antarctic plates. The largest is the Pacific plate, which is located mostly within the Pacific ocean. Notice from Figure 16.8 that several of the large plates include an entire continent plus a large area of seafloor (for example, the South American plate). This is a major departure from Wegener's continental drift hypothesis, which proposed that the continents moved through the ocean floor, not with it. Note also that

none of the plates are defined entirely by the margins of a continent.

Intermediate-size plates include the Caribbean, Nazca, Philippine, Arabian, Cocos, and Scotia plates. In addition, there are over a dozen smaller plates that have been identified but are not shown in Figure 16.8.

The lithospheric plates move relative to each other at a very slow but continuous rate that averages about 5 centimeters (2 inches) per year. This movement is ultimately driven by the unequal distribution of heat within Earth. Hot material found deep in the mantle moves slowly upward and serves as one part of our planet's internal convection system. Concurrently, cooler, denser slabs of oceanic lithosphere descend into the mantle, setting Earth's rigid outer shell into motion. Ultimately, the titanic, grinding movements of Earth's lithospheric plates generate earthquakes, create volcanoes, and deform large masses of rock into mountains.

Plate Boundaries

Lithospheric plates move as coherent units relative to all other plates. Although the interiors of plates may experience some deformation, all major interactions among individual plates (and therefore most deformation) occur along their *boundaries*. In fact, the first attempts to outline plate boundaries were made using locations of earthquakes. Later work showed that plates are bounded by three distinct types of boundaries, which are differentiated by the type of movement they exhibit. These boundaries are depicted at the bottom of Figure 16.8 and are briefly described here:

1. **Divergent plate boundaries** (*constructive margins*)—where two plates move apart, resulting in upwelling of material from the mantle to create new seafloor (Figure 16.8A).
2. **Convergent plate boundaries** (*destructive margins*)—where two plates move together, resulting in oceanic lithosphere being thrust beneath an overriding plate, eventually to be reabsorbed into the mantle (Figure 16.8B). Convergence can also result in the collision of two continental plates to create a mountain system.
3. **Transform fault boundaries** (*conservative margins*)—where two plates grind past each other without the production or destruction of lithosphere (Figure 16.8C).

Each plate is bounded by a combination of these three types of boundaries. For example, as shown in Figure 16.8, the Nazca plate has a divergent zone on the west, a convergent boundary on the east, and numerous transform faults, which offset segments of the divergent boundary. Although the total surface area of Earth does not change, individual plates may diminish or grow in area depending on the distribution of convergent and divergent boundaries. The Antarctic and African plates are almost entirely bounded by spreading centers and hence are growing larger. By contrast, the Pacific plate is being consumed into the mantle along its northern and western flanks and is therefore diminishing in size.

Furthermore, new plate boundaries can be created in response to changes in the forces acting on these rigid slabs. For example, a relatively new divergent boundary is located in Africa, in a region known as the East African Rift Valleys. If spreading continues there, the African plate will split into two plates separated by a new ocean basin. At other locations, plates carrying continental crust are presently moving toward each other. Eventually, these continents may collide and be sutured together. Thus, the boundary that once separated two plates disappears as the plates become one. The result of such a continental collision is a majestic mountain range such as the Himalayas.

In the following sections we will briefly summarize the nature of the three types of plate boundaries.

Divergent Plate Boundaries

Internal Processes
↳ Plate Tectonics

Most *divergent plate boundaries* are located along the crests of oceanic ridges and can be thought of as *constructive plate margins* since this is where new oceanic lithosphere is generated (Figure 16.9). Here, as the plates move away from the ridge axis, the fractures created are filled with molten rock that wells up from the hot mantle below. Gradually, this magma cools to produce new slivers of seafloor. In a continuous manner, spreading and upwelling of magma adds oceanic lithosphere between the diverging plates. As we shall see later, divergent boundaries can also form on the continents.

Oceanic Ridges and Seafloor Spreading

Along well-developed divergent plate boundaries, the seafloor is elevated forming the **oceanic ridge.** The interconnected oceanic ridge system is the longest topographic feature on Earth's surface, exceeding 70,000 kilometers (43,000 miles) in length with crests commonly 2 to 3 kilometers higher than the adjacent ocean basins. Representing 20 percent of Earth's surface, the oceanic ridge system winds through all major ocean basins in a manner similar to the seam on a baseball. The term "ridge" may be misleading as these features are not narrow but have widths from 1000 to 4000 kilometers. Further, along the axis of some segments are deep down-faulted structures called **rift valleys.**

The mechanism that operates along the oceanic ridge system to create new seafloor is appropriately called **seafloor spreading.** Typical rates of spreading average around 5 centimeters (2 inches) per year. Comparatively slow spreading rates of 2 centimeters per year are found in the North Atlantic, whereas spreading rates exceeding 15 centimeters (6 inches) have been measured along sections of the East Pacific Rise. Although these rates of lithospheric production are slow on a human time scale, they are nevertheless rapid enough so that all of Earth's ocean basins could have been generated within the last 200 million years. In fact, none of the ocean floor that has been dated exceeds 180 million years in age.

During seafloor spreading, the magma that is injected into newly developed fractures forms dikes that tend to cool from their outer borders inward toward their centers. Because the warm interiors of these newly formed dikes are weak, continued spreading produces new fractures that split these young rocks roughly in

Figure 16.8 A mosaic of rigid plates constitutes Earth's outer shell. (After W. B. Hamilton, U.S. Geological Survey)

North American
plate

Canadian Shield

Iceland

Eurasian plate

Alps

Juan de Fuca
plate

Rocky Mountains

Basin
and
Range

Appalachian Mts.

San Andreas
Fault

Caribbean
plate

African plate

Mid-Atlantic Ridge

Pacific plate

Cocos
plate

Galapagos
Ridge

Antilles
Arc

East Pacific Rise

South American
plate

Nazca
plate

Andes Mountains

Chile Ridge

Scotia plate

Antarctic plate

A. Divergent boundary

B. Convergent boundary

C. Transform fault boundary

Figure 16.9 Most divergent plate boundaries are situated along the crests of oceanic ridges.

half. As a result, new material is added about equally to the two diverging plates. Consequently, new ocean floor grows symmetrically on each side of a centrally located ridge crest. Indeed, the ridge systems of the Atlantic and Indian oceans are located near the middle of these water bodies and as a consequence are called mid-ocean ridges. However, the East Pacific Rise is situated far from the center of the Pacific Ocean. Despite uniform spreading along the East Pacific Rise, much of the Pacific Basin that once lay east of this spreading center has been overridden by the westward migration of the American plates.

Spreading Rates and Ridge Topography

When various segments of the oceanic ridge system were studied in detail, some topographic differences came to light. Most of these differences appear to be controlled by spreading rates. At comparatively slow spreading rates of 1 to 5 centimeters per year, such as occur at the Mid-Atlantic and Mid-Indian ridges, a prominent rift valley develops along the ridge crest. This structure is usually 30 to 50 kilometers across and between 1500 to 3000 meters deep. Here, the displacement of large slabs of oceanic crust along nearly vertical faults and outpourings of pillow lavas contribute to the characteristically rugged topography of these rift valleys.

Along the Galapagos ridge and the northern-most section of the East Pacific Rise, an intermediate spreading rate of 5 to 9 centimeters per year is the norm. In these settings the rift valleys that develop are shallow, often less than 200 meters deep, and their topography is relatively smooth.

At faster spreading rates (greater than 9 centimeters per year), such as those which occur along much of the East Pacific Rise, no median rift valleys develop. Here oceanic ridges are usually narrow (roughly 10 kilometers wide) topographic highs. These elevated structures are extensively faulted and exhibit topography consisting of numerous horsts and grabens (see Chapter 17). Faulting is the primary cause of topographic relief along fast spreading centers, whereas the buildup of volcanic structures is significant at slow spreading centers.

Continental Rifts

Spreading centers can also develop within a continent, in which case the landmass may split into two or more smaller segments, as Alfred Wegener had proposed for the breakup of the Pangaea. Examples of active continental rifts include the East African rift valleys, the Baikal Rift (south central Siberia), the Rhine Valley (Northwest Europe), the Rio Grande Rift, and the Basin and Range Province in the western United States. Whether any of these rifts will continue to develop and eventually split a continent is a matter of much speculation.

The most widely accepted model for continental splitting suggests that extensional forces must be acting on the lithospheric plate. These forces are thought to arise from the "pull" of cold lithoshperic plates as they subduct along the margins of a continent. It appears that by themselves these extensional forces are not great enough to actually tear the lithosphere apart. Rather, the rupture of the lithosphere is initiated only in those settings where plumes of hot rock rise from the mantle. The effect of this hot-spot activity is weakening of the lithosphere and doming of the crust directly above the hot rising plume. Uplifting stretches and thins the crust, as shown in Figure 16.10A. Extension is accompanied by alternating episodes of faulting and volcanism that result in the development of a rift valley that resem-

Figure 16.10   **A.** Rising magma forces the crust upward, causing numerous cracks in the rigid lithosphere. **B.** As the crust is pulled apart, large slabs of rock sink, generating a rift zone. **C.** Further spreading generates a narrow sea. **D.** Eventually, an expansive ocean basin and ridge system are created.

Figure 16.11 East African rift valleys and associated features.

bles those found along the axes of some ridges (Figure 16.10B).

The East African rift valleys may represent the initial stage in the breakup of a continent as described above (Figure 16.11). Large volcanic mountains such as Kilimanjaro and Mount Kenya exemplify the extensive volcanic activity that accompanies continental rifting.

In those settings where the extensional forces are maintained, the rift valley will lengthen and deepen, eventually extending out to the margin of the plate and

BOX 16.1 The Breakup of a Supercontinent

A great deal of evidence has been gathered to support the fact that Alfred Wegener's supercontinent of Pangaea began to break apart about 200 million years ago as he had originally suggested (Figure 16.A, part A). An important consequence of this continental rifting was the creation of a "new" ocean basin: the Atlantic. As you can see in Figure 16.A, part B, the rifting that eventually resulted in the Atlantic basin occurred over an extended period of time. The first rift developed between North America and Africa and generated great volumes of Jurassic-age basalts. These basalts are present as weathered flows and sills along the Eastern Seaboard of the United States but primarily buried beneath rocks that form the continental shelf. Radiometric dating of this material indicates that rifting began in various stages between 200 million and 165 million years ago. This time span can be used as the birth date for this section of the North Atlantic.

Rifting that formed the South Atlantic began about 130 million years ago near the tips of what are now Africa and South America. As this rift propagated northward, it gradually opened the South Atlantic (compare Figure 16.A, parts B and C). Continued rifting of the southern landmass of Gondwanaland sent India on a northward journey. By the early Cenozoic, about 50 million years ago, Australia began to separate from Antarctica, and the South Atlantic had emerged as a full-fledged ocean (Figure 16.A, part D).

A modern map (Figure 16.A, part E) shows that India eventually collided with Asia, an event that began about 45 million years ago and created the Himalayas as well as the Tibetan Highlands. About the same time, the separation of Greenland from Eurasia completed the breakup of the northern landmass (Laurasia). Also, notice the recent formation of the Baja Peninsula along with the Gulf of California. This event began less than 10 million years ago.

split it in two (Figure 16.10C). At this point the rift becomes a narrow linear sea with an outlet to the ocean, similar to the Red Sea (Figure 16.11). The Red Sea formed when the Arabian Peninsula rifted from Africa, an event that began about 20 million years ago. Consequently, the Red Sea provides oceanographers with a view of how the Atlantic Ocean may have looked in its infancy (see Box 16.1).

Not all rift valleys develop into full-fledged spreading centers. Running through the central United States is a failed rift extending from Lake Superior to Oklahoma. This once active rift valley is filled with volcanic rock that was extruded onto the crust more than a billion years ago. Why one rift valley develops into an active spreading center while others are abandoned is not yet known.

A. 200 Million Years Ago

B. 150 Million Years Ago
(Jurassic Period)

C. 100 Million Years Ago
(Cretaceous Period)

D. 50 Million Years Ago
(Early Cenozoic)

E. Present

Figure 16.A   Several views of the breakup of Pangaea over a period of 200 million years.

Convergent Plate Boundaries

Internal Processes
↳ Plate Tectonics

Although new lithosphere is constantly being added at the oceanic ridges, our planet is not growing larger—its total surface area remains constant. To accommodate the newly created lithosphere, older portions of oceanic plates return to the mantle along *convergent plate boundaries*. (Because lithosphere is "destroyed" at convergent boundaries, they are also called *destructive plate margins*.) As two plates slowly converge, the leading edge of one is bent downward, allowing it to slide beneath the other. The surface expression produced by the descending plate is an ocean **trench,** like the Peru–Chile trench (Figure 16.12A). Trenches formed in this manner may be thousands of kilometers long, 8 to 12 kilometers deep, and between 50 and 100 kilometers wide (Figure 16.13). Destructive plate margins where oceanic crust is being consumed in the mantle are called **subduction zones.**

The average angle at which oceanic lithosphere descends into the mantle is about 45 degrees. However, depending on its buoyancy, a plate may descend at an angle as small as a few degrees or it may plunge vertically (90 degrees) into the mantle. When a spreading center is located near a subduction zone, the lithosphere is young and therefore warm and buoyant. Consequently, the angle of descent is small. This is the situation along parts of the Peru–Chile trench. Low dip angles are usually associated with a strong coupling between the descending slab and the overriding plate. As a result, these regions experience great earthquakes. By contrast, some subduction zones, such as the Mariana trench, have steep angles of descent and few strong earthquakes.

Although all convergent zones have the same basic characteristics, they are highly variable features. Each is controlled by the type of crustal material involved and the tectonic setting. Convergent boundaries can form between two oceanic plates, one oceanic and one continental plate, or two continental plates. All three situations are illustrated in Figure 16.12.

Oceanic-Continental Convergence

Whenever the leading edge of a continental plate converges with an oceanic plate, the buoyant continental plate remains "floating," while the denser oceanic slab sinks into the asthenosphere (Figure 16.12A). When a descending plate reaches a depth of about 100 to 150 kilometers, heat drives water and other volatile components from the subducted sediments into the overlying mantle. These substances act as a flux does at a foundry, inducing partial melting of mantle rocks at re-

duced temperatures. The partial melting of mantle rock generates magmas having a basaltic or occasionally an andesitic composition. The newly formed magma, being less dense than the rocks of the mantle, will buoyantly rise. Often the magma, being more dense than continental rocks, will pond beneath the overlying continental crust, where it may melt some silica-enriched rocks. Eventually some of this silica-rich magma may migrate to the surface, where it can give rise to volcanic eruptions, some of which are explosive.

The volcanoes of the Andean arc located along the western flank of South America are the product of magma generated as the Nazca plate descends beneath the continent (see Figure 16.8). In the central section of the Southern Andes, the subduction angle is very shallow, which probably accounts for the lack of volcanism in that area. As the South American plate moves westward, it overruns the Nazca plate. The result is a seaward migration of the Peru–Chile trench and a reduction in the size of the Nazca plate.

Mountains such as the Andes, which are produced in part by volcanic activity associated with the subduction of oceanic lithosphere, are called **continental volcanic arcs.** Another active continental volcanic arc is located in the western United States. The Cascade Range of Washington, Oregon, and California consists of several well-known volcanic mountains, including Mount Rainier, Mount Shasta, and Mount St. Helens. As the continuing activity of Mount St. Helens testifies, the Cascade Range is still active. The magma here arises from melting triggered by the subduction of the Juan de Fuca plate.

A remnant of a formerly extensive continental volcanic arc is California's Sierra Nevada, in which Yosemite National Park is located. The Sierra Nevada is much older than the Cascade Range and has been inactive for several million years, as evidenced by the absence of volcanic cones. Here erosion has stripped away most of the obvious traces of volcanic activity and left exposed the large, crystallized magma chambers that once fed lofty volcanoes.

Oceanic-Oceanic Convergence

When two oceanic slabs converge, one descends beneath the other, initiating volcanic activity in a manner similar to that which occurs at an oceanic-continental boundary. In this case, however, the volcanoes form on the ocean floor rather than on a continent (Figure 16.12B). If this activity is sustained, it will eventually build a chain of volcanic structures that emerge as islands. The volcanic islands are spaced about 80 kilometers apart and are built upon submerged ridges of volcanic material a few hundred kilometers wide. This newly formed land consisting of an arc-shaped chain of small volcanic islands is called a **volcanic island arc.**

Continental
volcanic arc

Figure 16.12 Zones of plate convergence. **A.** Oceanic-continental **B.** Oceanic-oceanic **C.** Continental-continental

A.

Volcanic island arc

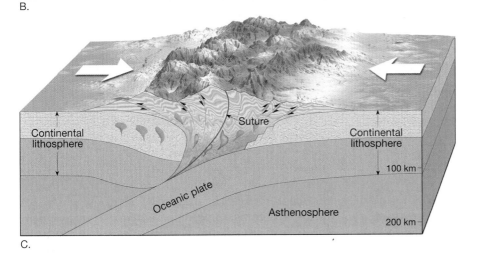

B.

C.

The Aleutian, Mariana, and Tonga islands are examples of volcanic island arcs. Island arcs such as these are generally located 200 to 300 kilometers from a trench axis. Located adjacent to the island arcs just mentioned are the Aleutian trench, Mariana trench, and the Tonga trench (Figure 16.13).

Only two volcanic island arcs are located in the Atlantic—the Lesser Antilles arc adjacent to the Caribbean Sea, and the Sandwich Islands in the South Atlantic. The Lesser Antilles are a product of the subduction of the Atlantic plate beneath the Caribbean plate. Located within this arc is the island of Martinique, where Mount Pelée erupted in 1902, destroying the town of St. Pierre and killing an estimated 28,000 people; and the island of Montserrat, where volcanic activity has occurred very recently.

Relatively young island arcs are fairly simple structures that are underlain by crust that is generally less

Figure 16.13 Distribution of the world's oceanic trenches, ridge system, and transform faults. Where transform faults offset ridge segments, they permit the ridge to change direction (curve), as can be seen in the Atlantic Ocean.

than 20 kilometers (12 miles) thick. Examples include the arcs of Tonga, the Aleutians, and the Lesser Antilles. By contrast, older island arcs are more complex and are underlain by crust that ranges in thickness from 20 to 35 kilometers. Examples include the Japanese and Indonesian arcs, which are built upon material generated by earlier episodes of subduction. In a few places, volcanic island arcs are gradually transformed into an Andean-type volcanic chain. For example, the western section of the Aleutian arc consists of numerous volcanic islands built on oceanic crust whereas the volcanoes at the eastern end of the chain are part of the Alaskan Peninsula—a piece of continental crust.

Continental-Continental Convergence

As you saw earlier, when an oceanic plate is subducted beneath continental lithosphere, an Andean-type volcanic arc develops along the margin of the continent.

However, if the subducting plate also contains continental lithosphere, continued subduction eventually brings the two continents together (Figure 16.12C). Whereas oceanic lithosphere is relatively dense and sinks into the asthenosphere, continental lithosphere is buoyant, which prevents it from being subducted to any great depth. The result is a collision between the two continental blocks (Figure 16.14).

Such a collision occurred when the subcontinent of India rammed into Asia and produced the Himalayas—the most spectacular mountain range on Earth (Figure 16.14). During this collision, the continental crust buckled, fractured, and was generally shortened and thickened. In addition to the Himalayas, several other major mountain systems, including the Alps, Appalachians, and Urals, formed during continental collisions.

Prior to a continental collision, the landmasses involved are separated by an ocean basin. As the continental blocks converge, the intervening seafloor is

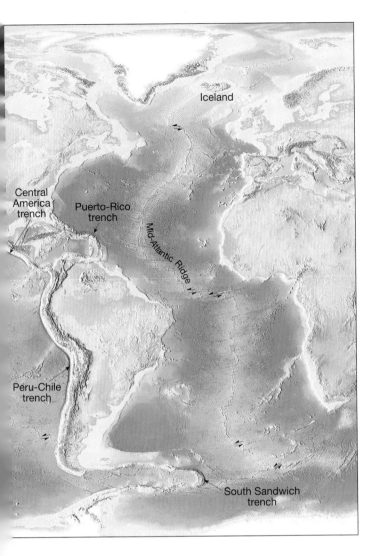

part for the high elevation of the Himalayas and may help explain the elevated Tibetan Plateau to the north.

Transform Fault Boundaries

The third type of plate boundary is the transform fault, where plates grind past one another without the production or destruction of lithosphere. Transform faults were first identified where they join offset segments of an oceanic ridge. At first it was erroneously assumed that the ridge system originally formed a long continuous chain that was later offset by horizontal displacement along these large faults. However, the displacement along these faults was found to be in the exact opposite direction required to produce the offset ridge segments.

The true nature of transform faults was discovered in 1965 by a Canadian researcher who suggested that these large faults connect the global active belts (convergent boundaries, divergent boundaries, and other transform faults) into a continuous network that divides Earth's outer shell into several rigid plates. Thus, he became the first to suggest that Earth was made of individual plates, while at the same time identifying the faults along which relative motion between the plates is made possible.

Most transform faults join two segments of a mid-ocean ridge (Figure 16.15). Here, they are part of prominent linear breaks in the oceanic crust known as *fracture zones*, which include both the transform faults and their inactive extension into the plate interior. These fracture zones are present approximately every 100 kilometers along the trend of a ridge axis. As shown in Figure 16.16, active transform faults lie only between the two offset ridge segments. Here seafloor produced at one ridge axis moves in the opposite direction as seafloor produced at an opposing ridge segment. Thus, between the ridge segments these adjacent slabs of oceanic crust are grinding past each other along a transform fault. Beyond the ridge crests are the inactive zones, which are preserved as linear topographic scars. These fracture zones tend to curve such that small segments roughly parallel the direction of plate motion at the time of their formation.

In another role, transform faults provide the means by which the oceanic crust created at ridge crests can be transported to a site of destruction, the deep-ocean trenches. Figure 16.16 illustrates this situation. Notice that the Juan de Fuca plate moves in a southeasterly direction, eventually being subducted under the western coast of the United States. The southern end of this relatively small plate is bounded by the Mendocino transform fault. This transform fault boundary connects the Juan de Fuca ridge to the Cascade subduction zone (Figure 16.16). Therefore, it facilitates the movement

subducted beneath one of the plates. Subduction initiates partial melting in the overlying mantle rocks, which in turn results in the growth of a volcanic arc.

Eventually, as the intervening seafloor is consumed, these continental masses collide. This folds and deforms the accumulation of sediments along the continental margin as if they were placed in a gigantic vise. The result is the formation of a new mountain range composed of deformed and metamorphosed sedimentary rocks, fragments of the volcanic arc, and possibly slivers of oceanic crust.

After continents collide, the subducted oceanic plate may separate from the continental block and continue its downward movement. However, because of its buoyancy, continental lithosphere cannot be carried very far into the mantle. In the case of the Himalayas, the leading edge of the Indian plate was forced partially under Asia, generating an unusually great thickness of continental lithosphere. This accumulation accounts in

Figure 16.14 The ongoing collision of India and Asia, starting about 45 million years ago, produced the majestic Himalayas.

of the crustal material created at the ridge crest to its destination beneath the North American continent (Figure 16.16).

Another example of a *ridge-trench* transform fault is found southeast of the tip of South America. Here transform faults on the north and south margins of the South Sandwich plate connect the trench to a short spreading axis (see Figure 16.8).

Although most transform faults are located within the ocean basins, a few cut through continental crust.

Two examples are the earthquake-prone San Andreas Fault of California and the Alpine Fault of New Zealand. Notice in Figure 16.16 that the San Andreas Fault connects a spreading center located in the Gulf of California to the Cascade subduction zone and the Mendocino transform fault located along the northwest coast of the United States. Along the San Andreas Fault, the Pacific plate is moving toward the northwest, past the North American plate. If this movement continues, that part of California west of the fault zone, in-

Figure 16.15 Illustration of a transform fault joining segments of an oceanic ridge.

Figure 16.16 The role of transform faults. The Mendocino transform fault permits seafloor generated at the Juan de Fuca ridge to move southeastward past the Pacific plate and beneath the North American plate. Thus, this transform fault connects a divergent boundary to a subduction zone. Furthermore, the San Andreas Fault, also a transform fault, connects two spreading centers—the Juan de Fuca ridge and a divergent zone located in the Gulf of California.

cluding the Baja Peninsula, will become an island off the western coast of the United States and Canada. It could eventually reach Alaska. However, a more immediate concern is the earthquake activity triggered by movements along this fault system.

Testing the Plate Tectonics Model

With the birth of the plate tectonics model, researchers from all of the Earth sciences began testing it. Some of the evidence supporting continental drift and seafloor spreading has already been presented. Some of the evidence that was instrumental in solidifying the support for this new concept follows. Note that some of the ev-

idence was not new; rather, it was a new interpretation of old data that swayed the tide of opinion.

Evidence: Paleomagnetism

Probably the most persuasive evidence to the geologic community for the plate tectonics theory comes from the study of Earth's magnetic field. Anyone who has used a compass to find direction knows that the magnetic field has a north pole and a south pole. These magnetic poles align closely, but not exactly, with the geographic poles. (The geographic poles are simply the top and bottom of the spinning sphere we live on, the points through which passes the imaginary axis of rotation.)

In many respects the magnetic field is very much like that produced by a simple bar magnet. Invisible lines of force pass through Earth and extend from one pole to the other. A compass needle, itself a small magnet free to move about, becomes aligned with these lines of force and thus points toward the magnetic poles.

The technique used to study ancient magnetic fields relies on the fact that certain rocks contain minerals that serve as fossil compasses. These iron-rich minerals, such as magnetite, are abundant in lava flows of basaltic composition. When heated above a certain temperature, called the *Curie point*, these magnetic minerals lose their magnetism.

However, when these iron-rich grains cool below their Curie point (about 580°C), they become magnetized in the direction parallel to the existing magnetic field. Once the minerals solidify, the magnetism they possess will remain "frozen" in this position. In this regard, they behave much like a compass needle inasmuch as they "point" toward the existing magnetic poles. Then, if the rock is moved, or if the magnetic pole changes position, the rock magnetism will, in most instances, retain its original alignment. Thus, rocks formed thousands or millions of years ago "remember" the location of the magnetic poles at the time of their formation and are said to possess fossil magnetism, or **paleomagnetism.**

Polar Wandering. A study of lava flows conducted in Europe in the 1950s led to an amazing discovery. The magnetic alignment in the iron-rich minerals in lava flows of different ages was found to vary widely. A plot of the apparent positions of the magnetic north pole revealed that during the past 500 million years, the location of the pole had gradually wandered from a spot near Hawaii northward through eastern Siberia and finally to its present site (Figure 16.17A). This was clear evidence that either the magnetic poles had migrated through time, an idea known as **polar wandering,** or that the lava flows had moved—in other words, the continents had drifted.

Although the magnetic poles are known to move, studies of the magnetic field indicated that the average positions of the magnetic poles correspond closely to the positions of the geographic poles. This is consistent with our knowledge of Earth's magnetic field, which is generated in part by the rotation of Earth about its axis. If the geographic poles do not wander appreciably, which we believe is true, neither can the magnetic poles. Therefore, a more acceptable explanation for the apparent polar wandering is provided by the plate tectonics theory. *If the magnetic poles remain stationary, their apparent movement was produced by the drifting of the continents.*

Further evidence for plate tectonics came a few years later when polar wandering curves were constructed for North America and Europe (Figure 16.17A). To nearly everyone's surprise, the curves for North America and Europe had similar paths, except that they

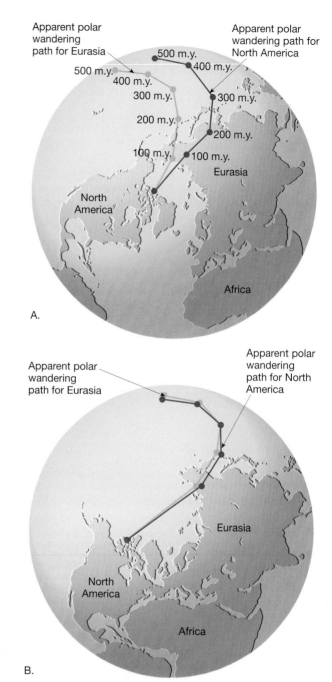

Figure 16.17 Simplified apparent polar wandering paths as established from North American and Eurasian paleomagnetic data. **A.** The more westerly path determined from North American data is thought to have been caused by the westward drift of North America by about 24 degrees from Eurasia. **B.** The positions of the wandering paths when the landmasses are reassembled in their predrift locations.

were separated by about 24 degrees of longitude. When these rocks solidified, could there have been two magnetic north poles that migrated parallel to each other? This is very unlikely. The differences in these migration paths, however, can be reconciled if the two presently separated continents are placed next to one another, as we now believe they were prior to the opening of the Atlantic Ocean (Figure 16.17B).

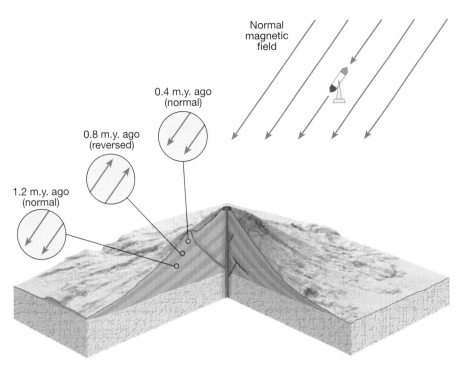

Figure 16.18 Schematic illustration of paleomagnetism preserved in lava flows of various ages. Data such as these from various locales were used to establish the time scale of polarity reversals shown in Figure 16.19.

Magnetic Reversals and Seafloor Spreading.

Another discovery came when geophysicists learned that Earth's magnetic field periodically reverses polarity; that is, the north magnetic pole becomes the south magnetic pole, and vice versa. A rock solidifying during one of the periods of reverse polarity will be magnetized with the polarity opposite that of rocks being formed today.

When rocks exhibit the same magnetism as the present magnetic field, they are said to possess **normal polarity,** while those rocks exhibiting the opposite magnetism are said to have **reverse polarity.** Evidence for magnetic reversals was obtained from lavas and sediments from around the world.

Once the concept of magnetic reversals was confirmed, researchers set out to establish a time scale for polarity reversals. There are many areas where volcanic activity has occurred sporadically for periods of millions of years (Figure 16.18). The task was to measure the directions of paleomagnetism in numerous lava flows of various ages. These data were collected from several places and were used to determine the dates when the polarity of Earth's magnetic field changed. Figure 16.19 shows the time scale of the polarity reversals established for the last few million years.

A significant relationship was uncovered between the magnetic reversals and the seafloor-spreading hypothesis. Very sensitive instruments called *magnetometers* were towed by research vessels across a segment of the ocean floor located off the western coast of the United States. Here workers from the Scripps Institution of Oceanography discovered alternating strips of high- and low-intensity magnetism that trended in roughly a north–south direction. This relatively simple

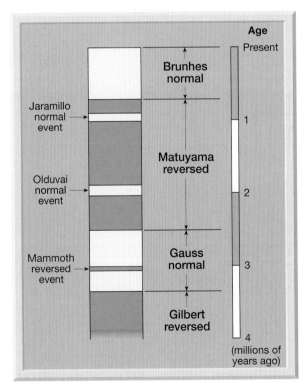

Figure 16.19 Time scale of Earth's magnetic field in the recent past. This time scale was developed by establishing the magnetic polarity for lava flows of known age. (Data from Allen Cox and G. B. Dalrymple)

pattern of magnetic variation defied explanation until 1963, when it was tied to the concept of seafloor spreading. The strips of high-intensity magnetism are regions where the paleomagnetism of the ocean crust is of the normal type.

Consequently, these positively magnetized rocks *enhance* the existing magnetic field. Conversely, the low-intensity strips represent regions where the ocean crust is polarized in the reverse direction and, therefore, *weaken* the existing magnetic field. But how do parallel strips of normally and reversely magnetized rock become distributed across the ocean floor?

As new basalt is added to the ocean floor at the oceanic ridges, it becomes magnetized according to the existing magnetic field (Figure 16.20). As new rock is added in approximately equal amounts to the trailing edges of both plates, we should expect strips of equal size and polarity to parallel both sides of the ocean ridges, as shown in Figure 16.20C. This explanation of the alternating strips of normal and reverse polarity, which lay as mirror images across the ocean ridges, was the strongest evidence so far presented in support of the concept of seafloor spreading.

Now that the dates of the most recent magnetic reversals have been established, the rate at which spreading occurs at the various ridges can be determined accurately. In the Pacific Ocean, for example, the magnetic strips are much wider for corresponding time intervals than are those of the Atlantic Ocean. Hence, we conclude that a faster spreading rate exists for the spreading center of the Pacific as compared to the Atlantic. When we apply numerical dates to these magnetic events, we find that the spreading rate for the North Atlantic Ridge is only 2 centimeters per year.

In contrast, the rate is somewhat faster for the South Atlantic. The spreading rates for the East Pacific Rise generally range between 6 and 12 centimeters annually, with a maximum rate exceeding 15 centimeters (6 inches) in one segment. Thus, we have a magnetic tape recorder that records changes in Earth's magnetic field. This recorder also permits us to determine the rate of seafloor spreading.

Evidence: Earthquake Patterns

By 1968, the basic outline of global tectonics was firmly established. In this same year, three seismologists published papers demonstrating how successfully the new plate tectonics model accounted for the global distribution of earthquakes (Figure 16.21). In particular, these scientists were able to account for the close association between deep-focus earthquakes and ocean trenches. Furthermore, the absence of deep-focus earthquakes along the oceanic ridge system was shown to be consistent with the new theory.

The close association between plate boundaries and earthquakes can be seen by comparing the distribution of earthquakes shown in Figure 16.21 with the map of

Figure 16.20 As new basalt is added to the ocean floor at the oceanic ridges, it is magnetized according to Earth's existing magnetic field. Hence, it behaves much like a tape recorder as it records each reversal of the planet's magnetic field.

A. Period of normal magnetism

B. Period of reverse magnetism

C. Period of normal magnetism

Figure 16.21   Distribution of shallow-, intermediate-, and deep-focus earthquakes. Note that deep-focus earthquakes only occur in association with subduction zones. (Data from NOAA)

plate boundaries in Figure 16.8 and trenches in Figure 16.13. In trench regions where dense slabs of lithosphere plunge into the mantle, this association is especially striking. When the depths of earthquake foci and their locations within the trench systems are plotted, an interesting pattern emerges. Figure 16.22, which shows the distribution of earthquakes in the vicinity of the Japan trench, is an example. Here most shallow-focus earthquakes occur within or adjacent to the trench,

whereas intermediate- and deep-focus earthquakes occur toward the mainland.

In the plate tectonics model, deep-ocean trenches are produced where cool, dense slabs of oceanic lithosphere plunge into the mantle. Shallow-focus earthquakes are produced as the descending plate interacts with the overriding lithosphere. As the slab descends farther into the asthenosphere, deeper-focus earthquakes are generated (Figure 16.22). Because the earth-

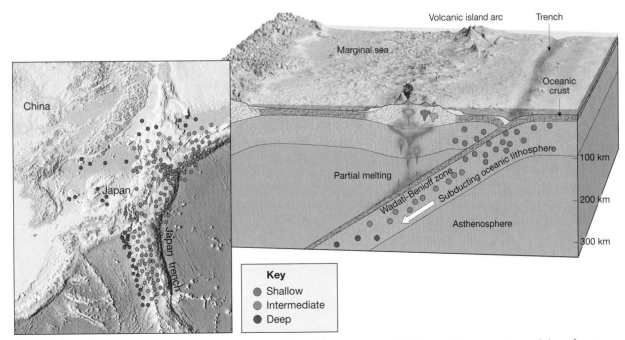

Figure 16.22   Distribution of earthquake foci in the vicinity of the Japan trench. Note that intermediate- and deep-focus earthquakes occur only within the sinking slab of oceanic lithosphere. (Data from NOAA)

quakes occur within the rigid subducting plate rather than in the plastic mantle, they provide a method for tracking the plate's descent. Very few earthquakes have been recorded below 700 kilometers (435 miles), possibly because the slab has been heated sufficiently to lose its rigidity.

Evidence: Ocean Drilling

Some of the most convincing evidence confirming the plate tectonics theory has come from drilling directly into ocean-floor sediment. From 1968 until 1983, the source of these important data was the Deep Sea Drilling Project, an international program sponsored by several major oceanographic institutions and the National Science Foundation. A new drilling ship, the *Glomar Challenger,* represented a significant technological breakthrough, because this vessel could lower drill pipe thousands of meters to the ocean floor and then drill hundreds of meters into the sediments and underlying basaltic crust.

Operations began in August 1968, in the South Atlantic. At several sites, holes were drilled through the entire thickness of sediments to the basaltic rock below. An important objective was to gather samples of sediment from just above the igneous crust as a means of dating the seafloor at each site. (Radiometric dates of the ocean crust itself are unreliable because seawater alters basalt.)

When the oldest sediment from each drill site was plotted against its distance from the ridge crest, it was revealed that the age of the sediment increased with increasing distance from the ridge. This finding agreed with the seafloor-spreading hypothesis, which predicted that the youngest oceanic crust would be found at the ridge crest and that the oldest oceanic crust would be at the continental margins.

Data from the Deep Sea Drilling Project also reinforced the idea that the ocean basins are geologically youthful, because no sediment with an age in excess of 180 million years was found. By comparison, some continental crust has been dated at 3.9 billion years.

During its 15 years of operation, the *Glomar Challenger* drilled 1092 holes and obtained more than 96 kilometers (60 miles) of invaluable core samples. The Ocean Drilling Program has succeeded the Deep Sea Drilling Project and, like its predecessor, is a major international program. A more technologically advanced drilling ship, the *JOIDES Resolution,* now continues the work of the *Glomar Challenger* (Figure 16.23). To learn more about the *JOIDES Resolution,* see Box 14.1 on p. 287.

Evidence: Hot Spots

Mapping of seafloor volcanoes called *seamounts* in the Pacific revealed a chain of volcanic structures extending from the Hawaiian Islands to Midway Island and then continuing northward toward the Aleutian trench (Figure 16.24). Radiometric dates of volcanoes in this chain showed that the volcanoes increase in age with increasing distance from Hawaii. Suiko Seamount, which is located near the Aleutian trench, is 65 million years old, Midway Island is 27 million years old, and the island of Hawaii was built up from the seafloor less than a million years ago (Figure 16.24).

Researchers are in agreement that a rising plume of mantle material is located below the island of Hawaii. Decompression melting of this hot rock as it enters the low-pressure environment near the surface generates a volcanic area or **hot spot.** As the Pacific plate moved over the hot spot, successive volcanic mountains have been built. The age of each volcano indicates the time when it was situated over the relatively stationary mantle plume.

This pattern is shown in Figure 16.24. Kauai is the oldest of the large islands in the Hawaiian chain. Five million years ago, when it was positioned over the hot spot, Kauai was the only Hawaiian Island in existence (Figure 16.24). Visible evidence of the age of Kauai can be seen by examining its extinct volcanoes, which have

Figure 16.23 The *JOIDES Resolution*, the drilling ship of the Ocean Drilling Program. This modern drilling ship has replaced the *Glomar Challenger* in the important work of sampling the floors of the world's oceans. JOIDES is an acronym for Joint Oceanographic Institutions for Deep Earth Sampling. (Photo courtesy of Ocean Drilling Program)

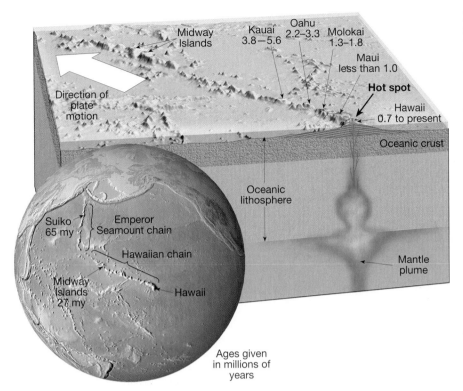

Figure 16.24 The chain of islands and seamounts that extends from Hawaii to the Aleutian trench results from the movement of the Pacific plate over an apparently stationary hot spot. Radiometric dating of the Hawaiian Islands shows that the volcanic activity decreases in age toward the island of Hawaii.

been eroded into jagged peaks and vast canyons. By contrast, the south slopes of the relatively youthful island of Hawaii consist of fresh lava flows, and two of Hawaii's volcanoes, Mauna Loa and Kilauea, remain active.

This evidence supports the fact that the plates do indeed move relative to Earth's interior. The hot spot tracks also trace the direction of plate motion. Notice, for example, in Figure 16.24 that the Hawaiian Island–Emperor Seamount chain bends. This particular bend in the trace occurred about 40 million years ago when the motion of the Pacific plate changed from nearly due north to a northwesterly path.

Measuring Plate Motion

A number of methods have been employed to establish the direction and rate of plate motion. As noted earlier, hot spot "tracks" like those of the Hawaiian Island–Emperor Seamount chain trace the movement of the Pacific Plate relative to the mantle below. Further, by measuring the length of this volcanic chain and the time interval between the formation of the oldest structure (Suiko Seamount) and youngest structure (Hawaii), an average rate of plate motion can be calculated. In this case the volcanic chain is roughly 3000 kilometers long and has formed over the past 65 million years—making the average rate of movement about 9 centimeters (4 inches) per year. The accuracy of this calculation hinges on the hot spot maintaining a fixed position in the mantle. Based on current evidence, this appears to be a reasonable assumption.

Recall that the magnetic stripes measured on the floor of the ocean also provide a method to measure rates of plate motion—at least as averaged over millions of years. Using paleomagnetism and other indirect techniques, researchers been able to work out relative plate velocities as shown on the map in Figure 16.25.

It is currently possible, with the use of space-age technology, to directly measure the relative motion between plates. This is accomplished by periodically establishing the exact locations, and hence the distance between two observing stations situated on opposite sides of a plate boundary. Two of the methods used for this calculation are *Very Long Baseline Interferometry* (VLBI) and a satellite positioning technique that employs the *Global Positioning System* (GPS). The Very Long Baseline Interferometry system utilizes large radio telescopes to record signals from very distant quasars (quasi-stellar objects) (Figure 16.26). Quasars lie billions of light-years from Earth, so they act as stationary reference points. The millisecond differences in the arrival times of the same signal at different earthbound observatories provide a means of establishing the precise distance between receivers. A typical survey may take a day to perform and involves two widely spaced radio telescopes observing perhaps a dozen quasars, 5 to 10 times each. This scheme provides an estimate of the distance between these observatories, which is accurate to about 2 centimeters. By repeating this experiment at a later date, researchers can establish the relative motion of these sites. This method has been particularly useful in establishing large-scale plate motions, such as the

Figure 16.25 This map illustrates directions and rates of plate motion in centimeters per year. Seafloor-spreading velocities (as shown with black arrows and labels) are based on the spacing of dated magnetic stripes (anomalies). The colored arrows show Very Long Baseline Interferometry (VLBI) data of plate motion at selected locations. The data obtained by these methods are typically consistent. (Seafloor data from DeMets and others, VLBI data from Ryan and others)

separation that is occurring between the United States and Europe (see Box 16.2).

You may be familiar with the Global Positioning System, which uses 21 satellites to accurately locate any individual who is equipped with a handheld receiver. By using two spaced receivers, signals obtained by these instruments can be used to calculate their relative positions with considerable accuracy. Techniques using GPS receivers have been shown to be useful in establishing small-scale crustal movements such as those that occur along local faults in regions known to be tectonically active.

Confirming data obtained from these and other techniques leave little doubt that real plate motion has been detected (Figure 16.25). Calculations show that Hawaii is moving in a northwesterly direction and approaching Japan at 8.3 centimeters per year. A site located in Maryland is retreating from one in England at a rate of about 1.7 centimeters per year—a rate that is close to the 2.3-centimeters-per-year spreading rate that was established from paleomagnetic evidence.

The Driving Mechanism

The plate tectonics theory *describes* plate motion and the effects of this motion. Therefore, acceptance of this model does not rely on the mechanism that drives plate motion. This is fortunate, because none of the driving mechanisms yet proposed can account for all major facets of plate tectonics. Nevertheless, researchers generally agree on the following:

1. Convective flow in the rocky 2900-kilometer-thick mantle—in which warm, less dense rock rises and cooler, more dense material sinks—is the basic driving force for plate movement.
2. Mantle convection and plate tectonics are part of the same system. Oceanic plates represent the cold downward-moving portion of convective flow.

Figure 16.26 Radio telescopes like this one located at Green Bank, West Virginia, are used to determine accurately the distance between two distant sites. Data collected by repeated measurements have detected relative plate motions of 1 to 15 centimeters per year between various sites worldwide. (Courtesy of National Radio Astronomy Observatory)

BOX 16.2

Plate Tectonics into the Future

Two geologists, Robert Dietz and John Holden, extrapolated present-day plate movements into the future. Figure 16.B illustrates where they envision Earth's landmasses will be 50 million years from now if present plate movements persist for this time span.

In North America we see that the Baja Peninsula and the portion of southern California that lies west of the San Andreas Fault will have slid past the North American plate. If this northward migration takes place, Los Angeles and San Francisco will pass each other in about 10 million years, and in about 60 million years Los Angeles will begin to descend into the Aleutian trench.

Significant changes are seen in Africa, where a new sea emerges as East Africa parts company with the mainland. In addition, Africa will have moved slowly into Europe, perhaps initiating the next major mountain-building stage on our dynamic planet. Meanwhile, the Arabian Peninsula continues to diverge from Africa, allowing the Red Sea to widen and causing the Persian Gulf to close.

In other parts of the world, Australia is now astride the equator and, along with New Guinea, is on a collision course with Asia. Meanwhile, North and South America are beginning to separate, while the Atlantic and Indian oceans continue to grow at the expense of the Pacific Ocean.

These projections into the future, although interesting, must be viewed with caution because many assumptions must be correct for these events to unfold as just described. Nevertheless, changes in the shapes and positions of continents that are equally profound will undoubtedly occur for millions of years to come.

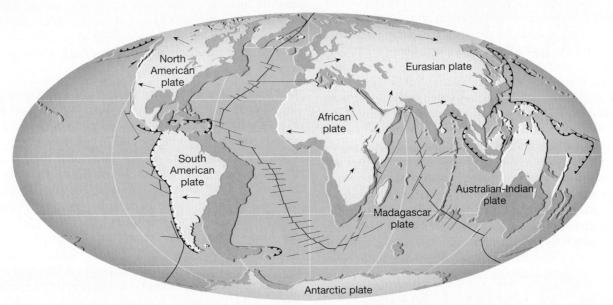

Figure 16.B   The world as it may look 50 million years from now. (From "The Breakup of Pangaea," Robert S. Dietz and John C. Holden. Copyright 1970 by Scientific American, Inc., All rights reserved.)

3. The slow movements of the plates and mantle are driven by the unequal distribution of heat within Earth's interior. This flow is the mechanism that transports heat away from Earth's interior.

What is not known with any large degree of certainty is the precise nature of this convective flow.

Some researchers have argued that the mantle is like a giant layer cake, divided at a depth of 660 kilometers. Convection operates in both layers, but mixing between layers is minimal. At the other end of the spectrum is a model that loosely resembles a pot of just barely boiling water, churning ever so slowly from top to bottom over eons of geologic time. Neither model fits all of the available data. We will first look at some of the processes that are thought to contribute to plate motion and then examine a few of the models that have been proposed to describe plate-mantle convection.

Slab-Pull, Ridge-Push, and Mantle Plumes

Several mechanisms generate forces that contribute to plate motion. One relies on the fact that old oceanic crust, which is relatively cool and dense, sinks into the asthenosphere and "pulls" the trailing lithosphere along. This mechanism, called **slab-pull,** is thought to be the primary downward arm of the convective flow operating in the mantle. By contrast, **ridge-push** results from the elevated position of the oceanic ridge system and causes oceanic lithosphere to gravitationally slide down the flanks of the ridge. Ridge-push, although apparently active in some spreading centers, is probably less important than slab-pull.

Most models suggest that hot, buoyant plumes of rock are the upward flowing arms in the convective mechanism at work in the mantle. These rising **mantle plumes** manifest themselves on Earth's surface as hot spots with their associated volcanic activity. Mapping of Earth's interior using seismic tomography indicates that at least some of these hot plumes extend upward from the vicinity of the mantle-core boundary. Others, however, appear to originate higher in the mantle. Further, some upward flow in the mantle is not related to plumes and occurs just below the ridge crests as a result of seafloor spreading.

Models of Plate-Mantle Convection

Any model describing mantle convection must be consistent with the observed physical and chemical properties of the mantle. In particular, these models must explain why basalts that erupt along the oceanic ridge are fairly homogeneous in composition and depleted in certain trace elements. It is assumed that these ridge basalts are derived from rocks located in the upper mantle that experienced an earlier period of chemical differentiation, in which trace elements were removed. By contrast, higher concentrations of these elements are evident in basaltic eruptions associated with hot-spot volcanism. Because basalts that erupted in different settings have different concentrations of trace elements, they are assumed to be derived from chemically distinct regions of the mantle. Basalts associated with mantle plumes are thought to come from a primitive (less differentiated) source, which more closely resembles the average chemical composition of the early mantle.

Layering at 660 Kilometers. Earlier we referred to the "layer cake" version of mantle convection. As shown in Figure 16.27A, one of these layered models has two zones of convection—a thin, convective layer above 660 kilometers and a thick one located below. This model successfully explains why the basalts that erupt along the oceanic ridges have a somewhat different composition than those that erupt in Hawaii as a result of hot-spot activity. The oceanic ridge basalts come from the upper convective layer, which is well mixed, whereas the mantle plume that feeds the Hawaiian volcanoes taps a deeper, more primitive source that resides in the lower convective layer.

Despite evidence that supports this model, recent seismic imaging has shown that subducting slabs of cold oceanic lithosphere are able to penetrate the 660-kilometer boundary. The subducting lithosphere serves to mix the upper and lower layers together. As a result, the layered mantle structure is lost.

Whole-Mantle Convection. Because of problems with the layered model, researchers began to favor whole-mantle convection. In a whole-mantle convection model, slabs of cold oceanic lithosphere descend into the lower mantle, providing the downward arm of convective flow (Figure 16.27B). Simultaneously, hot mantle plumes originating near the mantle-core boundary transport heat toward the surface. It was also suggested that at extremely slow rates of convection, primitive (undepleted) mantle rock would exist at depth in quantities sufficient to feed the rising mantle plumes.

Recent work, however, has shown that whole mantle mixing would cause the lower mantle to become appreciably mixed in a matter of a few hundred million years. This mixing would tend to eliminate the source of primitive magma observed in hot-spot volcanism.

Deep-Layer Model. A remaining possibility is layering deeper in the mantle. One deep-layer model has been described as analogous to a lava lamp on a low setting. As shown in Figure 16.27C, the lower perhaps one-third of the mantle is like the colored fluid in the bottom layer of a lava lamp. Like a lava lamp on low, heat from Earth's interior causes the two layers to slowly swell and shrink in complex patterns without substantial mixing. A small amount of material from the lower layer flows upward as mantle plumes to generate hot-spot volcanism at the surface.

This model provides the two chemically different mantle sources for basalt that are required by observational data. Further, it is compatible with seismic images that show cold lithospheric plates sinking deep into the mantle. Despite its attractiveness, there is little seismic evidence to suggest that a deep mantle layer of this nature exists, except for the very thin layer located at the mantle-core boundary.

Although there is still much to be learned about the mechanisms that cause plates to move, some facts are clear. The unequal distribution of heat in Earth generates some type of thermal convection that ultimately drives plate-mantle motion. Furthermore, the descending lithospheric plates are active components of downwelling, and they serve to transport cold material into the mantle. Exactly how this convective flow operates is yet to be determined.

A.

B.

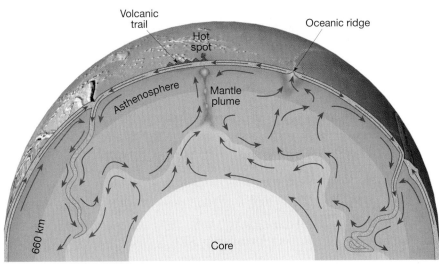

C.

Figure 16.27 Proposed models for mantle convection. **A.** The model shown in this illustration consists of two convection layers—a thin, convective layer above 660 kilometers and a thick one below. **B.** In this whole-mantle convection model, cold oceanic lithosphere descends into the lowermost mantle while hot mantle plumes transport heat toward the surface. **C.** This deep-layer model suggests that the mantle operates similar to a lava lamp on a low setting. Earth's heat causes these layers of convection to slowly swell and shrink in complex patterns without substantial mixing. Some material from the lower layer flows upward as mantle plumes.

The Chapter in Review

- In the early 1900s *Alfred Wegener* set forth his *continental drift* hypothesis. One of its major tenets was that a supercontinent called *Pangaea* began breaking apart into smaller continents about 200 million years ago. The smaller continental fragments then "drifted" to their present positions. To support the claim that the now separate continents were once joined, Wegener and others used the *fit of South America and Africa, fossil evidence, rock types and structures*, and *ancient climates*. One of the main objections to the continental drift hypothesis was its inability to provide an acceptable mechanism for the movement of continents.

- The theory of *plate tectonics*, a far more encompassing theory than continental drift, holds that Earth's outer shell, called the *lithosphere*, consists of seven major and numerous smaller segments called *plates* that are in motion relative to each other. Most of Earth's *seismic activity, volcanism*, and *mountain building* occur along the dynamic margins of these plates.

- A major departure of the plate tectonics theory from the continental drift hypothesis is that large plates contain both continental and ocean crust, and the entire plate moves. By contrast, in continental drift, Wegener proposed that the sturdier continents drifted by breaking through the oceanic crust, much like ice breakers cut through ice.

- *Divergent plate boundaries* occur where plates move apart, resulting in upwelling of material from the mantle to create new seafloor. Most divergent boundaries occur along the axis of the oceanic ridge system and are associated with seafloor spreading, which occurs at rates of 2 to 15 centimeters per year. New divergent boundaries may form within a continent (for example, the East African rift valleys) where they may fragment a landmass and develop a new ocean basin.

- *Convergent plate boundaries* occur where plates move together, resulting in the subduction of oceanic lithosphere into the mantle along a deep oceanic trench. Convergence between an oceanic and continental block results in subduction of the oceanic slab and the formation of a *continental volcanic arc* such as the Andes of South America. Oceanic–oceanic convergence results in an arc-shaped chain of volcanic islands called a *volcanic island arc*. When two plates carrying continental crust converge, both plates are too buoyant to be subducted. The result is a "collision" resulting in the formation of a mountain belt such as the Himalayas.

- *Transform fault boundaries* occur where plates grind past each other without the production or destruction of lithosphere. Most transform faults join two segments of an oceanic ridge. Others connect spreading centers to subduction zones and thus facilitate the transport of oceanic crust created at a ridge crest to its site of destruction, at a deep-ocean trench. Still others, like the San Andreas Fault, cut through continental crust.

- The theory of plate tectonics is supported by (1) *paleomagnetism*, the direction and intensity of Earth's magnetism in the geologic past; (2) the global distribution of *earthquakes* and their close association with plate boundaries; (3) the ages of *sediments* from the floors of the deep-ocean basins; and (4) the existence of island groups that formed over *hot spots* and provide a frame of reference for tracing the direction of plate motion.

- Three basic models for mantle convection are currently being evaluated. Mechanisms that contribute to this convective flow are slab-pull, ridge-push, and mantle plumes. *Slab-pull* occurs where cold, dense oceanic lithosphere is subducted and pulls the trailing lithosphere along. *Ridge-push* results when gravity sets the elevated slabs astride oceanic ridges in motion. Hot, buoyant *mantle plumes* are considered the upward flowing arms of mantle convection. One model suggests that mantle convection occurs in two layers separated at a depth of 660 kilometers. Another model proposes whole-mantle convection that stirs the entire 2900-kilometer-thick rocky mantle. Yet another model suggests that the bottom third of the mantle gradually bulges upward in some areas and sinks in others without appreciable mixing.

Key Terms

asthenosphere (p. 336)
continental drift (p. 331)
continental volcanic arc (p. 334)
convergent plate boundary (p. 337)
divergent plate boundary (p. 337)

hot spot (p. 354)
lithosphere (p. 336)
mantle plumes (p. 358)
normal polarity (p. 351)
oceanic ridge (p. 337)
paleomagnetism (p. 350)
Pangaea (p. 331)

plate (p. 336)
plate tectonics (p. 336)
polar wandering (p. 350)
reverse polarity (p. 351)
ridge-push (p. 357)
rift (rift valley) (p. 337)
seafloor spreading (p. 337)

slab-pull (p. 357)
subduction zone (p. 344)
transform fault boundary (p. 337)
trench (p. 344)
volcanic island arc (p. 344)

Questions for Review

1. Who is credited with developing the continental drift hypothesis?

2. What was probably the first evidence that led some to suspect that the continents were once connected?

3. What was Pangaea?

4. List the evidence that Wegener and his supporters gathered to substantiate the continental drift hypothesis.

5. Explain why the discovery of the fossil remains of *Mesosaurus* in both South America and Africa, but nowhere else, supports the continental drift hypothesis.

6. What was the prevailing view early in the twentieth century of how land animals migrated across vast expanses of ocean?

7. How did Wegener account for the existence of glaciers in the southern landmasses, while at the same time areas in North America, Europe, and Siberia supported lush tropical swamps?

8. On what basis were plate boundaries first established?

9. What are the three major types of plate boundaries? Describe the relative plate motion at each of these boundaries.

10. What is seafloor spreading? Where is active seafloor spreading occurring today?

11. What is a subduction zone? With what type of plate boundary is it associated?

12. Where is lithosphere being consumed? Why must the production and destruction of lithosphere be going on at approximately the same rate?

13. Briefly describe how the Himalaya Mountains formed.

14. Differentiate between transform faults and the other two types of plate boundaries.

15. Some predict that California will sink into the ocean. Is this idea consistent with the theory of plate tectonics?

16. Define the term *paleomagnetism*.

17. How does the continental drift hypothesis account for the apparent wandering of Earth's magnetic poles?

18. Describe the distribution of earthquake epicenters and foci depths as they relate to oceanic trench systems.

19. What is the age of the oldest sediments recovered by deep-ocean drilling? How do the ages of these sediments compare to the ages of the oldest continental rocks?

20. How do hot spots and the plate tectonics theory account for the fact that the Hawaiian Islands vary in age?

21. With what type of plate boundary are the following places or features associated (be as specific as possible): Himalayas, Aleutian Islands, Red Sea, Andes Mountains, San Andreas Fault, Iceland, Japan, Mount St. Helens?

22. Briefly describe the three models proposed for mantle-plate convection. What is lacking in each of these models?

Web Resources

The *Essentials of Geology* Website uses the resources and flexibility of the Internet to aid in your study of the topics in this chapter. Written and developed by geology instructors, this site will help improve your understanding of geology. Visit **www.prenhall.com/lutgens** and click on the cover of *Essentials of Geology, 8e* to find:

- On-line review quizzes.
- Web-based critical thinking and writing exercises.
- Links to chapter-specific Web resources.
- Internet-wide key term searches.

http://www.prenhall.com/lutgens

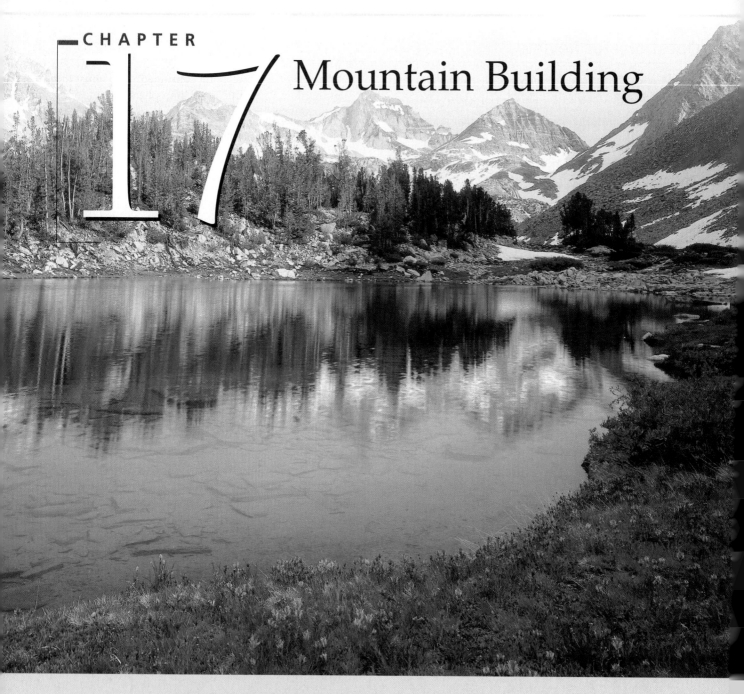

CHAPTER 17 Mountain Building

Focus on Learning

To assist you in learning the important concepts in this chapter, you will find it helpful to focus on the following questions:

- What are the two basic types of rock deformation? What factors influence how rock deforms?

- What are the most common types of folds and faults? How does each type form?

- How are Aleutian- and Andean-type mountain building similar? How are they different?

- How does continental accretion relate to mountain building?

- How is the concept of isostasy related to mountain building?

The Sierra Nevada in California's John Muir Wilderness. (Photo by Carr Clifton)

Mountains provide some of the most spectacular scenery on our planet (Figure 17.1). This splendor has been captured by poets, painters, and songwriters alike. Geologists believe that at some time all continental regions were mountainous masses and have concluded that the continents grow by the addition of mountains to their flanks. Consequently, as geologists unravel the secrets of mountain formation, they also gain a deeper understanding of the evolution of Earth's continents. If continents do indeed grow by adding mountains to their flanks, how do geologists explain the existence of mountains (the Urals, for example) that are located in the interior of a landmass? To answer this and related questions, this chapter attempts to piece together the sequence of events believed to generate these lofty structures.

Deformation

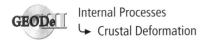

Internal Processes
↳ Crustal Deformation

Every body of rock, no matter how strong, has a point at which it will fracture or flow. **Deformation** is a general term that refers to all changes in the original shape and/or size of a rock body. Most crustal deformation occurs along plate margins. Recall from Chapter 16 that the lithosphere consists of large segments (plates) that move relative to one another. Plate motions and the interactions along plate boundaries generate forces that cause rock units to deform.

Figure 17.2   Rocks in Colorado exhibiting the results of ductile behavior. These rocks were deformed at great depth and were subsequently exposed at the surface. (Photo by Martin Miller/Visuals Unlimited)

When rocks are subjected to forces (stresses) greater than their own strength, they begin to deform, usually by folding, flowing, or fracturing (Figure 17.2). It is easy to visualize how rocks break, because we normally think of them as being brittle. But how can rock masses be bent into intricate folds without being broken during the process? To answer this question, geologists performed laboratory experiments in which rocks were subjected to forces under conditions that simulated those existing at various depths within the crust.

Figure 17.1   Uplift along high angle faults, Mount Sneffels, Colorado Rockies. (Photo by Art Wolfe, Inc.)

Although each rock type deforms somewhat differently, the general characteristics of rock deformation were determined from these experiments. Geologists discovered that when stress is gradually applied, rocks first respond by deforming elastically. Changes that result from *elastic deformation* are recoverable; that is, like a rubber band, the rock will return to nearly its original size and shape when the force is removed. (As you saw in Chapter 15, the energy for most earthquakes comes from stored elastic energy that is released as rock snaps back to its original shape.)

Once the elastic limit (strength) of a rock is surpassed, it either flows (*ductile deformation*) or fractures (*brittle deformation*). The factors that influence the strength of a rock and thus how it will deform include temperature, confining pressure, rock type, and time.

Temperature and Confining Pressure

Rocks near the surface, where temperatures and confining pressure are low, tend to behave like a brittle solid and fracture once their strength is exceeded. This type of deformation is called **brittle failure** or **brittle deformation.** From our everyday experience, we know that glass objects, wooden pencils, china plates, and even our bones exhibit brittle failure once their strength is surpassed. By contrast at depth, where temperatures and confining pressures are high, rocks exhibit *ductile* behavior. **Ductile deformation** is a type of solid-state flow that produces a change in the size and shape of an object without fracturing. Ordinary objects that display ductile behavior include modeling clay, bee's wax, caramel candy, and most metals. For example, a copper penny placed on a railroad track will be flattened and deformed (without breaking) by the force applied by a passing train.

Ductile deformation of a rock—strongly aided by high temperature and high confining pressure—is somewhat similar to the deformation of a penny flattened by a train. Rocks that display evidence of ductile flow usually were deformed at great depth and may exhibit contorted folds that give the impression that the strength of the rock was akin to soft putty (Figure 17.2).

Rock Type

In addition to the physical environment, the mineral composition and texture of a rock greatly influence how it will deform. For example, crystalline rocks, such as granite, basalt, and quartzite, that are composed of minerals that have strong internal molecular bonds tend to fail by brittle fracture. By contrast, sedimentary rocks that are weakly cemented, or metamorphic rocks that contain zones of weakness such as foliation, are more susceptible to ductile flow. Rocks that are weak and thus most likely to behave in a ductile manner when subjected to differential forces include rock salt, gypsum,

and shale, whereas limestone, schist, and marble are of intermediate strength. In fact, rock salt is so weak that it deforms under small amounts of differential stress and rises like stone pillars through beds of sediment that lie in and around the Gulf of Mexico.

Time

One key factor that researchers are unable to duplicate in the laboratory is how rocks respond to small amounts of force applied over long spans of *geologic time*. However, insights into the effects of time on deformation are provided in everyday settings. For example, marble benches have been known to sag under their own weight over a span of a hundred years or so, and wooden bookshelves may bend after being loaded with books for a relatively short period. In nature, small stresses applied over long time spans surely play an important role in the deformation of rock. Forces that are unable to deform rock when initially applied may cause rock to flow if the force is maintained over an extended period of time.

Folds

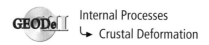

GEODe
Internal Processes
↳ Crustal Deformation

During mountain building, flat-lying sedimentary and volcanic rocks are often bent into a series of wavelike undulations called **folds.** Folds in sedimentary strata are much like those that would form if you were to hold the ends of a sheet of paper and then push them together. In nature, folds come in a wide variety of sizes and configurations. Some folds are broad flexures in which rock units hundreds of meters thick have been slightly warped. Others are very tight microscopic structures found in metamorphic rocks. Size differences notwithstanding, most folds are the result of *compressional forces* that result in the shortening and thickening of the crust.

Types of Folds

The two most common types of folds are anticlines and synclines (Figure 17.3). An **anticline** is most commonly formed by the upfolding, or arching, of rock layers.* Anticlines are sometimes spectacularly displayed where highways have been cut through deformed strata

---

*By strict definition, an anticline is a structure in which the oldest strata are found in the center. This most typically occurs when strata are upfolded. Further, a syncline is strictly defined as a structure in which the youngest strata are found in the center. This occurs most commonly when strata are downfolded.

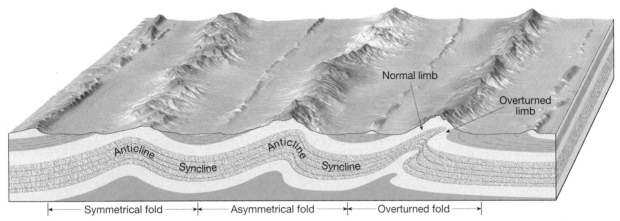

Figure 17.3 Block diagram of principal types of folded strata. The upfolded, or arched, structures are anticlines. The downfolds, or troughs, are synclines. Notice that the limb of an anticline is also the limb of the adjacent syncline.

(Figure 17.4). Often found in association with anticlines are downfolds, or troughs, called **synclines.** Notice in Figure 17.4 that the limb of an anticline is also a limb of the adjacent syncline.

Depending on their orientation, these basic folds are described as *symmetrical* when the limbs are mirror images of each other and *asymmetrical* when they are not. An asymmetrical fold is said to be *overturned* if one limb is tilted beyond the vertical (Figure 17.3). An overturned fold can also "lie on its side" so that a plane extending through the axis of the fold would be horizontal. These *recumbent* folds are common in mountainous regions such as the Alps.

Folds do not continue forever; rather, their ends die out much like the wrinkles in cloth. Some folds *plunge* because the axis of the fold penetrates into the ground (Figure 17.5). As the figure shows, both anticlines and synclines can plunge. Figure 17.6 shows an example of a plunging anticline and the pattern produced when erosion removes the upper layers of the structure and exposes its interior. Note that the outcrop pattern of an anticline points in the direction it is plunging, whereas

the opposite is true for a syncline. A good example of the kind of topography that results when erosional forces attack folded sedimentary strata is found in the Valley and Ridge Province of the Appalachians.

Although we have separated our discussion of folds and faults, in the real world, folds are generally intimately coupled with faults. Examples of this close association are broad, regional features called *monoclines*. Particularly prominent features of the Colorado Plateau, **monoclines** are large, steplike folds in otherwise horizontal sedimentary strata (Figure 17.7). These folds appear to be the result of the reactivating of steeply dipping fault zones located in basement rocks beneath the plateau. As large blocks of basement rock were displaced upward along ancient faults, the comparatively ductile sedimentary strata above responded by folding. On the Colorado Plateau, monoclines display a narrow zone of steeply inclined beds that flatten out to form the uppermost layers of large elevated areas, including the Zuni Uplift, Echo Cliffs Uplift, and San Rafael Swell (Figure 17.7). Displacement along these reactivated faults often exceeds 1 kilometer (0.6 mile).

Figure 17.4 Syncline (left) and anticline (right) share a common limb. (Photo by E. J. Tarbuck)

Figure 17.5   Plunging folds.
A. Idealized view of plunging folds in which a horizontal surface has been added. **B.** View of plunging folds as they might appear after extensive erosion. Notice that in a plunging anticline the outcrop pattern "points" in the direction of the plunge, while the opposite is true of plunging synclines.

Figure 17.6   Sheep Mountain, a doubly plunging anticline. Note that erosion has cut the flanking sedimentary beds into low ridges that make a "V" pointing in the direction of plunge. (Photo by John S. Shelton)

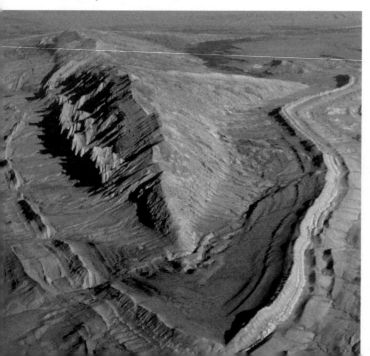

Domes and Basins

Broad upwarps in basement rock may deform the overlying cover of sedimentary strata and generate large folds. When this upwarping produces a circular or elongated structure, the feature is called a **dome.** Downwarped structures having a similar shape are termed **basins.**

The Black Hills of western South Dakota is a large domed structure thought to be generated by upwarping. Here erosion has stripped away the highest portions of the upwarped sedimentary beds, exposing older igneous and metamorphic rocks in the center (Figure 17.8). Remnants of these once continuous sedimentary layers are visible, flanking the crystalline core of these mountains.

Several large basins exist in the United States (Figure 17.9). The basins of Michigan and Illinois have very gently sloping beds similar to saucers. They are thought to be the result of large accumulations of sediment, whose weight caused the crust to subside.

Because large basins usually contain sedimentary beds sloping at very low angles, they are usually identified by the age of the rocks composing them. The

Figure 17.7 Monocline. **A.** The San Rafael monocline, Utah. (Photo by Stephen Trimble) **B.** Monocline consisting of bent sedimentary beds that were deformed by faulting in the bedrock below.

youngest rocks are found near the center, and the oldest rocks are at the flanks. This is just the opposite order of a domed structure, such as the Black Hills, where the oldest rocks form the core.

Faults

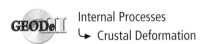
GEODe Internal Processes
↳ Crustal Deformation

Faults are fractures in the crust along which appreciable displacement has taken place. Occasionally, small faults can be recognized in road cuts where sedimentary beds have been offset a few meters, as shown in Figure 17.10. Faults of this scale usually occur as single discrete breaks. By contrast, large faults, like the San Andreas Fault in California, have displacements of hundreds of kilometers and consist of many interconnecting fault surfaces. These *fault zones* can be several kilometers wide and are often easier to identify from high-altitude photographs than at ground level.

Dip-Slip Faults

Faults in which the movement is primarily parallel to the inclination, or dip, of the fault surface are called **dip-slip faults.** Vertical displacements along dip-slip faults

may produce long, low cliffs called **fault scarps.** Fault scarps, such as the one shown in Figure 17.11, are produced by displacements that generate earthquakes.

It has become common practice to call the rock surface that is immediately above the fault the *hanging wall,* and to call the rock surface below, the *footwall.* This nomenclature arose from prospectors and miners who excavated shafts and tunnels along fault zones because these are frequently sites of ore deposits. In these tunnels, the miners would walk on the rocks below the mineralized fault zoned (the footwall) and hang their lanterns on the rocks above (the hanging wall).

Two major types of dip-slip faults are *normal faults* and *reverse faults.*

Normal Faults. Dip-slip faults are classified as **normal faults** when the hanging wall block moves down relative to the footwall block (Figure 17.12A). Most normal faults have steep dips of about 60°, which tend to flatten out with depth. However, some dip-slip faults have much lower dips, with some approaching horizontal. Because of the downward motion of the hanging wall block, normal faults accommodate lengthening, or extension, of the crust.

Most normal faults are small, having displacements of only a meter or so, like the one shown in the road cut in Figure 17.10. Others extend for tens of kilometers where they may sinuously trace the boundary of a

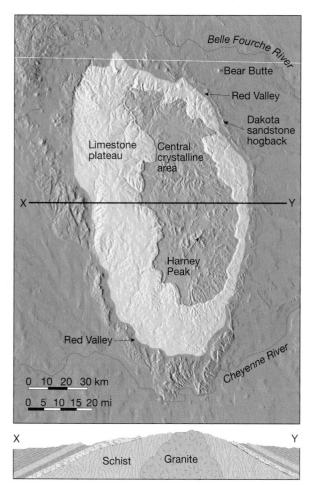

Figure 17.8 The Black Hills of South Dakota, a large domal structure with resistant igneous and metamorphic rocks exposed in the core.

mountain front. In the western United States, large-scale normal faults like these are associated with structures called **fault-block mountains.**

Examples of fault-block mountains include the Teton Range of Wyoming (see Figure 17.17, p. 376) and the Sierra Nevada of California. Both are faulted along their eastern flanks, which were uplifted as the blocks tilted downward to the west. These precipitous mountain fronts were produced over a period of 5 million to 10 million years by many irregularly spaced episodes of faulting. Each event was responsible for just a few meters of displacement.

Normal faulting is also prevalent at spreading centers where plate divergence occurs. Here, a central block called a **graben** is bounded by normal faults and drops as the plates separate (Figure 17.13). These grabens produce an elongated valley bounded by relatively uplifted structures called **horsts.** The Rift Valley of East Africa is made up of several large grabens, above which tilted horsts produce a linear mountainous topography. This valley, nearly 6000 kilometers (3600 miles) long, contains the excavation sites of some of the earliest human fossils.

An excellent example of horst and graben topography is found in the Basin and Range Province, a region that encompasses Nevada and portions of surrounding states (Figure 17.14). Here the crust has been elongated and broken to create more than 200 relatively small mountain ranges. Averaging about 80 kilometers in length, the ranges rise 900 to 1500 meters above the adjacent down-faulted basins. Notice in Figure 17.14 that slopes of the normal faults in the Basin and Range Province decrease with depth and join together to form a nearly horizontal fault called a **detachment fault.** These detachment faults extend for several kilometers below the surface. Here they form a major boundary between the rocks below, which exhibit ductile deformation, and the rocks above, which demonstrate brittle deformation via faulting.

Fault motion provides geologists with a method of determining the nature of the forces at work within Earth. Normal faults indicate the existence of tensional forces that pull the crust apart. This "pulling apart" can be accomplished either by uplifting, which causes the surfaces to stretch and break or by opposing horizontal forces.

Reverse and Thrust Faults. **Reverse faults** and **thrust faults** are dip-slip faults in which the hanging wall block moves up relative to the footwall block (Figure 17.12B and C). Reverse faults have dips greater than 45°, and thrust faults have dips less than 45°. Because the hanging wall block moves up and over the footwall block, reverse and thrust faults accommodate shortening of the crust.

Most high-angle reverse faults are small and accommodate local displacements in regions dominated by other types of faulting. Thrust faults, on the other hand, exist at all scales. In mountainous regions such as the Alps, Northern Rockies, Himalayas, and Appalachians, thrust faults have displaced strata as far as 50 kilometers over adjacent rock units. The result of this large-scale movement is that older strata end up overlying younger rocks.

Whereas normal faults occur in tensional environments, reverse and thrust faults result from strong compressional stresses. In these settings, crustal blocks are displaced *toward* one another, with the hanging wall being displaced upward relative to the footwall. Thrust faulting is most pronounced in subduction zones and other convergent boundaries where plates are colliding. Compressional forces generally produce folds as well as faults and result in a thickening and shortening of the material involved.

Strike-Slip Faults

Faults in which the dominant displacement is horizontal and parallel to the trend, or strike, of the fault surface are called **strike-slip faults** (Figure 17.12D). Because of their large size and linear nature, many strike-slip faults produce a trace that is visible over a great distance

Figure 17.9 The bedrock geology of the Michigan Basin. Notice that the youngest rocks are centrally located, while the oldest beds flank this structure.

(Figure 17.15). Rather than a single fracture along which movement takes place, large strike-slip faults consist of a zone of roughly parallel fractures. The zone may be up to several kilometers wide. The most recent movement, however, is often along a strand only a few meters wide, which may offset features such as stream channels (see Figure 17.A). Furthermore, crushed and broken rocks produced during faulting are more easily eroded, often producing linear valleys or troughs that mark the locations of strike-slip faults.

The earliest scientific records of strike-slip faulting were made following surface ruptures that produced large earthquakes. One of the most noteworthy of these was the great San Francisco earthquake of 1906. During this strong earthquake, structures such as fences that were built across the San Andreas Fault were displaced

Figure 17.10 Faulting caused the vertical displacement of these beds located near Kanab, Utah. Arrows show relative motion of rock units. (Photo by Tom Bean/DRK Photo)

Figure 17.11 A fault scarp located near Joshua Tree National Monument, California. (Photo by A. P. Trujillo/APT Photos)

A. Normal fault (tensional)

B. Reverse fault (compressional)

C. Thrust fault (compressional)

D. Strike-slip fault (shear)

Figure 17.12 Block diagrams of four types of faults. **A.** Normal fault. **B.** Reverse fault. **C.** Thrust fault. **D.** Strike-slip fault.

as much as 4.7 meters (15 feet). Because the movement along the San Andreas causes the crustal block on the opposite side of the fault to move to the right as you face the fault, it is called a *right-lateral* strike-slip fault. The Great Glen fault in Scotland is a well-known example of a *left-lateral* strike-slip fault, which exhibits the opposite sense of displacement.

Many major strike-slip faults cut through the lithosphere and accommodate motion between two large crustal plates. This special kind of strike-slip fault is called a **transform fault.** Numerous transform faults cut the oceanic lithosphere and link offset segments of oceanic ridges. Others accommodate displacement between continental plates that move horizontally with

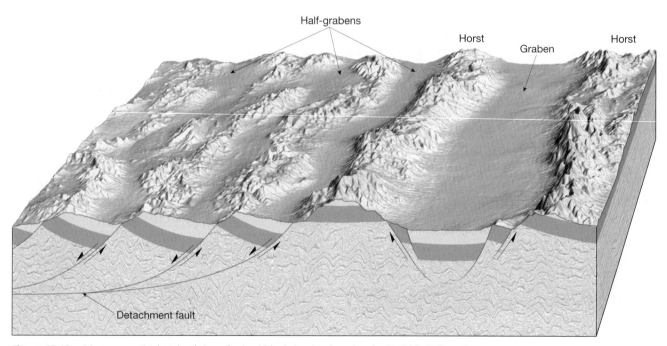

Figure 17.13 Diagrammatic sketch of downfaulted block (graben) and unfaulted block (horst).

Figure 17.14 Normal faulting in the Basin and Range Province. Here, tensional stresses have elongated and fractured the crust into numerous blocks. Movement along these fractures has tilted the blocks producing parallel mountain ranges called fault-block mountains. (Photo by Michael Collier)

Tensional stresses

Normal faults

Detachment fault

Figure 17.15 Aerial view of strike-slip (right-lateral) fault in Southern Nevada. (Photo by Martin G. Miller)

respect to each other. One of the best-known transform faults is California's San Andreas Fault (see Box 17.1). This plate-bounding fault can be traced for about 950 kilometers (600 miles) from the Gulf of California to a point along the Pacific Coast north of San Francisco, where it heads out to sea. Ever since its formation, about 29 million years ago, displacement along the San Andreas Fault has exceeded 560 kilometers (340 miles). This movement has accommodated the northward displacement of southwestern California and the Baja Peninsula of Mexico in relation to the remainder of North America.

Joints

 GEODe

Internal Processes
↳ Crustal Deformation

Among the most common rock structures are fractures called joints. Unlike faults, **joints** are fractures along which *no appreciable displacement* has occurred. Although some joints have a random orientation, most occur in roughly parallel groups (Figure 17.16).

The San Andreas Fault System

The great San Francisco earthquake and fire of 1906 first attracted the public's attention to what is now California's best-known fault: the San Andreas (Figure 17.A). It is North America's largest fault system. Geologic studies after the quake discovered that a displacement up to 5 meters (16 feet) along the fault caused the tremor. It is now known that this event was just one of thousands caused by repeated slippage along the San Andreas throughout its 29-million-year history.

The San Andreas runs for nearly 1300 kilometers (780 miles) through much of western California (Figure 17.B). To the south, it connects with a spreading center in the Gulf of California. To the north, the fault enters the Pacific Ocean at Point Arena, where it is thought to continue northwestward, eventually joining the Mendocino fracture zone.

In its central section, the San Andreas is relatively simple and straight. However, at its north and south ends, several branches spread from the main trace, so that in some areas the fault zone exceeds 100 kilometers (60 miles) in width.

With the development of the plate tectonics theory, geologists began to recognize the significance of this great fault system. The San Andreas Fault is a transform boundary separating two crustal plates. The Pacific plate, on the west, is

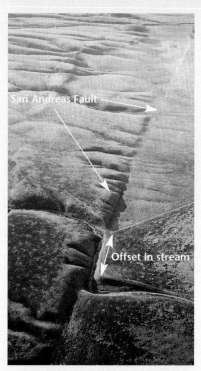

Figure 17.A   Aerial view showing the displacement of stream channels along the trace of the San Andreas Fault. (Photo by Michael Collier/DRK Photo)

slipping and lurching northwestward relative to the North American plate on the east, causing earthquakes along the fault.

Over much of its extent, a linear trough reveals the presence of the San Andreas Fault. When viewed from the air, linear scars, offset stream channels, and elongated ponds strikingly mark its trace (Figure 17.A). On the ground, however, surface expressions of the fault are much more difficult to detect.

The San Andreas is undoubtedly the most studied fault system on Earth. Geologists have learned that each fault segment exhibits somewhat different behavior. Some portions creep slowly with little noticeable seismic activity. Other segments regularly slip, producing small earthquakes. Still other segments seem to store elastic energy for hundreds of years and then rupture to generate great earthquakes. This knowledge is useful when assigning a potential earthquake hazard to a given segment of the fault zone.

Because of its great length and complexity, the San Andreas is more appropriately called a "fault system." It consists of the San Andreas Fault and several major branches, including the Hayward, Calaveras, San Jacinto, and Elsinore faults. These major segments, plus a vast number of smaller faults that include the Imperial, San Fernando, and Santa Monica faults, collectively accommodate the relative motion between the North American and Pacific plates. Geologists have determined that the total accumulated displacement from earthquakes and creep exceeds 560 kilometers (340 miles).

We have already considered two types of joints. Earlier we learned that *columnar joints* form when igneous rocks cool and develop shrinkage fractures that produce elongated, pillarlike columns (see Figure 4.25, p. 91). Also recall that sheeting produces a pattern of gently curved joints that develop more or less parallel to the surface of large exposed igneous bodies such as batholiths. Here the jointing results from the gradual expansion that occurs when erosion removes the overlying load (see Figure 5.4, p. 104).

In contrast to the situations just described, most joints are produced when rocks in the outermost crust are deformed. Here forces associated with crustal movements cause the rock to fail by brittle fracture. For example, when folding occurs, rocks situated at the axes of the folds are elongated and pulled apart to produce tensional joints. Extensive joint patterns can also develop in response to relatively subtle and often barely perceptible regional upwarping and downwarping of the crust. In many cases, the

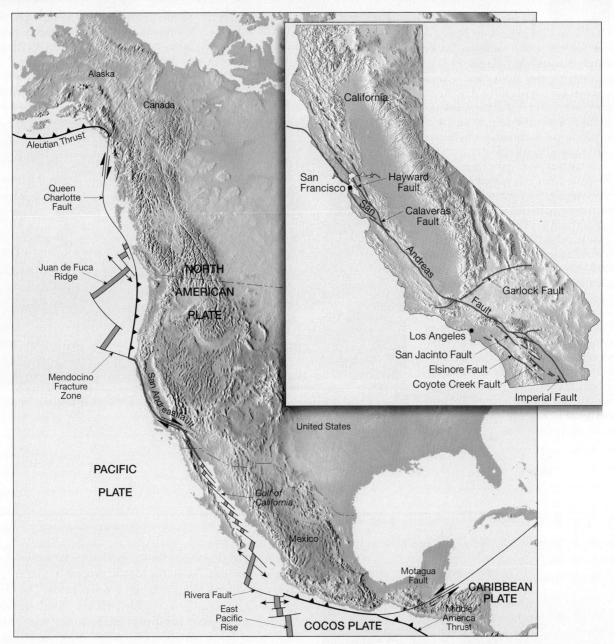

Figure 17.B   **The San Andreas fault system, part of the boundary between the Pacific and North American plates. Inset shows a few of the many splinter faults that are part of this great fault system.**

cause for jointing at a particular locale is not readily apparent.

Many rocks are broken by two or even three sets of intersecting joints that slice the rock into numerous regularly shaped blocks. These joint sets often exert a strong influence on other geologic processes. For example, chemical weathering tends to be concentrated along joints, and in many areas groundwater movement and the resulting dissolution in soluble rocks is controlled by the joint pattern (Figure 17.16). Moreover, a system of joints can influence the direction that stream courses follow. The rectangular drainage pattern described in Chapter 9 is such a case.

Mountain Belts

Like other people, geologists have been inspired more by Earth's mountains than by any other landform (Figure 17.17). Through extensive scientific exploration over the last 150 years, much has been learned about

BOX 17.2

Mountain Building Away from Plate Margins

In the American West, extending from the Front Range of the southern Rocky Mountains across the Colorado Plateau and through the Basin and Range Province, the topography consists of lofty peaks and elevated plateaus (Figure 17.C). According to the plate tectonics model, mountain belts are produced along continental margins in association with convergent plate boundaries. But this region of mountainous topography extends inland almost 1600 kilometers (1000 miles), far from the nearest plate boundary.

The portion of the Rocky Mountains that extends from southern Montana to New Mexico was produced by a period of deformation known as the *Laramide Orogeny*. This event, which created some of the most picturesque scenery in the United States, peaked about 60 million years ago. The mountain ranges generated during the Laramide Orogeny include Colorado's Front Range, the Sangre de Cristo of New Mexico and Colorado, and Wyoming's Bighorns (see Figure 17.20).

The event that may have triggered this period of uplift started with the nearly horizontal subduction of the Farallon plate eastward beneath North America as far inland as the Black Hills of South Dakota (Figure 17.D). As the subducted slab scraped beneath the continent, compressional forces initiated a period of tectonic activity. About 65 million years ago the Farallon plate began to sink into the mantle. As this relatively cool plate gradually separated

Figure 17.C The spectacular Maroon Bells are part of the Colorado Rockies. (Photo by Peter Saloutos/The Stock Market)

Figure 17.16 Chemical weathering is enhanced along joints in sandstone, southeastern Utah. (Photo by Martin Miller)

the internal processes that generate these often spectacular terrains. The name for the processes that collectively produce a mountain belt is **orogenesis.** The rocks comprising mountains provide striking visual evidence of the enormous compressional forces that have deformed large sections of Earth's crust and subsequently elevated them to their present positions. Although folding is often the most conspicuous sign of these forces, thrust faulting, metamorphism, and igneous activity are always present in varying degrees.

Mountain building has occurred during the recent geologic past in several locations around the world. These young mountainous belts include the American Cordillera, which runs along the western margin of the Americas from Cape Horn to Alaska and includes the Andes and Rocky mountains; the Alpine-Himalaya chain, which extends from the Mediterranean through Iran to northern India and into Indochina; and the mountainous terrains of the western Pacific, which include volcanic island arcs such as Japan, the Philippines, from the lithosphere above, it was replaced by hot rock that upwelled from the mantle (Figure 17.D). Thus, according to this scenario, the hot mantle provided the buoyancy to raise the southern Rockies, as well as the Colorado Plateau and the Basin and Range Province.

In the southern Rockies, this event uplifted large blocks of ancient basement rocks along high-angle faults to produce mountainous topography separated by large basins that became filled with sediment as the mountains subsequently eroded. The upwelling that is associated with the Basin and Range Province started about 50 million years ago and remains active today. Here the buoyancy of the warm material caused upwarping and rifting that elongated the overlying crust by 200 to 300 kilometers. The lower crust is ductile and easily stretched. The upper crust, on the other hand, is brittle and deforms by faulting (see Figure 17.14). The extension and faulting broke the uplifted crust, causing individual blocks to shift. The high portions of these tilted blocks make up the mountain ranges, whereas their low areas form the basins, now partially filled with sediment.

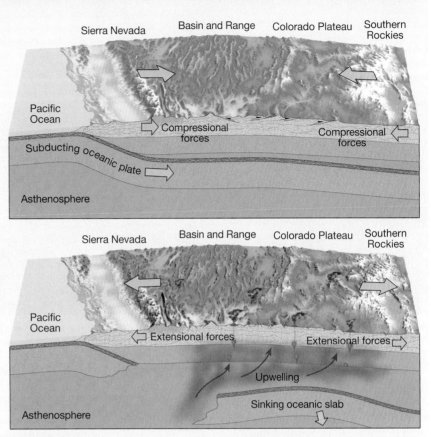

Figure 17.D **One possible mechanism for the uplift that led to the formation of the southern Rockies, Colorado Plateau, and Basin and Range Province. A. Nearly horizontal subduction of an oceanic plate initiated a period of tectonic activity. B. Sinking of this oceanic slab allowed for upwelling of hot mantle material that buoyantly raised the crust.**

and Sumatra. Most of these young mountain belts have come into existence within the last 100 million years. Some, including the Himalayas, began their growth as recently as 45 million years ago.

In addition to these young mountain belts, several chains of Paleozoic- and Precambrian-age mountains exist on Earth as well. Although these older structures are deeply eroded and topographically less prominent, they clearly possess the same structural features found in younger mountains. Typical of this older group are the Appalachians in the eastern United States and the Urals in Russia.

Over the years, several hypotheses have been put forward regarding the formation of Earth's major mountain belts. One early proposal suggested that mountains are simply wrinkles in Earth's crust, produced as the planet cooled from its original semimolten state. As Earth lost heat, it contracted and shrank. In response to this process, the crust was deformed similar to when the peel of an orange wrinkles as the fruit dries

out. However, neither this nor any other early hypothesis was able to withstand careful scrutiny.

Mountain Building at Convergent Boundaries

With the development of the theory of plate tectonics, a widely accepted model for orogenesis became available. According to this model, most mountain building occurs at convergent plate boundaries. Here colliding plates provide the compressional forces that fold, fault, and metamorphose the thick accumulations of sediments that have been deposited along the flanks of landmasses. In addition, partial melting of mantle rock provides a source of magma that intrudes and further deforms these deposits.

To unravel the events that produce mountains, researchers examine ancient mountain structures as well as sites where orogenesis is currently active. Of partic-

Figure 17.17 The Grand Tetons of Wyoming are an example of fault-block mountains. (Photo by David Muench)

ular interest are active subduction zones, where lithospheric plates are converging. At most modern-day subduction zones, volcanic arcs are forming. This situation is typified by Alaska's Aleutian Islands and by the Andean arc of western South America.

Aleutian-Type Mountain Building

Aleutian-type convergent zones occur where two oceanic plates converge and one is subducted beneath the other (Figure 17.18A). This activity results in partial melting of the mantle wedge located above the subducting plate and eventually leads to the growth of a volcanic island arc on the ocean floor. Because they are associated with subducting oceanic lithosphere, island arcs are typically found on the margins of a shrinking ocean basin, such as the Pacific—where the majority of volcanic island arcs are found. Examples of active island arcs include the Mariana, New Hebrides, Tonga, and Aleutian arcs.

Island arcs result from the steady subduction of oceanic lithosphere and tend to be relatively long-lived features. Here, somewhat sporadic volcanic activity and the emplacement of magma at depth, as well as the accumulation of sediment that is scraped off the subducting plate, increase the volume of the crust (Figure 17.18B). Some volcanic island arcs, such as Japan, appear to have been built up by two or three different episodes of subduction or possibly by the accretion of two island arcs.

As illustrated by Japan, the continued development of a volcanic island arc can result in the formation of mountainous topography consisting of belts of igneous and metamorphic rocks. This activity, however, is viewed as just one phase in the development of a major mountain belt. As you will see later, these volcanic arcs eventually become positioned along the margin of a large continental block, where sooner or later they will be involved in a major mountain-building episode.

Andean-Type Mountain Building

Mountain building along continental margins involves the convergence of an oceanic plate and a plate whose leading edge contains continental crust. Exemplified by the Andes Mountains, an *Andean-type convergent zone* results in the formation of a continental volcanic arc and related tectonic features inland of the continental margin. Both Andean- and Aleutian-type plate margins have many broadly similar features, reflecting their comparable tectonic settings. There are a number of differences, however, related to the type of overriding plate.

Passive Margins. The first stage in the development of an idealized Andean-type mountain belt occurs prior to the formation of the subduction zone. During this period the continental margin is a **passive continental margin;** that is, it is not a plate boundary but a part of the same plate as the adjoining oceanic crust. The East Coast of the United States provides a present-

376

Figure 17.18 The development of a volcanic island arc by the convergence of two oceanic plates. Continuous subduction along these Aleutian-type convergent zones results in the development of thick units of continental-type crust.

day example of a passive continental margin. Here, as at other passive continental margins surrounding the Atlantic, deposition of sediment on the continental shelf is producing a thick wedge of shallow-water sandstones, limestones, and shales (Figure 17.19A). Beyond the continental shelf, turbidity currents are depositing sediments on the continental slope and rise (see Chapter 14).

Active Continental Margins. At some point the continental margin becomes active. A subduction zone forms and the deformation process begins (Figure 17.19B). A good place to examine an **active continental margin** is the west coast of South America. Here the Nazca plate is being subducted beneath the South American plate along the Peru–Chile trench (see Figure 16.8, pp. 338–339). This subduction zone probably formed prior to the breakup of the supercontinent Pangaea.

In an idealized Andean-type subduction, convergence of the continental block and the subducting oceanic plate leads to deformation and metamorphism of the continental margin. Once the oceanic plate descends to about 100 kilometers, partial melting of mantle rock above the subducting slab generates magma that migrates upward (Figure 17.19B). During the development of this continental volcanic arc, sediment derived from the land and scraped from the subducting plate is plastered against the landward side of the trench. This chaotic accumulation of sedimentary and metamorphic rocks with occasional scraps of ocean crust is called an **accretionary wedge** (Figure 17.19B). Prolonged subduction can build an accretionary wedge that is large enough to stand above sea level (Figure 17.19C).

Andean-type mountain belts are composed of two roughly parallel zones. The volcanic arc develops on the continental block. It consists of volcanoes and large intrusive bodies intermixed with high-temperature metamorphic rocks. The seaward segment is the accretionary wedge. It consists of folded, faulted, and metamorphosed sediments and volcanic debris (Figure 17.19C).

Sierra Nevada and Coast Ranges. One of the best examples of an inactive Andean-type orogenic belt is found in the western United States. It includes the Sierra Nevada and the Coast Ranges in California (Figure 17.20). These parallel mountain belts were produced by the subduction of a portion of the Pacific Basin under the western edge of the North American plate. The Sierra Nevada batholith is a remnant of a portion of the continental volcanic arc that was produced by several surges of magma over tens of millions of years. Subsequent uplifting and erosion have removed most of the evidence of past volcanic activity and exposed a core of crystalline, igneous, and associated metamorphic rocks.

In the trench region, sediments scraped from the subducting plate, plus those provided by the eroding continental volcanic arc, were intensely folded and faulted into an accretionary wedge. This chaotic mixture of rocks presently constitutes the Franciscan Formation of California's Coast Ranges. Uplifting of the Coast Ranges took place only recently, as evidenced by the young unconsolidated sediments that still mantle portions of these highlands.

Continental Collisions

So far, we have discussed the formation of mountain belts where the leading edge of just one of the two lithospheric plates was capped by continental crust. However, it is possible that both plates may be carrying continental crust. Because continental crust is apparently too buoyant to undergo any appreciable subduction, a collision between the continental fragments results (Figure 17.21).

Figure 17.19 Orogenesis along an Andean-type subduction zone. **A.** Passive continental margin with extensive wedge of sediments. **B.** Plate convergence generates a subduction zone, and partial melting produces a developing continental volcanic arc. **C.** Continued convergence and igneous activity further deform and thicken the crust, elevating the mountain belt, while the accretionary wedge grows.

An example of such a collision began about 45 million years ago when India collided with the Eurasian plate. Prior to this event, India, which was once part of Antarctica, had split from that continent and over the course of millions of years slowly moved a few thousand kilometers due north (Figure 17.21A). The result of the collision was the formation of the spectacular Himalaya Mountains and the Tibetan Plateau (Figure 17.21B).

Although most of the oceanic crust that separated these landmasses prior to the collision was subducted, some was caught up in the squeeze, along with sediment that lay offshore. Today these sedimentary rocks and slivers of oceanic crust are elevated high above sea level. Geologists believe that following such a collision, the subducted oceanic plate decouples from the rigid continental plate and continues its descent into the mantle.

The spreading center that propelled India northward is still active; hence, India continues to be thrust into Asia at an estimated rate of a few centimeters each year. However, numerous earthquakes recorded off the southern coast of India indicate that a new subduction zone may be in the making. If formed, it would provide a subduction site for the floor of the Indian Ocean, which is continually being produced at a spreading center located to the southwest. Should this occur, India's northward journey, relative to Asia, would come

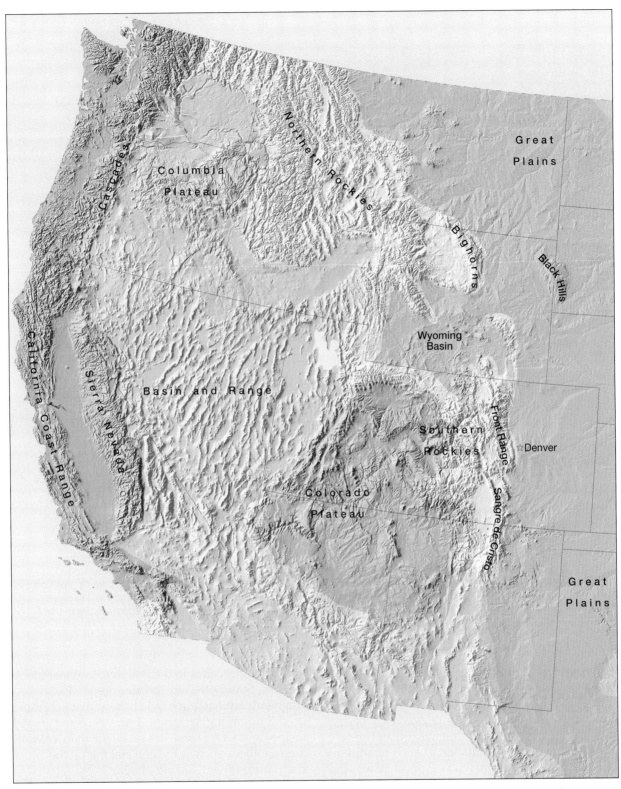

Figure 17.20   Map of mountains and landform regions of the western United States. (After Thelin and Pike, USGS)

to an end and the growth of the Himalayas would cease.

A similar but much older collision is believed to have taken place when the European continent collided with the Asian continent to produce the Ural Mountains, which extend in a north–south direction through Russia. Prior to

the discovery of plate tectonics, geologists had difficulty explaining the existence of mountain ranges such as the Urals, which are located deep within continental interiors. How could thousands of meters of marine sediment be deposited and then become highly deformed while situated in the middle of a large stable landmass?

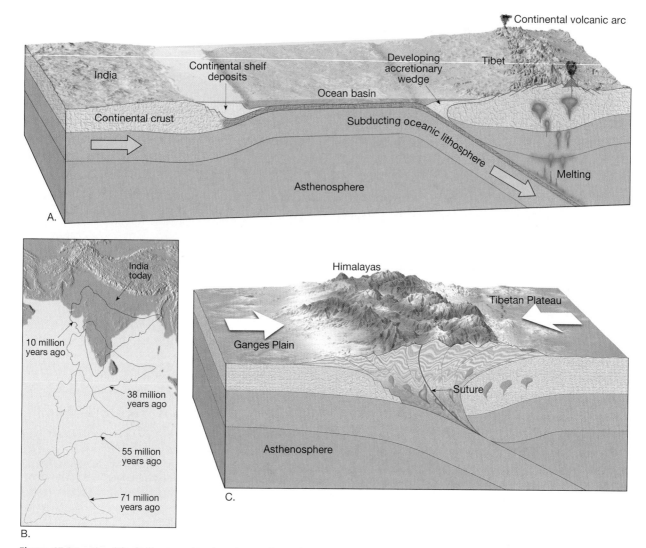

Figure 17.21 Simplified diagrams showing the northward migration and collision of India with the Eurasian plate. **A.** Converging plates generated a subduction zone, while partial melting of the subducting oceanic slab produced a continental volcanic arc. Sediments scraped from the subducting plate were added to the accretionary wedge. **B.** Position of India in relation to Eurasia at various times. **C.** Eventually the two landmasses collided, deforming and elevating the accretionary wedge and continental shelf deposits. In addition, slices of the Indian crust were thrust up onto the Indian plate. (Modified after Peter Molnar)

Continental Accretion and Mountain Building

Plate tectonics theory originally suggested two mechanisms for orogenesis: (1) Continental collisions were proposed to explain the formation of such mountainous terrains as the Alps, Himalayas, Appalachians, and Urals; and (2) subduction of oceanic lithosphere was thought to be the underlying tectonic process for many circum-Pacific mountain chains, as typified by the Andes, Aleutians, and Japan. Subsequent investigations, however, indicated a third mechanism of orogenesis: Smaller crustal fragments collide and merge with continental margins, and through this process of collision and accretion, many of the mountainous regions rimming the Pacific have been generated.

Geologists refer to these accreted crustal blocks as terranes. Simply, the term **terrane** designates any crustal fragment that has a geologic history distinct from that of the adjoining terranes. Terranes come in varied shapes and sizes; some are no larger than volcanic islands. Others, such as the one making up the entire Indian subcontinent, are very large.

What is the nature of these crustal fragments, and where do they come from? Research suggests that prior to their accretion to a continental block, some of the fragments may have been microcontinents similar to the present-day island of Madagascar, located in the Indian Ocean east of Africa. Many others were island arcs like Japan, the Philippines, and the Aleutian Islands. Still others may have been submerged crustal fragments, such as those occuring on the floor of the western Pacific (Figure 17.22). Over a hundred of these *"oceanic plateaus"* are presently known to exist. Their origins vary; some are thought to be submerged continental fragments, whereas others are extinct volcanic islands, such as those of the Hawaiian Island–Emperor Seamount chain. Others are submerged volcanic plateaus created by massive outpourings of lava associated with hot-spot activity.

Accretion of Foreign Terranes.   The widely accepted view is that as oceanic plates move, they carry the embedded oceanic plateaus, volcanic island arcs, and microcontinents to an Andean-type subduction zone. Eventually a collision between the crustal fragment and the continent occurs. Relatively small crustal pieces are peeled from the subducting plate and thrust in relatively thin sheets upon the adjacent continental block. This newly added material increases the width and thickness of the continental crust. The material may later be overridden and/or displaced farther inland by the addition of other fragments.

The accretion of larger crustal fragments, such as a mature island arc, may result in a mountain range—although much smaller than those resulting from a continent–continent collision (Figure 17.23). Because of its buoyancy, an island arc will not subduct beneath the continental plate and will therefore plow into the continent, deforming both blocks. When no further convergence is possible, a new trench may develop on the seaward side of the accreted volcanic arc.

Accretion and Orogenesis.   The idea that orogenesis occurs in association with the accretion of crustal fragments to a continental mass arose principally from studies conducted in the northern portion of the North

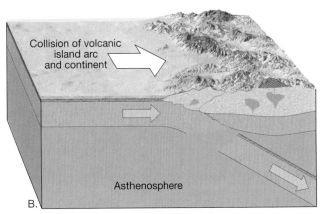

Figure 17.23   This sequence illustrates the collision of an inactive volcanic island arc with an Andean-type plate margin.

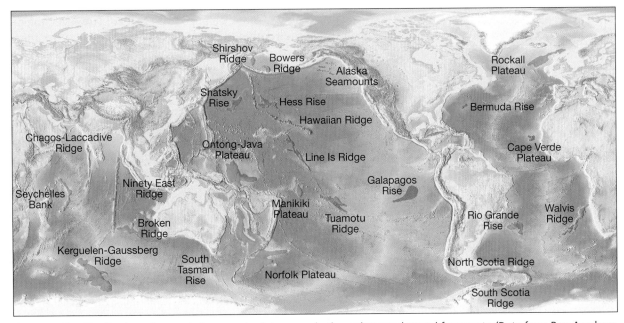

Figure 17.22   Distribution of present-day oceanic plateaus and other submerged crustal fragments. (Data from Ben-Avraham and others)

American Cordillera (Figure 17.24). Here it was determined that some mountainous areas, principally those in the orogenic belts of Alaska and British Columbia, contain fossil and magnetic evidence indicating that these strata once lay nearer the equator.

It is now assumed that many other terranes found in the North American Cordillera were once scattered throughout the eastern Pacific, much as we find island arcs and oceanic plateaus distributed in the western Pacific today (see Figure 17.22). Since the breakup of Pangaea 200 million years ago, North America has been migrating westward, overriding the Pacific Basin. Apparently, this activity resulted in the piecemeal addition of fragments to the entire Pacific margin of the continent, from the Baja Peninsula to northern Alaska (Figure 17.24). In a like manner, many modern microcontinents will eventually be accreted to active continental margins, producing new orogenic belts.

Isostasy and Crustal Uplift

In addition to the horizontal movements of lithospheric plates, gradual up-and-down motions of the crust are observed at many locations around the globe. Although much of this vertical movement occurs along plate margins and is associated with episodes of mountain building, some of it is not. Rather, it occurs in the interiors of continents far from plate boundaries.

Evidence for crustal uplift occurs along the West Coast of the United States. When the elevation of a coastal area remains unchanged for an extended period, a wave-cut platform develops (see Figure 13.9, p. 269). In parts of California, ancient wave-cut platforms can now be found as terraces hundreds of meters above sea level (Figure 17.25). Such evidence of crustal uplift is easy to find; unfortunately, the reason for uplift is not always as easy to establish.

Early workers discovered that Earth's less-dense crust floats on top of the denser and deformable rocks of the mantle. The concept of a floating crust in gravitational balance is called **isostasy.** One way to grasp the concept of isostasy is to envision a series of wooden blocks of different heights floating in water, as shown in Figure 17.26. Note that the thicker wooden blocks float higher than the thinner blocks. In a similar manner, many mountain belts stand high above the surface because they have buoyant (less dense) crustal "roots" that extend deep into the denser mantle that supports them from below.

Now visualize what would happen if another small block of wood were placed atop one of the blocks in Figure 17.26. The combined block would sink until a new isostatic balance was reached. However, the top of the combined block would actually be higher than be-

Figure 17.24 Map showing terranes thought to have been added to western North America during the past 200 million years. (Redrawn after D. R. Hutchinson and others)

Figure 17.25 Former wave-cut platforms now exist as a series of elevated terraces on the western side of San Clemente Island off the southern California coast. Once at sea level, the highest terraces are now about 400 meters (1320 feet) above it. (Photo by John S. Shelton)

Figure 17.26 This drawing illustrates how wooden blocks of different thicknesses float in water. In a similar manner, thick sections of crustal material float higher than thinner crustal slabs.

fore, and the bottom would be lower. This process of establishing a new level of gravitational equilibrium is called **isostatic adjustment.**

Applying the concept of isostasy, we should expect that when weight is added to the crust, it will respond by subsiding, and when weight is removed, the crust will rebound. (Visualize what happens to a ship as cargo is being loaded and unloaded.) Evidence for crustal subsidence followed by crustal rebound is provided by Ice Age glaciers. When continental ice sheets occupied portions of North America during the Pleistocene epoch, the added weight of 3-kilometer-thick masses of ice caused downwarping of Earth's crust by hundreds of meters. In the 8000 years since the last ice sheet melted, uplifting of as much as 330 meters (1000 feet) has occurred in Canada's Hudson Bay region, where the thickest ice had accumulated.

As this example illustrates, crustal buoyancy can account for considerable vertical movement. Recall that mountain building causes the crust to shorten and thicken. Thus, because of isostasy, deformed strata experience regional uplift both during mountain building and for an extended period afterward. Usually, little additional deformation is associated with the latter ac-

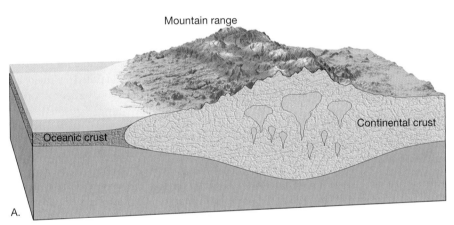

Mountain range

Continental crust

Oceanic crust

A.

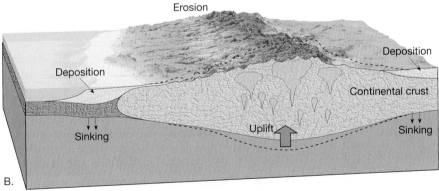

Erosion

Deposition

Deposition

Continental crust

Sinking

Uplift

Sinking

B.

Figure 17.27 This sequence illustrates how the combined effect of erosion and isostatic adjustment results in a thinning of the crust in mountainous regions. **A.** When mountains are young, the continental crust is thickest. **B.** As erosion lowers the mountains, the crust rises in response to the reduced load. **C.** Erosion and uplift continue until the mountains reach "normal" crustal thickness.

Mountain remnant

Continental crust

C.

tivity. As the crust rises, the processes of erosion accelerate, carving the deformed strata into a mountainous landscape.

One of the consequences of the isostatic adjustment is that as erosion lowers the summits of mountains, the crust will rise in response to the reduced load (Figure 17.27). However, each episode of isostatic uplift is somewhat less than the elevation loss due to erosion.

The processes of uplifting and erosion will continue until the mountain block reaches "normal" crustal thickness. When this occurs, the mountains will be eroded to near sea level, and the once deeply buried interior of the mountain will be exposed at the surface. In addition, as mountains are worn down, the eroded sediment is deposited on adjacent landscapes, causing these areas to subside (Figure 17.27).

The Chapter in Review

- *Deformation* refers to changes in the shape and/or volume of a rock body and is most pronounced along plate margins. Rocks deform differently depending on the environment (temperature and confining pressure), the composition and texture of the rock, and the length of time stress is maintained. Rocks first respond by deforming *elastically* and will return to their original shape when the stress is removed. Once their elastic limit (strength) is surpassed, rocks either deform by ductile flow or they fracture. *Ductile deformation* is a solid-state flow that results in a change in size and shape of an object without fracturing. Ductile deformation occurs in a high-temperature/high-pressure environment. In a near-surface environment, most rocks deform by *brittle failure*.

- The most basic geologic structures associated with rock deformation are *folds* (layers of rocks bent into a series of wavelike undulations) and *faults*. The two most common types of folds are *anticlines*, formed by the upfolding, or arching, of rock layers, and *synclines*, which are downfolds. Most folds are the result of horizontal *compressional forces*. Folds can be *symmetrical, asymmetrical*, or, if one limb has been tilted beyond the vertical, *overturned*. *Domes* (upwarped structures) and *basins* (downwarped structures) are circular or somewhat elongated folds formed by vertical displacements of strata.

- Faults are fractures in the crust along which appreciable displacement has occurred. Faults in which the movement is primarily vertical are called *dip-slip faults*. Dip-slip faults include both *normal* and *reverse faults*. Low-angle reverse faults are called *thrust faults*. Normal faults indicate *tensional forces* that pull the crust apart. Along the spreading centers of plates, divergence can cause a central block called a *graben*, bounded by normal faults, to drop as the plates separate.

- Reverse and thrust faulting indicate that *compressional forces* are at work. Large *thrust faults* are found along subduction zones and other convergent boundaries where plates are colliding. In mountainous regions such as the Alps, Northern Rockies, Himalayas, and Appalachians, thrust faults have displaced strata as far as 50 kilometers over adjacent rock units.

- *Strike-slip faults* exhibit mainly horizontal displacement parallel to the fault surface. Large strike-slip faults, called

transform faults, accommodate displacement between plate boundaries. Most transform faults cut the oceanic lithosphere and link spreading centers. The San Andreas Fault cuts the continental lithosphere and accommodates the northward displacement of southwestern California.

- *Joints* are fractures along which no appreciable displacement has occurred. Joints generally occur in groups with roughly parallel orientations and are the result of brittle failure of rock units located in the outermost crust.

- The name for the processes that collectively produce a mountain system is *orogenesis*. Most mountains consist of roughly parallel ridges of folded and faulted sedimentary and volcanic rocks, portions of which have been strongly metamorphosed and intruded by younger igneous bodies.

- Major mountain systems form along *convergent plate boundaries*. *Andean-type mountain building* along continental margins involves the convergence of an oceanic plate and a plate whose leading edge contains continental crust. At some point in the formation of Andean-type mountains a *subduction zone* forms along with a *continental volcanic arc*. Sediment from the land, as well as material scraped from the subducting plate, becomes plastered against the landward side of the trench, forming an *accretionary wedge*. One of the best examples of an inactive Andean-type mountain belt is found in the western United States and includes the Sierra Nevada and the Coast Range in California.

- *Continental collisions*, in which both plates are carrying continental crust, have resulted in the formation of the Himalaya Mountains and the Tibetan Plateau. Continental collisions also formed many other mountain belts, including the Alps, Urals, and Appalachians.

- Recent investigations indicate that *accretion*, a third mechanism of orogenesis, takes place where *small crustal fragments collide and accrete to continental margins* along plate boundaries. The accreted crustal blocks are referred to as *terranes*. The mountainous topography of Alaska and the westernmost regions of Canada, the United States, and Mexico formed as the result of the accretion of terranes to North America.

- Earth's less dense crust floats on top of the denser and deformable rocks of the mantle, much like wooden blocks floating in water. The concept of a floating crust in gravitational balance is called *isostasy*. Most mountainous topography is located where the crust has been shortened and thickened. Therefore, mountains have deep crustal roots that isostatically support them. As erosion lowers the peaks, *isostatic adjustment* gradually raises the mountains in response. The processes of uplifting and erosion will continue until the mountain block reaches "normal" crustal thickness.

Key Terms

accretionary wedge
  (p. 377)
active continental margin
  (p. 377)
anticline (p. 364)
basin (p. 366)
brittle failure
  (deformation) (p. 364)
deformation (p. 363)

detachment fault (p. 368)
dip-slip fault (p. 367)
dome (p. 366)
ductile deformation
  (p. 364)
fault (p. 367)
fault scarp (p 367)
fault-block mountains
  (p. 368)

fold (p. 364)
graben (p. 368)
horst (p. 368)
isostasy (p. 382)
isostatic adjustment
  (p. 383)
joint (p. 371)
monocline (p. 365)
normal fault (p. 367)

orogenesis (p. 374)
passive continental margin
  (p. 376)
reverse fault (p. 368)
strike-slip fault (p. 368)
syncline (p. 365)
terrane (p. 380)
thrust fault (p. 368)
transform fault (p. 370)

Questions for Review

1. What is rock deformation?

2. How is brittle deformation different from ductile deformation?

3. List three factors that determine how rocks will behave when exposed to stresses that exceed their strength. Briefly explain the role of each.

4. Distinguish between anticlines and synclines, domes and basins, anticlines and domes.

5. How is a monocline different from an anticline?

6. The Black Hills of South Dakota are a good example of what type of structural feature?

7. Contrast the movements that occur along normal and reverse faults. What type of force is indicated by each fault?

8. Is the fault shown in Figure 17.10 (p. 369) a normal or reverse fault?

9. Describe a horst and a graben. Explain how a graben valley forms, and name one.

10. What type of faults are associated with fault-block mountains?

11. How are reverse faults different from thrust faults? In what way are they the same?

12. The San Andreas Fault is an excellent example of a _____ fault.

13. How are joints different from faults?

14. In the plate tectonics model, which type of plate boundary is most directly associated with mountain building?

15. Briefly describe the development of a volcanic island arc.

16. The formation of mountainous topography at a volcanic island arc, such as Japan, is considered just one phase in the development of a major mountain belt. Explain.

17. What is an accretionary wedge? Briefly describe its formation.

18. What is a passive continental margin? Give an example. Give an example of an active continental margin.

19. In what way are the Sierra Nevada and the western Andes similar?

20. Compare mountain building along an Andean-type convergent boundary and a continent-continent convergent boundary.

21. How does the plate tectonics theory help explain the existence of fossil marine life in rocks atop the Ural Mountains?

22. Define the term *terrane*. How is it different from the term *terrain*?

23. List three structures that fall under the heading "oceanic plateaus." In addition to oceanic plateaus, what other features are thought to be carried by the oceanic lithosphere and eventually accreted to a continent?

24. List some evidence that supports the concept of crustal uplift.

25. What happens to a floating object when weight is added? Subtracted? How does this principal apply to changes in the elevation of mountains? What term is applied to the adjustment that causes crustal uplift of this type?

26. How does the formation and melting of Pleistocene ice sheets support the idea that the lithosphere tries to remain in isostatic balance?

Web Resources

The *Essentials of Geology* Website uses the resources and flexibility of the Internet to aid in your study of the topics in this chapter. Written and developed by geology instructors, this site will help improve your understanding of geology. Visit **www.prenhall.com/lutgens** and click on the cover of *Essentials of Geology, 8e* to find:

- On-line review quizzes.
- Web-based critical thinking and writing exercises.
- Links to chapter-specific Web resources.
- Internet-wide key term searches.

http://www.prenhall.com/lutgens

Geologic Time

Focus on Learning

To assist you in learning the important concepts in this chapter, you will find it helpful to focus on the following questions:

- What are the two types of dates used by geologists to interpret Earth history?
- What are the laws, principles, and techniques used to establish relative dates?
- What are fossils? What are the conditions that favor the preservation of organisms as fossils?
- How are fossils used to correlate rocks of similar ages that are in different places?
- What is radioactivity, and how are radioactive isotopes used in radiometric dating?
- What is the geologic time scale, and what are its principal subdivisions?
- Why is it difficult to assign reliable numerical dates to samples of sedimentary rock?

The strata exposed in the Grand Canyon contain clues to hundreds of millions of years of Earth history. This view is from the North Rim. (Photo by David Muench)

I n the late eighteenth century, James Hutton recognized the immensity of Earth history and the importance of time as a component in all geological processes. In the nineteenth century, others effectively demonstrated that Earth had experienced many episodes of mountain building and erosion, which must have required great spans of geologic time. Although these pioneering scientists understood that Earth was very old, they had no way of knowing its true age. Was it tens of millions, hundreds of millions, or even billions of years old? Rather, a geologic time scale was developed that showed the sequence of events based on relative dating principles. What are these principles? What part do fossils play? With the discovery of radioactivity and radiometric dating techniques, geologists now can assign fairly accurate dates to many of the events in Earth history. What is radioactivity? Why is it a good "clock" for dating the geologic past?

Geology Needs a Time Scale

In 1869, John Wesley Powell, who was later to head the U.S. Geological Survey, led a pioneering expedition down the Colorado River and through the Grand Canyon (Figure 18.1). Writing about the rock layers that were exposed by the downcutting of the river, Powell said that "the canyons of this region would be a Book of Revelations in the rock-leaved Bible of geology." He was undoubtedly impressed with the millions of years of Earth history exposed along the walls of the Grand Canyon (see chapter-opening photo).

Powell realized that the evidence for an ancient Earth is concealed in its rocks. Like the pages in a long and complicated history book, rocks record the geological events and changing life forms of the past. The book, however, is not complete. Many pages, especially in the early chapters, are missing. Others are tattered, torn, or smudged. Yet enough of the book remains to allow much of the story to be deciphered.

Interpreting Earth history is a prime goal of the science of geology. Like a modern-day sleuth, the geologist must interpret clues found preserved in the rocks. By studying rocks, especially sedimentary rocks, and the features they contain, geologists can unravel the complexities of the past.

Geological events by themselves, however, have little meaning until they are put into a time perspective. Studying history, whether it be the Civil War or the Age of Dinosaurs, requires a calendar. Among geology's major contributions to human knowledge is the *geologic time scale* and the discovery that Earth history is exceedingly long.

Relative Dating—Key Principles

GEODe

Geologic Time
↳ Relative Dating

The geologists who developed the geologic time scale revolutionized the way people think about time and

A.

B.

Figure 18.1 **A.** Start of the expedition from Green River station. A drawing from Powell's 1875 book. **B.** Major John Wesley Powell, pioneering geologist and the second director of the U.S. Geological Survey. (Courtesy of the U.S. Geological Survey, Denver)

perceive our planet. They learned that Earth is much older than anyone had previously imagined and that its surface and interior have been changed over and over again by the same geological processes that operate today.

During the late 1800s and early 1900s, various attempts were made to determine Earth's age. Although some of the methods appeared promising at the time, none proved to be reliable. What these scientists were seeking was a **numerical date.** Such dates specify the actual number of years that have passed since an event occurred. Today our understanding of radioactivity allows us to accurately determine numerical dates for rocks that represent important events in Earth's distant past. We will study radioactivity later in this chapter. Prior to the discovery of radioactivity, geologists had no accurate and dependable method of numerical dating and had to rely solely on relative dating.

Relative dating means placing rocks in their proper *sequence of formation*—which formed first, second, third, and so on. Relative dating cannot tell us how long ago something took place, only that it followed one event and preceded another. The relative dating techniques that were developed are valuable and still widely used. Numerical dating methods did not replace these techniques; they simply supplemented them. To establish a relative time scale, a few basic principles or rules had to be discovered and applied. Although they may seem obvious to us today, they were major breakthroughs in thinking at the time, and their discovery and acceptance was an important scientific achievement.

Law of Superposition

Nicolaus Steno, a Danish anatomist, geologist, and priest (1636–1686), is credited with being the first to recognize a sequence of historical events in an outcrop of sedimentary rock layers. Working in the mountains of western Italy, Steno applied a very simple rule that has come to be the most basic principle of relative dating—the **law of superposition.** The law simply states that in an underformed sequence of sedimentary rocks, each bed is older than the one above it and younger than the one below. Although it may seem obvious that a rock layer could not be deposited unless it had something older beneath it for support, it was not until 1669 that Steno clearly stated the principle.

This rule also applies to other surface-deposited materials, such as lava flows and beds of ash from volcanic eruptions. Applying the law of superposition to the beds exposed in the upper portion of the Grand Canyon (Figure 18.2), you can easily place the layers in their proper order. Among those that are shown, the sedimentary rocks in the Supai Group must be the oldest, followed in order by the Hermit Shale, Coconino Sandstone, Toroweap Formation, and Kaibab Limestone.

Principle of Original Horizontality

Steno is also credited with recognizing the importance of another basic principle, called the **principle of original horizontality.** Simply stated, it means that layers of sediment are generally deposited in a horizontal position. Thus, if we observe rock layers that are flat, it

A.

B.

Figure 18.2 Applying the law of superposition to these layers exposed in the upper portion of the Grand Canyon, the Supai Group is oldest, and the Kaibab Limestone is youngest. (Photo by E. J. Tarbuck)

Figure 18.3 Most layers of sediment are deposited in a nearly horizontal position. Thus, when we see rock layers that are inclined, we can assume that they must have been moved into that position by crustal disturbances after their deposition. Hartland Quay, Devon, England. (Photo by Tom Bean/DRK Photo)

means they have not been disturbed, and they still have their *original* horizontality. The layers in the Grand Canyon illustrate this in Figure 18.2. But if they are folded or inclined at a steep angle, they must have been moved into that position by crustal disturbances sometime after their deposition (Figure 18.3).

Principle of Cross-Cutting Relationships

When a fault cuts through other rocks, or when magma intrudes and crystallizes, we can assume that the fault or intrusion is younger than the rocks affected. For example, in Figure 18.4, the faults and dikes clearly must have occurred *after* the sedimentary layers were deposited.

This is the **principle of cross-cutting relationships.** By applying the cross-cutting principle, you can see that fault *A* occurred *after* the sandstone layer was deposited, because it "broke" the layer. However, fault *A* occurred *before* the conglomerate was laid down, because that layer is unbroken.

We can also state that dike *B* and its associated sill are older than dike *A*, because dike *A* cuts the sill. In the same manner, we know that the batholith was emplaced after movement occurred along fault *B*, but before dike *B* was formed. This is true because the batholith cuts across fault *B* and dike *B* cuts across the batholith.

Inclusions

Sometimes inclusions can aid the relative dating process. **Inclusions** are pieces of one rock unit that are contained within another. The basic principle is logical and straightforward. The rock mass adjacent to the one containing the inclusions must have been there first in order to provide the rock fragments. Therefore, the rock mass containing inclusions is the younger of the two. Figure 18.5 provides an example. Here the inclusions of intrusive igneous rock in the adjacent sedimentary layer indicate that the sedimentary layer was deposited on top of a weathered igneous mass rather than being intruded from below by magma that later crystallized.

Unconformities

When we observe layers of rock that have been deposited essentially without interruption, we call them **conformable.** Particular sites exhibit conformable beds representing certain spans of geologic time. However, no place on Earth has a complete set of conformable strata.

Throughout Earth history, the deposition of sediment has been interrupted again and again. All such breaks in the rock record are termed *unconformities*. An **unconformity** represents a long period during which deposition ceased, erosion removed previously formed rocks, and then deposition resumed. In each case, up-

Figure 18.4 Cross-cutting relationships represent one principle used in relative dating. An intrusive rock body is younger than the rocks it intrudes. A fault is younger than the rock layers it cuts.

lift and erosion are followed by subsidence and re-newed sedimentation. Unconformities are important features because they represent significant geologic events in Earth history. Moreover, their recognition helps us identify what intervals of time are not represented by strata and thus are missing from the geologic record.

The rocks exposed in the Grand Canyon of the Colorado River represent a tremendous span of geologic history. It is a wonderful place to take a trip through time. The canyon's colorful strata record a long history of sedimentation in a variety of environments—advancing seas, rivers and deltas, tidal flats, and sand dunes. But the record is not continuous. Unconformities represent vast amounts of time that have not been recorded in the canyon's layers. Figure 18.6 is a geologic cross section of the Grand Canyon. Refer to it as you read about the three basic types of unconformities: angular unconformities, disconformities, and nonconformities.

Angular Unconformity.   Perhaps the most easily recognized unconformity is an **angular unconformity.** It consists of tilted or folded sedimentary rocks that are overlain by younger, more flat-lying strata. An angular

unconformity indicates that during the pause in deposition, a period of deformation (folding or tilting) and erosion occurred (Figure 18.7).

When James Hutton studied an angular unconformity in Scotland more than 200 years ago, it was clear to him that it represented a major episode of geologic activity.* He also appreciated the immense time span implied by such relationships. When a companion later wrote of their visit to the site, he stated that, "The mind seemed to grow giddy by looking so far into the abyss of time."

Disconformity.   When contrasted with angular unconformities, **disconformities** are more common, but usually far less conspicuous because the strata on either side are essentially parallel. For example, look at the disconformities in the cross section of the Grand Canyon in Figure 18.6. Many disconformities are difficult to identify because the rocks above and below are similar and there is little evidence of erosion. Such a break often resembles an ordinary bedding plane.

*This pioneering geologist is discussed in the section on "The Birth of Modern Geology" in Chapter 1.

A. Intrusive
igneous rock

B. Exposure and
weathering of intrusive igneous rock

C. Deposition of
sedimentary layers

Figure 18.5 These three diagrams illustrate one way that inclusions can form, and a type of unconformity termed a *nonconformity*. In the third diagram we know the igneous rock must be older because pieces of it are included in the overlying sedimentary bed. When older intrusive igneous rocks are overlain by younger sedimentary layers, a nonconformity is said to exist. The photo shows an inclusion of dark igneous rock in a lighter-colored and younger host rock. (Photo by Tom Bean)

Other disconformities are easier to identify because the ancient erosion surface is cut deeply into the older rocks below.

Nonconformity. The third basic type of unconformity is a **nonconformity.** Here the break separates older metamorphic or intrusive igneous rocks from younger sedimentary strata (Figures 18.5 and 18.6). Just as angular unconformities and disconformities

imply crustal movements, so too do nonconformities. Intrusive igneous masses and metamorphic rocks originate far below the surface. Thus, for a nonconformity to develop, there must be a period of uplift and the erosion of overlying rocks. Once exposed at the surface, the igneous or metamorphic rocks are subjected to weathering and erosion prior to subsidence and the renewal of sedimentation.

Using Relative Dating Principles

If you apply the principles of relative dating to the hypothetical geologic cross section in Figure 18.8, you can place in proper sequence the rocks and the events they represent. The statements within the figure summarize the logic used to interpret the cross section.

In this example, we establish a relative time scale for the rocks and events in the area of the cross section. Remember that this method gives us no idea of how many years of Earth history are represented, for we have no numerical dates. Nor do we know how this area compares to any other.

Correlation of Rock Layers

To develop a geologic time scale that is applicable to the entire Earth, rocks of similar age in different regions must be matched up. Such a task is referred to as **correlation.**

Within a limited area, correlating the rocks of one locality with those of another may be done simply by walking along the outcropping edges. However, this may not be possible when the rocks are mostly concealed by soil and vegetation. Correlation over short distances is often achieved by noting the position of a distinctive rock layer in a sequence of strata. Or, a layer may be identified in another location if it is composed of very distinctive or uncommon minerals.

By correlating the rocks from one place to another, a more comprehensive view of the geologic history of a region is possible. Figure 18.9, for example, shows the correlation of strata at three sites on the Colorado Plateau in southern Utah and northern Arizona. No single locale exhibits the entire sequence, but correlation reveals a more complete picture of the sedimentary rock record.

Many geologic studies involve relatively small areas. Such studies are important in their own right, but their full value is realized only when the rocks are correlated with those of other regions. Although the methods just described are sufficient to trace a rock formation over relatively short distances, they are not adequate for matching rocks that are separated by great distances. When correlation between widely separated areas or between continents is the objective, geologists must rely on fossils.

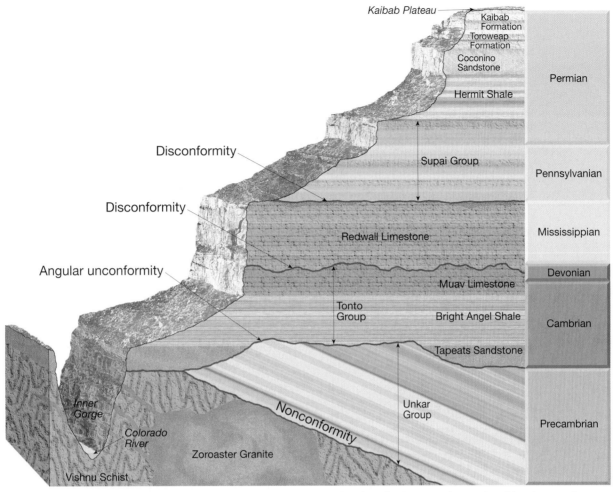

Figure 18.6 This cross-section through the Grand Canyon illustrates the three basic types of unconformities. An angular unconformity can be seen between the tilted Precambrian Unkar Group and the Cambrian Tapeats Sandstone. Two disconformities are marked, above and below the Redwall Limestone. A nonconformity occurs between the igneous and metamorphic rocks of the Inner Gorge and the sedimentary strata of the Unkar Group.

Fossils: Evidence of Past Life

Fossils, the remains or traces of prehistoric life, are important inclusions in sediment and sedimentary rocks. They are important tools for interpreting the geologic past. Knowing the nature of the life forms that existed at a particular time helps researchers understand past environmental conditions. Further, fossils are important time indicators and play a key role in correlating rocks of similar ages that are from different places.

Types of Fossils

Fossils are of many types. The remains of relatively recent organisms may not have been altered at all. Such objects as teeth, bones, and shells are common examples. Far less common are entire animals, flesh included, that have been preserved because of rather unusual circumstances. Remains of prehistoric elephants called *mammoths*, which were frozen in the Arctic tundra

of Siberia and Alaska, are examples, as are the mummified remains of sloths preserved in a dry cave in Nevada.

Given enough time, the remains of an organism are likely to be modified. Often fossils become *petrified* (literally, "turned into stone"), meaning that the small internal cavities and pores of the original structure are filled with precipitated mineral matter (Figure 18.10A). In other instances *replacement* may occur. Here the cell walls and other solid material are removed and replaced with mineral matter. Sometimes the microscopic details of the replaced structure are faithfully retained.

Molds and casts constitute another common class of fossils. When a shell or other structure is buried in sediment and then dissolved by underground water, a *mold* is created. The mold faithfully reflects only the shape and surface marking of the organism; it does not reveal any information concerning its internal structure. If these hollow spaces are subsequently filled with mineral matter, *casts* are created (Figure 18.10B).

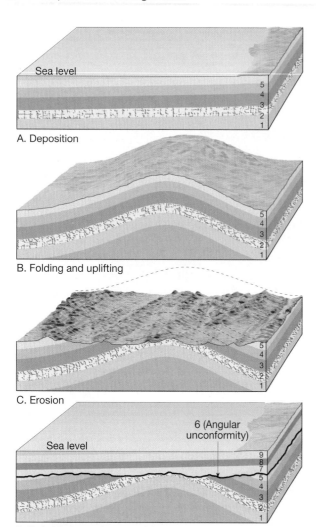

A. Deposition

B. Folding and uplifting

C. Erosion

6 (Angular unconformity)

Sea level

D. Subsidence and renewed deposition

E.

Figure 18.7 Formation of an angular unconformity. An angular unconformity represents an extended period during which deformation and erosion occurred. Part E shows an angular unconformity at Siccar Point, Scotland, that was first described by James Hutton more than 200 years ago. (Photo by A. P. Trujillo/APT Photos)

A type of fossilization called *carbonization* is particularly effective in preserving leaves and delicate animal forms. It occurs when fine sediment encases the remains of an organism. As time passes, pressure squeezes out the liquid and gaseous components and leaves behind a thin residue of carbon (Figure 18.10C). Black shales deposited as organic-rich mud in oxygen-poor environments often contain abundant carbonized remains. If the film of carbon is lost from a fossil preserved in fine-grained sediment, a replica of the surface, called an *impression*, may still show considerable detail (Figure 18.10D).

Delicate organisms, such as insects, are difficult to preserve and consequently are relatively rare in the fossil record. Not only must they be protected from decay, but they must not be subjected to any pressure that would crush them. One way in which some insects have been preserved is in *amber*, the hardened resin of ancient trees. The fly in Figure 18.10E was preserved after being trapped in a drop of sticky resin. Resin sealed off the insect from the atmosphere and protected the remains from damage by water and air. As the resin hardened, a protective, pressure-resistant case was formed.

In addition to the fossils already mentioned, there are numerous other types, many of them only traces of prehistoric life. Examples of such indirect evidence include:

1. Tracks—animal footprints made in soft sediment that was later lithified (Figure 18.10F).
2. Burrows—tubes in sediment, wood, or rock made by an animal. These holes may later become filled with mineral matter and preserved. Some of the oldest-known fossils are believed to be worm burrows.
3. Coprolites—fossil dung and stomach contents that can provide useful information pertaining to food habits of organisms.
4. Gastroliths—highly polished stomach stones that were used in the grinding of food by some extinct reptiles.

Conditions Favoring Preservation

Only a tiny fraction of the organisms that have lived during the geologic past have been preserved as fossils. Normally, the remains of an animal or plant are destroyed. Under what circumstances are they preserved? Two special conditions appear to be necessary: rapid burial and the possession of hard parts.

When an organism perishes, its soft parts usually are quickly eaten by scavengers or decomposed by bacteria. Occasionally, however, the remains are buried by sediment. When this occurs, the remains are protected

Interpretation:

1. Applying the *law of superposition*, beds *A*, *B*, *C*, and *E* were deposited in that order.

2. Bed *D* is a sill (a concordant igneous intrusion). Applying the principle of cross-cutting relationships, sill *D* must be younger than the rocks that were intruded. Further evidence that sill *D* is younger than beds *C* and *E* are the *inclusions* in the sill of fragments from these beds. If this igneous mass contains pieces of adjacent strata, then the adjacent strata must have been there first.

3. Following the intrusion of sill *D*, the intrusion of dike *F* occurred. Because the dike cuts through beds *A* through *E*, it must be younger than all of them (principle of cross-cutting relationships).

4. Next, the rocks were tilted and eroded. The tilting happened first because the upturned ends of the strata have been eroded. The tilting and erosion, followed by further deposition, produced an *angular unconformity*.

5. Beds, *G*, *H*, *I*, *J*, and *K* were deposited in that order, again using the *law of superposition*. Although the lava flow (bed *H*) is not a sedimentary rock layer, it is a surface-deposited layer, and thus superposition may be applied.

6. Finally, the irregular surface and the stream valley indicate that another gap in the rock record is being produced by erosion.

Figure 18.8   Geologic cross-section of a hypothetical region.

from the environment where destructive processes operate. Rapid burial therefore is an important condition favoring preservation.

In addition, animals and plants have a much better chance of being preserved as part of the fossil record if they have hard parts. Although traces and imprints of soft-bodied animals such as jellyfish, worms, and insects exist, they are not common. Flesh usually decays so rapidly that preservation is exceedingly unlikely. Hard parts such as shells, bones, and teeth predominate in the record of past life.

Because preservation is contingent on special conditions, the record of life in the geologic past is biased. The fossil record of those organisms with hard parts

that lived in areas of sedimentation is quite abundant. However, we get only an occasional glimpse of the vast array of other life forms that did not meet the special conditions favoring preservation.

Fossils and Correlation

The existence of fossils had been known for centuries, yet it was not until the late 1700s and early 1800s that their significance as geologic tools was made evident. During this period an English engineer and canal builder, William Smith, discovered that each rock formation in the canals he worked on contained fossils unlike those in the beds either above or below. Further, he

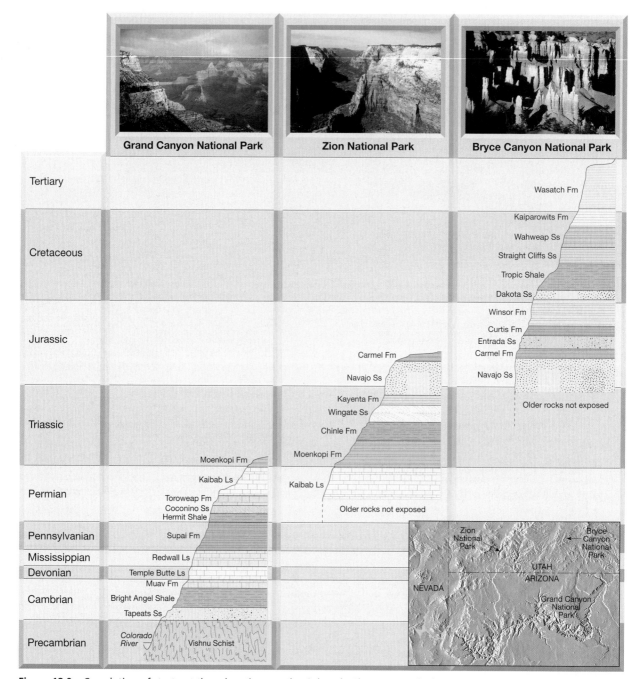

Figure 18.9   Correlation of strata at three locations on the Colorado Plateau reveals the total extent of sedimentary rocks in the region. (After U.S. Geological Survey; Photos by E. J. Tarbuck)

noted that sedimentary strata in widely separated areas could be identified and correlated by their distinctive fossil content.

Based on Smith's classic observations and the findings of many geologists who followed, one of the most important and basic principles in historical geology was formulated: *Fossil organisms succeed one another in a definite and determinable order, and therefore any time period can be recognized by its fossil content.* This has come to be known as the **principle of fossil succession.** In other words, when fossils are arranged according to their age

by applying the law of superposition to the rocks in which they are found, they do not present a random or haphazard picture. To the contrary, fossils document the evolution of life through time.

For example, an Age of Trilobites is recognized quite early in the fossil record. Then, in succession, paleontologists recognize an Age of Fishes, an Age of Coal Swamps, an Age of Reptiles, and an Age of Mammals. These "ages" pertain to groups that were especially plentiful and characteristic during particular time periods. Within each of the "ages," there are many subdivisions

Figure 18.10 There are many types of fossilization. Six examples are shown here. **A.** Petrified wood in Petrified Forest National Park, Arizona. **B.** Natural casts of shelled invertebrates. **C.** A fossil bee preserved as a thin carbon film. **D.** Impressions are common fossils and often show considerable detail. **E.** Insect in amber. **F.** Dinosaur footprint in fine-grained limestone near Tuba City, Arizona. (Photo A by David Muench; Photos B, D, and F by E. J. Tarbuck; Photo C courtesy of the National Park Service; Photo E by Breck P. Kent)

based, for example, on certain species of trilobites and certain types of fish, reptiles, and so on. This same succession of dominant organisms, never out of order, is found on every continent.

Once fossils were recognized as time indicators, they became the most useful means of correlating rocks of similar age in different regions. Geologists pay particular attention to certain fossils called **index fossils.** These fossils are widespread geographically and are limited to a short span of geologic time, so their presence provides an important method of matching rocks of the same age. Rock formations, however, do not always contain a specific index fossil. In such situations, groups of fossils are used to establish the age of the bed. Figure 18.11 illustrates how a group of fossils can be used to date rocks more precisely than could be accomplished by the use of only one of the fossils.

In addition to being important and often essential tools for correlation, fossils are important environmental indicators. Although much can be deduced about past environments by studying the nature and characteristics of sedimentary rocks, a close examination of any fossils present can usually provide a great deal more information.

For example, when the remains of certain clam shells are found in limestone, the geologist can assume that the region was once covered by a shallow sea, because that is where clams live today. Also, by using what we know of living organisms, we can conclude that fos-

sil animals with thick shells capable of withstanding pounding and surging waves must have inhabited shorelines. In contrast, animals with thin, delicate shells probably indicate deep, calm offshore waters. Hence, by looking closely at the types of fossils, the approximate position of an ancient shoreline may be identified.

Further, fossils can indicate the former temperature of the water. Certain present-day corals require warm and shallow tropical seas like those around Florida and the Bahamas. When similar corals are found in ancient limestones, they indicate that a Florida-like marine environment must have existed when they were alive. These examples illustrate how fossils can help unravel the complex story of Earth history.

Dating with Radioactivity

GEODe II   Geologic Time
↳ Radiometric Dating

In addition to establishing relative dates by using the principles described in the preceding sections, it is also possible to obtain reliable numerical dates for events in the geologic past. For example, we know that Earth is about 4.5 billion years old and that the dinosaurs became extinct about 65 million years ago. Dates that are

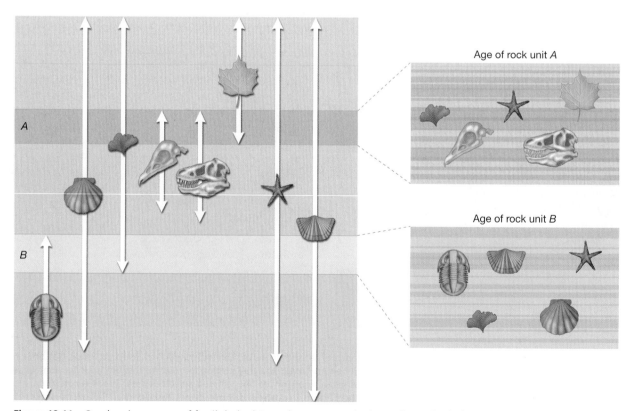

Figure 18.11   Overlapping ranges of fossils help date rocks more exactly than using a single fossil.

expressed in millions and billions of years truly stretch our imagination because our personal calendars involve time measured in hours, weeks, and years. Nevertheless, the vast expanse of geologic time is a reality, and it is radiometric dating that allows us to measure it. In this section you will learn about radioactivity and its application in radiometric dating.

Reviewing Basic Atomic Structure

Recall from Chapter 2 that each atom has a *nucleus* containing protons and neutrons and that the nucleus is orbited by electrons. *Electrons* have a negative electrical charge, and *protons* have a positive charge. A *neutron* is actually a proton and an electron combined, so it has no charge (it is neutral).

The *atomic number* (each element's identifying number) is the number of protons in the nucleus. Every element has a different number of protons and thus a different atomic number (hydrogen = 1, carbon = 6, oxygen = 8, uranium = 92, etc.). Atoms of the same element always have the same number of protons, so the atomic number stays constant.

Practically all of an atom's mass (99.9%) is in the nucleus, indicating that electrons have virtually no mass at all. So, by adding the protons and neutrons in an atom's nucleus, we derive the atom's *mass number*. The number of neutrons can vary, and these variants, or *isotopes*, have different mass numbers.

To summarize with an example, uranium's nucleus always has 92 protons, so its atomic number always is 92. But its neutron population varies, so uranium has three isotopes: uranium-234 (protons + neutrons = 234), uranium-235, and uranium-238. All three isotopes are mixed in nature. They look the same and behave the same in chemical reactions.

Radioactivity

The forces that bind protons and neutrons together in the nucleus usually are strong. However, in some isotopes, the nuclei are unstable because the forces binding protons and neutrons together are not strong enough. As a result, the nuclei spontaneously break apart (decay), a process called **radioactivity.**

What happens when unstable nuclei break apart? Three common types of radioactive decay are illustrated in Figure 18.12 and are summarized as follows:

1. Alpha particles (α particles) may be emitted from the nucleus. An alpha particle is composed of 2 protons and 2 neutrons. Consequently, the emission of an alpha particle means that the mass number of the isotope is reduced by 4 and the atomic number is decreased by 2.

2. When a beta particle (β particle), or electron, is given off from a nucleus, the mass number remains unchanged because electrons have practically no mass. However, because the electron has come from a neutron (remember, a neutron is a combination of a proton and an electron), the nucleus contains one more proton than before. Therefore, the atomic number increases by 1.

3. Sometimes an electron is captured by the nucleus. The electron combines with a proton and forms an additional neutron. As in the last example, the mass number remains unchanged. However, as the nucleus now contains one less proton, the atomic number decreases by 1.

An unstable (radioactive) isotope of an element is called the *parent*. The isotopes resulting from the decay of the parent are the *daughter products*. Figure 18.13 provides an example of radioactive decay. Here it can be seen that when the radioactive parent, uranium-238 (atomic number 92, mass number 238) decays, it follows a number of steps, emitting 8 alpha particles and 6 beta particles before finally becoming the stable daughter product lead-206 (atomic number 82, mass number 206). One of the unstable daughter products produced during this decay series is radon. Box 18.1 examines the hazards associated with this radioactive gas.

Certainly among the most important results of the discovery of radioactivity is that it provided a reliable means of calculating the ages of rocks and minerals that contain particular radioactive isotopes. The procedure is called **radiometric dating.** Why is radiometric dating reliable? Because the rates of decay for many isotopes have been precisely measured and do not vary under the physical conditions that exist in Earth's outer layers. Therefore, each radioactive isotope used for dating has been decaying at a fixed rate ever since the formation of the rocks in which it occurs, and the products of decay have been accumulating at a corresponding rate.

For example, when uranium is incorporated into a mineral that crystallizes from magma, there is no lead (the stable daughter product) from previous decay. The radiometric "clock" starts at this point. As the uranium in this newly formed mineral disintegrates, atoms of the daughter product are trapped and measurable amounts of lead eventually accumulate.

Half-Life

The time required for half of the nuclei in a sample to decay is called the **half-life** of the isotope. Half-life is a common way of expressing the rate of radioactive disintegration. Figure 18.14 illustrates what occurs when a radioactive parent decays directly into its stable daughter product. When the quantities of parent and daughters are equal (ratio 1:1), we know that one

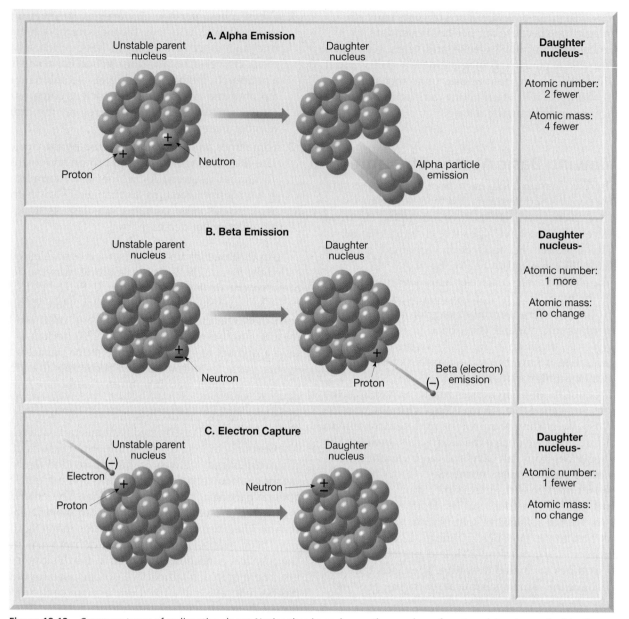

Figure 18.12 Common types of radioactive decay. Notice that in each case the number of protons (atomic number) in the nucleus changes, thus producing a different element.

half-life has transpired. When one quarter of the original parent atoms remain and three quarters have decayed to the daughter product, the parent/daughter ratio is 1:3, and we know that two half-lives have passed. After three half-lives, the ratio of parent atoms to daughter atoms is 1:7 (one parent for every seven daughter atoms).

If the half-life of a radioactive isotope is known and the parent/daughter ratio can be measured, the age of the sample can be calculated. For example, assume that the half-life of a hypothetical unstable isotope is 1 million years and the parent/daughter ratio in a sample is 1:15. Such a ratio indicates that four half-lives have passed and that the sample must be 4 million years old.

Radiometric Dating

Notice that the *percentage* of radioactive atoms that decay during one half-life is always the same: 50 percent. However, the *actual number* of atoms that decay with the passing of each half-life continually decreases. Thus, as the percentage of radioactive parent atoms declines, the proportion of stable daughter atoms rises, with the increase in daughter atoms just matching the drop in parent atoms. This fact is the key to radiometric dating.

Of the many radioactive isotopes that exist in nature, five have proved particularly useful in providing radiometric ages for ancient rocks (Table 18.1). Rubidium-87, thorium-232, and the two isotopes of uranium

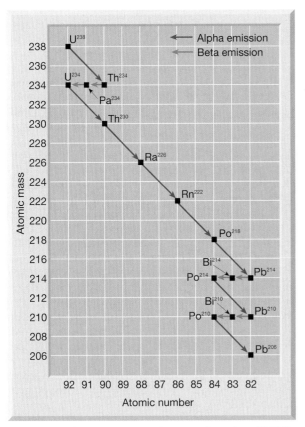

Figure 18.13 The most common isotope of uranium (U-238) is an example of a radioactive decay series. Before the stable end product (Pb-206) is reached, many different isotopes are produced as intermediate steps.

are used only for dating rocks that are millions of years old, but potassium-40 is more versatile.

Potassium-Argon. Although the half-life of potassium-40 is 1.3 billion years, analytical techniques make possible the detection of tiny amounts of its stable daughter product, argon-40, in some rocks that are younger than 100,000 years. Another important reason for its frequent use is that potassium is an abundant constituent of many common minerals, particularly micas and feldspars.

Although potassium (K) has three natural isotopes—K^{39}, K^{40}, and K^{41}—only K^{40} is radioactive. When K^{40} decays, it does so in two ways. About 11 percent changes to argon-40 (Ar^{40}) by means of electron capture (see Figure 18.12C). The remaining 89 percent of K^{40} decays to calcium-40 (Ca^{40}) by beta emission (see Figure 18.12B). The decay of K^{40} to Ca^{40}, however, is not useful for radiometric dating, because the Ca^{40} produced by radioactive disintegration cannot be distinguished from calcium that may have been present when the rock formed.

The potassium-argon clock begins when potassium-bearing minerals crystallize from a magma or form within a metamorphic rock. At this point the new min-

erals will contain K^{40} but will be free of Ar^{40}, because this element is an inert gas that does not chemically combine with other elements.

As time passes, the K^{40} steadily decays by electron capture. The Ar^{40} produced by this process remains trapped within the mineral's crystal lattice. Because no Ar^{40} was present when the mineral formed, all of the daughter atoms trapped in the mineral must have come from the decay of K^{40}. To determine a sample's age, the K^{40}/Ar^{40} ratio is measured precisely and the known half-life for K^{40} applied.

Sources of Error. It is important to realize that an accurate radiometric date can be obtained only if the mineral remained a closed system during the entire period since its formation. A correct date is not possible unless there was neither the addition nor loss of parent or daughter isotopes.

However, this is not always the case. In fact, an important limitation of the potassium-argon method arises from the fact that argon is a gas and it may leak from minerals, throwing off measurements. Indeed, losses can be significant if the rock is subjected to relatively high temperatures. Of course, a reduction in the amount of Ar^{40} leads to an underestimation of the rock's age. Sometimes temperatures are high enough for a sufficiently long period that all argon escapes. When this happens, the potassium-argon clock is reset, and dating the sample will give only the time of thermal resetting, not the true age of the rock. For other radiometric clocks, a loss of daughter atoms can occur if the rock has been subjected to weathering or leaching. To avoid such a problem, one simple safeguard is to use only fresh, unweathered material and not samples that may have been chemically altered.

If parent/daughter ratios are not always reliable, how can meaningful radiometric dates be obtained? One common precaution against unknown errors is the use of cross checks. Often this simply involves subjecting a sample to two different radiometric methods. If the two dates agree, the likelihood is high that the date is reliable. Conversely, if an appreciable difference is found between the two dates, other cross checks must be employed to determine which, if either, is correct.

Dating with Carbon-14

To date very recent events, carbon-14 is used. Carbon-14 is the radioactive isotope of carbon. The process is often called **radiocarbon dating.** Because the half-life of carbon-14 is only 5730 years, it can be used for dating events from the historic past as well as those from very recent geologic history (see Box 18.2). In some cases carbon-14 can be used to date events as far back as 75,000 years.

BOX 18.1

Radon

Richard L. Hoffmann, Ph.D.*

Radioactivity is defined as the spontaneous emission of atomic particles and/or electromagnetic waves from unstable atomic nuclei. For example, in a sample of uranium-238, unstable nuclei decay and produce a variety of radioactive progeny or daughter products, as well as energetic forms of radiation (Table 18.A). One of its radioactive decay products is radon—a colorless, odorless, invisible gas.

Radon gained public attention in 1984 when a worker in a Pennsylvania nuclear power plant set off radiation alarms—not when he left work, but as he first arrived there. His clothing and hair were contaminated with radon decay products. Investigation revealed that his basement at home had a radon level 2800 times the average level in indoor air. The home was located along a geological formation known as the Reading Prong—a mass of uranium-bearing rock that runs from near Reading, Pennsylvania, to near Trenton, New Jersey.

Originating in the radio decay of traces of uranium and thorium found in almost all soils, radon isotopes (Rn-222 and Rn-220) are continually renewed in an ongoing, natural process. Geologists estimate that the top 6 feet of soil from an average acre of land contains about 50 pounds of uranium (about 2 to 3 parts per million); some

types of rocks contain more. Radon is continually generated by the gradual decay of this uranium. Because uranium has a half-life of about 4.5 billion years, radon will be with us forever.

Radon itself decays, having a half-life of only about four days. Its decay products (except lead-206) are all radioactive solids that adhere to dust particles, many of which we inhale. During prolonged exposure to a radon-contaminated environment, some decay will occur while the gas is in the lungs, thereby placing the radioactive radon progeny in direct contact with delicate lung tissue. Steadily accumulating evidence indicates radon to be a significant cause of lung cancer second only to smoking.

A house with a radon level of 4.0 picocuries per liter of air has about eight to nine atoms of radon decaying every minute in every liter of air. The Environmental Protection Agency (EPA) suggests that indoor radon levels be kept below this level. The EPA risk estimates are conservative; they are based on an assumption that one would spend 75 percent of a 70-year time span (about 52 years) in the contaminated space, which most people would not.

Once radon is produced in the soil, it diffuses throughout the tiny spaces between soil particles. Some radon ultimately reaches the soil surface, where it dissipates into the air. Radon enters buildings and homes through holes and cracks in basement floors and walls. Radon's density is greater than air, so it tends to remain in basements during its short decay cycle.

The source of radon is as enduring as its generation mechanism within Earth; radon will never go away. However, cost-effective mitigation strategies are available to reduce radon to acceptable levels, generally without great expense.

*Dr. Hoffmann is Professor of Chemistry, Emeritus, Illinois Central College.

Table 18.A Decay products of uranium-238.

Some Decay Products of Uranium-238	Decay Particle Produced	Half-Life
Uranium-238	alpha	4.5 billion years
Radium-226	alpha	1600 years
Radon-222	alpha	3.82 days
Polonium-218	alpha	3.1 minutes
Lead-214	beta	26.8 minutes
Bismuth-214	beta	19.7 minutes
Polonium-214	alpha	1.6×10^{-4} second
Lead-210	beta	20.4 years
Bismuth-210	beta	5.0 days
Polonium-210	alpha	138 days
Lead-206	none	stable

Carbon-14 is continuously produced in the upper atmosphere as a consequence of cosmic-ray bombardment. Cosmic rays, which are high-energy particles, shatter the nuclei of gas atoms, releasing neutrons. Some of the neutrons are absorbed by nitrogen atoms (atomic number 7), causing their nuclei to emit a proton. As a result, the atomic number decreases by 1 (to 6), and a different element, carbon-14, is created (Figure 18.15A). This isotope of carbon quickly becomes incorporated into carbon dioxide, which circulates in the atmosphere and is absorbed by living matter. As a result, all organisms contain a small amount of carbon-14, including yourself.

While an organism is alive, the decaying radiocarbon is continually replaced, and the proportions of carbon-14 and carbon-12 remain constant. Carbon-12 is the stable and most common isotope of carbon. However, when any plant or animal dies, the amount of carbon-14 gradually decreases as it decays to nitrogen-14 by beta emission (Figure 18.15B). By comparing the proportions of carbon-14 and carbon-12 in a sample, radiocarbon dates can be determined.

Although carbon-14 is useful in dating only the last small fraction of geologic time, it has become a very valuable tool for anthropologists, archeologists, and historians, as well as for geologists who study very recent

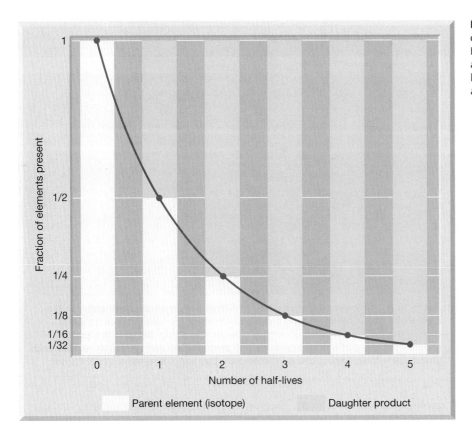

Figure 18.14   The radioactive-decay curve shows change that is exponential. Half of the radioactive parent remains after one half-life. After a second half-life, one quarter of the parent remains, and so forth.

Earth history. In fact, the development of radiocarbon dating was considered so important that the chemist who discovered this application, Willard F. Libby, received a Nobel Prize.

Importance of Radiometric Dating

Bear in mind that although the basic principle of radiometric dating is simple, the actual procedure is quite complex. The analysis that determines the quantities of parent and daughter must be painstakingly precise. In addition, some radioactive materials do not decay directly into the stable daughter product. As you saw in Figure 18.13, uranium-238 produces 13 intermediate unstable daughter products before the 14th and final daughter product, the stable isotope lead-206, is produced.

Radiometric dating methods have produced literally thousands of dates for events in Earth history. Rocks from several localities have been dated at more than 3 billion years, and geologists realize that still older rocks exist. For example, a granite from South Africa has been dated at 3.2 billion years—and it contains inclusions of quartzite. (Remember that inclusions are older than the rock containing them.) Quartzite itself is a metamorphic rock that originally was the sedimentary rock sandstone. Sandstone, in turn, is the product of the lithification of sediments produced by the weathering of existing rocks. Thus, we have a positive indication that much older rocks existed.

Radiometric dating has vindicated the ideas of James Hutton, Charles Darwin, and others who inferred that geologic time must be immense. Indeed, modern dating methods have proved that there has been enough time for the processes we observe to have accomplished tremendous tasks.

Table 18.1 Radioactive isotopes frequently used in radiometric dating.

Radioactive Parent	Stable Daughter Product	Currently Accepted Half-life Values
Uranium-238	Lead-206	4.5 billion years
Uranium-235	Lead-207	713 million years
Thorium-232	Lead-208	14.1 billion years
Rubidium-87	Strontium-87	47.0 billion years
Potassium-40	Argon-40	1.3 billion years

The Geologic Time Scale

 Geologic Time
↳ Geologic Time Scale

Geologists have divided the whole of geologic history into units of varying magnitude. Together they com-

BOX 18.2

Using Tree Rings to Date and Study the Recent Past

If you look at the top of a tree stump or at the end of a log, you will see that it is composed of a series of concentric rings. Each of these *tree rings* becomes larger in diameter outward from the center (Figure 18.A). Every year in temperate regions trees add a layer of new wood under the bark. Characteristics of each tree ring, such as size and density, reflect the environmental conditions (especially climate) that prevailed during the year when the ring formed. Favorable growth conditions produce a wide ring; unfavorable ones produce a narrow ring. Trees growing at the same time in the same region show similar tree-ring patterns.

Because a single growth ring is usually added each year, the age of the tree when it was cut can be determined by counting the rings. If the year of cutting is known, the age of the tree and the year in which each ring formed can be determined by counting back from the outside ring.* This procedure can be used to determine the dates of recent

*Scientists are not limited to working with trees that have been cut down. Small, nondestructive core samples can be taken from living trees.

Figure 18.A   Each year a growing tree produces a layer of new cells beneath the bark. If the tree is felled and the trunk examined (or if a core is taken, to avoid cutting the tree), each year's growth can be seen as a ring. Because the amount of growth (thickness of a ring) depends upon precipitation and temperature, tree rings are useful records of past climates. (Photo by Stephen J. Krasemann/DRK Photo)

geologic events. For example, the minimum number of years since a new land surface was created by a landslide or a flood. The dating and study of annual rings in trees is called *dendrochronology*.

To make the most effective use of tree rings, extended patterns known as ring chronologies are established. They are

produced by comparing the patterns of rings among trees in an area. If the same pattern can be identified in two samples, one of which has been dated, the second sample can be dated from the first by matching the ring pattern common to both. This technique, called *cross dating*, is illustrated in Figure 18.B. Tree-ring chronologies extending back for thou-

Figure 18.15   **A.** Production and **B.** decay of carbon-14. These sketches represent the nuclei of the respective atoms.

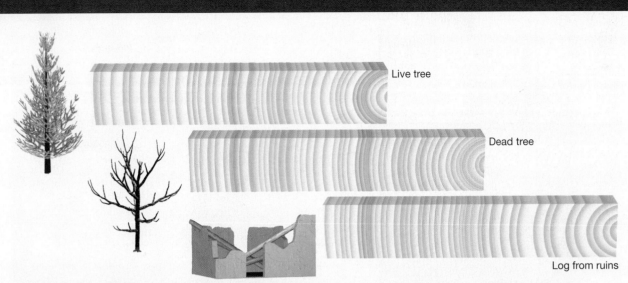

Live tree

Dead tree

Log from ruins

Figure 18.B   Cross dating is a basic principle in dendrochronology. Here it was used to date an archaeological site by correlating tree-ring patterns for wood from trees of three different ages. First, a tree-ring chronology for the area is established using cores extracted from living trees. This chronology is extended further back in time by matching overlapping patterns from older, dead trees. Finally, cores taken from beams inside the ruin are dated using the chronology established from the other two sites.

sands of years have been established for some regions. To date a timber sample of unknown age, its ring pattern is matched against the reference chronology.

Tree-ring chronologies are unique archives of environmental history and have important applications in such disciplines as climate, geology, ecology, and archaeology. For example, tree rings are used to reconstruct climate variations within a region for spans of thousands of years prior to human historical records. Knowledge of such long-term variations is of great value in making judgments regarding the recent record of climate change.

In summary, dendrochronology provides useful numerical dates for events in the historic and recent prehistoric past. Moreover, because tree rings are a storehouse of data, they are a valuable tool in the reconstruction of past environments.

prise the **geologic time scale** of Earth history (Figure 18.16). The major units of the time scale were delineated during the nineteenth century, principally by workers in Western Europe and Great Britain. Because radiometric dating was unavailable at that time, the entire time scale was created using methods of relative dating. It was only in the twentieth century that radiometric dating permitted numerical dates to be added.

Structure of the Time Scale

The geologic time scale subdivides the 4.5-billion-year history of Earth into many different units and provides a meaningful time frame within which the events of the geologic past are arranged. As shown in Figure 18.16, **eons** represent the greatest expanses of time. The eon that began about 540 million years ago is the **Phanerozoic,** meaning *visible life.* It is an appropriate description because the rocks and deposits of the Phanerozoic eon contain abundant fossils that document major evolutionary trends.

Another glance at the time scale reveals that the Phanerozoic eon is divided into **eras.** The three eras within the Phanerozoic are the **Paleozoic** ("ancient life"), the **Mesozoic** ("middle life"), and the **Cenozoic** ("recent life"). As the names imply, the eras are bounded by profound worldwide changes in life forms. Each era is subdivided into **periods.** The Paleozoic has seven, the Mesozoic three, and the Cenozoic two. Each of these 12 periods is characterized by a somewhat less profound change in life forms as compared with the eras. The eras and periods of the Phanerozoic, with brief explanations of each, are shown in Table 18.2.

Finally, periods are divided into still smaller units called **epochs.** As you can see in Figure 18.16, seven epochs have been named for the periods of the Cenozoic. The epochs of other periods, however, are not usually referred to by specific names. Instead, the terms *early, middle,* and *late* are generally applied to the epochs of these earlier periods.

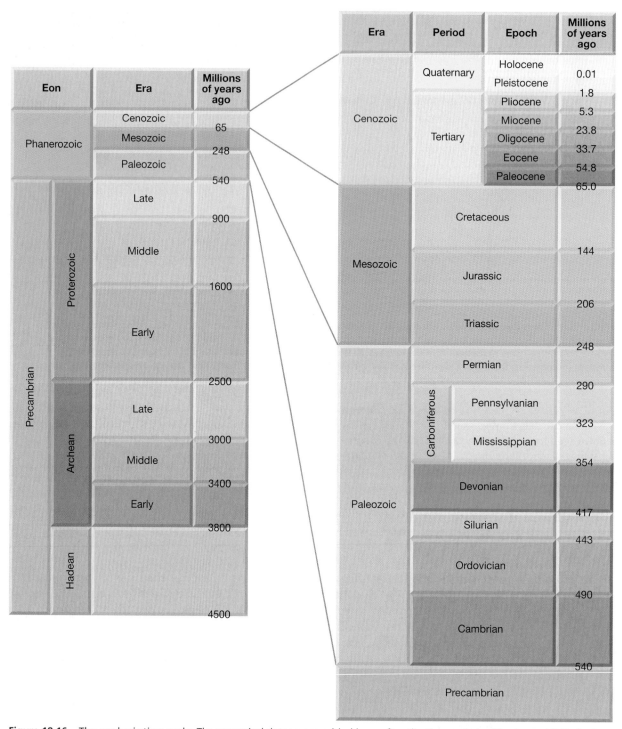

Figure 18.16 The geologic time scale. The numerical dates were added long after the time scale had been established using relative dating techniques. (Data from Geological Society of America)

Precambrian Time

Notice that the detail of the geologic time scale does not begin until about 540 million years ago, the date for the beginning of the Cambrian period. The nearly 4 billion years prior to the Cambrian is divided into three eons, the *Hadean,* the *Archean,* and the *Proterozoic.* It is also common for this vast expanse of time to simply be referred to as the **Precambrian.** Although it represents about 88 percent of Earth history, the Precambrian is not divided into nearly as many smaller time units as the Phanerozoic eon.

The quantity of information that geologists have deciphered about Earth's past is somewhat analogous

Table 18.2 Major divisions of geologic time.

Cenozoic Era (Age of Recent Life)	Quaternary period	The several geologic eras were originally named Primary, Secondary, Tertiary, and Quaternary. The first two names are no longer used; Tertiary and Quaternary have been retained but used as period designations.
	Tertiary period	
Mesozoic Era (Age of Middle Life)	Cretaceous period	Derived from Latin word for chalk (creta) and first applied to extensive deposits that form white cliffs along the English Channel (see Figure 6.11, p. 130).
	Jurassic period	Named for the Jura Mountains, located between France and Switzerland, where rocks of this age were first studied.
	Triassic period	Taken from word "trias" in recognition of the threefold character of these rocks in Europe.
Paleozoic Era (Age of Ancient Life)	Permian period	Named after the province of Perm, Russia, where these rocks were first studied.
	Pennsylvanian period*	Named for the state of Pennsylvania where these rocks have produced much coal.
	Mississippian period*	Named for the Mississippi River Valley where these rocks are well exposed.
	Devonian period	Named after Devonshire County, England, where these rocks were first studied.
	Silurian period	Named after Celtic tribes, the Silures and the Ordovices, that lived in Wales during the Roman Conquest.
	Ordovician period	
	Cambrian period	Taken from Roman name for Wales (Cambria), where rocks containing the earliest evidence of complex forms of life were first studied.
Precambrian		The time between the birth of the planet and the appearance of complex forms of life. About 88 percent of Earth's estimated 4.5 billion years fall into this span.

Outside of North America, the Mississippian and Pennsylvanian periods are combined into the Carboniferous period.
SOURCE: U.S. Geological Survey.

to the detail of human history. The further back we go, the less we know. Certainly more data and information exist about the past 10 years than for the first decade of the twentieth century; the events of the nineteenth century have been documented much better than the events of the first century A.D.; and so on. So it is with Earth history. The more recent past has the freshest, least disturbed, and most observable record. The further back in time the geologist goes, the more fragmented the record and clues become.

Difficulties in Dating the Geologic Time Scale

Although reasonably accurate numerical dates have been worked out for the periods of the geologic time scale (see Figure 18.16), the task is not without difficulty. The primary problem in assigning numerical dates to units of time is the fact that not all rocks can be dated by radiometric methods. Recall that for a radiometric date to be useful, all minerals in the rock must have formed at about the same time. For this reason, radioactive isotopes can be used to determine when minerals in an ig-

neous rock crystallized and when pressure and heat created new minerals in a metamorphic rock.

However, samples of sedimentary rock can only rarely be dated directly by radiometric means. A sedimentary rock might include particles that contain radioactive isotopes, but the rock's age cannot be accurately determined, because the grains making up the rock are not the same age as the rock in which they occur. Rather, the sediments have been weathered from rocks of diverse ages.

Radiometric dates obtained from metamorphic rocks might also be difficult to interpret, because the age of a particular mineral in a metamorphic rock does not necessarily represent the time when the rock initially formed. Instead, the date could indicate any one of a number of subsequent metamorphic phases.

If samples of sedimentary rocks rarely yield reliable radiometric ages, how can numerical dates be assigned to sedimentary layers? Usually the geologist must relate them to datable igneous masses, as in Figure 18.17. In this example, radiometric dating has determined the ages of the volcanic ash bed within the Morrison Formation and the dike cutting the Mancos Shale and Mesaverde Formation. The sedimentary beds below the

Figure 18.17 Numerical dates for sedimentary layers are usually determined by examining their relationship to igneous rocks. (After U.S. Geological Survey)

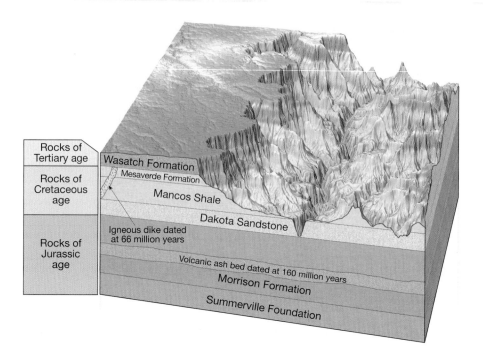

Rocks of Tertiary age

Rocks of Cretaceous age

Rocks of Jurassic age

Wasatch Formation

Mesaverde Formation

Mancos Shale

Dakota Sandstone

Igneous dike dated at 66 million years

Volcanic ash bed dated at 160 million years

Morrison Formation

Summerville Foundation

ash are obviously older than the ash, and all the layers above the ash are younger (principle of superposition). The dike is younger than the Mancos Shale and the Mesaverde Formation but older than the Wasatch Formation because the dike does not intrude the Tertiary rocks (cross-cutting relationships).

From this kind of evidence, geologists estimate that a part of the Morrison Formation was deposited about 160 million years ago, as indicated by the ash bed. Further, they conclude that the Tertiary period began after the intrusion of the dike, 66 million years ago. This is one example of literally thousands that illustrates how datable materials are used to *bracket* the various episodes in Earth history within specific time periods. It shows the necessity of combining laboratory dating methods with field observations of rocks.

The Chapter in Review

- The two types of dates used by geologists to interpret Earth history are (1) *relative dates*, which put events in their *proper sequence of formation*, and (2) *numerical dates*, which pinpoint the *time in years* when an event occurred.

- Relative dates can be established using the *law of superposition* (in an underformed sequence of sedimentary rocks or surface-deposited igneous rocks, each bed is older than the one above, and younger than the one below), *principle of original horizontality* (most layers are deposited in a horizontal position), *principle of cross-cutting relationships* (when a fault or intrusion cuts through another rock, the fault or intrusion is younger than the rocks cut through), and *inclusions* (the rock mass containing the inclusion is younger than the rock that provided the inclusion).

- *Unconformities* are gaps in the rock record. Each represents a long period during which deposition ceased, erosion removed previously formed rocks, and then deposition resumed. The three basic types of unconformities are *angular unconformities* (tilted or folded sedimentary rocks that are overlain by younger, more flat-lying strata), *disconformities*

(the strata on either side of the unconformity are essentially parallel), and *nonconformities* (where a break separates older metamorphic or intrusive igneous rocks from younger sedimentary strata).

- *Correlation*, matching rocks of similar age in different areas, is used to develop a geologic time scale that applies to the entire Earth.

- *Fossils* are the remains or traces of prehistoric life. The special conditions that favor preservation are *rapid burial* and the possession of *hard parts* such as shells, bones, or teeth.

- Fossils are used to *correlate* sedimentary rocks from different regions by using the rocks' distinctive fossil content and applying the *principle of fossil succession*. It is based on the work of *William Smith* in the late 1700s and states that fossil organisms succeed one another in a definite and determinable order, and therefore any time period can be recognized by its fossil content. The use of *index fossils*, those that are widespread geographically and are limited to a short span of geologic time, provides an important method for matching rocks of the same age.

- Each atom has a nucleus containing *protons* (positively charged particles) and *neutrons* (neutral particles). Orbiting the nucleus are negatively charged *electrons*. The *atomic number* of an atom is the number of protons in the nucleus. The *mass number* is the number of protons plus the number of neutrons in an atom's nucleus. *Isotopes* are variants of the same atom, but with a different number of neutrons, and hence a different mass number.

- *Radioactivity* is the spontaneous breaking apart (decay) of certain unstable atomic nuclei. Three common types of radioactive decay are (1) emission of *alpha particles* from the nucleus, (2) emission of *beta particles* (or electrons) from the nucleus, and (3) *capture of electrons* by the nucleus.

- An unstable *radioactive isotope*, called the *parent*, will decay and form stable *daughter products*. The length of time for half of the nuclei of a radioactive isotope to decay is called the *half-life* of the isotope. If the half-life of the isotope is known, and the parent/daughter ratio can be measured, the age of a sample can be calculated. An accurate radiometric date can only be obtained if the mineral containing the radioactive isotope remained a closed system during the entire period since its formation.

- The *geologic time scale* divides Earth's history into units of varying magnitude. It is commonly presented in chart form, with the oldest time and event at the bottom and the youngest at the top. The principal subdivisions of the geologic time scale, called *eons*, include the *Hadean, Archean, Proterozoic* (together, these three eons are commonly referred to as the *Precambrian*), and, beginning about 540 million years ago, the *Phanerozoic*. The Phanerozoic (meaning "visible life") eon is divided into the following *eras*: *Paleozoic* ("ancient life"), *Mesozoic* ("middle life"), and *Cenozoic* ("recent life").

- A significant problem in assigning numerical dates to units of time is that *not all rocks can be radiometrically dated*. A sedimentary rock may contain particles of many ages that have been weathered from different rocks that formed at various times. One way geologists assign numerical dates to sedimentary rocks is to relate them to datable igneous masses, such as dikes and volcanic ash beds.

Key Terms

angular unconformity (p. 391)
Cenozoic era (p. 405)
conformable (p. 390)
correlation (p. 392)
cross-cutting relationships, principle of (p. 390)
disconformity (p. 391)
eon (p. 405)

epoch (p. 405)
era (p. 405)
fossil (p. 393)
fossil succession, principle of (p. 396)
geologic time scale (p. 405)
half-life (p. 399)
inclusion (p. 390)
index fossil (p. 398)

Mesozoic era (p. 405)
nonconformity (p. 392)
numerical date (p. 389)
original horizontality, principle of (p. 389)
Paleozoic era (p. 405)
period (p. 405)
Phanerozoic eon (p. 405)
Precambrian (p. 406)

radioactivity (p. 399)
radiocarbon dating (p. 401)
radiometric dating (p. 399)
relative dating (p. 389)
superposition, law of (p. 389)
unconformity (p. 390)

Questions for Review

1. Distinguish between numerical and relative dating.

2. What is the law of superposition? How are cross-cutting relationships used in relative dating?

3. When you observe an outcrop of steeply inclined sedimentary layers, what principle allows you to assume that the beds were tilted *after* they were deposited?

4. Refer to Figure 18.4 and answer the following questions:
   a. Is fault *A* older or younger than the sandstone layer?
   b. Is dike *A* older or younger than the sandstone layer?
   c. Was the conglomerate deposited before or after fault *A*?
   d. Was the conglomerate deposited before or after fault *B*?
   e. Which fault is older, *A* or *B*?
   f. Is dike *A* older or younger than the batholith?

5. A mass of granite is in contact with a layer of sandstone. Using a principle described in this chapter, explain how you might determine whether the sandstone was deposited on top of the granite, or whether it was intruded from below after the sandstone was deposited.

6. Distinguish among angular unconformity, disconformity, and nonconformity.

7. What is meant by the term *correlation*?

8. Describe several types of fossils. What organisms have the best chance of being preserved as fossils?

9. Describe William Smith's important contribution to the science of geology.

10. Why are fossils such useful tools in correlation?

11. In addition to being important aids in dating and correlating rocks, how else are fossils helpful in geologic investigations?

12. If a radioactive isotope of thorium (atomic number 90, mass number 232) emits 6 alpha particles and 4 beta par-

ticles during the course of radioactive decay, what is the atomic number and mass number of the stable daughter product?

13. Why is radiometric dating a reliable method of dating the geologic past?

14. A hypothetical radioactive isotope has a half-life of 10,000 years. If the ratio of radioactive parent to stable daughter product is 1:3, how old is the rock containing the radioactive material?

15. Briefly describe why tree rings might be helpful in studying the recent geologic past (see Box 18.2).

16. In order to provide a reliable radiometric date, a mineral must remain a closed system from the time of its formation until the present. Why is this true?

17. What precautions are taken to ensure reliable radiometric dates?

18. To make calculations easier, let us round the age of Earth to 5 billion years.
    a. What fraction of geologic time is represented by recorded history (assume 5000 years for the length of recorded history)?
    b. The first abundant fossil evidence does not appear until the beginning of the Cambrian period (540 million years ago). What percentage of geologic time is represented by abundant fossil evidence?

19. What subdivisions make up the geologic time scale?

20. Briefly describe the difficulties in assigning numerical dates to layers of sedimentary rock.

21. Figure 18.18 is a block diagram of a hypothetical area in the American Southwest. Place the lettered features in the proper sequence, from oldest to youngest. Identify an angular unconformity and a nonconformity.

Figure 18.18 Use this block diagram in conjunction with Review Question 21.

Web Resources

 The *Essentials of Geology* Website uses the resources and flexibility of the Internet to aid in your study of the topics in this chapter. Written and developed by geology instructors, this site will help improve your understanding of geology. Visit **www.prenhall.com/lutgens** and click on the cover of *Essentials of Geology, 8e* to find:

- On-line review quizzes.
- Web-based critical thinking and writing exercises.
- Links to chapter-specific Web resources.
- Internet-wide key term searches.

http://www.prenhall.com/lutgens

Earth History: A Brief Summary

Focus on Learning

To assist you in learning the important concepts in this chapter, you will find it helpful to focus on the following questions:

• How did the planets in the solar system originate?

• How did Earth's atmosphere evolve and change through time?

• What were the principal geological and biological events of Earth history?

Dinosaur fossils in New York's Museum of Natural History. (Photo by Gail Mooney/Corbis)

Earth has a long and complex history. The splitting and colliding of continents has resulted in the formation of new ocean basins and the creation of Earth's great mountain ranges. Furthermore, biological evolution has spurred a constant changing of Earth's life forms.

Changes on planet Earth occur at a snail's pace, generally too slowly for people to perceive. Thus, human awareness of evolutionary change is fairly recent. Evolution is not confined to life forms, for all of Earth's spheres have evolved together: the atmosphere, hydrosphere, lithosphere, and biosphere. Examples are evolutionary changes in the air we breathe, evolution of the world oceans, the rise of mountains, ponderous movements of crustal plates, the comings and goings of immense ice sheets, and the evolution of a vast array of life forms. As each facet of Earth has evolved, it has powerfully influenced the others.

As we saw in Chapter 18, geologists have many tools at their disposal for interpreting the clues about Earth's past. Using these tools, and clues that are contained in the rock record, geologists have been able to unravel many of the complex events of the geological past. The goal of this chapter is to provide a brief overview of the history of our planet and its life forms (Figure 19.1). We will begin nearly 5 billion years ago with the formation of Earth and its atmosphere. Then we will describe how our physical world assumed its present form and how Earth's inhabitants changed through time.

Early Evolution of Earth

Earth is one of nine planets that, along with several dozen moons and numerous smaller bodies, revolve around the Sun. The orderly nature of our solar system leads most researchers to conclude that Earth and the other planets formed at essentially the same time and from the same primordial material as the Sun. The **nebular hypothesis** suggests that the bodies of our solar system evolved from an enormous rotating cloud called the *solar nebula*, composed mostly of hydrogen and helium, with a small percentage of the heavier elements (Figure 19.2).

About 5 billion years ago this huge cloud of gases and minute particles began to gravitationally contract. As this slowly spiraling nebula contracted, it rotated faster and faster for the same reason ice skaters do when they draw their arms toward their bodies. Eventually the inward pull of gravity came into balance with the outward force caused by the rotational motion of the nebula (Figure 19.2B and C). By this time the once vast cloud had assumed a flat disk shape with a large concentration of material at its center called the *protosun* (pre-Sun). During the collapse, gravitational energy was converted to thermal energy (heat), causing the temperature of the inner portion of the nebula to dramatically rise.

The formation of the Sun marked the end of the period of contraction and thus the end of gravitational heating. Temperatures in the region where the inner planets now reside began to decline. The decrease in

Figure 19.1 Paleontologists study ancient life. Here scientists are excavating the remains of *Albertasaurus*, a large carnivore similar to *Tyrannosaurus rex*, that lived during the last Cretaceous Period. The excavation site is near Red Deer River, Alberta, Canada. (Photo by Photo Researchers, Inc.)

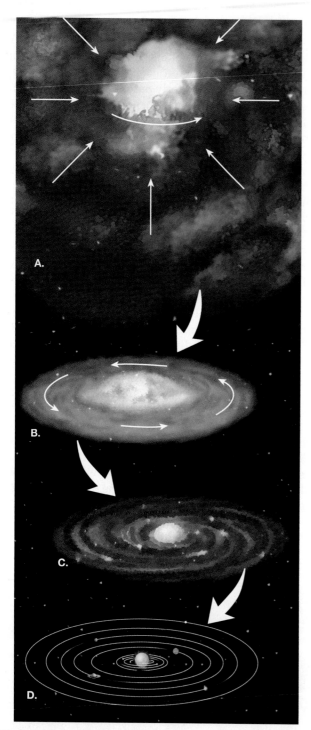

Figure 19.2 Nebular hypothesis. **A.** The nebula, a huge rotating cloud of dust and gases, began to contract. **B.** Most of the material was gravitationally swept toward the center, producing the Sun. However, owing to rotational motion, some dust and gases remained, orbiting the central body as a flattened disk. **C.** The planets began to accrete from the material that was orbiting within the flattened disk. **D.** In time, most of the remaining debris either coalesced to form the nine planets and their moons, or was swept into space by the solar wind.

temperature caused those substances with high melting points to coalesce (join together). Materials such as iron and nickel and the elements of which the rock-forming minerals are composed—silicon, calcium, sodium, and so forth—formed metallic and rocky clumps that orbited the Sun. Repeated collisions caused these fragments to coalesce into larger asteroid-size bodies, which in a few tens of millions of years accreted into the four inner planets we call Mercury, Venus, Earth, and Mars (Figure 19.2D). Not all of these clumps of matter were incorporated into the *protoplanets* (pre-planets). Those that remained in orbit are called *meteorites* when they survive an impact with Earth.

As more and more material was swept up by the protoplanets, the high-velocity impact of nebular debris caused the temperature of these bodies to rise. Because of their relatively high temperatures and weak gravitational fields, the inner planets were unable to accumulate much of the lighter components of the nebular cloud. The lightest of these, hydrogen and helium, were eventually whisked from the inner solar system by the solar winds. However, Earth, Mars, and Venus were able to retain various amounts of the heavier gases, including water vapor and carbon dioxide. Mercury, the smallest of the inner planets, and Earth's moon, were unable to retain even the heaviest of these gases.

At the same time that the inner planets were forming, the larger, outer planets (Jupiter, Saturn, Uranus, and Neptune), along with their extensive satellite systems, were also developing. Because of the frigid temperatures existing far from the Sun, the fragments from which these planets formed contained a high percentage of ices—water, carbon dioxide, ammonia, and methane. The two most massive planets, Jupiter and Saturn, had a surface gravity sufficient to attract and hold even the lightest elements—hydrogen and helium.

As material accumulated to form Earth (and for a short period afterward), the high-velocity impact of nebular debris and the decay of radioactive elements caused the temperature of our planet to steadily increase. During this time of intense heating, Earth became hot enough that iron and nickel began to melt. Melting produced liquid blobs of heavy metal that sank toward the center of the planet. This process occurred rapidly on the scale of geologic time and produced Earth's dense iron-rich core.

The early period of heating resulted in another process of chemical differentiation, whereby melting formed buoyant masses of molten rock that rose toward the surface where they solidified to produce a primitive crust. These rocky materials were enriched in oxygen and "oxygen-seeking" elements, particularly silicon and aluminum, along with lesser amounts of calcium, sodium, potassium, iron, and magnesium. In addition, some heavy metals such as gold, lead, and uranium, which have low melting points or were highly soluble in the ascending molten masses, were scavenged from

Earth's interior and concentrated in the developing crust. This early period of chemical segregation established the three basic divisions of Earth's interior—the iron-rich *core*, the thin *primitive crust*, and Earth's largest layer, called the *mantle*, which is located between the core and crust.

An important consequence of this early period of chemical differentiation is that large quantities of gaseous materials were allowed to escape from Earth's interior, as happens today during volcanic eruptions. By this process a primitive atmosphere gradually evolved. It is on this planet, with this atmosphere, that life as we know it came into existence.

Earth's Atmosphere Evolves

Today, the air you breathe is a stable mixture of 78 percent nitrogen, 21 percent oxygen, about 1 percent argon (an inert gas), and trace gases like carbon dioxide and water vapor. But our planet's original atmosphere, several billion years ago, was far different.

Earth's very earliest atmosphere probably was swept into space by the *solar wind*, a vast stream of particles emitted by the Sun. As Earth slowly cooled, a more enduring atmosphere formed. The molten surface solidified into a crust, and gases that had been dissolved in the molten rock were gradually released, a process called **outgassing.** Outgassing continues today from hundreds of active volcanoes worldwide. Thus, geologists hypothesize that Earth's original atmosphere was made up of gases similar to those released in volcanic emissions today: water vapor, carbon dioxide, nitrogen, and several trace gases.

As the planet continued to cool, the water vapor condensed to form clouds, and great rains commenced. At first, the water evaporated in the hot air before reaching the ground, or quickly boiled away upon contacting the surface, just like water sprayed on a hot grill. This accelerated the cooling of Earth's crust. When the surface had cooled below water's boiling point (100°C or 212°F), torrential rains slowly filled low areas, forming the oceans. This reduced not only the water vapor in the air but the amount of carbon dioxide as well, for it became dissolved in the water. What remained was a nitrogen-rich atmosphere.

If Earth's primitive atmosphere resulted from volcanic outgassing, we have a problem, because volcanoes do not emit free oxygen. Where did the very significant percentage of oxygen in our present atmosphere (nearly 21 percent) come from?

The major source of oxygen is green plants. Put another way, *life itself* has strongly influenced the composition of our present atmosphere. Plants did not just adapt to their environment; they actually influenced it, dramatically altering the composition of the entire planet's atmosphere by using carbon dioxide and releasing oxygen. This is a good example of how Earth operates as a giant system in which living things interact with their environment.

How did plants come to alter the atmosphere? The key is the way in which plants create their own food. They employ *photosynthesis*, in which they use light energy to synthesize food sugars from carbon dioxide and water. The process releases a waste gas, oxygen. Those of us in the animal kingdom rely on oxygen to metabolize our food, and we in turn exhale carbon dioxide as a waste gas. The plants use this carbon dioxide for more photosynthesis, and so on, in a continuing system.

The first life forms on Earth, probably bacteria, did not need oxygen. Their life processes were geared to the earlier, oxygenless atmosphere. Even today, many *anaerobic* bacteria thrive in environments that lack free oxygen. Later, primitive plants evolved that used photosynthesis and released oxygen. Slowly, the oxygen content of Earth's atmosphere increased. The Precambrian rock record suggests that much of the first free oxygen did not remain free, because it combined with (oxidized) other substances dissolved in water, especially iron. Iron has tremendous affinity for oxygen, and the two elements combine to form iron oxides (rust) at any opportunity.

Then, once the available iron satisfied its need for oxygen, substantial quantities of oxygen accumulated in the atmosphere. By the beginning of the Paleozoic era, about 4 billion years into Earth's existence (after seven-eighths of Earth's history had elapsed), the fossil record reveals abundant ocean-dwelling organisms that require oxygen to live. Hence, the composition of Earth's atmosphere has evolved together with its life forms, from an oxygenless envelope to today's oxygen-rich environment.

With this background on the evolution of the solar system, Sun, Earth, and atmosphere, we will look at Earth's history.

Precambrian Time: Vast and Enigmatic

The Precambrian encompasses immense geological time, from Earth's distant beginnings 4.5 billion years ago until the start of the Cambrian period, about 4 billion years later. Thus, the Precambrian spans about 88 percent of Earth's history. Our knowledge of this ancient time is sketchy, for much of the early rock record has been obscured by the very Earth processes you have been studying, especially plate tectonics, erosion, and deposition.

Untangling the long, complex Precambrian rock record is a formidable task, and it is far from complete. Most Precambrian rocks are devoid of fossils, which hinders correlation of rocks. Rocks of this great age are metamorphosed and deformed, extensively eroded, and obscured by overlying strata. Consequently, this least-understood span of Earth's history has not been successfully divided into briefer time units, as have later intervals. Indeed, Precambrian history is written in scattered, speculative episodes, like a long book with many missing chapters.

Precambrian Rocks

Looking upon Earth from the Space Shuttle, astronauts see plenty of ocean (71 percent) and much less land area (29 percent). Over large expanses of the continents, the orbiting space scientists gaze upon many Paleozoic, Mesozoic, and Cenozoic rock surfaces, but fewer Precambrian surfaces. This demonstrates the law of superposition: Precambrian rocks in these regions are buried from view beneath varying thicknesses of more recent rocks. Here, Precambrian rocks peek through the surface where younger strata are extensively eroded, as in the Grand Canyon and in some mountain ranges. However, on each continent, large core areas of Precambrian rocks dominate the surface mostly as deformed metamorphic rocks. These areas are called **shields** because they roughly resemble a warrior's shield in shape.

Figure 19.3 shows these shield areas of Precambrian rocks worldwide. In North America (including Greenland), the Canadian Shield encompasses 7.2 million square kilometers (2.8 million square miles), the equivalent of about 10 states of Texas put together.

Much of what we know about Precambrian rocks comes from mining the ores that some contain. The mining of iron, nickel, gold, silver, copper, chromium, uranium, and diamonds has provided Precambrian rock samples for study, and surveys to locate valuable ore deposits have revealed much about the rocks.

Noteworthy are extensive iron-ore deposits. Rocks from the middle Precambrian (1.2 billion to 2.5 billion years ago) contain most of Earth's iron ore, mainly as the mineral hematite (Fe_2O_3). These iron-rich sedimentary rocks probably represent the time when oxygen became sufficiently abundant to react with iron dissolved in shallow lakes and seas. Later, after much of the iron was oxidized and deposited on lake and sea bottoms, formation of these iron-rich deposits declined and oxygen levels in the ocean and atmosphere began to increase. Because most of Earth's free oxygen results from plant photosynthesis, the formation of extensive Precambrian iron ore deposits is linked to life in the sea.

Notably absent in the Precambrian are fossil fuels (coal, oil, natural gas). The reason is clear—a virtual absence of land plants to form coal swamps and of certain animals to form petroleum. Fossil fuels are from a later time.

Precambrian Fossils

A century ago, the earliest fossils known dated from the Cambrian period, about 540 million years ago. None were known from the Precambrian. This created a major problem for science at that time: How could complex organisms like trilobites abruptly appear in the geologic record? The answer, of course, was that there were fossils in Precambrian rocks, but they were rare, small, obscure, and simply had not been discovered yet. Today our knowledge of Precambrian life, although far from complete, is quite extensive.

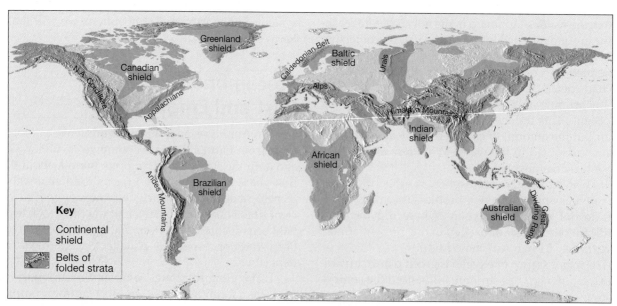

Figure 19.3   Today's continents look very different from the Precambrian. Remnants of Precambrian rocks are the continental shields, composed largely of metamorphic rocks.

A.   **B.**

Figure 19.4 Stromatolites are among the most common Precambrian fossils. **A.** Precambrian fossil stromatolites composed of calcium carbonate deposited by algae in the Helena Formation, Glacier National Park. (Photo by Ken M. Johns/Photo Researchers, Inc.) **B.** Modern stromatolites growing in shallow saline area, western Australia. (Photo by Bill Bachman/Photo Researchers, Inc.)

Precambrian fossils are disappointing if you are expecting to see fascinating plants and large animals, for these had not yet evolved. Instead, the most common Precambrian fossils are **stromatolites.** These are distinctively layered mounds or columns of calcium carbonate (Figure 19.4). Stromatolites are not the remains of actual organisms, but are material deposited by algae. They are indirect evidence of algae because they closely resemble similar deposits made by modern algae.

Stromatolites did not become common until the middle Precambrian, about 2 billion years ago. Stromatolites are large, but most actual organisms preserved in Precambrian rocks are microscopic. Well-preserved remains of many tiny organisms have been discovered, extending the record of life back beyond 3.5 billion years.

Many of these most ancient fossils are preserved in *chert*, a hard, dense chemical sedimentary rock. Chert must be very thinly sliced and studied under powerful microscopes to observe bacteria and algae fossils within it.

Microfossils have been found at several locations worldwide. Two notable areas are in southern Africa, where the rocks date to more than 3.1 billion years, and in the Gunflint Chert (named for its use in flintlock rifles) of Lake Superior, which dates to 1.7 billion years. In both places, bacteria and blue-green algae have been discovered. The fossils are of the most primitive organisms, *prokaryotes.* Their cells lack organized nuclei, and they reproduce asexually.

More advanced organisms, *eukaryotes*, have cells that contain nuclei. Eukaryotes are among billion-year-old fossils discovered at Bitter Springs in Australia, such as green algae. Unlike prokaryotes, eukaryotes reproduce sexually, which means that genetic material is exchanged between organisms. This reproductive mode permits greatly increased genetic variation. Thus, development of eukaryotes may have dramatically increased the rate of evolutionary change.

Plant fossils date from the middle Precambrian, but animal fossils came a bit later, in the late Precambrian. Many of these fossils are *trace fossils*, meaning that they are not of the animals themselves, but of their activities, such as trails and worm holes. Areas in Australia and Newfoundland have yielded hundreds of fossil impressions of soft-bodied creatures. Most, if not all, of the Precambrian fauna lacked shells, which would develop as protective armor during the Paleozoic.

As the Precambrian came to a close, the fossil record disclosed diverse and complete multicelled organisms. This set the stage for more complex plants and animals to evolve at the dawn of the Paleozoic era.

Paleozoic Era: Life Explodes

Following the long Precambrian, the most recent 540 million years of Earth history are divided into three eras: Paleozoic, Mesozoic, and Cenozoic. The Paleozoic era encompasses about 292 million years and is by far the longest of the three. Seven periods make up the Paleozoic era (Figure 19.5).

Before the Paleozoic, life forms possessed no hard parts, such as shells, scales, bones, or teeth. The beginning of the Paleozoic is marked by the appearance of the first life forms with hard parts (Figure 19.6). Hard parts greatly enhanced their chance of being preserved as part of the fossil record. Therefore, our knowledge of life's diversification improves greatly from the Paleozoic onward. This diversity is demonstrated in Figure 19.7.

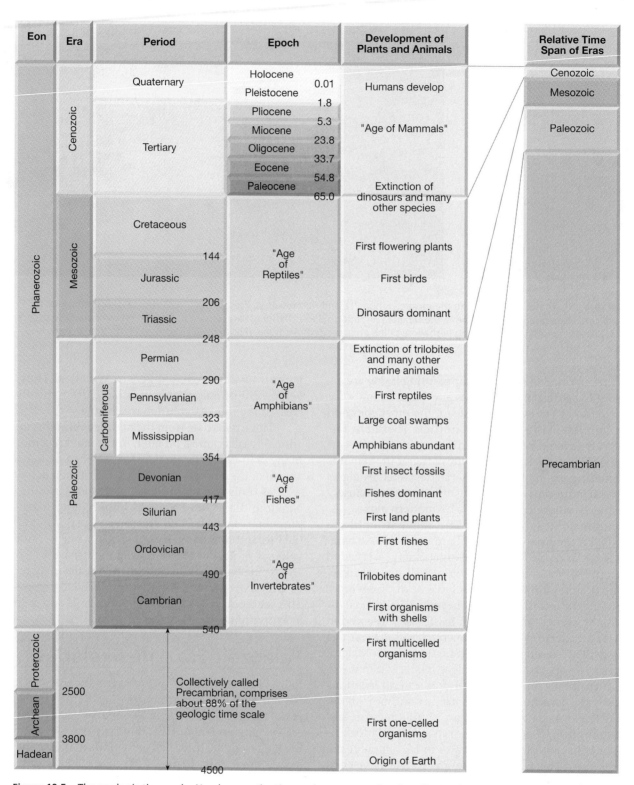

Figure 19.5   The geologic time scale. Numbers on the time scale represent time in millions of years before the present. These dates were added long after the time scale had been established using relative dating techniques. The Precambrian accounts for about 88 percent of geologic time. (Data from Geological Society of America)

A.

B.

Figure 19.6   Fossils of common Paleozoic life forms. **A.** Natural cast of a trilobite. Trilobites dominated the Paleozoic ocean, scavenging food from the bottom. **B.** Extinct coiled cephalopods. Like their modern descendants, these were highly developed marine organisms. (Photos by E. J. Tarbuck)

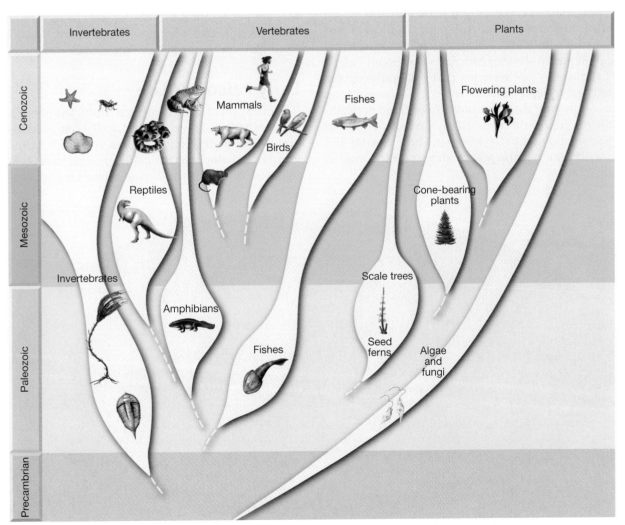

Figure 19.7   This chart indicates the times of appearance and relative abundance of major groups of organisms. The wider the band, the more dominant the group.

Abundant fossils have allowed geologists to construct a far more detailed time scale for the last one-eighth of geologic time than for the preceding seven-eighths, the Precambrian. Moreover, because every organism is associated with a particular environment, the greatly improved fossil record provided invaluable information for deciphering ancient environments. To facilitate our brief tour of the Paleozoic, we divide it into early Paleozoic (Cambrian, Ordovician, Silurian periods) and late Paleozoic (Devonian, Mississippian, Pennsylvanian, Permian periods).

Early Paleozoic History

The early Paleozoic consists of a 123-million-year span that embraces the Cambrian, Ordovician, and Silurian periods. Anyone approaching Earth from space at this time would have seen the familiar blue planet with plentiful white clouds, but the arrangement of continents would have looked very different from today (Figure 19.8). At this time, the vast southern continent of Gondwanaland encompassed five continents (South America, Africa, Australia, Antarctica, India, and perhaps China). Evidence of an extensive continental glaciation places western Africa near the South Pole!

Landmasses that were not part of Gondwanaland existed as five separate units and some scattered fragments. Although the exact position of these northern continents is uncertain, ancestral North America and Europe are thought to have been near the equator and separated by a narrow sea, as shown on the map in Figure 19.8.

As the Paleozoic opened, North America was a land with no living things, plant or animal. There were no Appalachian or Rocky Mountains; rather, the continent was largely a barren lowland. Several times during the Cambrian and Ordovician periods, shallow seas moved inland and then receded from the interior of the continent. Deposits of clean sandstones, used today to make glass, mark the edge of these shallow seas in the midcontinent.

Early in the Paleozoic, a mountain-building event affected eastern North America from the present-day central Appalachians to Newfoundland. The mountains produced during this event, the Taconic Orogeny, have since eroded away, leaving behind deformed strata and a large volume of detrital sedimentary rocks that were derived from the weathering of these mountains.

During the Silurian period, much of North America was once again inundated by shallow seas. This time large barrier reefs restricted circulation between shallow marine basins and the open ocean. Water in these basins evaporated, causing deposition of large quantities of rock salt and gypsum. Today these thick *evaporite beds* are important resources for the chemical, rubber, plasterboard, and photographic industries in Ohio, Michigan, and western New York State.

Early Paleozoic Life

Life in early Paleozoic time was restricted to the seas. Vertebrates had not yet evolved, so life consisted of several invertebrate groups (shown in Figure 19.9). The Cambrian period was the golden age of *trilobites*. More than 600 genera of these mud-burrowing scavengers

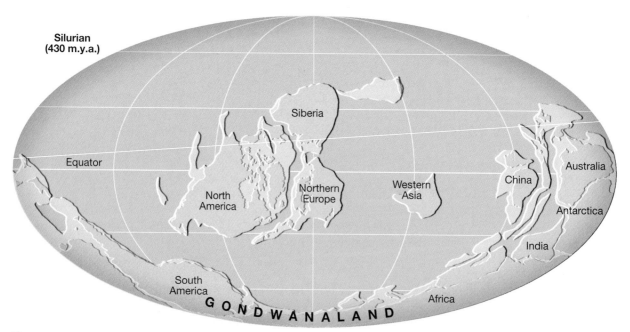

Figure 19.8 Reconstruction of Earth as it may have appeared in early Paleozoic time. The southern continents were joined into a single landmass called Gondwanaland. Four of the five landmasses that would later join to form the northern continent of Laurasia lay scattered roughly along the equator. (After C. Scotese, R. K. Bambach, C. Barton, R. VanderVoo, and A. Ziegler)

Figure 19.9 During the Ordovician period (490 million to 443 million years ago), the shallow waters of an inland sea over central North America contained an abundance of marine invertebrates. Shown in this reconstruction are straight-shelled cephalopods, trilobites, brachiopods, snails, and corals. (© The Field Museum, Neg. #GEO80820c, Chicago)

flourished worldwide. By Ordovician times, *brachiopods* outnumbered the trilobites. Brachiopods are among the most widespread Paleozoic fossils and, except for one modern group, are now extinct. Although the adults lived attached to the seafloor, the young larvae were free-swimming. This mobility accounts for the group's wide geographic distribution.

The Ordovician also marked the appearance of abundant *cephalopods*—mobile, highly developed mollusks that became the major predators of their time. The descendants of cephalopods include the modern squid, octopus, and nautilus. Cephalopods were the first truly large organisms on Earth (Figure 19.9). Whereas the largest trilobites seldom exceeded 30 centimeters (12 inches) in length and the biggest brachiopods were no more than about 20 centimeters (8 inches) across, one species of cephalopod reached a length of nearly 10 meters (30 feet).

The beginning of the Cambrian period marks an important event in animal evolution. For the first time, organisms appeared that secreted material that formed *hard parts*, such as shells. Why several diverse life forms began to develop hard parts about the same time remains unanswered. One proposal suggests that because an external skeleton provides protection from predators, hard parts evolved for survival. Yet, the fossil record does not seem to support this hypothesis. Organisms with hard parts were plentiful in the Cambrian period, whereas predators such as cephalopods were

not abundant until the Ordovician period, some 70 million years later.

Whatever the answer, hard parts clearly served many useful purposes and aided adaptations to new ways of life. Sponges, for example, developed a network of fine interwoven silica spicules that allowed them to grow larger and more erect, capable of extending above the surface in search for food. Mollusks (clams and snails) secreted external shells of calcium carbonate that protected them and allowed body organs to function in a more controlled environment. The successful trilobites developed an exoskeleton of a protein called *chitin*, which permitted them to burrow through soft sediment in search of food (Figure 19.10).

Late Paleozoic History

The late Paleozoic consists of four periods—the Devonian, Mississippian, Pennsylvanian, and Permian—that span about 160 million years. Tectonic forces reorganized Earth's landmasses during this time, culminating with the formation of the supercontinent *Pangaea* (Figure 19.11).

Forming Pangaea, the Supercontinent. As ancestral North America collided with Africa, the narrow sea that separated these landmasses began to close slowly (compare Figures 19.11B and 19.11C). Strong compressional forces from this collision deformed the

421

Figure 19.10 These trilobites date from the Devonian period, 417 million to 354 million years ago. (Photo by Phil Degginger/Tony Stone Images)

rocks to produce the original northern Appalachian Mountains of eastern North America.

During the fusion of North America and Africa, the other northern continents began to converge (Figure 19.11). By the Permian period, this newly formed landmass had collided with western Asia and the Siberian landmass along the line of the Ural Mountains. Through this union, the northern continent of *Laurasia* was born, encompassing present-day North America, Europe, western Asia, Siberia, and perhaps China.

As Laurasia was forming, Gondwanaland migrated northward. By the Pennsylvanian period, Gondwanaland collided with Laurasia, forming a mountainous belt through central Europe. Simultaneously, a collision between the African fragment of Gondwanaland and the southeastern edge of North America rumpled up the southern Appalachian Mountains.

By the close of the Paleozoic, all the continents had fused into the supercontinent of Pangaea (Figure 19.11). With only a single vast continent, the world's climate became very seasonal, having extremes far greater than those we experience today. These altered climatic conditions may have contributed to one of the most dramatic biological declines in all of Earth history.

Late Paleozoic Life

During most of the late Paleozoic, organisms diversified dramatically. Some 400 million years ago, plants that had adapted to survive at the water's edge began to move inland, becoming *land plants*. These earliest land plants were leafless, vertical spikes about the size of your index finger. However, by the end of the Devonian, 40 million years later, the fossil record indicates the existence of forests with trees tens of meters high.

In the oceans, armor-plated fishes that had evolved during the Ordovician continued to adapt. Their armor plates thinned to lightweight scales that increased their speed and mobility (Figure 19.12). Other fishes evolved during the Devonian, including primitive sharks that had a skeleton made of cartilage and bony fishes, the groups to which virtually all modern fishes belong. Because of this, the Devonian period is often called the "age of fishes."

By late Devonian time, two groups of bony fishes—the lung fish and the lobe-finned fish—became adapted to land environments. Not unlike their modern relatives, these fishes had primitive lungs that supplemented their breathing through gills. It is believed that the lobe-finned fish occupied tidal flats or small ponds, and that in times of drought they may have used their bony fins to "walk" from dried-up pools in search of other ponds. Through time, the lobe-finned fish began to rely more on their lungs and less on their gills. By late Devonian time, they had evolved into true air-breathing amphibians with fishlike heads and tails. It should be noted that insects had already invaded the land.

Modern amphibians, like frogs, toads, and salamanders, are small and occupy limited biological niches. But conditions during the remainder of the Paleozoic were ideal for these newcomers to the land. Plants and insects, which were their main diet, already were very abundant and large. Having only minimal competition from other land dwellers, the amphibians rapidly diversified. Some groups took on roles and forms that were more similar to modern reptiles, such as crocodiles, than to modern amphibians (Figure 19.13).

By the Pennsylvanian period, large tropical swamps extended across North America, Europe, and Siberia (Figure 19.14). Trees approached 30 meters (100 feet), with trunks more than 1 meter across. The coal deposits that fueled the Industrial Revolution and which provide a substantial portion of our electric power today, originated in these vast swamps. Further, it was in the lush coal swamp environment of the late Paleozoic that the amphibians evolved quickly into a variety of species.

Mesozoic Era: Age of the Dinosaurs

Spanning about 183 million years, the Mesozoic era is divided into three periods: the Triassic, Jurassic, and Cretaceous. The Mesozoic era witnessed the beginning of the breakup of the supercontinent Pangaea. Also in this era, organisms that had survived the great Permian extinction began to diversify in spectacular ways (see

A. Devonian (410 m.y.a.)

B. Mississippian (330 m.y.a.)

C. Permian (260 m.y.a.)

Figure 19.11 During the late Paleozoic, plate movements were joining together the major landmasses to produce the supercontinent of Pangaea. (After C. Scotese, R. K. Bambach, C. Barton, R. VanderVoo, and A. Ziegler)

Figure 19.12 These placoderms or "plate-skinned" fish were abundant during the Devonian (417 million to 354 million years ago). (Drawing after A. S. Romer)

Box 19.1). On land, dinosaurs became dominant and remained unchallenged for over 100 million years. Because of this, the Mesozoic era is often called the "age of dinosaurs."

Early geologists recognized a profound difference between the fossils in Permian strata and those in younger Triassic rocks, as if someone had drawn a bold line to separate the two time periods. Clearly one half of the fossil groups that occurred in late Paleozoic rocks were missing in Mesozoic rocks. On this basis, it was decided to separate the Paleozoic and Mesozoic at the Permian–Triassic boundary.

Mesozoic History

The Mesozoic era began with much of the world's land above sea level. In fact, in North America no period exhibits a more meager marine sedimentary record than the Triassic period. Of the exposed Triassic strata, most are red sandstones and mudstones that lack

Figure 19.13 Permian scene (290 million to 248 million years ago) shows the large amphibian *Eryops*, which exceeded 2 meters (6 feet) in length. Eryops was a carnivore, as you can see from the teeth.

Figure 19.14 Restoration of a Pennsylvania coal swamp (323 million to 290 million years ago). Shown are scale trees (left), seed ferns (lower left), and scouring rushes (right). Also note the large dragonfly. (© The Field Museum, Neg. #GEO85637c, Chicago. Photographer John Weinstein.)

fossils and contain features indicating a terrestrial environment.

As the Jurassic period opened, the sea invaded western North America. Adjacent to this shallow sea, extensive continental sediments were deposited on what is now the Colorado Plateau. The most prominent is the Navajo Sandstone, a windblown, white quartz sandstone that in places approaches a thickness of 300 meters (1000 feet). These massive dunes indicate that a

major desert occupied much of the American Southwest during early Jurassic times.

A well-known Jurassic deposit is the Morrison Formation, within which is preserved the world's richest storehouse of dinosaur fossils. Included are fossilized bones of huge dinosaurs such as *Apatosaurus* (formerly *Brontosaurus*), *Brachiosaurus*, and *Stegosaurus*.

As the Jurassic period gave way to the Cretaceous, shallow seas once again invaded much of western

BOX 19.1

The Great Paleozoic Extinction

The Paleozoic ended with the Permian period, a time when Earth's major landmasses joined to form the supercontinent Pangaea. This redistribution of land and water and changes in the elevations of landmasses brought pronounced changes in world climates. Broad areas of the northern continents became elevated above sea level, and the climate grew drier. These changes apparently triggered extinctions of many species on land and sea.

By the close of the Permian, 75 percent of the amphibian families had disappeared, and plants had declined in number and variety. Although many amphibian groups became extinct, their descendants, the reptiles, would become the most successful and advanced animals of Earth. Marine life was not spared. At least 80 percent, and perhaps as much as 95 percent, of marine life disappeared. Many marine invertebrates that had been dominant during the Paleozoic, including all the remaining trilobites as well as some types of corals and brachiopods, failed to adapt to the widespread environmental changes.

The late Paleozoic extinction was the greatest of at least five mass extinctions to occur over the past 600 million years. Each extinction wreaked havoc with the existing biosphere, wiping out large numbers of species. In each case, however, the survivors formed new biological communities that were more diverse than their predecessors. Thus, mass extinctions actually invigorated life on Earth, as the few hardy survivors eventually filled more niches than the ones left by the victims.

The cause of the great Paleozoic extinction is uncertain. The climate changes from the formation of Pangaea and the associated drop in sea level undoubtedly stressed many species. In addition, at least 2 million cubic kilometers of lava flowed across Siberia to produce what is called the Siberian Traps. Perhaps debris from these eruptions blocked incoming sunlight, or perhaps enough sulfuric acid was emitted to make the seas virtually uninhabitable. Some recent research suggests that an impact with an extraterrestrial body may have contributed. Whatever caused the late Paleozoic extinction, it is clear that without it a very different population of organisms would today inhabit this planet.

North America, the Atlantic, and Gulf coastal regions. This created great swamps like those of the Paleozoic era, forming Cretaceous coal deposits that are very important economically in the western United States and Canada. For example, on the Crow Indian reservation in Montana, there exists nearly 20 billion tons of high-quality coal of Cretaceous age.

A major event of the Mesozoic era was the breakup of Pangaea (Figure 19.15). A rift developed between what is now the eastern United States and western Africa, marking the birth of the Atlantic Ocean. It also represents the beginning of the breakup of Pangaea, a process that continued for 200 million years, through the Mesozoic and into the Cenozoic.

As Pangaea fragmented, the westward-moving North American plate began to override the Pacific plate. This tectonic event began a continuous wave of deformation that moved inland along the entire western margin of the continent. By Jurassic times, subduction of the Pacific plate had begun to produce the chaotic mixture of rocks that exists today in the Coast Ranges of California. Further inland, igneous activity was widespread, and for nearly 60 million years, huge masses of magma rose to within a few miles of the surface, where these masses cooled and solidified. The remnants of this intrusive activity included the granitic rocks of the Sierra Nevada, the Idaho batholith, and the Coast Range batholith of British Columbia.

Tectonic activity that began in the Jurassic continued throughout the Cretaceous, ultimately forming the vast mountains of western North America (Figure 19.16). Compressional forces moved huge rock units in a shinglelike fashion toward the east. Throughout much of the western margin of North America, older rocks were thrust eastward over younger strata, for a distance exceeding 150 kilometers (90 miles).

Toward the end of the Mesozoic, the middle and southern ranges of the Rocky Mountains formed. This mountain-building event, called the Laramide Orogeny, resulted when tectonic forces uplifted large blocks of Precambrian rocks in Colorado and Wyoming.

Mesozoic Life

As the Mesozoic era dawned, its life forms were survivors of the great Paleozoic extinction. These survivors diversified in many new ways to fill the biological voids created at the close of the Paleozoic. On land, conditions favored those that could adapt to drier climates. Among plants, the gymnosperms were one such group. Unlike the first plants to invade the land, the seed-bearing gymnosperms did not depend on free-standing water for fertilization. Consequently, these plants were not restricted to a life near the water's edge.

The gymnosperms quickly became the dominant trees of the Mesozoic. They included the cycads, the conifers, and the ginkgoes. The cycads resembled a large pineapple plant. The ginkgoes had panshaped leaves, much like their modern relatives. Largest were the conifers, whose modern descendants include the pines, firs, and junipers. The best-known fossil occurrence of these ancient trees is in northern Arizona's Petrified Forest National Park. Here, huge petrified logs lie exposed at the surface, having been weathered from rocks of the Triassic Chinle Formation (Figure 19.17).

The Shelled Egg. Among the animals, reptiles readily adapted to the drier Mesozoic environment. They were the first true terrestrial animals. Unlike amphibians, reptiles have shell-covered eggs that can be laid on land. The elimination of a water-dwelling stage (like the tadpole stage in frogs) was an important evolutionary step. Of interest is the fact that the watery fluid within the reptilian egg closely resembles seawater in chemical composition. Because the reptile embryo develops in this watery environment, the shelled egg has been characterized as a "private aquarium" in which the embryos of these land vertebrates spend their water-dwelling stage of life.

Dinosaurs Dominate. With the perfection of the shelled egg, reptiles quickly became the dominant land animals. They continued this dominance for more than 160 million years. Most awesome of the Mesozoic reptiles were the dinosaurs. Some of the huge dinosaurs were carnivorous (*Tyrannosaurus*), whereas others were herbivorous (like the ponderous *Apatosaurus*, formerly *Brontosaurus*). The extremely long neck of *Apatosaurus* may have been an adaptation for feeding on tall conifer trees. However, not all dinosaurs were large. In fact, certain small forms closely resembled modern fleet-footed lizards. Further, evidence indicates that some dinosaurs, unlike their present-day reptile relatives, were warm-blooded.

The reptiles made one of the most spectacular adaptive radiations in all of Earth history (Figure 19.18). One group, the pterosaurs, took to the air. These "dragons of the sky" possessed huge membranous wings that allowed them rudimentary flight (Figure 19.19). Another group of reptiles, exemplified by the fossil *Archaeopteryx*, led to more successful flyers: the birds. Whereas some reptiles took to the skies, others returned to the sea, including the fish-eating plesiosaurs and ichthyosaurs. These reptiles became proficient swimmers, but they retained their reptilian teeth and breathed by means of lungs.

At the close of the Mesozoic, many reptile groups became extinct. Only a few types survived to recent times, including the turtles, snakes, crocodiles, and lizards. The huge land-dwelling dinosaurs, the marine plesiosaurs, and the flying pterosaurs all are known only through the fossil record. What caused this great extinction (see Box 19.2)?

A. 200 Million Years Ago (Early Jurassic)

B. 150 Million Years Ago (Late Jurassic)

C. 100 Million Years Ago (Cretaceous)

D. 50 Million Years Ago (Early Cenozoic)

Figure 19.15   The breakup of the supercontinent of Pangaea began and continued throughout the Mesozoic era. (After D. Walsh and C. Scotese)

Figure 19.16 Uplifted and deformed sedimentary strata in the Canadian Rockies near Lake Louise, Alberta, Canada. (Photo by Carr Clifton/Minden Pictures)

Cenozoic Era: Age of Mammals

The Cenozoic era, or "era of recent life," encompasses the past 65 million years of Earth history. It is the "post-dinosaur era," the time of mammals, including humans. It is during this span that the physical landscapes and life forms of our modern world came into being. The Cenozoic era represents a much smaller fraction of geologic time than either the Paleozoic or the Mesozoic. Although shorter, it nevertheless possesses a rich history, because the completeness of the geologic record improves as time approaches the present. The rock formations of this time span are more widespread and less disturbed than those of any preceding time.

The Cenozoic era is divided into two periods of very unequal duration—the Tertiary period and the Quaternary period. The Tertiary period includes five epochs and embraces about 63 million years, practically all of the Cenozoic era. The Quaternary period consists of two epochs that represent only the last 2 million years of geologic time.

Cenozoic North America

Most of North America was above sea level throughout the Cenozoic era. However, the eastern and western margins of the continent experienced markedly contrasting events, because of their different relationships with plate boundaries. The Atlantic and Gulf coastal regions, far removed from an active plate boundary, were tectonically stable. Western North America, in contrast, was the leading edge of the North American plate. As a result, plate interactions during the Cenozoic gave rise to many events of mountain building, volcanism, and earthquakes in the West.

Eastern North America. The stable continental margin of eastern North America was the site of abundant marine sedimentation. The most extensive deposition

Figure 19.17 Petrified logs that have been weathered out of the Triassic Chinle Formation in the Petrified Forest of Arizona. (Photo by David Muench)

428

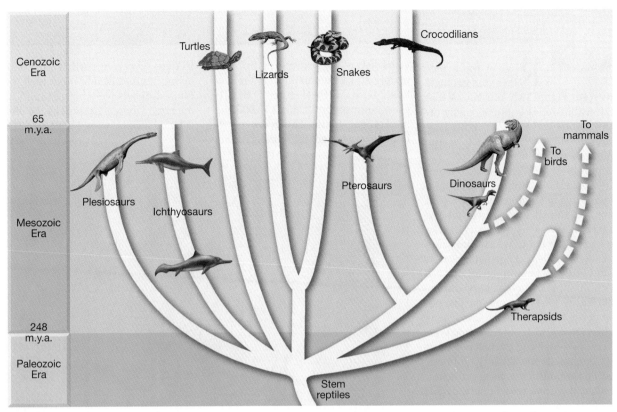

Figure 19.18 Adaptive radiation of major reptile groups during the Mesozoic era—"the golden age of reptiles." Dinosaurs, pterosaurs (flying reptiles), and large marine reptiles all became extinct by the end of the Mesozoic.

surrounded the Gulf of Mexico, from the Yucatan Peninsula to Florida. Here, the great buildup of sediment caused the crust to downwarp and produced numerous faults. In many instances, the faults created traps in which oil and natural gas accumulated. Today, these and other petroleum traps are the most econom-

Figure 19.19 Fossils of the great flying Pteranodon have been recovered from Cretaceous chalk deposits located in Kansas. Although Pteranodon had a wingspan of 7 meters (22 feet), flying reptiles with twice this wingspan have been discovered in strata of west Texas.

ically important resource in Cenozoic strata of the Gulf Coast, as evidenced by the Gulf's well-known offshore drilling platforms.

By early Cenozoic time, most of the original Appalachians had been eroded to a low plain. Then, by the mid-Cenozoic, isostatic adjustments raised the region once again, rejuvenating its rivers. Streams eroded with renewed vigor, gradually sculpturing the surface into its present-day topography (Figure 19.20). The sediments from all of this erosion were deposited along the eastern margin of the continent, where they attained a thickness of many kilometers. Today, portions of the strata deposited during the Cenozoic are exposed as the gently sloping Atlantic and Gulf coastal plains.

Western North America. In the West, the Laramide Orogeny that built the Rocky Mountains was coming to an end. As erosion lowered the mountains, the basins between uplifted ranges filled with sediments. Eastward, a great wedge of sediment from the eroding Rockies was building, creating the Great Plains.

Beginning in the Miocene epoch, a broad region from northern Nevada into Mexico experienced crustal movements that formed more than 150 fault-block mountain ranges. They rise abruptly above the adjacent basins, creating the Basin and Range Province (see Chapter 17).

BOX 19.2

Demise of the Dinosaurs

Review for a moment the geologic time scale, Figure 19.5. You can see that its divisions represent times of significant geological and/or biological change. Of special interest is the boundary between the Mesozoic era ("middle life") and Cenozoic era ("recent life"), about 65 million years ago. Around this time, more than half of all plant and animal species died out in a *mass extinction*. This boundary marks the end of the era in which dinosaurs and other reptiles dominated the landscape and the beginning of the era when mammals become very important (Figure 19.A). Because the last period of the Mesozoic is the Cretaceous (abbreviated K to avoid confusion with other C periods), and the first period of the Cenozoic is the Tertiary (abbreviated T), the time of this mass extinction is called the *Cretaceous-Tertiary*, or *KT, boundary*.

The extinction of the dinosaurs is generally attributed to this group's inability to adapt to some radical change in environmental conditions. What event could have triggered the rapid extinction of the dinosaurs—one of the most successful groups of land animals ever to have lived?

The most strongly supported hypothesis proposes that about 65 million years ago a large meteorite approximately 10 kilometers in diameter collided with Earth (Figure 19.B). The speed at impact was about 70,000 kilometers per hour! The force of the impact vaporized the meteorite and trillions of tons of Earth's crust. Huge quantities of dust and other metamorphosed debris were blasted high into the atmosphere. For months the encircling dust cloud would have greatly restricted the sunlight reaching Earth's

Figure 19.A Dinosaurs dominated the Mesozoic landscape until their extinction at the close of the Cretaceous period. This skeleton of *Tyrannosaurus* stands on display in Chicago's Field Museum. (Photo by Don and Pat Valenti/DRK Photo)

As the Basin and Range Province was forming, the entire western interior of the continent was gradually uplifted. This uplift re-elevated the Rockies and rejuvenated many of the West's major rivers. As the rivers became incised, many spectacular gorges were formed, including the Grand Canyon of the Colorado River, the Grand Canyon of the Snake River, and the Black Canyon of the Gunnison River. The present topography of the Rocky Mountains is in large measure the result of this late Tertiary uplift and the subsequent excavation of the early Tertiary basin deposits by rejuvenated streams (Figure 19.21).

Volcanic activity was common in the West during much of the Cenozoic. Beginning in the Miocene epoch, great volumes of fluid basaltic lava flowed from fissures in portions of present-day Washington, Oregon, and Idaho. These eruptions built the extensive (1.3 million square kilometers) Columbia Plateau. Immediately west of the Columbia Plateau, volcanic activity was quite different. Here, thicker magmas with a higher silica content erupted explosively, creating a chain of stratovolcanoes from northern California to the Canadian border, some of which remain active—like Mount St. Helens (Figure 19.22).

A final episode of folding occurred in the West in late Tertiary time, creating the Coast Ranges that stretch along the Pacific Coast. Meanwhile, the Sierra Nevada became fault-block mountains as they were uplifted along their eastern flank, creating the imposing mountain front we know today.

As the Tertiary period drew to a close, the effects of mountain building, volcanic activity, isostatic adjustments, and extensive erosion and sedimentation had created a physical landscape very similar to the configuration of today. All that remained of Cenozoic time was the final 2-million-year episode called the Quaternary period. During this most recent (and current) phase of Earth history, in which humans evolved, the action of glacial ice and other erosional agents added the finishing touches.

surface. With insufficient sunlight for photosynthesis, delicate food chains would have collapsed. Large, plant-eating dinosaurs would be affected more adversely than smaller life forms because of the tremendous volume of vegetation they consumed. Next would come the demise of large carnivores. When the sunlight returned, more than half of the species on Earth, including numerous marine organisms, had become extinct.

This period of mass extinction appears to have affected all land animals larger than dogs. Supporters of this catastrophic-event scenario suggest that small, ratlike mammals were able to survive a breakdown in their food chains lasting perhaps several months. Large dinosaurs, they contend, could not survive this event. The loss of these large reptiles opened habitats for the small mammals that remained. These new habitats, along with evolutionary forces, led to the development of the large mammals that occupy our modern world.

Figure 19.B Chicxulub crater is a giant impact crater that formed about 65 million years ago and has since been filled with sediments. About 180 kilometers in diameter, Chicxulub crater is regarded by some researchers to be the impact site that resulted in the demise of the dinosaurs.

What evidence points to such a catastrophic collision 65 million years ago? First, a thin layer of sediment nearly 1 centimeter thick has been discovered at the KT boundary worldwide. This sediment contains a high level of the element *iridium*, rare in Earth's crust but found in high proportions in stony meteorites. Could this layer be the scattered remains of the meteorite that was responsible for the environmental changes that led to the demise of many reptile groups?

Despite its growing support, some scientists disagree with the impact hypothesis. They suggest instead that huge volcanic eruptions led to the breakdown in the food chain. To support their hypothesis they site enormous outpourings of lavas in the Deccan Plateau of northern India about 65 million years ago.

Whatever caused the KT extinction, we now have a greater appreciation of the role of catastrophic events in shaping the history of our planet and the life that occupies it. Could a catastrophic event having similar results occur today? This possibility may explain why an event that occurred 65 million years ago has captured the interest of so many.

Cenozoic Life

Mammals replaced reptiles as the dominant land animals in the Cenozoic. Angiosperms (flowering plants with covered seeds) replaced gymnosperms as the dominant land plants. Marine invertebrates took on a modern look. Microscopic animals called *foraminifera* became especially important. Today, foraminifera are among the most intensely studied of all fossils because their widespread occurrence makes them invaluable in correlating Tertiary sediments. Tertiary strata are very important to the modern world, for they yield more oil than rocks of any other age.

The Cenozoic is often called the "age of mammals," because these animals came to dominate land life. It could also be called the "age of flowering plants," for the angiosperms enjoy a similar status in the plant world. As a result of advances in seed fertilization and dispersal, angiosperms experienced rapid development and expansion as the Mesozoic drew to a close. Thus, as the Cenozoic era began, angiosperms were already the dominant land plants.

Development of the flowering plants, in turn, strongly influenced the evolution of both birds and mammals. Birds that feed on seeds and fruits, for example, evolved rapidly during the Cenozoic in close association with the flowering plants. During the middle Tertiary, grasses developed rapidly and spread over the plains. This fostered the emergence of herbivorous (plant-eating) mammals that were mainly grazers. In turn, the development and spread of grazing animals established the setting for the evolution of the carnivorous mammals that preyed upon them.

Mammals Replace Reptiles. Back in the Mesozoic, an important evolutionary event was the appearance of primitive mammals in the late Triassic, about the same time that the dinosaurs emerged. Yet throughout the period of dinosaur dominance, mammals remained in the background as small and inconspicuous

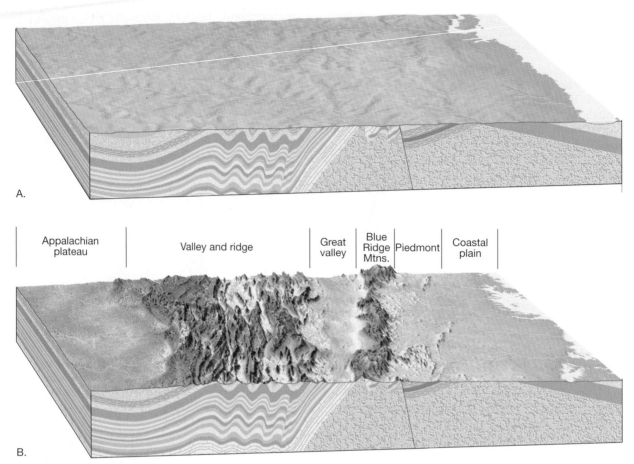

A.

| Appalachian plateau | Valley and ridge | Great valley | Blue Ridge Mtns. | Piedmont | Coastal plain |

B.

Figure 19.20 The formation of the modern Appalachian Mountains. **A.** The original Appalachians eroded to a low plain. **B.** The recent upwarping and erosion of the Appalachians began nearly 30 million years ago to produce the present topography.

Figure 19.21 Wyoming's spectacular Teton Range is part of the Rocky Mountains. They were formed during the late Tertiary uplift about 10 million years ago. (Photo by Tom Till)

Figure 19.22 Mount Hood, Oregon. This dormant volcano is one of several large composite cones that comprise the Cascade Range. (Photo by John M. Roberts/The Stock Market)

animals. By the close of the Mesozoic era, dinosaurs and other reptiles no longer dominated the land. It was only after these large reptiles became extinct that mammals came into their own as the dominant land animals. The transition is a major example in the fossil record of the replacement of one large group by another.

Mammals are distinct from reptiles in important respects. Mammalian young are born live, and mammals maintain a steady body temperature—that is, they are warm-blooded. This latter adaptation allowed mammals to lead more active and diversified lives than reptiles because they could survive in cold regions and search for food during any season or time of day. Other mammalian adaptations included the development of insulating body hair and more efficient heart and lungs.

It is worth noting that perfect boundaries cannot be drawn between mammalian and reptilian traits. One minor group of mammals, the *monotremes*, still lay eggs. The two species in this group, the duck-billed platypus and the spiny anteater, are found only in Australia. Moreover, although modern reptiles are cold-blooded, some paleontologists believe that dinosaurs may have been warm-blooded.

With the demise of most Mesozoic reptiles, Cenozoic mammals diversified rapidly. The many forms that exist today evolved from small primitive mammals that were characterized by short legs, flat five-toed feet, and small brains. Their development and specialization took four principal directions: (1) increase in size, (2) increase in brain capacity, (3) specialization of teeth to better accommodate a particular diet, and (4) specialization of

limbs to better equip the animal for life in a particular environment.

Marsupials, Placentals, and Diversity. Following the reptilian extinctions at the close of the Mesozoic, two groups of mammals, the marsupials and the placentals, evolved and expanded to dominate the Cenozoic. The groups differ principally in their modes of reproduction. Young marsupials are born live but at a very early stage of development. After birth, the tiny and immature young crawl into the mother's external stomach pouch to complete their development. Examples are kangaroos, opossums, and koalas.

Placental mammals, conversely, develop within the mother's body for a much longer period, so that birth occurs after the young are relatively mature and independent. Most mammals are placental, including humans.

Today, marsupials are found primarily in Australia, where they underwent a separate evolutionary expansion during the Cenozoic, largely isolated from placental mammals.

In South America, both primitive marsupials and placentals coexisted before that landmass became completely isolated during the breakup of Pangaea. Evolution and specialization of both groups continued undisturbed for approximately 40 million years until the close of the Pliocene epoch, when the Central American land bridge emerged, connecting the two American continents. Then an invasion of advanced carnivores from North America brought the extinction of many hoofed mammals that had persisted in South America for millions of years. The marsupials, except for opos-

433

Figure 19.23 Partially exposed skeleton of an extinct sabre-toothed cat. This lion-size cat roamed the White River Badlands of South Dakota during the Oligocene epoch. (Photo by T. A. Wiewandt/DRK Photo)

sums, also could not compete and thus became extinct. Both Australia and South America provide excellent examples of how isolation caused by the separation of continents increased the diversity of animals in the world.

Large Mammals and Extinction. As we have seen, mammals diversified quite rapidly during the Cenozoic era. One tendency was for some groups to become very large. For example, by the Oligocene epoch a hornless rhinoceros that stood nearly 5 meters high (16 feet) had evolved. It is the largest land mammal known to have existed. As time approached the present, many other types evolved to a large size as well; more, in fact, than now exist. Many of these large forms were common as recently as 11,000 years ago.

However, a wave of late Pleistocene extinctions rapidly eliminated these animals from the landscape.

In North America, the mastodon and mammoth, both huge relatives of the elephant, became extinct. In addition, saber-toothed cats, giant beavers, large ground sloths, horses, camels, giant bison, and others died out (Figure 19.23). In Europe, late Pleistocene extinctions included woolly rhinos, large cave bears, and the Irish elk. The reason for this recent wave of large animal extinctions puzzles scientists. These animals had survived several major glacial advances and interglacial periods, so it is difficult to ascribe these extinctions to climatic change. Some scientists believe that early humans hastened the decline of these mammals by selectively hunting large forms. Although this hypothesis is preferred by many, it is not yet accepted by all.

The Chapter in Review

- The *nebular hypothesis* describes the formation of the solar system. The planets and Sun began forming about 5 billion years ago from a large cloud of dust and gases composed of hydrogen and helium, with only a small percentage of all the other heavier elements. As the cloud contracted, it began to rotate and assume a disk shape. Material that was gravitationally pulled toward the center became the *protosun*. Within the rotating disk, small centers called *protoplanets* swept up more and more of the cloud's debris. Because of their high temperatures and weak gravitational fields, the inner planets (Mercury, Venus, Earth, and Mars) were unable to accumulate and retain many of the lighter components (hydrogen, helium, ammonia, methane, and water). However, because of the very cold temperatures existing far from the Sun, the fragments from which the large outer planets (Jupiter, Saturn, Uranus, and Neptune) formed consisted of huge amounts of lighter materials. These gaseous substances account for the comparatively large sizes and low densities of the outer planets.

- The decay of radioactive elements and heat released by colliding particles aided the melting of Earth's interior, allowing the denser elements, principally iron and nickel, to sink to its center. As a result of this *differentiation*, Earth's interior consists of shells or spheres of materials, each having distinct properties.

- Earth's primitive atmosphere probably consisted of water vapor, carbon dioxide, nitrogen, and several trace gases that were released in volcanic emissions, a process called *outgassing*. The first life forms on Earth, probably *anaerobic bacteria*, did not need oxygen. As life evolved, plants, through the process of *photosynthesis*, used carbon dioxide and water and released oxygen into the atmosphere. Once the available iron on Earth was oxidized (combined with oxygen), substantial quantities of oxygen accumulated in the atmosphere. About 4 billion years into Earth's existence, the fossil record reveals abundant ocean-dwelling organisms that require oxygen to live.

- The *Precambrian* spans about 88 percent of Earth's history, beginning with the formation of Earth about 4.5 billion years ago and ending approximately 540 million years ago with the diversification of life, which marks the start of the Paleozoic era. It is the least understood span of Earth's history. On each continent there is a core area of Precambrian rocks called a *shield*. The iron-ore deposits of Precambrian age represent the time when oxygen became abundant and combined with iron to form iron oxide. The most common middle Precambrian fossils are *stromatolites*. Microfossils of bacteria and blue-green algae, both primitive *prokaryotes* whose cells lack organized nuclei, have been found in *chert*, a hard, dense, chemical sedimentary rock in southern Africa (3.1 billion years of age) and near Lake Superior (1.7 billion years of age). *Eukaryotes*, with cells containing organized nuclei, are among billion-year-old fossils discovered in Australia. Plant fossils date from the middle Precambrian, but animal fossils came a bit later, in the late Precambrian. Many of these fossils are *trace fossils*.

- The *Paleozoic era* extends from 540 million years ago to about 248 million years ago. The beginning of the Paleozoic is marked by the *appearance of the first life forms with hard parts* such as shells. Therefore, abundant Paleozoic fossils occur, and a far more detailed record of Paleozoic events can be constructed. During the early Paleozoic (the Cambrian, Ordovician, and Silurian periods), the vast southern continent of *Gondwanaland* existed. Seas inundated and receded from North America several times, leaving thick evaporate beds of rock salt and gypsum. Life in the early Paleozoic was restricted to the seas and consisted of several invertebrate groups. During the late Paleozoic (the Devonian, Mississippian, Pennsylvanian, and Permian periods), ancestral North America collided with Africa to produce the original northern Appalachian Mountains, and the northern continent of *Laurasia* formed. By the close of the Paleozoic, all the continents had fused into the supercontinent of *Pangaea*. During most of the late Paleozoic, organisms diversified dramatically. Insects and plants moved onto the land, and amphibians evolved and diversified quickly. By the Pennsylvanian period, large tropical swamps, which became the major coal deposits of today, extended across North America, Europe, and Siberia. At the close of the Paleozoic, altered climatic conditions contributed to one of the most dramatic biological declines in all of Earth's history.

- The *Mesozoic era*, literally the *"era of middle life,"* is often called the "age of dinosaurs." It began about 248 million years ago and ended approximately 65 million years ago. Early in the Mesozoic, much of the land was above sea level. However, by the middle Mesozoic, seas invaded western North America. As Pangaea began to break up, the westward-moving North American plate began to override the Pacific plate, causing crustal deformation along the entire western margin of the continent. Organisms that had survived extinction at the end of the Paleozoic began to diversify in spectacular ways. *Gymnosperms* (cycads, conifers, and ginkgoes) became the dominant trees of the Mesozoic because they could adapt to the drier climates. Reptiles became the dominant land animals, with one group eventually becoming the birds. The most awesome of the Mesozoic reptiles were the *dinosaurs*. At the close of the Mesozoic, many reptile groups, including the dinosaurs, became extinct.

- The *Cenozoic era*, or *"era of recent life,"* begins approximately 65 million years ago and continues today. It is the *time of mammals*, including humans. The widespread, less disturbed rock formations of the Cenozoic provide a rich geologic record. Most of North America was above sea level throughout the Cenozoic. Owing to their different relations with tectonic plate boundaries, the eastern and western margins of the continent experienced contrasting events. The stable eastern margin was the site of abundant sedimentation as isostatic adjustment raised the Appalachians, causing streams to erode with renewed vigor and deposit their sediment along the continental margin. In the West, building of the Rocky Mountains (the *Laramide Orogeny*) was coming to an end, the Basin and Range Province was forming, and volcanic activity was extensive. The Cenozoic is often called *the age of mammals* because these animals replaced the reptiles as the dominant land life. Two groups of mammals, the marsupials and the placentals, evolved and expanded to dominate the era. One tendency was for some mammal groups to become very large. However, a wave of late *Pleistocene* extinctions rapidly eliminated these animals from the landscape. Some scientists believe that early humans hastened their decline by selectively hunting the larger animals. The Cenozoic could also be called the *"age of flowering plants."* As a source of food, flowering plants (angiosperms) strongly influenced the evolution of both birds and herbivorous (plant-eating) mammals throughout the Cenozoic era.

Key Terms

nebular hypothesis (p. 413)   outgassing (p. 415)   shields (p. 416)   stromatolites (p. 417)

Questions for Review

1. Briefly describe the events believed to have led to the formation of the solar system.

2. What is the major source of free oxygen in Earth's atmosphere?

3. Explain why Precambrian history is more difficult to decipher than more recent geological history.

4. Match the following words and phrases to the most appropriate time span. Select among the following: Precambrian, early Paleozoic, late Paleozoic, Mesozoic, Cenozoic.
   a. Pangaea came into existence.
   b. Bacteria and blue-green algae preserved in chert.
   c. The era that encompasses the least amount of time.
   d. Shields.
   e. Age of dinosaurs.
   f. Formation of the original northern Appalachian Mountains.
   g. Mastodons and mammoths.
   h. Extensive deposits of rock salt.
   i. Triassic, Jurassic, and Cretaceous.
   j. Coal swamps extended across North America, Europe, and Siberia.
   k. Gulf Coast oil deposits formed.
   l. Formation of most of the world's major iron-ore deposits.
   m. Massive sand dunes covered large portions of the Colorado Plateau region.
   n. The "age of fishes" occurred during this span.
   o. Cambrian, Ordovician, and Silurian.
   p. Pangaea began to break apart.
   q. "Age of mammals."
   r. Animals with hard parts first appeared in abundance.
   s. Gymnosperms were the dominant trees.
   t. Columbia Plateau formed.
   u. Stromatolites are among its more common fossils.
   v. "Golden age of trilobites" occurred during this span.
   w. Fault-block mountains form in the Basin and Range region.

5. Briefly discuss two proposals that attempt to explain why several groups developed hard parts at the beginning of the Cambrian period. Do these proposals appear to provide a satisfactory explanation? Why or why not?

6. List some differences between amphibians and reptiles. List differences between reptiles and mammals.

7. Describe one hypothesis that attempts to explain the extinction of the dinosaurs.

8. Contrast the eastern and western margins of North America during the Cenozoic era in terms of their relationships to plate boundaries.

Web Resources

The *Essentials of Geology* Website uses the resources and flexibility of the Internet to aid in your study of the topics in this chapter. Written and developed by geology instructors, this site will help improve your understanding of geology. Visit **www.prenhall.com/lutgens** and click on the cover of *Essentials of Geology, 8e* to find:

- On-line review quizzes.
- Web-based critical thinking and writing exercises.
- Links to chapter-specific Web resources.
- Internet-wide key term searches.

http://www.prenhall.com/lutgens

Metric and English Units Compared

Units

$$1 \text{ kilometer (km)} = 1000 \text{ meters (m)}$$
$$1 \text{ meter (m)} = 100 \text{ centimeters (cm)}$$
$$1 \text{ centimeter (cm)} = 0.39 \text{ inch (in.)}$$
$$1 \text{ mile (mi)} = 5280 \text{ feet (ft)}$$
$$1 \text{ foot (ft)} = 12 \text{ inches (in.)}$$
$$1 \text{ inch (in.)} = 2.54 \text{ centimeters (cm)}$$
$$1 \text{ square mile (mi}^2) = 640 \text{ acres (a)}$$
$$1 \text{ kilogram (kg)} = 1000 \text{ grams (g)}$$
$$1 \text{ pound (lb)} = 16 \text{ ounces (oz)}$$
$$1 \text{ fathom} = 6 \text{ feet (ft)}$$

Conversions

When you want
  to convert: | Multiply by: | To find:

Length

inches	2.54	centimeters
centimeters	0.39	inches
feet	0.30	meters
meters	3.28	feet
yards	0.91	meters
meters	1.09	yards
miles	1.61	kilometers
kilometers	0.62	miles

Area

square inches	6.45	square centimeters
square centimeters	0.15	square inches
square feet	0.09	square meters
square meters	10.76	square feet
square miles	2.59	square kilometers
square kilometers	0.39	square miles

Volume

cubic inches	16.38	cubic centimeters
cubic centimeters	0.06	cubic inches
cubic feet	0.028	cubic meters
cubic meters	35.3	cubic feet
cubic miles	4.17	cubic kilometers
cubic kilometers	0.24	cubic miles
liters	1.06	quarts
liters	0.26	gallons
gallons	3.78	liters

Masses and Weights

ounces	28.33	grams
grams	0.035	ounces
pounds	0.45	kilograms
kilograms	2.205	pounds

Temperature

When you want to convert degrees Fahrenheit (°F) to degrees Celsius (°C), subtract 32 degrees and divide by 1.8.

When you want to convert degrees Celsius (°C) to degrees Fahrenheit (°F), multiply by 1.8 and add 32 degrees.

When you want to convert degrees Celsius (°C) to kelvins (K), delete the degree symbol and add 273. When you want to convert kelvins (K) to degrees Celsius (°C), add the degree symbol and subtract 273.

Figure A.1 A comparison of Fahrenheit and Celsius temperature scales.

Topographic Maps

A map is a representation on a flat surface of all or a part of Earth's surface drawn to a specific scale. Maps are often the most effective means for showing the locations of both natural and human structures, their sizes, and their relationships to one another. Like photographs, maps readily display information that would be impractical to express in words.

While most maps show only the two horizontal dimensions, geologists, as well as other map users, often require that the third dimension—elevation—be shown on maps. Maps that show the shape of the land are called **topographic maps.** Although various techniques may be used to depict elevations, the most accurate method involves the use of contour lines.

Contour Lines

A **contour line** is a line on a map representing a corresponding imaginary line on the ground that has the same elevation above sea level along its entire length. While many map symbols are pictographs, resembling the objects they represent, a contour line is an abstraction that has no counterpart in nature. It is, however, an accurate and effective device for representing the third dimension on paper.

Some useful facts and rules concerning contour lines are listed as follows. This information should be studied in conjunction with Figure B.1.

1. Contour lines bend upstream or upvalley. The contours form V's that point upstream, and in the upstream direction the successive contours represent higher elevations. For example, if you were standing on a stream bank and wished to get to the point at the same elevation directly opposite you on the other bank without stepping up or down, you would need to walk upstream along the contour at that elevation to where it crosses the stream bed, cross the stream, and then walk back downstream along the same contour.
2. Contours near the upper parts of hills form closures. The top of a hill is higher than the highest closed contour.
3. Hollows (depressions) without outlets are shown by closed, hatched contours. Hatched contours are contours with short lines on the inside pointing downslope.
4. Contours are widely spaced on gentle slopes.
5. Contours are closely spaced on steep slopes.
6. Evenly spaced contours indicate a uniform slope.
7. Contours usually do not cross or intersect each other, except in the rare case of an overhanging cliff.
8. All contours eventually close, either on a map or beyond its margins.
9. A single high contour never occurs between two lower ones, and vice versa. In other words, a change in slope direction is always determined by the repetition of the same elevation either as two different contours of the same value or as the same contour crossed twice.
10. Spot elevations between contours are given at many places, such as road intersections, hill summits, and lake surfaces. Spot elevations differ from control elevation stations, such as benchmarks, in not being permanently established by permanent markers.

Relief

Relief refers to the difference in elevation between any two points. Maximum relief refers to the difference in elevation between the highest and lowest points in the area being considered. Relief determines the **contour interval,** which is the difference in elevation between succeeding contour lines that is used on topographic maps. Where relief is low, a small contour interval, such as 10 or 20 feet, may be used. In flat areas, such as wide river valleys or broad, flat uplands, a contour interval of 5 feet is often used. In rugged mountainous terrain, where relief is many hundreds of feet, contour intervals as large as 50 or 100 feet are used.

Scale

Map **scale** expresses the relationship between distance or area on the map to the true distance or area on Earth's surface. This is generally expressed as a ratio or fraction, such as 1:24,000 or 1/24,000. The numerator, usually 1, represents map distance, and the denominator, a large number, represents ground distance. Thus, 1:24,000 means that a distance of 1 unit on the map represents a distance of 24,000 such units on the surface of Earth. It does not matter what the units are.

Elevation in feet above mean sea level

Figure B.1 Perspective view of an area and a contour map of the same area. These illustrations show how features are depicted on a topographic map. The upper illustration is a perspective view of a river valley and the adjoining hills. The river flows into a bay, which is partly enclosed by a hooked sandbar. On either side of the valley are terraces through which streams have cut gullies. The hill on the right has a smoothly eroded form and gradual slopes, whereas the one on the left rises abruptly in a sharp precipice, from which it slopes gently, and forms an inclined plateau traversed by a few shallow gullies. A road provides access to a church and the two houses situated across the river from a highway that follows the seacoast and curves up the river valley. The lower illustration shows the same features represented by symbols on a topographic map. The contour interval (vertical distance between adjacent contours) is 20 feet. (After U.S. Geological Survey)

Often, the graphic or bar scale is more useful than the fractional scale, because it is easier to use for measuring distances between points. The graphic scale (Figure B.2) consists of a bar divided into equal segments, which represent equal distances on the map. One segment on the left side of the bar is usually divided into smaller units to permit more accurate estimates of fractional units.

Topographic maps, which are also referred to as *quadrangles*, are generally classified according to publication scale. Each series is intended to fulfill a specific type of map need. To select a map with the proper scale for a particular use, remember that large-scale

maps show more detail and small-scale maps show less detail. The sizes and scales of topographic maps published by the U.S. Geological Survey are shown in Table B.1.

Color and Symbol

Each color and symbol used on U.S. Geological Survey topographic maps has significance. Common topographic map symbols are shown in Figure B.3. The meaning of each color is as follows:

Blue — water features

Figure B.2 Graphic scale.

Black — works of humans, such as homes, schools, churches, roads, and so forth

Brown — contour lines

Green — woodlands, orchards, and so forth

Red — urban areas, important roads, public land subdivision lines

Table B.1 National topographic maps.

Series	Scale	1 inch Represents	Standard Quadrangle Size (latitude-longitude)	Quadrangle Area (square miles)	Paper Size E-W N-S Width Length (inches)
7 $^1/_2$-minute	1:24,000	2000 feet	7$^1/_2$′ × 7$^1/_2$′	49–70	22 × 27
Puerto Rico 7 $^1/_2$-minute	1:20,000	about 1667 feet	7$^1/_2$′ × 7$^1/_2$′	71	29$^1/_2$ × 32$^1/_2$
15-minute	1:62,500	nearly 1 mile	15′ × 15′	197–282	17 × 21
Alaska 1:63,360	1:63,360	1 mile	15′ × 20′–36′	207–281	18 × 21
U.S. 1:250,000	1:250,000	nearly 4 miles	1° × 2°	4580–8669	34 × 22
U.S. 1:1,000,000	1:1,000,000	nearly 16 miles	4° × 6°	73,734–102,759	27 × 27

Source: U.S. Geological Survey

TOPOGRAPHIC MAP SYMBOLS

VARIATIONS WILL BE FOUND ON OLDER MAPS

Primary highway, hard surface . . . .

Secondary highway, hard surface . . . .

Light-duty road, hard or improved surface . . . .

Unimproved road . . . .

Road under construction, alinement known . . . .

Proposed road . . . .

Dual highway, dividing strip 25 feet or less . . . .

Dual highway, dividing strip exceeding 25 feet . . . .

Trail . . . .

Railroad: single track and multiple track . . . .

Railroads in juxtaposition . . . .

Narrow gage: single track and multiple track . . . .

Railroad in street and carline . . . .

Bridge: road and railroad . . . .

Drawbridge: road and railroad . . . .

Footbridge . . . .

Tunnel: road and railroad . . . .

Overpass and underpass . . . .

Small masonry or concrete dam . . . .

Dam with lock . . . .

Dam with road . . . .

Canal with lock . . . .

Buildings (dwelling, place of employment, etc.) . . . .

School, church, and cemetery . . . .

Buildings (barn, warehouse, etc.) . . . .

Power transmission line with located metal tower . . . .

Telephone line, pipeline, etc. (labeled as to type) . . . .

Wells other than water (labeled as to type) . . . .   oOil . . . . . oGas

Tanks: oil, water, etc. (labeled only if water) . . . .   ● ● ⊘Water

Located or landmark object; windmill . . . .   o . . . . 웃

Open pit, mine, or quarry; prospect   x

Shaft and tunnel entrance . . . .   Y

Horizontal and vertical control station:

    Tablet, spirit level elevation . . . .   BM △ 5653

    Other recoverable mark, spirit level elevation . . . .   △ 5455

Horizontal control station: tablet, vertical angle elevation   VABM △ 95/9

    Any recoverable mark, vertical angle or checked elevation   △ 3775

Vertical control station: tablet, spirit level elevation . . . . .   BM X 957

    Other recoverable mark, spirit level elevation . . . .   X 954

Spot elevation . . . .   X 7369   X 7369

Water elevation   670   670

Boundaries: National . . . .

    State . . . .

    County, parish, municipio . . . .

    Civil township, precinct, town, barrio . . . .

    Incorporated city, village, town, hamlet . . . .

    Reservation, National or State . . . .

    Small park, cemetery, airport, etc. . . . .

    Land grant . . . .

Township or range line, United States land survey . . . . . .

Township or range line, approximate location . . . .

Section line, United States land survey . . . .

Section line, approximate location . . . .

Township line, not United States land survey . . . .

Section line, not United States land survey . . . .

Found corner: section and closing . . . .

Boundary monument: land grant and other . . . .

Fence or field line . . . .

Index contour . . . .   Intermediate contour . .

Supplementary contour   Depression contours . .

Fill . . . .   Cut . . . .

Levee . . . .   Levee with road . . . . . .

Mine dump . . . .   Wash . . . .

Tailings . . . .   Tailings pond . . . .

Shifting sand or dunes   Intricate surface . . . . .

Sand area . . . .   Gravel beach . . . .

Perennial streams . . . . .   Intermittent streams . . .

Elevated aqueduct . . . . .   Aqueduct tunnel . . . . . .

Water well and spring . . .   Glacier . . . .

Small rapids . . . .   Small falls . . . .

Large rapids . . . .   Large falls . . . .

Intermittent lake . . . .   Dry lake bed . . . .

Foreshore flat . . . .   Rock or coral reef . . . .

Sounding, depth curve . .   10   Piling or dolphin . . . . .

Exposed wreck: . . . .   Sunken wreck . . . .

Rock, bare or awash; dangerous to navigation . . . .

Marsh (swamp) . . . .   Submerged marsh . . . .

Wooded marsh . . . .   Mangrove . . . .

Woods or brushwood . . .   Orchard . . . .

Vineyard . . . .   Scrub . . . .

Land subject to
controlled inundation . . . .   Urban area

Figure B.3   U.S. Geological Survey topographic map symbols. (Variations will be found on older maps.)

Appendix

C

Landforms of the Conterminous United States

Figure C.1 Outline map showing major physiographic provinces of the United States.

Landforms of the Conterminous United States

Figure C.2 Digital shaded relief landform map of the United States. (Data provided by the U.S. Geological Survey)

Glossary

Aa A type of lava flow that has a jagged, blocky surface.

Ablation A general term for the loss of ice and snow from a glacier.

Abrasion The grinding and scraping of a rock surface by the friction and impact of rock particles carried by water, wind, or ice.

Abyssal plain Very level area of the deep-ocean floor, usually lying at the foot of the continental rise.

Accretionary wedge A large wedge-shaped mass of sediment that accumulates in subduction zones. Here sediment is scraped from the subducting oceanic plate and accreted to the overriding crustal block.

Active continental margin Usually narrow and consisting of highly deformed sediments. They occur where oceanic lithosphere is being subducted beneath the margin of a continent.

Active layer The zone above the permafrost that thaws in summer and refreezes in winter.

Aftershock A smaller earthquake that follows the main earthquake.

Alluvial fan A fan-shaped deposit of sediment formed when a stream's slope is abruptly reduced.

Alluvium Unconsolidated sediment deposited by a stream.

Alpine glacier See *Valley glacier*.

Andesitic Igneous rocks having a mineral makeup between that of granite and basalt, after the common volcanic rock andesite. See also *Intermediate*.

Angle of repose The steepest angle at which loose material remains stationary without sliding downslope.

Angular unconformity An unconformity in which the older strata dip at an angle different from that of the younger beds.

Anthracite A hard, metamorphic form of coal that burns clean and hot.

Anticline A fold in sedimentary strata that resembles an arch.

Aphanitic texture A texture of igneous rocks in which the crystals are too small for individual minerals to be distinguished with the unaided eye.

Aquifer Rock or sediment through which groundwater moves easily.

Aquitard An impermeable bed that hinders or prevents groundwater movement.

Archean eon The second eon of Precambrian time. The eon following the Hadean eon and preceding the Proterozoic eon. It extends between about 3.8 and 2.5 billion years ago.

Arête A narrow, knifelike ridge separating two adjacent glaciated valleys.

Arkose A feldspar-rich sandstone.

Artesian well A well in which the water rises above the level where it was initially encountered.

Asthenosphere A subdivision of the mantle situated below the lithosphere. This zone of weak material exists below a depth of about 100 kilometers and in some regions extends as deep as 700 kilometers. The rock within this zone is easily deformed.

Atmosphere The gaseous portion of a planet; the planet's envelope of air. One of the traditional subdivisions of Earth's physical environment.

Atoll A continuous or broken ring of coral reef surrounding a central lagoon.

Atom The smallest particle that exists as an element.

Atomic mass unit A mass unit equal to exactly one-twelfth the mass of a carbon-12 atom.

Atomic number The number of protons in the nucleus of an atom.

Atomic weight The average of the atomic masses of isotopes of a given element.

Aureole A zone or halo of contact metamorphism found in the host rock surrounding an igneous intrusion.

Back swamp A poorly drained area on a floodplain, resulting when natural levees are present.

Bar Common term for sand and gravel deposits in a stream channel.

Barchan dune A solitary sand dune shaped like a crescent with its tips pointed downwind.

Barchanoid dunes Dunes forming scalloped rows of sand oriented at right angles to the wind. This form is intermediate between isolated barchans and extensive waves of transverse dunes.

Barrier island A low, elongated ridge of sand that parallels the coast.

Basal slip A mechanism of glacial movement in which the ice mass slides over the surface below.

Basalt An aphanitic igneous rock of mafic composition.

Basaltic Term used to describe igneous rocks that contain abundant dark (ferromagnesian) minerals and about 50 percent silica.

Base level The level below which a stream cannot erode.

Basin A circular downfolded structure.

Batholith A large mass of igneous rock that formed when magma was emplaced at depth, crystallized, and was subsequently exposed by erosion.

Baymouth bar A sandbar that completely crosses a bay, sealing it off from the main body of water.

Beach drift The transport of sediment in a zigzag pattern along a beach. It is caused by the uprush of water from obliquely breaking waves.

Beach nourishment Large quantities of sand are added to the beach system to offset losses caused by wave erosion. By building beaches seaward, beach quality and storm protection are both improved.

Bed See *Strata*.

Bedding plane A nearly flat surface separating two beds of sedimentary rock. Each bedding plane marks the end of one deposit and the beginning of another having different characteristics.

Bed load   Sediment rolled along the bottom of a stream by moving water, or particles robed along the ground surface by wind.

Belt of soil moisture   A zone in which water is held as a film on the surface of soil particles and may be used by plants or withdrawn by evaporation. The uppermost subdivision of the zone of aeration.

Biochemical   Describing a type of chemical sediment that forms when material dissolved in water is precipitated by water-dwelling organisms. Shells are common examples.

Biogenous sediment   Seafloor sediments consisting of material of marine-organic origin.

Bituminous coal   The most common form of coal, often called soft, black coal.

Blowout (deflation hollow)   A depression excavated by wind in easily eroded materials.

Body wave   A seismic wave that travels through Earth's interior.

Bottomset bed   A layer of fine sediment deposited beyond the advancing edge of a delta and then buried by continuous delta growth.

Bowen's reaction series   A concept proposed by N. L. Bowen that illustrates the relationship between magma and the minerals crystallizing from it during the formation of igneous rock.

Braided stream   A stream consisting of numerous intertwining channels.

Breakwater   A structure protecting a nearshore area from breaking waves.

Breccia   A sedimentary rock composed of angular fragments that were lithified.

Brittle deformation   Deformation that involves the fracturing of rock. Associated with rocks near the surface.

Burial metamorphism   Low-grade metamorphism that occurs in the lowest layers of very thick accumulations of sedimentary strata.

Caldera   A large depression typically caused by collapse of the summit area of a volcano following a violent eruption.

Caliche   A hard layer, rich in calcium carbonate, that forms beneath the *B* horizon in soils of arid regions.

Calving   Wastage of a glacier that occurs when large pieces of ice break off into water.

Capacity   The total amount of sediment a stream is able to transport.

Capillary fringe   A relatively narrow zone at the base of the zone of aeration. Here water rises from the water table in tiny threadlike openings between grains of soil or sediment.

Cap rock   A necessary part of an oil trap. The cap rock is impermeable and hence keeps upwardly mobile oil and gas from escaping at the surface.

Catastrophism   The concept that Earth was shaped by catastrophic events of a short-term nature.

Cavern   A naturally formed underground chamber or series of chambers most commonly produced by solution activity in limestone.

Cementation   One way in which sedimentary rocks are lithified. As material precipitates from water that percolates through the sediment, open spaces are filled and particles are joined into a solid mass.

Cenozoic era   A time span on the geologic time scale beginning about 65 million years ago following the Mesozoic era.

Chemical sedimentary rock   Sedimentary rock consisting of material that was precipitated from water by either inorganic or organic means.

Chemical weathering   The processes by which the internal structure of a mineral is altered by the removal and/or addition of elements.

Cinder cone   A rather small volcano built primarily of pyroclastics ejected from a single vent.

Cirque   An amphitheater-shaped basin at the head of a glaciated valley produced by frost wedging and plucking.

Clastic   A sedimentary rock texture consisting of broken fragments of preexisting rock.

Cleavage   The tendency of a mineral to break along planes of weak bonding.

Col   A pass between mountain valleys where the headwalls of two cirques intersect.

Color   A phenomenon of light by which otherwise identical objects may be differentiated.

Column   A feature found in caves that is formed when a stalactite and stalagmite join.

Columnar joints   A pattern of cracks that forms during cooling of molten rock to generate columns.

Compaction   A type of lithification in which the weight of overlying material compresses more deeply buried sediment. It is most important in fine-grained sedimentary rocks such as shale.

Competence   A measure of the largest particle a stream can transport; a factor dependent on velocity.

Composite cone   A volcano composed of both lava flows and proclastic material.

Compound   A substance formed by the chemical combination of two or more elements in definite proportions and usually having properties different from those of its constituent elements.

Concordant   A term used to describe intrusive igneous masses that form parallel to the bedding of the surrounding rock.

Conduit   A pipelike opening through which magma moves toward Earth's surface. It terminates at a surface opening called a vent.

Cone of depression   A cone-shaped depression immediately surrounding a well.

Conformable layers   Rock layers that were deposited without interruption.

Conglomerate   A sedimentary rock consisting of rounded, gravel-size particles.

Contact metamorphism   Changes in rock caused by the heat of a nearby magma body.

Continental drift   A hypothesis, credited largely to Alfred Wegener, that suggested all present continents once existed as a single supercontinent. Further, beginning about 200 million years ago, the supercontinent began breaking into smaller continents, which then drifted to their present positions.

Continental margin   See *Active continental margin* and *Passive continental margin*.

Continental rise   The gently sloping surface at the base of the continental slope.

Continental shelf   The gently sloping submerged portion of the continental margin, extending from the shoreline to the continental slope.

Continental slope The steep gradient that leads to the deep-ocean floor and marks the seaward edge of the continental shelf.

Continental volcanic arc Mountains formed in part by igneous activity associated with the subduction of oceanic lithosphere beneath a continent. Examples include the Andes and the Cascades.

Convergent plate boundary A boundary in which two plates move together, resulting in oceanic lithosphere being thrust beneath an overriding plate, eventually to be reabsorbed into the mantle. It can also involve the collision of two continental plates to create a mountain system.

Core Located beneath the mantle, it is Earth's innermost layer. The core is divided into an outer core and an inner core.

Correlation Establishing the equivalence of rocks of similar age in different areas.

Covalent bond A chemical bond produced by the sharing of electrons.

Crater The depression at the summit of a volcano, or that which is produced by a meteorite impact.

Creep The slow downhill movement of soil and regolith.

Crevasse A deep crack in the brittle surface of a glacier.

Cross-bedding Structure in which relatively thin layers are inclined at an angle to the main bedding. Formed by currents of wind or water.

Cross-cutting A principle of relative dating. A rock or fault is younger than any rock (or fault) through which it cuts.

Crust The very thin, outermost layer of Earth.

Crystal An orderly arrangement of atoms.

Crystal form The external appearance of a mineral as determined by its internal arrangement of atoms.

Crystallization The formation and growth of a crystalline solid from a liquid or gas.

Curie point The temperature above which a material loses its magnetization.

Cut bank The area of active erosion on the outside of a meander.

Cutoff A short channel segment created when a river erodes through the narrow neck of land between meanders.

Dark silicate Silicate minerals containing ions of iron and/or magnesium in their structure. They are dark in color and have a higher specific gravity than nonferromagnesian silicates.

Daughter product An isotope resulting from radioactive decay.

Debris flow A relatively rapid type of mass wasting that involves a flow of soil and regolith containing a large amount of water. Also called *mudflows*.

Decompression melting Melting that occurs as rock ascends due to a drop in confining pressure.

Deep-focus earthquake An earthquake focus at a depth of more than 300 kilometers.

Deep-ocean basin The portion of seafloor that lies between the continental margin and the oceanic ridge system. This region comprises almost 30 percent of Earth's surface.

Deep-ocean trench See *Trench*.

Deflation The lifting and removal of loose material by wind.

Deformation General term for the processes of folding, faulting, shearing, compression, or extension of rocks as the result of various natural forces.

Delta An accumulation of sediment formed where a stream enters a lake or ocean.

Dendritic pattern A stream system that resembles the pattern of a branching tree.

Density The weight per unit volume of a particular material.

Desert One of the two types of dry climate; the drier of the dry climates.

Desert pavement A layer of coarse pebbles and gravel created when wind removes the finer material.

Detachment fault A low-angle fault that represents a major boundary between unfaulted rocks below that exhibit ductile deformation and rocks above that exhibit brittle deformation via faulting.

Detrital sedimentary rocks Rocks that form from the accumulation of materials that originate and are transported as solid particles derived from both mechanical and chemical weathering.

Diagenesis A collective term for all the chemical, physical, and biological changes that take place after sediments are deposited and during and after lithification.

Dike A tabular-shaped intrusive igneous feature that cuts through the surrounding rock.

Dip The angle at which a rock layer is inclined from the horizontal. The direction of dip is at a right angle to the strike.

Dip-slip fault A fault in which the movement is parallel to the dip of the fault.

Discharge The quantity of water in a stream that passes a given point in a given period of time.

Disconformity A type of unconformity in which the beds above and below are parallel.

Discontinuity A sudden change with depth in one or more of the physical properties of the material making up Earth's interior. The boundary between two dissimilar materials in Earth's interior as determined by the behavior of seismic waves.

Discordant A term used to describe plutons that cut across existing rock structures, such as bedding planes.

Disseminated deposit Any economic mineral deposit in which the desired mineral occurs as scattered particles in the rock but in sufficient quantity to make the deposit an ore.

Dissolved load The portion of a stream's load carried in solution.

Distributary A section of a stream that leaves the main flow.

Diurnal tide A tide characterized by a single high and low water height each tidal day.

Divergent plate boundary A boundary in which two plates move apart, resulting in upwelling of material from the mantle to create new seafloor.

Divide An imaginary line that separates the drainage of two streams; often found along a ridge.

Dome A roughly circular, upfolded structure.

Drainage basin The land area that contributes water to a stream.

Drawdown The difference in height between the bottom of a cone of depression and the original height of the water table.

Drift See *Glacial drift*.

Drumlin A streamlined asymmetrical hill composed of glacial till. The steep side of the hill faces the direction from which the ice advanced.

Dry climate A climate in which the yearly precipitation is less than the potential loss of water by evaporation.

Ductile deformation   A type of solid-state flow that produces a change in the size and shape of a rock body without fracturing. Occurs at depths where temperatures and confining pressures are high.

Dune   A hill or ridge of wind-deposited sand.

Earthflow   The downslope movement of water-saturated, clay-rich sediment. Most characteristic of humid regions.

Earthquake   Vibration of Earth produced by the rapid release of energy.

Echo sounder   An instrument used to determine the depth of water by measuring the time interval between emission of a sound signal and the return of its echo from the bottom.

Elastic deformation   Nonpermanent deformation in which rock returns to its original shape when the stress is released.

Elastic rebound   The sudden release of stored strain in rocks that results in movement along a fault.

Electron   A negatively charged subatomic particle that has a negligible mass and is found outside the atom's nucleus.

Element   A substance that cannot be decomposed into simpler substances by ordinary chemical or physical means.

Eluviation   The washing out of fine soil components from the *A* horizon by downward-percolating water.

Emergent coast   A coast where land formerly below the sea level has been exposed either by crustal uplift or a drop in sea level or both.

End moraine   A ridge of till marking a former position of the front of the glacier.

Energy-level shell   The region occupied by electrons with a specific energy level.

Eon   The largest time unit on the geologic time scale, next in order of magnitude above *era*.

Epicenter   The location on Earth's surface that lies directly above the focus of an earthquake.

Epoch   A unit of the geological time scale that is a subdivision of a *period*.

Era   A major division on the geologic time scale; eras are divided into shorter units called *periods*.

Erosion   The incorporation and transportation of material by a mobile agent, such as water, wind, or ice.

Eruption column   Buoyant plumes of hot ash–laden gases that can extend thousands of meters into the atmosphere.

Esker   Sinuous ridge composed largely of sand gravel deposited by a stream flowing in a tunnel beneath a glacier near its terminus.

Estuary   A partially enclosed coastal water body that is connected to the ocean. Salinity here is measurably reduced by the freshwater flow of rivers.

Evaporite   A sedimentary rock formed of material deposited from solution by evaporation of water.

Evapotranspiration   The combined effect of evaporation and transpiration.

Exfoliation dome   Large, dome-shaped structure, usually composed of granite, formed by sheeting.

Exotic stream   A permanent stream that traverses a desert and has its source in well-watered areas outside the desert.

External process   Process such as weathering, mass wasting, or erosion that is powered by the Sun and transforms solid rock into sediment.

Extrusive   Igneous activity that occurs at Earth's surface.

Fall   A type of movement common to mass-wasting processes that refers to the free falling of detached individual pieces of any size.

Fault   A break in a rock mass along which movement has occurred.

Fault-block mountain   A mountain formed by the displacement of rock along a fault.

Fault creep   Slow, gradual displacement along a fault that occurs relatively smoothly and with little noticeable seismic activity.

Fault scarp   A cliff created by movement along a fault. It represents the exposed surface of the fault prior to modification by weathering and erosion.

Felsic   A term derived from *feld*spar and *sil*ica (quartz). It is a term used to describe granitic igneous rocks.

Ferromagnesian silicate   See *Dark silicate*.

Fetch   The distance that the wind has traveled across the open water.

Fiord   A steep-sided inlet of the sea formed when a glacial trough was partially submerged.

Fissure   A crack in rock along which there is a distinct separation.

Fissure eruption   An eruption in which lava is extruded from narrow fractures or cracks in the crust.

Flood   The overflow of a stream channel that occurs when discharge exceeds the channel's capacity. The most common and destructive geologic hazard.

Flood basalts   Flows of basaltic lava that issue from numerous cracks or fissures and commonly cover extensive areas to thicknesses of hundreds of meters.

Floodplain   The flat, low-lying portion of a stream valley subject to periodic inundation.

Flow   A type of movement common to mass-wasting processes in which water-saturated material moves downslope as a viscous fluid.

Fluorescence   The absorption of ultraviolet light, which is reemitted as visible light.

Focus (earthquake)   The zone within Earth where rock displacement produces an earthquake.

Fold   A bent layer or series of layers that were originally horizontal and subsequently deformed.

Foliated   A texture of metamorphic rocks that gives the rock a layered appearance.

Foliation   A term for a linear arrangement of textural features often exhibited by metamorphic rocks.

Foreset bed   An inclined bed deposited along the front of a delta.

Foreshocks   Small earthquakes that often precede a major earthquake.

Fossil   The remains or traces of organisms preserved from the geologic past.

Fossil fuel   General term for any hydrocarbon that may be used as a fuel, including coal, oil, natural gas, bitumen from tar sands, and shale oil.

Fossil succession   Fossil organisms succeed one another in a definite and determinable order, and any time period can be recognized by its fossil content.

Fractional crystallization   The process that separates magma into components having varied compositions and melting points.

Fracture (mineral)   One of the basic physical properties of minerals. It relates to the breakage of minerals when there are no planes of weakness in the crystalline structure. Examples include conchoidal, irregular, and splintery.

Fracture (rock)   Any break or rupture in rock along which no appreciable movement has taken place.

Fragmental texture   See *Pyroclastic texture*.

Frost wedging   The mechanical breakup of rock caused by the expansion of freezing water in cracks and crevices.

Fumarole   A vent in a volcanic area from which fumes or gases escape.

Gaining stream   Streams that gain water from the inflow of groundwater through the streambed.

Geology   The science that examines Earth, its form and composition, and the changes it has undergone and is undergoing.

Geothermal energy   Natural steam used for power generation.

Geothermal gradient   The gradual increase in temperature with depth in the crust. The average is 30°C per kilometer in the upper crust.

Geyser   A fountain of hot water ejected periodically from the ground.

Glacial budget   The balance, or lack of balance, between accumulation at the upper end of a glacier, and loss at the other end.

Glacial drift   An all-embracing term for sediments of glacial origin, no matter how, where, or in what shape they were deposited.

Glacial erratic   An ice-transported boulder that was not derived from the bedrock near its present site.

Glacial striations   Scratches and grooves on bedrock caused by glacial abrasion.

Glacial trough   A mountain valley that has been widened, deepened, and straightened by a glacier.

Glacier   A thick mass of ice originating on land from the compaction and recrystallization of snow. The ice shows evidence of past or present flow.

Glass (volcanic)   Natural glass produced when molten lava cools too rapidly to permit crystallization. Volcanic glass is a solid composed of unordered atoms.

Glassy texture   A term used to describe the texture of certain igneous rocks, such as obsidian, that contain no crystals.

Gneissic texture   The texture displayed by the metamorphic rock *gneiss* in which dark and light silicate minerals have separated, giving the rock a banded appearance.

Gondwanaland   The southern portion of Pangaea consisting of South America, Africa, Australia, India, and Antarctica.

Graben   A valley formed by the downward displacement of a fault-bounded block.

Graded bed   A sediment layer characterized by a decrease in sediment size from bottom to top.

Graded stream   A stream that has the correct channel characteristics to maintain the exact velocity required to transport the material supplied to it.

Gradient   The slope of a stream; generally measured in feet per mile.

Granitic   Igneous rocks composed mainly of light-colored silicates (quartz and feldspar) are said to have this composition.

Greenhouse effect   Carbon dioxide and water vapor in a planet's atmosphere absorb and reradiate infrared wavelengths, effectively trapping solar energy and raising the temperature.

Groin   A short wall built at a right angle to the seashore to trap moving sand.

Groundmass   The matrix of smaller crystals within an igneous rock that has porphyritic texture.

Ground moraine   An undulating layer of till deposited as the ice front retreats.

Groundwater   Water in the zone of saturation.

Guyot   A submerged flat-topped seamount.

Hadean eon   The first eon on the geologic time scale. The eon ending 3.8 billion years ago that preceded the Archean eon.

Half-life   The time required for one half of the atoms of a radioactive substance to decay.

Hanging valley   A tributary valley that enters a glacial trough at a considerable height above the floor of the trough.

Hardness   A mineral's resistance to scratching and abrasion.

Head   The vertical distance between the recharge and discharge points of a water table. Also, the source area or beginning of a valley.

Headward erosion   The extension upslope of the head of a valley due to erosion.

Historical geology   A major division of geology that deals with the origin of Earth and its development through time. Usually involves the study of fossils and their sequence in rock beds.

Hogback   A narrow, sharp-crested ridge formed by the upturned edge of a steeply dipping bed of resistant rock.

Horn   A pyramid-like peak formed by glacial action in three or more cirques surrounding a mountain summit.

Horst   An elongated, uplifted block of crust bounded by faults.

Hot spot   A proposed concentration of heat in the mantle capable of introducing magma that in turn extrudes onto Earth's surface. The intraplate volcanism that produced the Hawaiian Islands is one example.

Hot spring   A spring in which the water is 6° to 9°C (10° to 15°F) warmer than the mean annual air temperature of its locality.

Humus   Organic matter in soil produced by the decomposition of plants and animals.

Hydrogenous sediment   Seafloor sediments consisting of minerals that crystallize from seawater. The principal example is manganese nodules.

Hydrologic cycle   The unending circulation of Earth's water supply. The cycle is powered by energy from the Sun and is characterized by continuous exchanges of water among the oceans, the atmosphere, and the continents.

Hydrolysis   A chemical-weathering process in which minerals are altered by chemically reacting with water and acids.

Hydrosphere   The water portion of our planet; one of the traditional subdivisions of Earth's physical environment.

Hydrothermal metamorphism   Chemical alterations that occur as hot, ion-rich water circulates through fractures in rock.

Hydrothermal solution   The hot, watery solution that escapes from a mass of magma during the latter stages of crystallization. Such solutions may alter the surrounding country rock and are frequently the source of significant ore deposits.

Hypothesis   A tentative explanation that is then tested to determine if it is valid.

Ice cap   A mass of glacial ice covering a high upland or plateau and spreading out radially.

Ice-contact deposit   An accumulation of stratified drift deposited in contact with a supporting mass of ice.

Ice sheet   A very large, thick mass of glacial ice flowing outward in all directions from one or more accumulation centers.

Igneous rock   A rock formed by the crystallization of molten magma.

Immature soil   A soil lacking horizons.

Impact metamorphism   Metamorphism that occurs when meteorites strike Earth's surface.

Incised meander   Meandering channel that flows in a steep, narrow valley. These meanders form either when an area is uplifted or when base level drops.

Inclusion   A piece of one rock unit contained within another. Inclusions are used in relative dating. The rock mass adjacent to the one containing the inclusion must have been there first in order to provide the fragment.

Index fossil   A fossil that is associated with a particular span of geologic time.

Index mineral   A mineral that is a good indicator of the metamorphic environment in which it formed. Used to distinguish different zones of regional metamorphism.

Inertia   Objects at rest tend to remain at rest, and objects in motion tend to stay in motion unless either is acted upon by an outside force.

Infiltration   The movement of surface water into rock or soil through crack and pore spaces.

Infiltration capacity   The maximum rate at which soil can absorb water.

Inner core   The solid, innermost layer of Earth, about 1216 kilometers (754 miles) in radius.

Inselberg   An isolated mountain remnant characteristic of the late stage of erosion in a mountainous region.

Intensity (earthquake)   A measure of the degree of earthquake shaking at a given locale based on the amount of damage.

Interior drainage   A discontinuous pattern of intermittent streams that do not flow to the ocean.

Intermediate   Compositional category for igneous rocks found near the middle of Bowen's reaction series, mainly amphibole and the intermediate plagioclase feldspars.

Intermediate focus   An earthquake focus at a depth of between 60 and 300 kilometers.

Internal process   A process such as mountain building or volcanism that derives its energy from Earth's interior and elevates Earth's surface.

Intraplate volcanism   Igneous activity that occurs within a tectonic plate away from plate boundaries.

Intrusive rock   Igneous rock that formed below Earth's surface.

Ion   An atom or molecule that possesses an electrical charge.

Ionic bond   A chemical bond between two oppositely charged ions formed by the transfer of valence electrons from one atom to another.

Island arc   See *Volcanic island arc.*

Isostasy   The concept that Earth's crust is floating in gravitational balance upon the material of the mantle.

Isotopes   Varieties of the same element that have different mass numbers; their nuclei contain the same number of protons but different numbers of neutrons.

Joint   A fracture in rock along which there has been no movement.

Kame   A steep-sided hill composed of sand and gravel originating when sediment collected in openings in stagnant glacial ice.

Kame terrace   A narrow, terracelike mass of stratified drift deposited between a glacier and an adjacent valley wall.

Karst   A topography consisting of numerous depressions called *sinkholes.*

Kettle holes   Depressions created when blocks of ice become lodged in glacial deposits and subsequently melt.

Laccolith   A massive, concordant igneous body intruded between preexisting strata.

Lahar   Mudflows on the slopes of volcanoes that result when unstable layers of ash and debris become saturated and flow downslope, usually following stream channels.

Laminar flow   The movement of water particles in straight-line paths that are parallel to the channel. The water particles move downstream without mixing.

Lateral moraine   A ridge of till along the sides of a valley glacier composed primarily of debris that fell to the glacier from the valley walls.

Laterite   A red, highly leached soil type found in the tropics and rich in oxides of iron and aluminum.

Laurasia   The northern portion of Pangaea consisting of North America and Eurasia.

Lava   Magma that reaches Earth's surface.

Lava dome   A bulbous mass associated with an old-age volcano, produced when thick lava is slowly squeezed from the vent. Lava domes may act as plugs to deflect subsequent gaseous eruptions.

Lava tube   Tunnel in hardened lava that acts as a horizontal conduit for lava flowing from a volcanic vent. Lava tubes allow fluid lavas to advance great distances.

Law of superposition   In any undeformed sequence of sedimentary rocks or surface-deposited igneous materials, each layer is older than the one above it and younger than the one below.

Leaching   The depletion of soluble materials from the upper soil by downward-percolating water.

Light silicate   Silicate minerals that lack iron and/or magnesium. They are generally lighter in color and have lower specific gravities than dark silicates.

Lithification   The process, generally by cementation and/or compaction, of converting sediments to solid rock.

Lithosphere   The rigid outer layer of Earth, including the crust and upper mantle.

Loess   Deposits of windblown silt, lacking visible layers, generally buff colored, and capable of maintaining a nearly vertical cliff.

Longitudinal dunes   Long ridges of sand oriented parallel to the prevailing wind; these dunes form where sand supplies are limited.

Longitudinal profile   A cross section of a stream channel along its descending course from the head to the mouth.

Longshore current   A nearshore current that flows parallel to the shore.

Losing stream   Streams that lose water to the groundwater system by outflow through the streambed.

Lower mantle   See *Mesosphere.*

Luster   The appearance or quality of light reflected from the surface of a mineral.

Mafic Because basaltic rocks contain a high percentage of ferromagnesian minerals, they are also called mafic (from *mag*nesium and *fer*rum, the latin name for iron).

Magma A body of molten rock found at depth, including any dissolved gases and crystals.

Magnetometer A sensitive instrument used to measure the intensity of Earth's magnetic field at various points.

Magnitude (earthquake) The total amount of energy released during an earthquake.

Manganese nodules A type of hydrogenous sediment scattered on the ocean floor, consisting mainly of manganese and iron, and usually containing small amounts of copper, nickel, and cobalt.

Mantle The 2885-kilometer (1789-mile) thick layer of Earth located below the crust.

Mantle plume A mass of hotter-than-normal mantle material that ascends toward the surface, where it may lead to igneous activity. These plumes of solid yet mobile material may originate as deep as the core-mantle boundary.

Mass number The sum of the number of neutrons and protons in the nucleus of an atom.

Mass wasting The downslope movement of rock, regolith, and soil under the direct influence of gravity.

Meander A looplike bend in the course of a stream.

Meander scar A floodplain feature created when an oxbow lake becomes filled with sediment.

Mechanical weathering The physical disintegration of rock, resulting in smaller fragments.

Medial moraine A ridge of till formed when lateral moraines from two coalescing valley glaciers join.

Melt The liquid portion of magma excluding the solid crystals.

Mesosphere The part of the mantle that extends from the core-mantle boundary to a depth of 660 kilometers. Also known as the lower mantle.

Mesozoic era A time span on the geologic time scale between the Paleozoic and Cenozoic eras—from about 248 million to 65 million years ago.

Metallic bond A chemical bond present in all metals that may be characterized as an extreme type of electron sharing in which the electrons move freely from atom to atom.

Metamorphic rock Rock formed by the alteration of preexisting rock deep within Earth (but still in the solid state) by heat, pressure, and/or chemically active fluids.

Metamorphism The changes in mineral composition and texture of a rock subjected to high temperature and pressure within Earth.

Mid-ocean ridge See *Oceanic ridge.*

Migmatite A rock exhibiting both igneous and metamorphic rock characteristics. Such rocks may form when light-colored silicate minerals melt and then crystallize, while the dark silicate minerals remain solid.

Mineral A naturally occurring, inorganic crystalline material with a unique chemical structure.

Mineral resource All discovered and undiscovered deposits of a useful mineral that can be extracted now or at some time in the future.

Modified Mercalli intensity scale A 12-point scale developed to evaluate earthquake intensity based on the amount of damage to various structures.

Mohorovičić discontinuity (Moho) The boundary separating the crust and the mantle, discernible by an increase in seismic velocity.

Mohs scale A series of 10 minerals used as a standard in determining hardness.

Moment magnitude A more precise measure of earthquake magnitude than the Richter scale that is derived from the amount of displacement that occurs along a fault zone.

Monocline A one-limbed flexure in strata. The strata are usually flat lying or very gently dipping on both sides of the monocline.

Mouth The point downstream where a river empties into another stream or water body.

Mud crack A feature in some sedimentary rocks that forms when wet mud dries out, shrinks, and cracks.

Mudflow See *Debris flow.*

Natural levees The elevated landforms composed of alluvium that parallel some streams and act to confine their waters, except during floodstage.

Neap tide The lowest tidal range, occurring near the times of the first and third quarters of the Moon.

Nebular hypothesis A model for the origin of the solar system that assumes a rotating nebula of dust and gases that contracted to produce the Sun and planets.

Neutron A subatomic particle found in the nucleus of an atom. The neutron is electrically neutral, with a mass approximately equal to that of a proton.

Nonclastic A term for the texture of sedimentary rocks in which the minerals form a pattern of interlocking crystals.

Nonconformity An unconformity in which older metamorphic or intrusive igneous rocks are overlain by younger sedimentary strata.

Nonferromagnesian silicate See *Light silicate.*

Nonfoliated Metamorphic rocks that do not exhibit foliation.

Nonmetallic mineral resource Mineral resource that is not a fuel or processed for the metals it contains.

Nonrenewable resource Resource that forms or accumulates over such long time spans that it must be considered as fixed in total quantity.

Normal fault A fault in which the rock above the fault plane has moved down relative to the rock below.

Normal polarity A magnetic field the same as that which presently exists.

Nucleus The small, heavy core of an atom that contains all of its positive charge and most of its mass.

Nuée ardente Incandescent volcanic debris that is buoyed up by hot gases and moves downslope in an avalanche fashion.

Numerical date Date that specifies the actual number of years that have passed since an event occurred.

Oceanic ridge A continuous mountainous ridge on the floor of all the major ocean basins and varying in width from 500 to 5000 kilometers (300 to 3000 miles). The rifts at the crests of these ridges represent divergent plate boundaries.

Octet rule Atoms combine in order that each may have the electron arrangement of a noble gas; that is, the outer energy level contains eight electrons.

Oil trap A geologic structure that allows for significant amounts of oil and gas to accumulate.

Ore Usually a useful metallic mineral that can be mined at a profit. The term is also applied to certain nonmetallic minerals such as fluorite and sulfur.

Original horizontality Layers of sediment are generally deposited in a horizontal or nearly horizontal position.

Orogenesis The processes that collectively result in the formation of mountains.

Outer core A layer beneath the mantle about 2270 kilometers (1410 miles) thick that has the properties of a liquid.

Outgassing The release of gases dissolved in molten rock.

Outlet glacier A tongue of ice normally flowing rapidly outward from an ice cap or ice sheet, usually through mountainous terrain to the sea.

Outwash plain A relatively flat, gently sloping plain consisting of materials deposited by meltwater streams in front of the margin of an ice sheet.

Oxbow lake A curved lake produced when a stream cuts off a meander.

Oxidation The removal of one or more electrons from an atom or ion. So named because elements commonly combine with oxygen.

Pahoehoe A lava flow with a smooth-to-ropy surface.

Paleomagnetism The natural remnant magnetism in rock bodies. The permanent magnetization acquired by rock that can be used to determine the location of the magnetic poles and the latitude of the rock at the time it became magnetized.

Paleontology The systematic study of fossils and the history of life on Earth.

Paleozoic era A time span on the geologic time scale between the Precambrian and Mesozoic eras—from about 540 million to 248 million years ago.

Pangaea The proposed supercontinent that 200 million years ago began to break apart and form the present landmasses.

Parabolic dune A sand dune similar in shape to a barchan dune except that its tips point into the wind. These dunes often form along coasts that have strong onshore winds, abundant sand, and vegetation that partly covers the sand.

Paradigm Theory that is held with a very high degree of confidence and is comprehensive in scope.

Parasitic cone A volcanic cone that forms on the flank of a larger volcano.

Parent material The material upon which a soil develops.

Parent rock The rock from which a metamorphic rock formed.

Partial melting The process by which most igneous rocks melt. Because individual minerals have different melting points, most igneous rocks melt over a temperature range of a few hundred degrees. If the liquid is squeezed out after some melting has occurred, a melt with a higher silica content results.

Passive continental margin A margin that consists of a continental shelf, continental slope, and continental rise. They are *not* associated with plate boundaries and therefore experience little volcanism and few earthquakes.

Pater noster lakes A chain of small lakes in a glacial trough that occupy basins created by glacial erosion.

Pedalfer Soil of humid regions characterized by the accumulation of iron oxides and aluminum-rich clays in the *B* horizon.

Pedocal Soil associated with drier regions and characterized by an accumulation of calcium carbonate in the upper horizons.

Pegmatite A very coarse-grained igneous rock (typically granite) commonly found as a dike associated with a large mass of plutonic rock that has smaller crystals. Crystallization in a water-rich environment is believed to be responsible for the very large crystals.

Pegmatitic texture A texture of igneous rocks in which the interlocking crystals are all larger than one centimeter in diameter.

Peneplain In the idealized cycle of landscape evolution in a humid region, an undulating plain near base level associated with old age.

Perched water table A localized zone of saturation above the main water table created by an impermeable layer (aquitard).

Peridotite An igneous rock of ultramafic composition thought to be abundant in the upper mantle.

Period A basic unit of the geologic calendar that is a subdivision of an *era*. Periods may be divided into smaller units called *epochs*.

Permafrost Any permanently frozen subsoil. Usually found in the subarctic and arctic regions.

Permeability A measure of a material's ability to transmit water.

Phaneritic texture An igneous rock texture in which the crystals are roughly equal in size and large enough so that individual minerals can be identified with the unaided eye.

Phanerozoic eon That part of geologic time represented by rocks containing abundant fossil evidence. The eon extending from the end of the Proterozoic eon (about 540 million years ago) to the present.

Phenocryst Conspicuously large crystals in a porphyry that are imbedded in a matrix of finergrained crystals (the groundmass).

Physical geology A major division of geology that examines the materials of Earth and seeks to understand the processes and forces acting upon Earth's surface from below.

Piedmont glacier A glacier that forms when one or more valley glaciers emerge from the confining walls of mountain valleys and spread out to create a broad sheet in the lowlands at the base of the mountains.

Pillow lava Basaltic lava that solidifies in an underwater environment and develops a structure that resembles a pile of pillows.

Pipe A vertical conduit through which magmatic materials have passed.

Placer Deposit formed when heavy minerals are mechanically concentrated by currents, most commonly streams and waves. Placers are sources of gold, tin, platinum, diamonds, and other valuable minerals.

Plastic deformation Permanent deformation that results in a change in size and shape through folding or flowing.

Plastic flow A type of glacial movement that occurs within the glacier, below a depth of approximately 50 meters, in which the ice is not fractured.

Plate One of numerous rigid sections of the lithosphere that moves as a unit over the material of the asthenosphere.

Plate tectonics The theory that proposes Earth's outer shell consists of individual plates, which interact in various ways and thereby produce earthquakes, volcanoes, mountains, and the crust itself.

Playa The flat central area of an undrained desert basin.

Playa lake A temporary lake in a playa.

Pleistocene epoch An epoch of the Quaternary period beginning about 1.6 million years ago and ending about 10,000 years ago. Best known as a time of extensive continental ice sheets.

Plucking The process by which pieces of bedrock are lifted out of place by a glacier.

Pluton A structure that results from the emplacement and crystallization of magma beneath Earth's surface.

Pluvial lake A lake formed during a period of increased rainfall. For example, this occurred in many nonglaciated areas during periods of ice advance elsewhere.

Point bar A crescent-shaped accumulation of sand and gravel deposited on the inside of a meander.

Polar wandering hypothesis As the result of paleomagnetic studies in the 1950s, researchers proposed that either the magnetic poles migrated greatly through time or the continents had gradually shifted their positions.

Polymorphs Two or more minerals having the same chemical composition but different crystalline structures. Exemplified by the diamond and graphite forms of carbon.

Porosity The volume of open spaces in rock or soil.

Porphyritic texture An igneous rock texture characterized by two distinctively different crystal sizes. The larger crystals are called *phenocrysts*, and the matrix of smaller crystals is termed the *groundmass*.

Porphyroblastic texture A texture of metamorphic rocks in which particularly large grains (porphyroblasts) are surrounded by a fine-grained matrix of other minerals.

Porphyry An igneous rock with a porphyritic texture.

Pothole A depression formed in a stream channel by the abrasive action of the water's sediment load.

Precambrian All geologic time prior to the Paleozoic era.

Principle of fossil succession Fossil organisms succeed one another in a definite and determinable order, and any time period can be recognized by its fossil content.

Principle of original horizontality Layers of sediment are generally deposited in a horizontal or nearly horizontal position.

Proterozoic eon The eon following the Archean and preceding the Phanerozoic eon. It extends between 2.5 billion and 540 million years ago.

Proton A positively charged subatomic particle found in the nucleus of an atom.

P wave The fastest earthquake wave; travels by compression and expansion of the medium.

Pyroclastic flow A highly heated mixture, largely of ash and pumice fragments, traveling down the flanks of a volcano or along the surface of the ground.

Pyroclastic material The volcanic rock ejected during an eruption. Pyroclastics include ash, bombs, and blocks.

Pyroclastic texture An igneous rock texture resulting from the consolidation of individual rock fragments that are ejected during a violent eruption.

Radial drainage A system of streams running in all directions away from a central elevated structure, such as a volcano.

Radioactivity The spontaneous decay of certain unstable atomic nuclei.

Radiocarbon (carbon-14) The radioactive isotope of carbon produced continuously in the atmosphere and used in dating events as far back as 75,000 years.

Radiometric dating The procedure of calculating the absolute ages of rocks and minerals containing certain radioactive isotopes.

Rainshadow desert A dry area on the lee side of a mountain range. Many middle-latitude deserts are of this type.

Rapids A part of a stream channel in which the water suddenly begins flowing more swiftly and turbulently because of an abrupt steepening of the gradient.

Recessional moraine An end moraine formed as the ice front stagnated during glacial retreat.

Rectangular pattern A drainage pattern that develops on jointed or fractured bedrock and is characterized by numerous right-angle bends.

Refraction A change in direction of waves as they enter shallow water. The portion of the wave in shallow water is slowed, which causes the wave to bend and align with the underwater contours.

Regional metamorphism Metamorphism associated with large-scale mountain building.

Regolith The layer of rock and mineral fragments that nearly everywhere covers Earth's land surface.

Relative dating Rocks are placed in their proper sequence or order. Only the chronological order of events is determined.

Renewable resource A resource that is virtually inexhaustible or that can be replenished over relatively short time spans.

Reserve Already identified deposits from which minerals can be extracted profitably.

Reservoir rock The porous, permeable portion of an oil trap that yields oil and gas.

Residual soil Soil developed directly from the weathering of the bedrock below.

Reverse fault A fault in which the material above the fault plane moves up in relation to the material below.

Reverse polarity A magnetic field opposite to that which presently exists.

Richter scale A scale of earthquake magnitude based on the motion of a seismograph.

Ridge push A mechanism that may contribute to plate motion. It involves the oceanic lithosphere sliding down the oceanic ridge under the pull of gravity.

Rift A region of Earth's crust along which divergence (separation) is taking place.

Ripple marks Small waves of sand that develop on the surface of a sediment layer by the action of moving water or air.

Roche moutonnée An asymmetrical knob of bedrock formed when glacial abrasion smoothes the gentle slope facing the advancing ice sheet and plucking steepens the opposite side as the ice overrides the knob.

Rock A consolidated mixture of minerals.

Rock avalanche The very rapid downslope movement of rock and debris. These rapid movements may be aided by a layer of air trapped beneath the debris, and they have been known to reach speeds in excess of 200 kilometers per hour.

Rock cleavage The tendency of rock to split along parallel, closely spaced surfaces. These surfaces are often highly inclined to the bedding planes in the rock.

Rock cycle A model that illustrates the origin of the three basic rock types and the interrelatedness of Earth's materials and processes.

Rock flour Ground-up rock produced by the grinding effect of a glacier.

Rockslide The rapid slide of a mass of rock downslope along planes of weakness.

Runoff Water that flows over the land rather than infiltrating into the ground.

Saltation Transportation of sediment through a series of leaps or bounces.

Salt flat A white crust on the ground produced when water evaporates and leaves its dissolved materials behind.

Schistosity A type of foliation characteristic of coarser-grained metamorphic rocks. Such rocks have a parallel arrangement of platy minerals such as the micas.

Scoria Hardened lava that has retained the vesicles produced by escaping gases.

Scoria cone See *Cinder cone*.

Sea arch An arch formed by wave erosion when caves on opposite sides of a headland unite.

Seafloor spreading The hypothesis first proposed in the 1960s by Harry Hess that suggested that new oceanic crust is produced at the crests of mid-ocean ridges, which are the sites of divergence.

Seamount An isolated volcanic peak that rises at least 1000 meters (3300 feet) above the deep-ocean floor.

Sea stack An isolated mass of rock standing just offshore, produced by wave erosion of a headland.

Seawall A barrier constructed to prevent waves from reaching the area behind the wall. Its purpose is to defend property from the force of breaking waves.

Secondary enrichment The concentration of minor amounts of metals that are scattered through unweathered rocks into economically valuable concentrations by weathering processes.

Secondary (S) wave A seismic wave that involves oscillation perpendicular to the direction of propagation.

Sediment Unconsolidated particles created by the weathering and erosion of rock, by chemical precipitations from solution in water, or from the secretions of organisms, and transported by water, wind, or glaciers.

Sedimentary rock Rock formed from the weathered products of preexisting rocks that have been transported, deposited, and lithified.

Seiche The rhythmic sloshing of water in lakes, reservoirs, and other smaller enclosed basins. Some seiches are initiated by earthquake activity.

Seismic sea wave A rapidly moving ocean wave generated by earthquake activity and capable of inflicting heavy damage in coastal regions.

Seismogram The record made by a seismograph.

Seismograph An instrument that records earthquake waves.

Seismology The study of earthquakes and seismic waves.

Settling velocity The speed at which a particle falls through a still fluid. The size, shape, and specific gravity of particles influence settling velocity.

Shadow zone The zone between 105 and 140 degrees distance from an earthquake epicenter that direct waves do not penetrate because of refraction by Earth's core.

Shallow-focus earthquake An earthquake focus at a depth of less than 60 kilometers.

Shear Stress that causes two adjacent parts of a body to slide past one another.

Sheet flow Runoff moving in unconfined thin sheets.

Sheeting A mechanical weathering process characterized by the splitting off of slablike sheets of rock.

Shelf break The point at which a rapid steepening of the gradient occurs, marking the outer edge of the continental shelf and the beginning of the continental slope.

Shield A large, relatively flat expanse of ancient metamorphic rock within the stable continental interior.

Shield volcano A broad, gently sloping volcano built from fluid basaltic lavas.

Silicate Any one of numerous minerals that have the silicon-oxygen tetrahedron as their basic structure.

Silicon-oxygen tetrahedron A structure composed of four oxygen atoms surrounding a silicon atom that constitutes the basic building block of silicate minerals.

Sill A tabular igneous body that was intruded parallel to the layering of preexisting rock.

Sinkhole A depression produced in a region where soluble rock has been removed by groundwater.

Slab-pull A mechanism that contributes to plate motion in which cool, dense oceanic crust sinks into the mantle and "pulls" the trailing lithosphere along.

Slaty cleavage The type of foliation characteristic of slates in which there is a parallel arrangement of fine-grained metamorphic minerals.

Slide A movement common to mass-wasting processes in which the material moving downslope remains fairly coherent and moves along a well-defined surface.

Slip face The steep, leeward surface of a sand dune that maintains a slope of about 34 degrees.

Slump The downward slipping of a mass of rock or unconsolidated material moving as a unit along a curved surface.

Snowfield An area where snow persists year-round.

Snowline Lower limit of perennial snow.

Soil A combination of mineral and organic matter, water, and air; that portion of the regolith that supports plant growth.

Soil horizon A layer of soil that has identifiable characteristics produced by chemical weathering and other soil-forming processes.

Soil profile A vertical section through a soil showing its succession of horizons and the underlying parent material.

Solifluction Slow, downslope flow of water-saturated materials common to permafrost areas.

Solum The *O*, *A*, and *B* horizons in a soil profile. Living roots and other plant and animal life are largely confined to this zone.

Solution The change of matter from the solid or gaseous state into the liquid state by its combination with a liquid.

Sorting The degree of similarity in particle size in sediment or sedimentary rock.

Specific gravity The ratio of a substance's weight to the weight of an equal volume of water.

Speleothem A collective term for the dripstone features found in caverns.

Spheroidal weathering Any weathering process that tends to produce a spherical shape from an initially blocky shape.

Spit An elongated ridge of sand that projects from the land into the mouth of an adjacent bay.

Spring A flow of groundwater that emerges naturally at the ground surface.

Spring tide The highest tidal range; occurs near the times of the new and full moons.

Stalactite The icicle-like structure that hangs from the ceiling of a cavern.

Stalagmite The columnlike form that grows upward from the floor of a cavern.

Star dune Isolated hill of sand that exhibits a complex form and develops where wind conditions are variable.

Steppe One of the two types of dry climate. A marginal and more humid variant of the desert that separates the desert from bordering humid climates.

Stock A pluton similar to but smaller than a batholith.

Strata Parallel layers of sedimentary rock.

Stratified drift Sediments deposited by glacial meltwater.

Stratovolcano See *Composite cone*.

Streak The color of a mineral in powdered form.

Stream A general term to denote the flow of water within any natural channel. Thus, a small creek and a large river are both streams.

Stress The force per unit area acting on any surface within a solid.

Striations The multitude of fine parallel lines found on some cleavage faces of plagioclase feldspars but not present on orthoclase feldspar.

Striations (glacial) Scratches or grooves in a bedrock surface caused by the grinding action of a glacier and its load of sediment.

Strike The compass direction of the line of intersection created by a dipping bed or fault and a horizontal surface. Strike is always perpendicular to the direction of dip.

Strike-slip fault A fault along which the movement is horizontal.

Stromatolite Structures that are deposited by algae and that consist of layered mounds or columns of calcium carbonate.

Subduction The process of thrusting oceanic lithosphere into the mantle along a convergent zone.

Subduction zone A long, narrow zone where one lithospheric plate descends beneath another.

Submarine canyon A seaward extension of a valley that was cut on the continental shelf during a time when sea level was lower, or a canyon carved into the outer continental shelf, slope, and rise by turbidity currents.

Submergent coast A coast whose form is largely the result of the partial drowning of a former land surface either due to a rise of sea level or subsidence of the crust, or both.

Subsoil A term applied to the *B* horizon of a soil profile.

Superposition, law of In any undeformed sequence of sedimentary rocks, each bed is older than the one above and younger than the one below.

Surf A collective term for breakers; also the wave activity in the area between the shoreline and the outer limit of breakers.

Surface soil The upper portion of a soil profile consisting of the *O* and *A* horizons.

Surface waves Seismic waves that travel along the outer layer of Earth.

Surge A period of rapid glacial advance; surges are typically sporadic and short-lived.

Suspended load The fine sediment carried within the body of flowing water or air.

S wave An earthquake wave, slower than a P wave, that travels only in solids.

Swells Wind-generated waves that have moved into an area of weaker winds or calm.

Syncline A linear downfold in sedimentary strata; the opposite of anticline.

System A group of interacting or interdependent parts that form a complex whole.

Tablemount See *Guyot*.

Tabular Describing a feature such as an igneous pluton having two dimensions that are much longer than the third.

Talus An accumulation of rock debris at the base of a cliff.

Tarn A small lake in a cirque.

Tectonics The study of the large-scale processes that collectively deform Earth's crust.

Temporary (local) base level The level of a lake, resistant rock layer, or any other base level that stands above sea level.

Terminal moraine The end moraine marking the farthest advance of a glacier.

Terrace A flat, benchlike structure produced by a stream that was left elevated as the stream cut downward.

Terrane A crustal block bounded by faults whose geologic history is distinct from the histories of adjoining crustal blocks.

Terrigenous sediment Seafloor sediment derived from terrestrial weathering and erosion.

Texture The size, shape, and distribution of the particles that collectively constitute a rock.

Theory A well-tested and widely accepted view that explains certain observable facts.

Thermal metamorphism See *Contact metamorphism*.

Thrust fault A low-angle reverse fault.

Tide Periodic change in the elevation of the ocean's surface.

Till Unsorted sediment deposited directly by a glacier.

Tillite A rock formed when glacial till is lithified.

Tombolo A ridge of sand that connects an island to the mainland or to another island.

Topset bed An essentially horizontal sedimentary layer deposited on top of a delta during floodstage.

Transform fault A major strike-slip fault that cuts through the lithosphere and accommodates motion between two plates.

Transform fault boundary A boundary in which two plates slide past one another without creating or destroying lithosphere.

Transpiration The release of water vapor to the atmosphere by plants.

Transported soil Soils that form on unconsolidated deposits.

Transverse dunes A series of long ridges oriented at right angles to the prevailing wind; these dunes form where vegetation is sparse and sand is very plentiful.

Travertine A form of limestone ($CaCO_3$) that is deposited by hot springs or as a cave deposit.

Trellis drainage A system of streams in which nearly parallel tributaries occupy valleys cut in folded strata.

Trench An elongate depression in the seafloor produced by bending of oceanic crust during subduction.

Tsunami The Japanese word for a seismic sea wave.

Turbidite Turbidity current deposit characterized by graded bedding.

Turbidity current A downslope movement of dense, sediment-laden water created when sand and mud on the continental shelf and slope are dislodged and thrown into suspension.

Turbulent flow The movement of water in an erratic fashion often characterized by swirling, whirlpool-like eddies. Most streamflow is of this type.

Ultimate base level Sea level; the lowest level to which stream erosion could lower the land.

Ultramafic Compositional category for igneous rocks made up almost entirely of ferromagnesian minerals (mostly olivine and pyroxene).

Unconformity A surface that represents a break in the rock record; caused by erosion or nondeposition.

Uniformitarianism The concept that the processes that have shaped Earth in the geologic past are essentially the same as those operating today.

Valence electron The electrons involved in the bonding process; the electrons occupying the highest principal energy level of an atom.

Valley glacier A glacier confined to a mountain valley, which in most instances had previously been a stream valley.

Valley train A relatively narrow body of stratified drift deposited on a valley floor by meltwater streams that issue from the terminus of a valley glacier.

Vein deposit A mineral filling a fracture or fault in a host rock. Such deposits have a sheetlike, or tabular, form.

Ventifact A cobble or pebble polished and shaped by the sandblasting effect of wind.

Vesicles Spherical or elongated openings on the outer portion of a lava flow that were created by escaping gases.

Vesicular texture A term applied to aphanitic igneous rocks that contain many small cavities, called *vesicles*.

Viscosity A measure of a fluid's resistance to flow.

Volatiles Gaseous components of magma dissolved in the melt. Volatiles will readily vaporize (form a gas) at surface pressures.

Volcanic Pertaining to the activities, structures, or rock types of a volcano.

Volcanic island arc A chain of volcanic islands generally located a few hundred kilometers from a trench where there is active subduction of one oceanic plate beneath another.

Volcanic bomb A streamlined pyroclastic fragment ejected from a volcano while the fragment is still molten.

Volcanic neck An isolated, steep-sided, erosional remnant consisting of lava that once occupied the vent of a volcano.

Volcano A mountain formed from lava and/or pyroclastics.

Wash A desert stream course that is typically dry except for brief periods immediately following rainfall.

Water gap A pass through a ridge or mountain in which a stream flows.

Water table The upper level of the saturated zone of groundwater.

Wave-cut cliff A seaward-facing cliff along a steep shoreline formed by wave erosion at its base and by mass wasting.

Wave-cut platform A bench or shelf along a shore at sea level, cut by wave erosion.

Wave height The vertical distance between the trough and crest of a wave.

Wave length The horizontal distance separating successive crests or troughs.

Wave of oscillation A water wave in which the wave form advances as the water particles move in circular orbits.

Wave of translation The turbulent advance of water created by breaking waves.

Wave period The time interval between the passage of successive crests at a stationary point.

Weathering The disintegration and decomposition of rock at or near the surface of the Earth.

Welded tuff A pyroclastic deposit composed of particles fused together by the combination of heat still contained in the deposit after it has come to rest and by the weight of overlying material.

Well An opening bored into the zone of saturation.

Wind gap An abandoned water gap. These gorges typically result from stream piracy.

Xenolith An inclusion of unmelted country rock in an igneous pluton.

Xerophyte A plant highly tolerant of drought.

Yazoo tributary A tributary that flows parallel to the main stream because a natural levee is present.

Zone of accumulation The part of a glacier characterized by snow accumulation and ice formation. The outer limit of this zone is the *snowline*.

Zone of aeration Area above the water table where openings in soil, sediment, and rock are not saturated but are filled mainly with air.

Zone of fracture The upper portion of a glacier consisting of brittle ice.

Zone of saturation Zone where all open spaces in sediment and rock are completely filled with water.

Index

Installation Instructions

Macintosh Information

Hardware and software requirements:

- Macintosh with 256 color graphics
- 13" or larger display (640×480 or higher)
- CD-ROM drive (double-speed)
- 6 MB of free RAM memory
- System 7.0 or newer

Refer to the Read Me file on the CD-ROM if you need information regarding how to run the program.

Windows Information

Hardware and software requirements:

- IBM compatible 486 or better
- SVGA (256 colors in 640×480)
- CD-ROM drive (double-speed)
- 16 MB RAM suggested
- Windows 3.1 (enhanced mode) or Windows 95/98/NT

Refer to the README.TXT file on the CD-ROM if you need information regarding how to run the program.

For technical support, you may write to:

Media Production
Engineering/Science/Math Division
Prentice-Hall, Inc.
1 Lake St.
Upper Saddle River, NJ 07458

GEODe III

Geologic Explorations on Disk

YOU SHOULD CAREFULLY READ THE TERMS AND CONDITIONS BEFORE USING THE DISKETTE PACKAGE. USING THIS DISKETTE PACKAGE INDICATES YOUR ACCEPTANCE OF THESE TERMS AND CONDITIONS.

Prentice-Hall, Inc. provides this program and licenses its use. You assume responsibility for the selection of the program to achieve your intended results, and for the installation, use, and results obtained from the program. This license extends only to use of the program in the United States or countries in which the program is marketed by authorized distributors.

License Grant

You hereby accept a nonexclusive, nontransferable, permanent license to install and use the program ON A SINGLE COMPUTER at any given time. You may copy the program solely for backup or archival purposes in support of your use of the program on the single computer. You may not modify, translate, disassemble, decompile, or reverse engineer the program, in whole or in part.

Term

The License is effective until terminated. Prentice-Hall, Inc. reserves the right to terminate this License automatically if any provision of the License is violated. You may terminate the License at any time. To terminate this License, you must return the program, including documentation, along with a written warranty stating that all copies in your possession have been returned or destroyed.

Limited Warranty

THE PROGRAM IS PROVIDED "AS IS" WITHOUT WARRANTY OF ANY KIND, EITHER EXPRESSED OR IMPLIED, INCLUDING, BUT NOT LIMITED TO, THE IMPLIED WARRANTIES OR MERCHANTABILITY AND FITNESS FOR A PARTICULAR PURPOSE. THE ENTIRE RISK AS TO THE QUALITY AND PERFORMANCE OF THE PROGRAM IS WITH YOU. SHOULD THE PROGRAM PROVE DEFECTIVE, YOU (AND NOT PRENTICE-HALL, INC. OR ANY AUTHORIZED DEALER) ASSUME THE ENTIRE COST OF ALL NECESSARY SERVICING, REPAIR, OR CORRECTION. NO ORAL OR WRITTEN INFORMATION OR ADVICE GIVEN BY PRENTICE-HALL, INC., ITS DEALERS, DISTRIBUTORS, OR AGENTS SHALL CREATE A WARRANTY OR INCREASE THE SCOPE OF THIS WARRANTY.

SOME STATES DO NOT ALLOW THE EXCLUSION OF IMPLIED WARRANTIES, SO THE ABOVE EXCLUSION MAY NOT APPLY TO YOU. THIS WARRANTY GIVES YOU SPECIFIC LEGAL RIGHTS AND YOU MAY ALSO HAVE OTHER LEGAL RIGHTS THAT VARY FROM STATE TO STATE.

Prentice-Hall, Inc. does not warrant that the functions contained in the program will meet your requirements or that the operation of the program will be uninterrupted or error-free.

However, Prentice-Hall, Inc. warrants the diskette(s) on which the program is furnished to be free from defects in material and workmanship under normal use for a period of ninety (90) days from the date of delivery to you as evidenced by a copy of your receipt.

The program should not be relied on as the sole basis to solve a problem whose incorrect solution could result in injury to person or property. If the program is employed in such a manner, it is at the user's own risk and Prentice-Hall, Inc. explicitly disclaims all liability for such misuse.

Limitations of Remedies

Prentice-Hall, Inc.'s entire liability and your exclusive remedy shall be:

1. the replacement of any diskette not meeting Prentice-Hall, Inc.'s "LIMITED WARRANTY" and that is returned to Prentice-Hall, or

2. if Prentice-Hall is unable to deliver a replacement diskette that is free of defects in materials and workmanship, you may terminate this agreement by returning the program.

IN NO EVENT WILL PRENTICE-HALL, INC. BE LIABLE TO YOU FOR ANY DAMAGES, INCLUDING ANY LOST PROFITS, LOST SAVINGS, OR OTHER INCIDENTAL OR CONSEQUENTIAL DAMAGES ARISING OUT OF THE USE OR INABILITY TO USE SUCH PROGRAM EVEN IF PRENTICE-HALL, INC. OR AN AUTHORIZED DISTRIBUTOR HAS BEEN ADVISED OF THE POSSIBILITY OF SUCH DAMAGES, OR FOR ANY CLAIM BY ANY OTHER PARTY.

SOME STATES DO NOT ALLOW FOR THE LIMITATION OR EXCLUSION OF LIABILITY FOR INCIDENTAL OR CONSEQUENTIAL DAMAGES, SO THE ABOVE LIMITATION OR EXCLUSION MAY NOT APPLY TO YOU.

General

You may not sublicense, assign, or transfer the License of the program. Any attempt to sublicense, assign or transfer any of the rights, duties, or obligations hereunder is void.

This Agreement will be governed by the laws of the State of New York.

Should you have any questions concerning this Agreement, you may contact Prentice-Hall, Inc. by writing to:

Director of Multimedia Development
Higher Education Division
Prentice-Hall, Inc.
1 Lake Street
Upper Saddle River, NJ 07458

YOU ACKNOWLEDGE THAT YOU HAVE READ THIS AGREEMENT, UNDERSTAND IT, AND AGREE TO BE BOUND BY ITS TERMS AND CONDITIONS. YOU FURTHER AGREE THAT IT IS THE COMPLETE AND EXCLUSIVE STATEMENT OF THE AGREEMENT BETWEEN US THAT SUPERSEDES ANY PROPOSAL OR PRIOR AGREEMENT, ORAL OR WRITTEN, AND ANY OTHER COMMUNICATIONS BETWEEN US RELATING TO THE SUBJECT MATTER OF THIS AGREEMENT.